Lecture Notes in Computer Science 9566

Commenced Publication in 1973
Founding and Former Series Editors:
Gerhard Goos, Juris Hartmanis, and Jan van Leeuwen

More information about this series at http://www.springer.com/series/7410

Orr Dunkelman · Liam Keliher (Eds.)

Selected Areas in Cryptography – SAC 2015

22nd International Conference
Sackville, NB, Canada, August 12–14, 2015
Revised Selected Papers

 Springer

Editors
Orr Dunkelman
University of Haifa
Haifa
Israel

Liam Keliher
Mount Allison University
Sackville, NB
Canada

ISSN 0302-9743 ISSN 1611-3349 (electronic)
Lecture Notes in Computer Science
ISBN 978-3-319-31300-9 ISBN 978-3-319-31301-6 (eBook)
DOI 10.1007/978-3-319-31301-6

Library of Congress Control Number: 2016933230

LNCS Sublibrary: SL4 – Security and Cryptology

Printed on acid-free paper

This Springer imprint is published by Springer Nature
The registered company is Springer International Publishing AG Switzerland

Preface

For the last 22 years, the Conference on Selected Areas in Cryptography (SAC) has been the leading Canadian venue for the presentation and publication of cryptographic research. The conference, which this year was held at Mount Allison University in Sackville, New Brunswick (for the second time; the first was in 2008), offers a relaxed and supportive atmosphere for researchers to present and discuss new results.

SAC has three regular themes:

- Design and analysis of symmetric key primitives and cryptosystems, including block and stream ciphers, hash functions, MAC algorithms, and authenticated encryption schemes
- Efficient implementations of symmetric and public key algorithms
- Mathematical and algorithmic aspects of applied cryptology

The following special (or focus) theme was added this year:

- Privacy- and anonymity-enhancing technologies and their analysis

A total of 91 submissions were received, out of which the Program Committee selected 29 papers for presentation (three of which were accepted as short papers). It is our pleasure to thank the authors of all the submissions for the high quality of their work. The review process was thorough (each submission received the attention of at least three reviewers, and at least five for submissions involving a Program Committee member).

There were two invited talks. The Stafford Tavares Lecture was given by Paul Syverson, who spoke about "Trust Aware Traffic Security," and the second invited talk was given by Gaëtan Leurent, who spoke on "Generic Attacks Against MAC Algorithms."

Finally, this year we expanded SAC in a new direction by introducing what we hope will become an annual tradition — the SAC Summer School (S3). S3 is intended to be a place where young researchers can increase their knowledge of cryptography through instruction by and interaction with leading researchers. This year, we were fortunate to have Kaisa Nyberg and Christian Rechberger presenting symmetric-key cryptanalysis, and Jan Camenisch and Paul Syverson presenting various aspects of privacy-enhancing technologies. We would like to express our sincere gratitude to these four presenters for dedicating their time and effort to this inaugural event.

Finally, the members of the Program Committee, especially the co-chairs, are extremely grateful to Thomas Baignères and Matthieu Finiasz for the iChair software, which facilitated a smooth and easy submission and review process.

August 2015

Orr Dunkelman
Liam Keliher

SAC 2015
The 22th Selected Areas in Cryptography Conference

Sackville, New Brunswick, Canada, August 12–14, 2015

Program Chairs

Orr Dunkelman   University of Haifa, Israel
Liam Keliher   Mount Allison University, Canada

Program Committee

Carlisle Adams	University of Ottawa, Canada
Elena Andreeva	KU Leuven, Belgium
Jean-Philippe Aumasson	Kudelski Security, Switzerland
Roberto Avanzi	Qualcomm Product Security, Germany
Paulo S.L.M. Barreto	University of Sao Paulo, Brazil
Josh Benaloh	Microsoft Research, USA
Daniel J. Bernstein	University of Illinois at Chicago, USA, and TU Eindhoven, The Netherlands
John Black	University of Colorado at Boulder, USA
Nikita Borisov	University of Illinois at Urbana-Champaign, USA
Itai Dinur	École Normale Supérieure, France
Orr Dunkelman	University of Haifa, Israel (Co-chair)
Guang Gong	University of Waterloo, Canada
Tim Güneysu	University of Bremen, Germany
Seda Gürses	Princeton University, USA
Michael Jacobson	University of Calgary, Canada
Antoine Joux	Paris 6 University, France
Nathan Keller	Bar Ilan University, Israel
Liam Keliher	Mount Allison University, Canada (Co-chair)
Tanja Lange	TU Eindhoven, The Netherlands
Gregor Leander	Ruhr University Bochum, Germany
Anja Lehmann	IBM Research Zurich, Switzerland
Petr Lisonek	Simon Fraser University, Canada
Florian Mendel	TU Graz, Austria
María Naya-Plasencia	Inria, France
Kaisa Nyberg	Aalto University, Finland
Christian Rechberger	Technical University of Denmark, Denmark
Dipanwita Roy Chowdhury	IIT Kharagpur, India
Palash Sarkar	Indian Statistical Institute, India
Meltem Sönmez Turan	NIST, USA

Douglas Stinson	University of Waterloo, Canada
Carmela Troncoso	Gradiant, Spain
Vanessa Vitse	Institut Fourier, Université de Grenoble I, France
Bo-Yin Yang	Academia Sinica, Taiwan
Amr Youssef	Concordia University, Canada
Moti Yung	Google, USA

Additional Reviewers

Ahmed AbdelKhalek	Virginie Lallemand
Christof Beierle	Kerstin Lemke-Rust
Begül Bilgin	Gaëtan Leurent
Céline Blondeau	Sebastian Lindner
Christina Boura	Atul Luykx
Billy Bob Brumley	Evan MacNeil
Yao Chen	Nicolas Meloni
Chitchanok Chuengsatiansup	Bart Mennink
Thomas De Cnudde	Amir Moradi
Christoph Dobraunig	Nicky Mouha
Manu Drijvers	Christophe Négre
Léo Ducas	Oscar Reparaz
Maria Eichlseder	Arnab Roy
Philippe Elbaz-Vincent	Enrique Argones Rua
Junfeng Fan	Dhiman Saha
Xinxin Fan	Pascal Sasdrich
Magnus Gausdal Find	Tobias Schneider
Benoît Gérard	Peter Schwabe
Shamit Ghosh	Nicolas Sendrier
Alejandro Hevia	Abhrajit Sengupta
Simon Hoerder	Valentin Suder
Philipp Jovanovic	Eduarda S.V. Freire
Marc Joye	Yin Tan
Elif Bilge Kavun	Cihangir Tezcan
John M. Kelsey	Tyge Tiessen
Aggelos Kiayias	Elmar Tischhauser
Stefan Kölbl	M. Tolba
Thorsten Kranz	Christine van Vredendaal
Stephan Krenn	Kerem Varıcı
Sukhendu Kuila	Alexander Wild
Thijs Laarhoven	Bo Zhu

Stafford Tavares Lecture

Trust Aware Traffic Security

Paul Syverson

U.S. Naval Research Laboratory
paul.syverson@nrl.navy.mil

Abstract. We will trace the development of trust-aware traffic security in onion routing networks 2009–2015. Beginning with the question of how to mathematically model diversity of trust across an onion routing network, we will also discuss defining adversaries for a trust-aware context, representing trust in a usable way, accounting for all the network elements to which trust might be assigned, and making use of trust to design more secure routing.

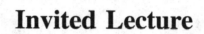
Invited Lecture

Generic Attacks Against MAC Algorithms

Gaëtan Leurent

Inria, France
Gaetan.Leurent@inria.fr

Message Authentication Codes (MACs) are important cryptographic constructions, used to ensure the authenticity of messages. MACs can be built from scratch (SipHash, Chaskey), from block ciphers (CBC-MAC, PMAC), from hash functions (HMAC), or from universal hash functions (GMAC, Poly1305). Constructions based on a lower level primitive are usually studied with a provable security approach: for commonly used MAC algorithms, security proofs rule out any attack on the MAC with complexity less than $2^{n/2}$ (the birthday bound), when using an ideal n-bit block cipher or compression function. On the other hand, a generic collision attack against iterated MACs by Preneel and van Oorschot [8] using $2^{n/2}$ queries show that the security proofs are tight.

The security of MAC algorithms seems to be well understood, but the generic attack of Preneel and van Oorschot is only an existential forgery attack, and stronger attacks (*e.g.* key-recovery attacks) are usually more expensive. In order to evaluate the security of common MAC algorithms below the birthday bound, we now focus on generic attacks, rather than on security proofs. The two approaches are complementary: generic attacks yields upper bounds on the security of a mode, and security proofs yield a lower bound. It is important to comin both, because constructions with a similar security proof can actually have a very different loss of security after the birthday bound.

For instance, there is a collision-based almost universal forgery attack against several CBC-MAC variants with birthday complexity [4]. There is also an attack with birthday complexity recovering the secret mask in PMAC using collisions [5]; this implies a universal forgery attack. More recently, a similar attack was shown against AEZ v3, but it yields a full key-recovery attack because of the way the secret mask is derived from the master key [2].

Hash-based MACs also have varying security beyond the birthday bound. The main constructions process the message with an unkeyed iteration, and use the key only in the initialization (secret-prefix MAC), in the finalization (secret-suffix MAC), or both in the initialization and finalization (HMAC, envelope MAC, sandwich MAC). All these constructions are good MACs when used with a random oracle, but they offer different levels of security in practice. Constructions using the key only in the finalization are much less secure, because collisions in the hash function directly lead to forgeries independently of the key. In particular, these collisions can be computed offline, and do not require any queries to a MAC oracle. Secret-suffix MAC and envelope MAC are also susceptible to a key-recovery attack based on collisions with partial blocks [9]. Interestingly, the key-recovery attack can be applied to envelope

MAC but not to sandwich MAC, although the construction are very close and have a similar security proof.

Over the last years, a series of papers have studied more complex generic attacks against HMAC and similar hash-based MAC algorithms [1, 3, 6, 7]. This proved that distinguishing-H, state-recovery, and universal forgery attacks against HMAC require less than 2^n operations, contrary to what was previously assumed. The first attacks used the structure of the cycle graph of random functions in an elegant way to build a distinguishing-H and state-recovery attack with complexity $2^{n/2}$ [6]. Variants with short messages were also given, using the entropy loss of random functions (with complexity $2^{2n/3}$), and later extended to HAIFA hash functions (with complexity $2^{4n/5}$) [1]. Surprisingly, extensions of those techniques also lead to universal forgery attacks against long messages [3, 7]. In addition, the state recovery attacks can be extended to key-recovery attacks for HMAC based on a hash function with an internal checksum. In particular, this gives a key-recovery attack with complexity 2^{192} against HMAC-GOST (with $n = 256$), and 2^{419} against HMAC-Streebog (with $n = 512$).

All those examples show a large variety in the attack techniques and complexity, with several key-recovery attacks more efficient than exhaustive search. This highlights the importance of studying generic attacks in addition to the provable security driven approach to the design of MAC algorithms.

References

1. Dinur, I., Leurent, G.: Improved generic attacks against hash-based MACs and HAIFA. In: Garay, J.A., Gennaro, R. (eds.) CRYPTO 2014. LNCS, vol. 8616, pp. 149–168. Springer, Berlin (2014)
2. Fuhr, T., Leurent, G., Suder, V.: Collision attacks against CAESAR candidates. In: Iwata, T., Cheon, J.H. (eds.) Advances in Cryptology - ASIACRYPT 2015. LNCS, vol. 9453, pp. 510–532. Springer, Berlin (2015)
3. Guo, J., Peyrin, T., Sasaki, Y., Wang, L.: Updates on generic attacks against HMAC and NMAC. In: Garay, J.A., Gennaro, R. (eds.) CRYPTO 2014. LNCS, vol. 8616, pp. 131–148. Springer, Berlin (2014)
4. Jia, K., Wang, X., Yuan, Z., Xu, G.: Distinguishing and second-preimage attacks on CBC-Like MACs. In: Garay, J.A., Miyaji, A., Otsuka, A. (eds.) CANS 2009. LNCS, vol. 5888, pp. 349–361. Springer, Berlin (2009)
5. Lee, C., Kim, J., Sung, J., Hong, S., Lee, S.: Forgery and key recovery attacks on PMAC and Mitchell's TMAC variant. In: Batten, L.M., Safavi-Naini, R. (eds.) ACISP 2006. LNCS, vol. 4058, pp. 421–431. Springer, Berlin (2006)
6. Leurent, G., Peyrin, T., Wang, L.: New generic attacks against hash-based MACs. In: Sako, K., Sarkar, P. (eds.) ASIACRYPT 2013. LNCS, vol. 8270, pp. 1–20. Springer, Berlin (2013)
7. Peyrin, T., Wang, L.: Generic universal forgery attack on iterative hash-based MACs. In: Nguyen, P.Q., Oswald, E. (eds.) EUROCRYPT 2014. LNCS, vol. 8441, pp. 147–164. Springer, Heidelberg (2014)
8. Preneel, B., van Oorschot, P.C.: MDx-MAC and building fast MACs from hash functions. In: Coppersmith, D. (ed.) CRYPTO 1995. LNCS, vol. 963, pp. 1–14. Springer, Berlin (1995)
9. Preneel, B., van Oorschot, P.C.: On the security of Two MAC algorithms. In: Maurer, U.M. (ed.) EUROCRYPT 1996. LNCS, vol. 1070, pp. 19–32. Springer, Berlin (1996)

Contents

Short Papers

Privacy Preserving Data Processing

Side Channel Attacks and Defenses

New Cryptographic Constructions

Privacy Enhancing Technologies

Formal Treatment of Privacy-Enhancing Credential Systems

Jan Camenisch[1], Stephan Krenn[2]([✉]), Anja Lehmann[1],
Gert Læssøe Mikkelsen[3], Gregory Neven[1], and Michael Østergaard Pedersen[4]

[1] IBM Research – Zurich, Rüschlikon, Switzerland
{jca,anj,nev}@zurich.ibm.com
[2] AIT Austrian Institute of Technology GmbH, Vienna, Austria
stephan.krenn@ait.ac.at
[3] Alexandra Institute, Aarhus, Denmark
gert.l.mikkelsen@alexandra.dk
[4] Miracle A/S, Aarhus, Denmark
mop@miracleas.dk

Abstract. Privacy-enhancing attribute-based credentials (PABCs) are
the core ingredients to privacy-friendly authentication systems. They
allow users to obtain credentials on attributes and prove possession of
these credentials in an unlinkable fashion while revealing only a subset of
the attributes. In practice, PABCs typically need additional features like
revocation, pseudonyms as privacy-friendly user public keys, or advanced
issuance where attributes can be "blindly" carried over into new credentials. For many such features, provably secure solutions exist in isolation,
but it is unclear how to securely combined them into a full-fledged PABC
system, or even which properties such a system should fulfill.

We provide a formal treatment of PABCs supporting a variety of
features by defining their syntax and security properties, resulting in
the most comprehensive definitional framework for PABCs so far. Unlike
previous efforts, our definitions are not targeted at one specific use-case;
rather, we try to capture generic properties that can be useful in a variety
of scenarios. We believe that our definitions can also be used as a starting point for diverse application-dependent extensions and variations of
PABCs. We present and prove secure a generic and modular construction
of a PABC system from simpler building blocks, allowing for a "plug-
and-play" composition based on different instantiations of the building
blocks. Finally, we give secure instantiations for each of the building
blocks.

Keywords: Privacy · Anonymous credentials · Provable security

This work was supported by the Horizon 2020 project PRISMACLOUD under grant
agreement no. 644962, and the FP7 projects FutureID and ABC4Trust under grant
agreement nos. 318424 and 257782. Parts of this work were done while the second
author was at IBM Research – Zurich. The full version is available online [1].

© Springer International Publishing Switzerland 2016
O. Dunkelman and L. Keliher (Eds.): SAC 2015, LNCS 9566, pp. 3–24, 2016.
DOI: 10.1007/978-3-319-31301-6_1

1   Introduction

Privacy-enhancing attribute-based credentials systems (aka PABCs, *anonymous credentials*, or *pseudonym systems*) allow for cryptographically strong user authentication while preserving the users' privacy by giving users full control over the information they reveal. There are three types of parties in a PABC system. *Issuers* assign sets of attribute values to *users* by issuing credentials for these sets. Users can present (i.e., prove possession of) their credentials to *verifiers* by revealing a subset of the attributes from one or more credentials. The verifiers can then check the validity of such presentations using the issuers' public keys, but they do not learn any information about the hidden attributes and cannot link different presentations by the same user. This basic functionality of a PABC system can be extended in a large number of ways, including pseudonyms, revocation of credentials, inspection, proving relations among attributes hidden in presented credentials, and key binding [2,3].

The importance of privacy and data minimization has been emphasized, e.g., by the European Commission in the European privacy standards [4,5] and by the US government in the National Strategy for Trusted Identities in Cyberspace (NSTIC) [6]. With IBM's Identity Mixer based on CL-signatures [7–10] and Microsoft's U-Prove based on Brands' signatures [11,12], practical solutions for PABCs exist and are currently deployed in several pilot projects [2,13–15]. In fact, numerous anonymous credential schemes as well as special cases thereof such as group signatures, direct anonymous attestation, or identity escrow have been proposed, offering a large variety of different features [8,10–12,16–24].

Despite this large body of work, a unified definitional framework for the security properties of a full-fledged PABC system is still missing. Existing schemes either have targeted security definitions for specific use cases [8,18–20] or do not provide provable security guarantees at all [12,21,22]. One possible reason for the lack of a generic framework is that dedicated schemes for specific scenarios are often more efficient than generic solutions. However, defining, instantiating, and re-proving tailored variants of PABCs is hard and error-prone. Clearly, it is desirable to have a unified definitional approach providing security definitions for a full-fledged PABC system. It turns out that achieving such definitions is far from trivial as they quickly become very complex, in particular, if one allows for relations between hidden attributes when issuing and presenting credentials, as we shall discuss. Nevertheless, in this paper we take a major step towards such a unified framework for PABCs by formally defining the most relevant features, detached from specific instantiations or use cases. We further provide a generic construction of a PABC system based on a number of simpler building blocks such as blind signatures or revocation schemes, and a formal proof that this construction meets our security definitions. Finally, we give concrete instantiations of these components, with detailed security proofs provided in the full version.

Considered Features. Our definition of PABC systems comprises the richest set of features integrated into a holistic PABC scheme so far. It supports credentials with any fixed number of attributes, of which any subset can be revealed during presentation. A single presentation can reveal attributes from *multiple credentials*. Users can *prove equality* of attributes, potentially across different credentials, without revealing their exact values. Users have *secret keys* from which arbitrarily many *scope-exclusive pseudonyms* can be derived. That is, for a given secret key and *scope*, only one unique pseudonym can be derived. Thus, by reusing the same scope in multiple presentations, users can intentionally create linkability between presentations; using different scopes yields mutually unlinkable pseudonyms.

Credentials can optionally be bound to the users' secret keys to prevent users from sharing their credentials. For a presentation that involves multiple credentials and/or a pseudonym, all credentials and the pseudonym must be bound to or derived from the same user secret key, respectively. Issuers can *revoke credentials*, so that they can no longer be used. During issuance, some attribute values may be hidden from the issuer or *"carried over"* from existing credentials. This latter *advanced issuance* means that the issuer does not learn their values but is guaranteed that they are equal to an attribute in an existing credential.

Our Contributions. We give formal security definitions for a full PABC system that incorporates all of the features mentioned above. We provide a generic construction from lower-level building blocks that satisfies our definitions and we present secure instantiations of the building blocks.

In terms of security, informally, we expect presentations to be *unforgeable*, i.e., users can only present attributes from legitimately obtained and unrevoked credentials, and to be *private*, i.e., they do not reveal anything more than intended. For privacy, we distinguish weak privacy, where presentations of a credential cannot be linked to a specific issuance session, and the strictly stronger notion of simulatable privacy, where in addition presentations of the same credential cannot be linked to each other. This allows us to cover (slight variants of) the most prevalent schemes used in practice, U-Prove and Identity Mixer.

Formally defining these properties is far from trivial because of the complexity of our envisaged system. For example, user can obtain credentials on (*i*) revealed, (*ii*) blindly carried-over, and (*iii*) fully blind attributes. Each type comes with different security expectations that must be covered by a single definition. Carried-over attributes, for example, present a challenge when defining unforgeability: While the issuer never learns the attributes, the adversary must not be able to present a value that was not previously issued as part of a pre-existing credential. For privacy, one challenge is to formalize the exact information that users intend to reveal, as they might reveal the same and possibly identifying attributes in different presentations. Revocation gives the issuer the power to exclude certain credentials from being used, which must be modeled without cementing trivial linking attacks into the model that would turn the definition moot.

As our definitions are rather complex, proving that a concrete scheme satisfies them from scratch can be a challenging and tedious task. Also, proofs for such monolithic definitions tend to be hard to verify. We thus further define building blocks, strongly inspired by existing work, and show how to generically compose them to build a secure PABC system. Our construction is efficient in the sense that its complexity is roughly the sum of the complexities of the building blocks. Additionally, this construction allows for simple changes of individual components (e.g., the underlying revocation scheme) without affecting any other building blocks and without having to reprove the security of the system. Finally, we give concrete instantiations for all our building blocks based on existing protocols.

Related Work. Our definitions are inspired by the work of Chase et al. [18,25], who provide formal, property-based definitions of delegatable anonymous credential systems and give a generic construction from so-called P-signatures [26]. However, their work focus on pseudonymous access control with delegation, but lacks additional features such as attributes, revocation, and advanced issuance.

PABCs were first envisioned by Chaum [16,27], and they have been a vivid area of research over the last decade. The currently most mature solutions are IBM's Identity Mixer based on CL-signatures [7–10] and Microsoft's U-Prove based on Brands' signatures [11,12]. A first formal definition [8] in the ideal-real world paradigm covered the basic functionalities without attributes and revocation. Their definition is stand-alone and does not allow composability as honest parties never output any cryptographic values such as a credentials or pseudonyms. This restriction makes it infeasible to use their schemes as building block in a larger system. These drawbacks are shared by the definition of Garman et al. [19].

A recent MAC-based credential scheme [20] allows for multiple attributes per credential, but requires issuer and verifier to be the same entity. It does not cover pseudonyms, advanced issuance, or revocation and provides rather informal definitions only. Similarly, Hanser and Slamanig [28] do not consider any of these features nor blind issuance, i.e., the issuer always learns all the user's attributes.

Baldimtsi and Lysyanskaya [29] define a blind signature scheme with attributes and claim that it yields an efficient *linkable* anonymous credential scheme, again without giving formal definitions. The scheme can be seen as a weakened version of our signature building block without unlinkability or extractability of hidden attributes – properties that are crucial for our PABC system.

Camenisch et al. [24] provide a UC-definition for anonymous credentials and a construction based on bilinear maps. However, their definition does not cover a full-blown credential systems but just a primitive to issue and verify signatures on attributes, i.e., a primitive we call privacy-enhancing signatures in this paper.

Finally, existing definitions of attribute-based signatures, e.g., [30–32] differ substantially from those of our building block for privacy-enhancing attribute-based signature (cf. Sect. 4.3), as again they do not consider, e.g., blind issuance.

2   Notation

Algorithms and parties are denoted by sans-serif fonts, e.g., A, B. For deterministic (probabilistic) algorithms we write $a \leftarrow \mathsf{A}(in)$ $(a \overset{\$}{\leftarrow} \mathsf{A}(in))$, if a is the output of A on inputs in. For an interactive protocol (A, B) let $(out_\mathsf{A}, out_\mathsf{B})$ $\leftarrow \langle \mathsf{A}(in_\mathsf{A}), \mathsf{B}(in_\mathsf{B}) \rangle$ denote that, on private inputs in_A to A and in_B to B, A and B obtained outputs out_A and out_B, respectively. For a set $\mathcal{S}$, $s \overset{\$}{\leftarrow} \mathcal{S}$ denotes that s is drawn uniformly at random from $\mathcal{S}$. We write $\Pr[\mathcal{E} : \Omega]$ to denote the probability of event $\mathcal{E}$ over the probability space Ω. We write vectors as $\vec{x} = (x_i)_{i=1}^{k} = (x_1, \ldots, x_k)$.

A function $\nu : \mathbb{N} \to \mathbb{R}$ is *negligible* if for every k and all sufficiently large n we have $\nu(n) < \frac{1}{n^k}$. Let $\pm\{0,1\}^k := [-2^k + 1, 2^k - 1] \cap \mathbb{Z}$, and $[n] := \{1, \ldots, n\}$.

Finally, κ is the main security parameter, and ε the empty string or list.

3   Privacy ABC Systems

A privacy-enhancing attribute-based credential system for an attribute space $\mathcal{AS}$ is a set of algorithms SPGen, UKGen, IKGen, Present, Verify, ITGen, ITVf, and Revoke and a protocol $\langle \mathcal{U}.\mathsf{Issue}, \mathcal{I}.\mathsf{Issue} \rangle$. Wherever possible (i.e., except for (advanced) issuance which is inherently interactive), we opted for non-interactive protocols. This is because rounds of interaction are an expensive resource in practice, and should thus be kept as few as possible.

Parties are grouped into issuers, users, and verifiers. The publicly available system parameters are generated using SPGen by a trusted party (in practice, this might be implemented using multiparty techniques). Each issuer can issue and revoke credentials that certify a list of attribute values under his issuer public key. Users hold secret keys that can be used to derive *pseudonyms* that are unique for a given scope string, but are unlinkable across scopes. Using Present, users can create non-interactive *presentation tokens* from their credentials that reveal any subset of attributes from any subset of their credentials, or prove that certain attributes are equal without revealing them. Presentation tokens can be publicly verified using Verify, on input the token and the issuers' public keys. To obtain a credential, a user generates an *issuance token* defining the attribute values of the new credential using ITGen. After the issuer verified the issuance token using ITVf, the user and the issuer run $\langle \mathcal{U}.\mathsf{Issue}, \mathcal{I}.\mathsf{Issue} \rangle$, at the end of which the user obtains a credential. Issuance can be combined with a presentation of existing credentials to hide some of the attribute values from the issuer, or to prove that they are equal to attributes in credentials that the user already owns. Hence, issuance tokens can be seen as an extension of presentation tokens. Credentials can optionally be bound to a user's secret key, meaning that knowledge of this key is required to prove possession of the credential. Now, if a pseudonym or multiple key-bound credentials are used in a presentation token, then all credentials and the pseudonym must be bound to the same key. Finally, an issuer can revoke a credential using the Revoke algorithm. This algorithm outputs some public revocation information RI that is published and should be used as input to the verification algorithm of presentation and issuance tokens.

Issuers and users agree on the parameters for issuance tokens including a revocation handle and the revealed attributes of the new credential (upon which the user has generated the issuance token) in a step preceding issuance. Issuers further verify the validity of these tokens before engaging in this protocol. There are no requirements on how revocation handles are chosen, but in practice they should be different for each credential an issuer issues.

3.1   Syntax

Before formalizing the security properties, we introduce the syntax of PABCs.

System parameter generation. The system parameters of a PABC-system are generated as $spar \xleftarrow{\$} \mathsf{SPGen}(1^\kappa)$. For simplicity we assume that the system parameters in particular contain an integer L specifying the maximum number of attributes that can be certified by one credential, as well as a description of the attribute space $\mathcal{AS}$. For the rest of this document, we assume that all honest parties only accept attributes from $\mathcal{AS}$ and abort otherwise.

The system parameters are input to all algorithms presented in the following. However, for notational convenience, we will sometimes not make this explicit.

User key generation. Each user generates a secret key as $usk \xleftarrow{\$} \mathsf{UKGen}(spar)$.

Issuer key generation. Each issuer generates a public/private issuer key pair and some initial revocation information as $(ipk, isk, RI) \xleftarrow{\$} \mathsf{IKGen}(spar)$. We assume that RI also defines the set of all supported revocation handles for this issuer, and that honest parties only accept such revocation handles and abort otherwise.

Presentation. A user generates a pseudonym nym and presentation token pt as

$$(nym, pt) \xleftarrow{\$} \mathsf{Present}\Big(usk, scope, \big(ipk_i, RI_i, cred_i, (a_{i,j})_{j=1}^{n_i}, R_i\big)_{i=1}^{k}, E, M\Big), \text{ where}$$

- usk is the user's secret key, which can be ε if $scope = \varepsilon$ and none of the credentials $(cred_i)_{i=1}^{k}$ is bound to a user secret key;
- $scope$ is the scope of the generated pseudonym nym, where $scope = \varepsilon$ if no pseudonym is to be generated (in which case $nym = \varepsilon$);
- $(cred_i)_{i=1}^{k}$ are k user-owned credentials that are involved in this presentation;
- ipk_i and RI_i are the public key and current revocation information of the issuer of $cred_i$;
- $(a_{i,j})_{j=1}^{n_i}$ is the list of attribute values certified in $cred_i$, where each $a_{i,j} \in \mathcal{AS}$;
- $R_i \subseteq [n_i]$ is the set of attribute indices for which the value is revealed;
- E induces an equivalence relation on $\{(i,j) : i \in [k] \;\wedge\; j \in [n_i]\}$, where $((i,j),(i',j')) \in E$ means that pt will prove that $a_{i,j} = a_{i',j'}$ without revealing the actual attribute values. That is, E enables one to prove equality predicates;
- $M \in \{0,1\}^*$ is a message to which the presentation token is to be bound. This might, e.g., be a nonce chosen by the verifier to prevent replay attacks in which a verifier uses a valid presentation token to impersonate a user.

If $k = 0$ and $scope \neq \varepsilon$, only a pseudonym nym is generated while $pt = \varepsilon$.

Presentation verification. A verifier can check the validity of a pseudonym *nym* and a presentation token *pt*:

$$\mathsf{accept/reject} \leftarrow \mathsf{Verify}\Big(nym, pt, scope, \big(ipk_i, RI_i, (a_{i,j})_{j \in R_i}\big)_{i=1}^{k}, E, M\Big),$$

where the inputs are as for presentation. For notational convenience from now on a term like $(a_{i,j})_{j \in R_i}$ implicitly also describes the set R_i.

Issuance token generation. Before issuing a credential, a user generates an *issuance token* defining the attributes of the credentials to be issued, where (some of) the attributes and the secret key can be hidden from the issuer and can be blindly "carried over" from credentials that the user already possesses (so that the issuer is guaranteed that hidden attributes were vouched for by another issuer). Similarly to a presentation token we have:

$$(nym, pit, sit) \xleftarrow{\$} \mathsf{ITGen}\Big(usk, scope, rh, \big(ipk_i, RI_i, cred_i, (a_{i,j})_{j=1}^{n_i}, R_i\big)_{i=1}^{k+1}, E, M\Big),$$

where most of the inputs and outputs are as before, but

- *pit* and *sit* are the public and secret parts of the issuance token,
- $cred_{k+1} = \varepsilon$ is the new credential to be issued,
- *rh* is the revocation handle for $cred_{k+1}$ (maybe chosen by the issuer before),
- ipk_{k+1} and RI_{k+1} are the public key and current revocation information of the issuer of the new credential,
- $(a_{k+1,j})_{j \in R_{k+1}}$ are the attributes of $cred_{k+1}$ that are revealed to the issuer,
- $(a_{k+1,j})_{j \notin R_{k+1}}$ are the attributes of $cred_{k+1}$ that remain hidden, and
- $((k+1,j),(i',j')) \in E$ means that the jth attribute of the new credential will have the same value as the j'^{th} attribute of the i'^{th} credential.

Issuance token verification. The issuer verifies an issuance token as follows:

$$\mathsf{accept/reject} \xleftarrow{\$} \mathsf{ITVf}\Big(nym, pit, scope, rh, \big(ipk_i, RI_i, (a_{i,j})_{j \in R_i}\big)_{i=1}^{k+1}, E, M\Big).$$

For $j \in [k]$ all inputs are as for Verify, but for $k+1$ they are for the new credential to be issued based on *pit*.

Issuance. Issuance of credentials is a protocol between a user and an issuer:

$$(cred, RI') \xleftarrow{\$} \langle\!\langle \mathcal{U}.\mathsf{Issue}(sit, pit); \mathcal{I}.\mathsf{Issue}(isk, pit, RI)\rangle.$$

The inputs are defined as before, and *pit* has been verified by the issuer before. The user obtains a credential as an output, while the issuer receives an updated revocation information RI'.

Revocation. To revoke a credential with revocation handle *rh*, the issuer runs: $RI' \xleftarrow{\$} \mathsf{Revoke}(isk, RI, rh)$ to generate the new revocation information RI' based on the issuer's secret key, the current revocation information, and the revocation handle to be revoked.

3.2   Oracles for Our Security Definitions

Our security definitions of PABC systems require a number of oracles, some of which are the same for different definitions. We therefore present them all in one place. The oracles are initialized with a set of honestly generated keys of n_I honest issuers $\{(ipk_i^*, isk_i^*, RI_i^*)\}_{i=1}^{n_I}$ and n_U users with keys $\{usk_i^*\}_{i=1}^{n_U}$, respectively. Let $IK^* = \{ipk_i^*\}_{i=1}^{n_I}$. The oracles maintain initially empty sets $\mathcal{C}$, $\mathcal{HP}$, $\mathcal{IT}$, $\mathcal{IRH}$, $\mathcal{RRH}$, $\mathcal{RI}$. Here, $\mathcal{C}$ contains all credentials that honest users have obtained as instructed by the adversary, while $\mathcal{HP}$ contains all presentation tokens generated by honest users. All public issuance tokens that the adversary used in successful issuance protocols with honest issuers are stored in $\mathcal{IT}$. The set $\mathcal{IRH}$ contains all issued revocation handles, i.e., the revocation handles of credentials issued by honest issuers, while $\mathcal{RRH}$ contains the revoked handles per issuer. Finally, $\mathcal{RI}$ contains the history of the valid revocation information of honest issuers at any point in time. Time is kept per issuer I through a counter $epoch_I^*$ that is initially 0 and increased at each issuance and revocation by I.

Honest Issuer Oracle $\mathcal{O}^{\text{issuer}}$. This oracle allows the adversary to obtain and revoke credentials from honest issuers. It provides the following interfaces:

- On input $(\text{issue}, nym, pit, scope, rh, (ipk_i, RI_i, (a_{i,j})_{j \in R_i})_{i=1}^{k+1}, E, M)$ the oracle checks that ITVf accepts the issuance token pit and that the revocation information of all honest issuers is authentic, i.e., that a tuple $(ipk_i, RI_i, \cdot)$ exists in $\mathcal{RI}$ for all honest issuers ipk_i. Further, it verifies that ipk_{k+1} is the public key of an honest issuer ipk_I^* with corresponding secret key isk_I^* and current revocation information $RI_I^* = RI_{k+1}$. If one of the checks fails, the oracle outputs $\perp$ and aborts.
  The oracle then runs $\mathcal{I}.\text{Issue}(isk_{k+1}, pit, RI_I^*)$ in interaction with A until the protocol outputs RI_{k+1}. It returns RI_{k+1} to A, sets $RI_I^* \leftarrow RI_{k+1}$, increases $epoch_I^*$, adds $(ipk_I^*, RI_I^*, epoch_I^*)$ to $\mathcal{RI}$. It adds $(nym, pit, scope, (ipk_i, RI_i, (a_{i,j})_{j \in R_i})_{i=1}^{k+1}, E, M)$ to $\mathcal{IT}$ and adds (ipk_I^*, rh) to the set $\mathcal{IRH}$.
- On input (revoke, I, rh) the oracle checks that the revocation handle rh has been the input of a successful issuance protocol with an honest issuer with key ipk_I^*, or returns $\perp$ otherwise. The oracle runs $RI_I^* \xleftarrow{\$} \text{Revoke}(isk_I^*, RI_I^*, rh)$, increases $epoch_I^*$, adds $(ipk_I^*, rh, epoch_I^*)$ to $\mathcal{RRH}$, adds $(ipk_I^*, RI_I^*, epoch_I^*)$ to $\mathcal{RI}$, and returns RI_I^* to the adversary.

Honest User Oracle $\mathcal{O}^{\text{user}}$. This oracle gives the adversary access to honest users, which he can trigger to obtain credentials and request presentation tokens on inputs of his choice. The adversary does not get the actual credentials, but only unique credential identifiers cid by which he can refer to the credentials. It provides the following interfaces:

- On input $(\text{present}, U, scope, (ipk_i, RI_i, cid_i, R_i)_{i=1}^{k}, E, M)$ the oracle checks if $U \in [n_U]$ and if, for all $i \in [k]$, a tuple $(U, cid_i, cred_i, (a_{i,j})_{j=1}^{n_i}) \in \mathcal{C}$ exists. For all credentials from honest issuers, $\mathcal{O}^{\text{user}}$ verifies if RI_i is authentic, i.e., if for all honest issuer public keys $ipk_i = ipk_I^*$ there exists $(ipk_i, RI_i, \cdot) \in \mathcal{RI}$.

If any check fails, $\mathcal{O}^{\text{user}}$ returns $\bot$, otherwise it computes $(nym, pt) \xleftarrow{\$}$ Present$\big(usk_U^*, scope, (ipk_i, RI_i, cred_i, (a_{i,j})_{j=1}^{n_i}, R_i)_{i=1}^k, E, M\big)$ and adds $\big(nym,$ $pt, scope, (ipk_i, RI_i, (a_{i,j})_{j\in R_i})_{i=1}^k, E, M\big)$ to $\mathcal{HP}$. It returns (nym, pt) to A.

- On input $\big($obtain, $U, scope, rh, (ipk_i, RI_i, cid_i, R_i)_{i=1}^{k+1}, E, M, (a_{k+1,j})_{j=1}^{n_{k+1}}\big)$ the oracle checks if $U \in [n_U]$ and, for all $i \in [k]$, a tuple $(U, cid_i, cred_i, (a_{i,j})_{j=1}^{n_i}) \in$ $\mathcal{C}$ exist. It further checks that the revocation information of honest issuers is authentic, i.e., that a tuple $(ipk_i, RI_i, \cdot) \in \mathcal{RI}$ exists for all honest issuers $ipk_i \in IK^*$. The oracle computes the issuance token $(nym, pit, sit) \xleftarrow{\$}$ ITGen$\big(usk_U^*, scope, rh, (ipk_i, RI_i, cred_i, (a_{i,j})_{j=1}^{n_i}, R_i)_{i=1}^{k+1}, E, M\big)$. If $ipk_{k+1} \notin$ IK^*, the oracle sends (nym, pit) to the adversary and runs $\mathcal{U}$.Issue(sit, pit) in interaction with the adversary until it returns a credential $cred$ and stores $(U, cid_{k+1}, cred, (a_{k+1,j})_{j=1}^{n_{k+1}})$ in $\mathcal{C}$. If $ipk_{k+1} = ipk_I^* \in IK^*$, the oracle runs $\mathcal{U}$.Issue(sit, pit) internally against $\mathcal{I}$.Issue(isk_I^*, pit) until they output a credential $cred$ and revocation information RI', respectively. The oracle adds $(U, cid_{k+1}, cred, (a_{k+1,j})_{j=1}^{n_{k+1}})$ to $\mathcal{C}$ and adds (ipk_I^*, rh) to $\mathcal{IRH}$. It further increases $epoch_I^*$, sets $RI_I^* \leftarrow RI'$, and adds $(ipk_I^*, RI_I^*, epoch_I^*)$ to $\mathcal{RI}$. Finally, the oracle chooses a fresh and unique credential identifier cid_{k+1} for the new credential and outputs it to A (note that A's choice of cid_{k+1} is ignored).

3.3 Security Definitions for PABCs

We now formalize the security properties required from PABC-systems.

Correctness. The correctness requirements are what one would expect, i.e., whenever all parties are honest and run all algorithms correctly, none of the algorithms aborts or outputs reject. That is, (i) if all inputs are correct, Issue always outputs a valid credential, (ii) holding valid credentials satisfying the specified relations allows a user to generate valid presentation- and issuance tokens, and (iii) issuers are able to revoke any attribute of their choice.

Pseudonym Collision-Resistance. On input $spar \xleftarrow{\$} \mathsf{SPGen}(1^\kappa)$, no PPT adversary can come up with two user secret keys $usk_0 \neq usk_1$ and a scope $scope$ such that $\mathsf{NymGen}(spar, usk_0, scope) = \mathsf{NymGen}(spar, usk_1, scope)$ with non-negligible probability.

Furthermore, we require that a pseudonym is a deterministic function of the system parameters, user secret and the scope. That is, we require that for some deterministic function NymGen, and for all $usk, scope, rh, E, E', M, M'$ and all honest tuples C, C' of the form $(ipk_i, RI_i, cred_i, (a_{i,j})_{j=1}^{n_i}, R_i)_{i=1}^k$ it holds that:

$$\Pr[nym = nym_{\mathrm{i}} = nym_{\mathrm{p}} : spar \xleftarrow{\$} \mathsf{SPGen}(1^\kappa), nym \leftarrow \mathsf{NymGen}(spar, usk, scope),$$
$$(nym_{\mathrm{i}}, pit, sit) \xleftarrow{\$} \mathsf{ITGen}(usk, scope, rh, C', E', M'),$$
$$(nym_{\mathrm{p}}, pt) \xleftarrow{\$} \mathsf{Present}(usk, scope, C, E, M)] = 1.$$

Unforgeability. In the unforgeability game, the adversary can request credentials from honest issuers, or trigger honest users to receive or present credentials, using the oracles from Sect. 3.2. After this interaction, the adversary outputs a number of presentation tokens and pseudonyms, i.e., a set $\mathcal{FT}$ of tuples $(nym, pt, scope, (ipk_i, RI_i, (a_{i,j})_{j \in R_i})_{i=1}^k), E, M)$ as defined in the syntax of the Verify algorithm. The adversary wins the game if at least one of the presentation tokens is a forgery or if at least one of the issuance tokens submitted to the honest issuer oracle was a forgery. Note that unforgeability of credentials is implied by this definition, as the adversary can always derive a forged presentation token from a forged credential.

Experiment $\mathsf{Forge}_A(1^\kappa, n_I, n_U)$:
$spar \xleftarrow{\$} \mathsf{SPGen}(1^\kappa)$
$(ipk_I^*, isk_I^*, RI_I^*) \xleftarrow{\$} \mathsf{IKGen}(spar)$ for $I = 1, \ldots, n_I$
$usk_U^* \xleftarrow{\$} \mathsf{UKGen}(spar)$ for $U = 1, \ldots, n_U$
$\mathcal{FT} \xleftarrow{\$} A^{\mathcal{O}^{\mathrm{issuer}}, \mathcal{O}^{\mathrm{user}}}(spar, (ipk_I^*, RI_I^*)_{I=1}^{n_I}, n_U)$
Return 1 if and only if:
$\quad \mathcal{FT}, \mathcal{IT}, \mathcal{HP}, \mathcal{IRH}, \mathcal{RRH}$, and $\mathcal{RI}$ are not consistent.

Fig. 1. $\mathsf{Forge}_A(1^\kappa, n_I, n_U)$

Informally, a forgery is an issuance or presentation token for which the corresponding credentials were not issued to the adversary or are not supposed to be valid w.r.t. the revocation information stated in the token. Now, as the issuer does not see all attributes nor the user secret key of issued credentials, it is often not clear whether or not a given issued credential is one of the credentials corresponding to a token. However, if we assume that we knew all hidden values for each credential issued (including the user secret key), then we can efficiently test whether or not a given issuance or presentation token is a forgery. Thus, if there is an assignment for all the hidden values of the issued credentials such that all the issuance and presentation tokens presented by the adversary correspond to valid credentials, then there is no forgery among the tokens. Or, in other words, if there is no such assignment, then the adversary has produced a forgery and wins the game. Regarding the validity of credentials, the adversary also wins if he outputs a valid token for a credential that was already revoked with respect to the revocation information specified by the adversary (which may not necessarily be the latest published revocation information).

Definition 1 (Unforgeability). *A* PABC-*scheme satisfies* unforgeability, *if for every PPT adversary* A *and all* $n_U, n_I \in \mathbb{N}$ *there exists a negligible function* ν *such that* $\Pr[\mathsf{Forge}_A = 1] \leq \nu$, *where the experiment is described in Fig. 1, and the oracles* $\mathcal{O}^{\mathrm{issuer}}$ *and* $\mathcal{O}^{\mathrm{user}}$ *are as in Sect. 3.2.*

We now define what *consistency* of the sets $\mathcal{FT}, \mathcal{IT}, \mathcal{HP}, \mathcal{IRH}, \mathcal{RRH}$, and $\mathcal{RI}$ means. First, the set $\mathcal{FT}$ must only contain "fresh" and valid presentation tokens, meaning that for each tuple $(nym, pt, scope, (ipk_i, RI_i, (a_{i,j})_{j \in R_i})_{i=1}^k)$,

E, M) in $\mathcal{FT}$ must pass Verify and that for all $i \in [k]$ where $ipk_i \in IK^*$ there exists a tuple $(ipk_i, RI_i, \cdot) \in \mathcal{RI}$. If any tuple in $\mathcal{FT}$ does not satisfy these conditions, then the adversary loses the game. Let $USK \subset \{0,1\}^*$ be a hypothetical set containing the user secret keys that the adversary may have used throughout the game, and let $CRED$ be a hypothetical set containing the credentials from honest issuers that the adversary may have collected during the game. In the latter set, we write $(usk, ipk_I^*, rh, (\alpha_1, \ldots, \alpha_n)) \in CRED$ if the adversary obtained a credential from issuer ipk_I^* with revocation handle rh and attribute values $(\alpha_1, \ldots, \alpha_n)$ bound to user secret usk. Now, we consider the sets $\mathcal{FT}, \mathcal{IT}$, $\mathcal{HP}, \mathcal{IRH}, \mathcal{RRH}$ and $\mathcal{RI}$ to be *consistent* if there exist sets USK and $CRED$ such that the following conditions hold:

1. *Each credential is the result of a successful issuance.* For all revocation handles rh and honest issuer public keys ipk_I^*, the number of tuples $(\cdot, ipk_I^*, rh, \cdot) \in CRED$ is at most the number of tuples $(ipk_I^*, rh) \in \mathcal{IRH}$.
2. *All presentation or issuance tokens correspond to honestly obtained unrevoked credentials.* For every tuple $\left(nym, pt, scope, (ipk_i, RI_i, epoch_i, (a_{i,j})_{j \in R_i})_{i=1}^k, E, M\right) \in \mathcal{FT} \cup \mathcal{IT}$ there exists a $usk \in USK$ and a set of credentials $\{ cred_i = (usk_i, ipk_i, rh_i, (\alpha_{i,j})_{j=1}^{n_i}) : ipk_i \in IK^* \} \subseteq CRED$ such that:
   (a) $usk_i \in \{usk, \varepsilon\}$ (all key-bound credentials are bound to the same key),
   (b) $nym = scope = \varepsilon$ or $nym = \mathsf{NymGen}(spar, usk, scope)$ (if there is a pseudonym, it is for usk and $scope$),
   (c) $\alpha_{i,j} = a_{i,j}$ for all $j \in R_i$ (the revealed attribute values are correct),
   (d) $\alpha_{i,j} = \alpha_{i',j'}$ for $((i,j),(i',j')) \in E$ with $ipk_{i'} \in IK^*$ (the attributes satisfy the equality relations to other credentials from honest issuers), and
   (e) there exists $(ipk_i, RI_i, epoch_i) \in \mathcal{RI}$ such that there exists no tuple $(ipk_i, rh_i, epoch_i') \in \mathcal{RRH}$ with $epoch_i' \leq epoch_i$ (the credentials were not revoked in the epoch where they were presented).

Thus, the adversary wins the game, if there do not exist sets USK and $CRED$ that satisfy all the information captured in $\mathcal{FT}, \mathcal{IT}, \mathcal{HP}, \mathcal{IRH}, \mathcal{RRH}$, and $\mathcal{RI}$.

We had to make a number of design decisions for our definitions, which we want to explain in the following:

- First, we do not require strong unforgeability, i.e., we do not consider a token a forgery if it contains the identical elements as a pt or pit generated by an honest user. In practice, tokens will be bound to messages M that uniquely identify the current session, and thus an adversary will neither be able to reuse pt or pit, nor will it be able to benefit from a weak forgery thereof.
- Next, we do not require that adversarially generated issuance tokens are satisfied until the issuance protocol finishes. That is, we consider forgery of issuance tokens to be a problem only if they are later used in a successful issuance protocol. This is acceptable, as an invalid issuance token does not give an adversary any meaningful power if he cannot use it to obtain a credential. Requiring the stronger property that issuance tokens are unforgeable by themselves is possible, but would further increase the complexity of our definitions – without providing stronger guarantees in practice.

- For blindly issued attributes we require that they satisfy E, but do not forbid, e.g., that an adversary runs the issuance protocol twice with the same issuance token, but with the resulting credentials containing different values for the blinded attributes as long as they satisfy E. This is not a real-world problem, as the adversary could otherwise just run multiple sessions for different issuance tokens, as the issuer will, by definition of blind attributes, not obtain any guarantees on those attributes apart from what E specifies.
- Finally, we allow presentation tokens for earlier epochs than the one underlying a credential to be generated. This makes sense for off-line verifiers who cannot update their revocation information continuously. However, our definition and construction could easily be modified to forbid such tokens.

Simulatable Privacy. For privacy, all issuance protocols and presentation tokens performed by honest users must be simulatable using only the public information that is explicitly revealed during the issuance or presentation. That is, the simulator is not given the credentials, values of hidden attributes, or even the index of the user that is supposed to perform the presentation or issuance, but must provide a view to A that is indistinguishable from a real user.

Experiment $\mathsf{Privacy}_A(1^\kappa, n_U)$:
$b \xleftarrow{\$} \{0, 1\}$
If $b = 0$:
    $spar \xleftarrow{\$} \mathsf{SPGen}(1^\kappa)$
    $usk^*_U \xleftarrow{\$} \mathsf{UKGen}(spar)$ for $U = 1, \dots, n_U$
    $USK^* = \{usk^*_U\}^{n_U}_{U=1}$
    $b' \xleftarrow{\$} A^{\mathcal{O}^{\mathsf{user}}}(spar, n_U)$
Return 1 if and only if $b = b'$.

If $b = 1$:
    $(spar, \tau) \xleftarrow{\$} S_1(1^\kappa)$

    $b' \xleftarrow{\$} A^{\mathcal{F}(n_U, \cdot) | S_2(\tau)}(spar, n_U)$

Fig. 2. $\mathsf{Privacy}_A(1^\kappa, n_U)$

Formalizing this is not straightforward, however. It does not suffice to require two separate simulators that work for issuance and presentation, respectively, because pseudonyms and revocation introduce explicit dependencies across different issuance and presentation queries that must also be reflected in the simulation. Moreover, the simulator must not generate presentation tokens that could not have been generated in the real world, e.g., because the user does not have the required credentials. But as the simulator does not see any user indices or hidden attribute values, it cannot know which queries can be satisfied.

We therefore define a game where an adversary A either runs in a real world with access to an honest user oracle performing the actual protocols, or runs in a simulated world, where oracle queries are first filtered by a *filter* $\mathcal{F}$ and then responded to by a stateful simulator S. The filter's role is to sanitize the queries from non-public information such as user indices, credential identifiers, etc., and to intercept queries that could not be satisfied in the real world. Note that the filter thereby enforces that the adversary can only obtain presentation tokens for valid inputs. This must be guaranteed by the credential system as well, otherwise the adversary could distinguish between both worlds.

Definition 2 (Privacy). *A PABC system is private, if there exist PPT algorithms* S_1, S_2 *such that for every PPT adversary* A *and every* $n_U \in \mathbb{N}$ *there exists a negligible function* ν *such that:* $\Pr[\mathsf{Privacy}_A(1^\kappa, n_U) = 1] \leq 1/2 + \nu(\kappa)$, *cf. (Fig. 2).*

  Here, $\mathcal{O}^{\mathtt{user}}$ *is as described in Sect. 3.2, while* $\mathcal{F}$ *maintains initially empty lists* C *and* P, *a counter* $ctr = 0$, *and internal state* $\mathsf{st_S} = \tau$, *and behaves as follows:*

- *On input* $(\mathtt{present}, U, scope, (ipk_i, RI_i, cid_i, R_i)_{i=1}^k, E, M)$, *the filter checks if* $U \in [n_U]$ *and if, for all* $i \in [k]$, *a tuple* $(U, cid_i, ipk_i, (a_{i,j})_{j=1}^{n_i}, \mathsf{rev}_i) \in C$ *exists. Here,* rev_i *is the code of an algorithm that on input* RI_i *outputs a bit indicating whether* $cred_i$ *is to be considered revoked.* $\mathcal{F}$ *checks if* $\mathsf{rev}_i(RI_i) = 0 \; \forall i \in [k]$ *and that* $a_{i,j} = a_{i',j'}$ *for all* $((i,j), (i',j')) \in E$. *If a check fails, the filter returns* $\perp$. *If* $scope \neq \varepsilon$ *and* $(U, scope, p) \notin P$ *then* $\mathcal{F}$ *sets* $ctr \leftarrow ctr + 1$, $p \leftarrow ctr$, *and adds* $(U, scope, p)$ *to* P. *It then executes* $(\mathsf{st_S}, nym, pt) \xleftarrow{\$} S_2(\mathsf{st_S}, \mathtt{present}, scope, p, (ipk_i, (a_{i,j})_{j \in R_i})_{i=1}^k, E, M)$. *Finally, it returns* (nym, pt) *to* A.
- *On input* $(\mathtt{obtain}, U, scope, rh, (ipk_i, RI_i, cid_i, R_i)_{i=1}^{k+1}, E, M, (a_{k+1,j})_{j=1}^{n_{k+1}})$, *the filter checks if* $U \in [n_U]$ *and if, for all* $i \in [k]$, *a tuple* $(U, cid_i, ipk_i, (a_{i,j})_{j=1}^{n_i}, \mathsf{rev}_i) \in C$ *exists. For all credentials,* $\mathcal{F}$ *checks if* $\mathsf{rev}_i(RI_i) = 0 \; \forall i \in [k]$ *and that* $a_{i,j} = a_{i',j'}$ *for all* $((i,j), (i',j')) \in E$. *If any of the checks fails, the filter returns* $\perp$. *The filter then looks up the same value* p *as in* $\mathcal{O}^{\mathtt{present}}$. *It sets* $(\mathsf{st_S}, nym, pit) \xleftarrow{\$} S_2(\mathsf{st_S}, \mathtt{obtain}, scope, p, (ipk_i, (a_{i,j})_{j \in R_i})_{i=1}^{k+1}, E, M, rh)$ *and returns* (nym, pit) *to* A. *For the subsequent flows in the issuance protocol,* $\mathcal{F}$ *answers each incoming message* M_{in} *from* A *by running* $(\mathsf{st_S}, M_{\mathrm{out}}) \xleftarrow{\$} S_2(\mathsf{st_S}, M_{\mathrm{in}})$. *At the last flow,* S_2 *returns a tuple* $(\mathsf{st_S}, M_{\mathrm{out}}, cid, \mathsf{rev})$. *If* $cid \neq \perp$, $\mathcal{F}$ *adds* $(U, cid, ipk_{k+1}, (a_{k+1,j})_{j=1}^{n_{k+1}}, \mathsf{rev})$ *to* C *and returns* M_{out} *to* A.

Weak Privacy. The definition above ensures a very strong notion of privacy that is not satisfied by all existing PABC schemes as they do not provide unlinkability across multiple presentation tokens that were derived from the same credential. For instance, for U-Prove [11,12] an arbitrary number of presentations cannot be linked to a specific issuance session, but any two presentations of the same credential can be linked to each other. We thus also introduce a strictly weaker privacy notion called *weak privacy*. Informally, we there give the simulator some more information to be able to generate "linkable" presentation tokens if the adversary requests multiple presentations tokens for some credential. This is done by giving S_2 "anonymized" pointers to credential identifiers as input, and thus, the simulator is aware if the same credential is used in multiple presentation sessions and can prepare the simulated token accordingly. Due to the anonymization, the simulator still does not learn the connection between an issued credential and a presentation token, thus untraceability (meaning presentation sessions cannot be linked to the actual issuance of the credential) still holds.

4 Building Blocks

We next introduce the syntax for the building blocks needed in our construction. We omit detailed security definitions of the building blocks here and refer to the full version [1], where we also present concrete instantiations of all components. Compared to PABC-systems, most of the security requirements presented in the following are relatively easy to formalize and prove for a specific instantiation. However, in Sect. 5 we will show that these properties are actually sufficient to obtain PABC-systems, by giving a generic construction for PABC-systems from these building blocks. We stress that the following definitions are heavily inspired by existing work, but have been adapted to facilitate the generic construction.

4.1 Global Setup

Global system parameters are parameters that are shared by all building blocks.

Global System Parameter Generation. The global system parameters are generated as $spar_g = (1^\kappa, \mathcal{AS}, \ell, L, spar_c, ck, spar'_g) \xleftarrow{\$} \mathsf{SPGen}_g(1^\kappa)$, where $\mathcal{AS} \subseteq \pm\{0,1\}^\ell$ is the message space of the signature scheme, and the revocation and pseudonym systems support inputs from at least $\pm\{0,1\}^\ell$. The integer L specifies the maximum number of attributes that can be signed by one signature. Furthermore, $spar_c \xleftarrow{\$} \mathsf{SPGen}_c(1^\kappa, \ell)$ and $ck \xleftarrow{\$} \mathsf{ComKGen}(spar_c)$ is a public master commitment key. Finally, $spar'_g$ potentially specifies further parameters.

4.2 Commitment Schemes

A commitment scheme is a tuple of six algorithms (SPGen_c, $\mathsf{ComKGen}$, Com, $\mathsf{ComOpenVf}$, ComPf, $\mathsf{ComProofVf}$), where SPGen_c generates commitment parameters $spar_c$. Taking these as inputs, $\mathsf{ComKGen}$ generates commitment keys ck, which can be used to compute a commitment/opening pair (c, o) to a message m using the commitment algorithm Com. This pair can be verified using $\mathsf{ComOpenVf}$. Furthermore, we require that one can generate a non-interactive proof of knowledge π of the content m of a commitment c, and the corresponding opening c, using ComPf. This proof can then be publicly verified using $\mathsf{ComProofVf}$.

4.3 Privacy-Enhancing Signatures

We next define the main building block: privacy-enhancing attribute-based signatures (PABS). Informally, parties are split into issuers signing attributes, users obtaining signatures, and verifiers checking whether users possess valid signatures on certain attributes. After setting up some PABS-specific system parameters, each issuer computes his signing/verification key pair, such that everybody can verify that keys are well-formed. At issuance time, users can reveal certain attributes to the issuer and get the remaining attributes signed blindly. Having received a signature, a user can verify its correctness. Presentation is then done in a non-interactive manner: users compute signature presentation tokens, potentially revealing certain attributes, and verifiers can check these tokens.

System Parameter Generation. The signature system parameters are generated as $spar_s = (spar_g, spar'_s) \xleftarrow{\$} \mathsf{SPGen_s}(spar_g)$, where the input are global system parameters and $spar'_s$ potentially specifies further parameters.

Similar to Sect. 3.1, we assume that the respective system parameters are input to all the other algorithms of the given scheme. However, for notational convenience, we will sometimes not make this explicit.

Key Generation. An issuer generates a key pair $(ipk, isk) \xleftarrow{\$} \mathsf{IKGen}(spar_s)$. We assume that the issuer public key implicitly also defines a maximum L of attributes a signature may contain.

Key Verification. An issuer public key ipk can be verified for correctness with respect to κ as $\mathsf{accept/reject} \leftarrow \mathsf{KeyVf}(ipk)$.

Signature Issuance. Issuance of a signature is a protocol between a user and a signature issuer:

$$(sig/\perp, \varepsilon) \xleftarrow{\$} \langle \mathcal{U}.\mathsf{Sign}(ipk, (c_j, o_j)_{j \notin R}, \vec{a}); \mathcal{I}.\mathsf{Sign}(isk, (a_i)_{i \in R}, (c_j, \pi_j)_{j \notin R}) \rangle, \text{where}$$

- $\vec{a} = (a_i)_{i=1}^{L} \in \mathcal{AS}^L$ are the attributes to be signed,
- R denotes indices of attributes that are revealed to the issuer
- c_j is a verified commitment to a_j, o_j is the associated opening information, and π_j is a non-interactive proof of knowledge of the opening of c_j. In particular, c_j might be re-used from a preceding signature presentation, allowing users to blindly carry over attributes into new credentials.

Signature Verification. The correctness of a signature can be verified using SigVf on input a signature sig, attributes $\vec{a}$ and a issuer public key ipk:

$$\mathsf{accept/reject} \leftarrow \mathsf{SigVf}(sig, \vec{a}, ipk).$$

Signature Presentation Token Generation. The user can compute a signature presentation token spt that proves that he possesses a signature for a set of revealed attributes R and committed attributes $(c_j, o_j)_{j \in C}$ where $C \cap R = \emptyset$. Furthermore, a signature presentation token can be bound to a specific message M specifying, e.g., some context information or random nonce to disable adversaries to re-use signature presentation tokens in subsequent sessions:

$$spt/\perp \xleftarrow{\$} \mathsf{SignTokenGen}(ipk, sig, \vec{a}, R, (c_j, o_j)_{j \in C}, M).$$

Signature Presentation Token Verification. A signature presentation token can be verified as $\mathsf{accept/reject} \xleftarrow{\$} \mathsf{SignTokenVf}(ipk, spt, (a_i)_{i \in R}, (c_j)_{j \in C}, M)$.

4.4   Revocation Schemes

Suppose a set of users, e.g., employees, who are granted access to some online resource. Then this set will often change over time. While adding users would be possible with the features presented so far, revoking access for specific users would not. In the following we thus define revocation for signature systems.

We chose a blacklisting approach rather than whitelisting, as whitelists require verifiers to update their local copy of the revocation information every time a new signature gets issued, which makes it harder to realize offline applications. For blacklists, different verifiers may obtain updates at different intervals, depending on their security policies. As a consequence, our generic construction also only supports blacklisting, while the interfaces and definitions from Sect. 3 would also support whitelisting or hybrid approaches.

After having set up system parameters, a revocation authority generates a secret revocation key, together with some public revocation key and revocation information. Using its secret key, the authority can revoke revocation handles by updating the revocation information accordingly. Proving that a revocation handle has not yet been revoked is again done non-interactively: a user can generate a token showing that some commitment contains an unrevoked handle. This token can later be publicly verified.

System Parameter Generation. On input global system parameters, revocation parameters are generated as $spar_r = (spar_g, \mathcal{RS}, spar'_r) \xleftarrow{\$} \mathsf{SPGen}_r(spar_g)$, where $\mathcal{RS}$ specifies the set of supported revocation handles, and $spar'_r$ potentially specifies further parameters.

Revocation Setup. The revocation authority (in the definitions of Sect. 3 this role is taken by the issuer) runs the revocation setup to obtain a secret key rsk, an associated public key rpk and a public revocation information RI:

$$(rsk, rpk, RI) \xleftarrow{\$} \mathsf{RKGen}(spar_r).$$

Attribute Revocation. A revocation handle rh can get revoked by a revocation authority by updating the public revocation information RI to incorporate rh:

$$RI' \xleftarrow{\$} \mathsf{Revoke}(rsk, RI, rh).$$

Revocation Token Generation. A user can generate a token proving that a certain revocation handle, committed to in c, has not been revoked before:

$$rt/\bot \xleftarrow{\$} \mathsf{RevTokenGen}(rh, c, o, RI, rpk).$$

In practice, a revocation presentation will always be tied to a signature presentation to prove that the signature presentation is valid. The value of c in the former will therefore be one of the commitments from the latter.

Revocation Token Verification. A revocation token is verified by:

$$\mathsf{accept}/\mathsf{reject} \leftarrow \mathsf{RevTokenVf}(rt, c, RI, rpk).$$

4.5   Pseudonyms

Users can be known to different issuers and verifiers under unlinkable pseudonyms.

System Parameter Generation. The parameters of a pseudonym system are generated as $spar_p = (spar_g, spar_p') \xleftarrow{\$} \mathsf{SPGen}_p(spar_g)$, where the input are global system parameters, and $spar_p'$ potentially specifies further pseudonym parameters.

User key generation. A user generates his secret key as $usk \xleftarrow{\$} \mathsf{UKGen}(spar_p)$.

Pseudonym generation. A pseudonym nym for given usk and $scope \in \{0,1\}^*$ is computed deterministically as $nym \leftarrow \mathsf{NymGen}(usk, scope)$.

Pseudonym presentation. On input a user's secret key usk, a commitment c to usk with opening information o, and a scope string $scope \in \{0,1\}^*$, the pseudonym presentation algorithm generates a pseudonym nym with a proof π:

$$(nym, \pi) \xleftarrow{\$} \mathsf{NymPres}(usk, c, o, scope).$$

Pseudonym verification. A pseudonym nym and proof π are verified for a commitment c and scope $scope$ as $\mathsf{accept/reject} \leftarrow \mathsf{NymVf}(spar_p, c, scope, nym, \pi)$.

5   Generic Construction of PABCs

We next present a generic construction of PABC systems from the building blocks introduced above. Our construction uses a global setup procedure, a commitment scheme, a PABS scheme, a revocation scheme, as well as a pseudonym scheme, where the syntax is as introduced in Sect. 4.

The idea underlying our construction is to use the pseudonym and revocation schemes unchanged to obtain the according properties for the PABC-system. Issuance and presentation are realized via the given PABS-scheme. However, instead of just signing the attributes, the issuer additionally signs the user secret key and the revocation handle whenever applicable. Similarly, whenever a user computes a presentation token for a set of credentials, it proves knowledge of the corresponding signature on the contained attributes, the revocation handle, and the user secret key, where the latter is always treated as an unrevealed attribute.

These independent components are linked together using commitments. For instance, to show that indeed the revocation handle contained in a credential was shown to be unrevoked, the same commitment/opening pair is used to generate revocation and signature presentation tokens.

5.1 Formal Description of the Construction

Let eq be a function mapping an attribute index (i,j) to its equivalence class as induced by E, i.e., $\mathrm{eq}(i,j) = \{(i,j)\} \cup \{(i',j') : ((i,j),(i',j')) \in E\}$. Let $\overline{E}$ be the set of all these equivalence classes, and $(a_{\overline{e}})_{\overline{e}\in\overline{E}}$ be the corresponding attribute values, i.e., $\mathrm{eq}(i,j) = \overline{e} \Rightarrow a_{i,j} = a_{\overline{e}}$. Finally, let $E_i = \{j : ((i,j),(i',j')) \in E\}$.

As some algorithm names from PABC schemes also appear in the building blocks, we stress that all algorithms called in the construction are those from the building blocks and never from the PABC scheme.

System parameter generation. This algorithm outputs $spar = (spar_g, spar_s, spar_r, spar_p)$, where the different parts are generated using the system parameter algorithms of the building blocks. We assume that the algorithms of all building blocks take their respective parameters as implicit inputs.

User key generation. Users generate their secret keys as $usk \xleftarrow{\$} \mathsf{UKGen}(1^\kappa)$.

Issuer key generation. Issuers generate signature keys $(ipk', isk') \xleftarrow{\$} \mathsf{IKGen}(spar_s)$ and revocation keys $(rsk, rpk, RI') \xleftarrow{\$} \mathsf{RKGen}(spar_r)$. The algorithm outputs $(ipk, isk, RI) = ((ipk', rpk), (isk', rsk), RI')$.

Presentation. On inputs $(usk, scope, (ipk_i, RI_i, cred_i, (a_{i,j})_{j=1}^{n_i}, R_i)_{i=1}^k, E, M)$, this algorithm outputs whatever $\mathsf{AuxPresent}$ (cf. Fig. 3) outputs, where $ipk_i = (ipk_i', rpk_i)$, $cred_i = (sig_i, rh_i)$, and $\hat{M} = \mathbf{pres}\|M$.

Presentation verification. On inputs $(nym, pt, scope, (ipk_i, RI_i, (a_{i,j})_{j\in R_i})_{i=1}^k, E, M)$, Verify outputs whatever $\mathsf{AuxVerify}$ described in Fig. 4 outputs.

Issuance token generation. An issuance token for inputs $(usk, scope, rh_{k+1}, (ipk_i, RI_i, cred_i, (a_{i,j})_{j=1}^{n_i}, R_i)_{i=1}^{k+1}, E, M)$, is generated as specified in Fig. 5.

$(c_{usk}, o_{usk}) \xleftarrow{\$} \mathsf{Com}(usk)$
$(nym, \pi_{nym}) = (\varepsilon, \varepsilon)$
if $scope \neq \varepsilon$:
  $(nym, \pi_{nym}) \xleftarrow{\$} \mathsf{NymPres}(usk, c_{usk}, o_{usk}, scope)$
$(c_{\overline{e}}, o_{\overline{e}}) \xleftarrow{\$} \mathsf{Com}(a_{\overline{e}}) \; \forall \overline{e} \in \overline{E}$
for $i = 1, \ldots, k$ do:
  $(c_{rh,i}, o_{rh,i}) \leftarrow (\varepsilon, \varepsilon)$
  if $cred_i$ is revocable:
    $(c_{rh,i}, o_{rh,i}) \xleftarrow{\$} \mathsf{Com}(rh_i)$
   $rt_i \xleftarrow{\$} \mathsf{RevTokenGen}(rh_i, c_{rh,i}, o_{rh,i}, RI_i, rpk_i)$
$\hat{M} = \hat{M}\|scope\|(ipk_i, (a_{i,j})_{j\in R_i})_{i=1}^k\|E$
for $i = 1, \ldots, k$ do:
  $spt_i \xleftarrow{\$} \mathsf{SignTokenGen}(ipk_i', sig_i, ((a_{i,j})_{j=1}^{n_i}, usk, rh_i), R_i,$
                $((c_{\mathrm{eq}(i,j)}, o_{\mathrm{eq}(i,j)})_{j\in E_i}, c_{usk}, o_{usk}, c_{rh,i}, o_{rh,i}), \hat{M})$
$pt = (c_{usk}, \pi_{nym}, (c_{\overline{e}})_{\overline{e}\in\overline{E}}, (c_{rh,i}, rt_i, spt_i)_{i=1}^k)$
if $rt_i \neq \bot$ and $spt_i \neq \bot$ for $i = 1, \ldots, k$:
  Output (nym, pt)
Output $(\bot, \bot)$

Fig. 3. $\mathsf{AuxPresent}(usk, scope, (ipk_i, RI_i, cred_i, (a_{i,j})_{j=1}^{n_i}, R_i)_{i=1}^k, E, \hat{M})$

$\hat{M} = \hat{M}\| scope \|(ipk_i, (a_{i,j})_{j \in R_i})_{i=1}^k \| E$
if $scope \neq \varepsilon \wedge \mathsf{NymVf}(c_{usk}, scope, nym, \pi_{nym}) = \mathsf{reject}$:
  Output reject
for $i = 1, \ldots, k$ do:
  if $\mathsf{RevTokenVf}(rt_i, c_{rh,i}, RI_i, rpk_i) = \mathsf{reject}$ or
    $\mathsf{SignTokenVf}(ipk_i, spt_i, (a_{i,j})_{j \in R_i}, ((c_{eq(i,j)})_{j \in E_i}, c_{usk}, c_{rh,i})_{i=1}^k, \hat{M}) = \mathsf{reject}$:
    Output reject
Output accept

Fig. 4. $\mathsf{AuxVerify}(nym, pt, scope, (ipk_i, RI_i, (a_{i,j})_{j \in R_i})_{i=1}^k, E, \hat{M})$

$(nym, pt) \xleftarrow{\$} \mathsf{AuxPresent}(usk, scope, (ipk_i, RI_i, cred_i, (a_{i,j})_{j=1}^{n_i}, R_i)_{i=1}^k, E, \mathsf{iss}\|M)$
  thereby saving the used $(c_{\bar{e}}, o_{\bar{e}})_{\bar{e} \in \bar{E}}$
$(c_j, o_j) \xleftarrow{\$} \mathsf{Com}(a_{k+1,j}) \;\forall j \notin R_{k+1} \cup E_{k+1}$
$\pi_j \xleftarrow{\$} \mathsf{ComPf}(c_{eq(k+1,j)}, o_{eq(k+1,j)}, a_{k+1,j}) \;\forall j \in E_{k+1}$
$\pi_j \xleftarrow{\$} \mathsf{ComPf}(c_j, o_j, a_{k+1,j}) \;\forall j \notin R_{k+1} \cup E_{k+1}$
$\pi_{usk} \xleftarrow{\$} \mathsf{ComPf}(c_{usk}, o_{usk}, usk)$
$pit = (pt, rh_{k+1}, (c_j)_{j \in R_{k+1} \cup E_{k+1}}, \pi_{usk}, (\pi_j)_{j \notin R_{k+1}})$
$sit = ((c_{eq(k+1,j)}, o_{eq(k+1,j)})_{j \in E_{k+1}}, (c_j, o_j)_{j \notin R_{k+1} \cup E_{k+1}}, c_{usk}, o_{usk}, ipk'_{k+1},$
$$(a_{k+1,j})_{j=1}^{n_{k+1}}, usk, rh_{k+1})$$
Output (pit, sit, nym)

Fig. 5. $\mathsf{ITGen}(usk, scope, rh_{k+1}, (ipk_i, RI_i, cred_i, (a_{i,j})_{j=1}^{n_i}, R_i)_{i=1}^{k+1}, E, M)$

Issuance token verification. To verify $pit = (pt, rh_{k+1}, (c_{k+1,j}, \pi_{k+1,j})_{j \notin R_{k+1}}, \pi_{usk})$, the verifier returns the output of:

$$\mathsf{AuxVerify}(nym, pt, scope, (ipk_i, RI_i, (a_{i,j})_{j \in R_i})_{i=1}^k, E, \mathsf{iss}\|M).$$

Issuance. For issuance, the user and the issuer run:

$\langle \mathcal{U}.\mathsf{Sign}(ipk'_{k+1}, ((c_{eq(k+1,j)}, o_{eq(k+1,j)})_{j \in E_{k+1}}, (c_j, o_j)_{j \notin R_{k+1} \cup E_{k+1}}, c_{usk}, o_{usk}),$
$((a_{k+1,j})_{j=1}^{n_{k+1}}, usk, rh_{k+1}));$
$\mathcal{I}.\mathsf{Sign}(isk', ((a_i)_{i \in R_{k+1}}, rh_{k+1}), ((c_{eq(k+1,j)}, \pi_j)_{j \in E_{k+1}}, (c_j, \pi_j)_{j \notin R_{k+1} \cup E_{k+1}},$
$(c_{usk}, \pi_{usk})))\rangle,$

where they extract their inputs from *sit*, and *isk* and *pit*, respectively. When the user's protocol returns *sig*, the user outputs $cred = (sig, rh_{k+1})$.

Revocation. On input an issuer secret key $isk = (isk', rsk)$, a revocation information RI and a revocation handle rh, the revocation algorithm returns $RI' \xleftarrow{\$} \mathsf{Revoke}(rsk, RI, rh)$.

The formal statements and proofs of the following theorem are given in [1].

Theorem 1 (informal). *Let the used building blocks satisfy all security properties introduced in Sect. 4. Then the* PABC *scheme resulting from the above construction is secure and simulatably private according Sect. 3.3. Furthermore, if the* PABS *scheme is weakly user private, the resulting scheme is secure and weakly private.*

6 Conclusion

We provided security definitions, a modular construction, and secure instantiations of a PABC system. Our framework encompasses a rich feature set including multi-attribute credentials, multi-credential presentations, key binding, pseudonyms, attribute equality proofs, revocation, and advanced credential issuance with carried-over attributes. Prior to our work, most of these features found provably secure instantiations in isolation, but their combination into a bigger PABC system was never proved secure, nor even defined formally.

Proving formal implications among existing definitions and ours might require substantial further research as for each related work, all definitions would first have to be reduced to the set of commonly considered features.

Finally, even though we think that our feature set is rich enough to cover a wide range of use cases (cf. Sect. 1), there are more features that can be added to our framework. Among these features are *inspection*, where presentation tokens can be de-anonymized by trusted inspectors, or *attribute predicates*, allowing to prove for example greater-than relations between attributes.

References

1. Camenisch, J., Krenn, S., Lehmann, A., Mikkelsen, G.L., Neven, G., Pedersen, M.O.: Formal Treatment of Privacy-Enhancing Credential Systems. ePrint, 2014/708 (2014)
2. ABC4Trust - Attribute-based Credentials for Trust: EU FP7 Project (2015). http://www.abc4trust.eu
3. Camenisch, J., Dubovitskaya, M., Lehmann, A., Neven, G., Paquin, C., Preiss, F.-S.: Concepts and languages for privacy-preserving attribute-based authentication. In: Fischer-Hübner, S., de Leeuw, E., Mitchell, C. (eds.) IDMAN 2013. IFIP AICT, vol. 396, pp. 34–52. Springer, Heidelberg (2013)
4. European Parliament and Council of the European Union: Regulation (EC) No 45/2001. Official Journal of the European Union (2001)
5. European Parliament and Council of the European Union: Directive 2009/136/EC. Official Journal of the European Union (2009)
6. Schmidt, H.A.: National strategy for trusted identities in cyberspace. Cyberwar-Resources Guide, Item 163 (2010)
7. Camenisch, J., Herreweghen, E.V.: Design and Implementation of the idemix Anonymous Credential System. In: Atluri, V. (ed.) ACM CCS 02, pp. 21–30. ACM (2002)
8. Camenisch, J.L., Lysyanskaya, A.: An efficient system for non-transferable anonymous credentials with optional anonymity revocation. In: Pfitzmann, B. (ed.) EUROCRYPT 2001. LNCS, vol. 2045, pp. 93–118. Springer, Heidelberg (2001)
9. Camenisch, J.L., Lysyanskaya, A.: A signature scheme with efficient protocols. In: Cimato, S., Galdi, C., Persiano, G. (eds.) SCN 2002. LNCS, vol. 2576, pp. 268–289. Springer, Heidelberg (2003)
10. Camenisch, J.L., Lysyanskaya, A.: Signature schemes and anonymous credentials from bilinear maps. In: Franklin, M. (ed.) CRYPTO 2004. LNCS, vol. 3152, pp. 56–72. Springer, Heidelberg (2004)

11. Brands, S.: Rethinking Public Key Infrastructure and Digital Certificates - Building in Privacy. Ph.D. thesis, Eindhoven Institute of Technology (1999)
12. Paquin, C., Zaverucha, G.: U-prove Cryptographic Specification v1.1 (Revision 2). Technical report, Microsoft Corporation (2013)
13. IRMA - I Reveal My Attributes: Research Project (2015). https://www.irmacard. org
14. IBM Research Security Team: Specification of the Identity Mixer Cryptographic Library. IBM Technical report RZ 3730 (99740) (2010)
15. Corporation, M.: Proof of Concept on integrating German Identity Scheme with U-Prove technology (2011). http://www.microsoft.com/mscorp/twc/endtoendtrust/vision/eid.aspx
16. Chaum, D.: Untraceable electronic mail, return addresses, and digital pseudonyms. Commun. ACM **24**(2), 84–88 (1981)
17. Verheul, E.R.: Self-blindable credential certificates from the weil pairing. In: Boyd, C. (ed.) ASIACRYPT 2001. LNCS, vol. 2248, p. 533. Springer, Heidelberg (2001)
18. Belenkiy, M., Camenisch, J., Chase, M., Kohlweiss, M., Lysyanskaya, A., Shacham, H.: Randomizable proofs and delegatable anonymous credentials. In: Halevi, S. (ed.) CRYPTO 2009. LNCS, vol. 5677, pp. 108–125. Springer, Heidelberg (2009)
19. Garman, C., Green, M., Miers, I.: Decentralized anonymous credentials. In: NDSS 2014. The Internet Society (2014)
20. Chase, M., Meiklejohn, S., Zaverucha, G.M.: Algebraic MACs and Keyed-Verification Anonymous Credentials. eprint, 2013/516 (2013)
21. Nguyen, L., Paquin, C.: U-Prove Designated-Verifier Accumulator Revocation Extension. Technical report MSR-TR-2013-87 (2013)
22. Zaverucha, G.: U-Prove ID escrow extension. Technical report MSR-TR-2013-86 (2013)
23. Baldimtsi, F., Lysyanskaya, A.: On the security of one-witness blind signature schemes. In: Sako, K., Sarkar, P. (eds.) ASIACRYPT 2013, Part II. LNCS, vol. 8270, pp. 82–99. Springer, Heidelberg (2013)
24. Camenisch, J., Dubovitskaya, M., Haralambiev, K., Kohlweiss, M.: Composable & modular anonymous credentials: definitions and practical constructions. In: Iwata, T., Jung, H.C. (eds.) ASIACRYPT 2015, PartII. LNCS, vol. 9453, pp. 262–288. Springer, Heidelberg (2015)
25. Chase, M.: Efficient Non-Interactive Zero-Knowledge Proofs for Privacy Applications. Ph.D. thesis, Brown University (2008)
26. Belenkiy, M., Chase, M., Kohlweiss, M., Lysyanskaya, A.: P-signatures and noninteractive anonymous credentials. In: Canetti, R. (ed.) TCC 2008. LNCS, vol. 4948, pp. 356–374. Springer, Heidelberg (2008)
27. Chaum, D.: Security without identification: transaction systems to make big brother obsolete. Commun. ACM **28**(10), 1030–1044 (1985)
28. Hanser, C., Slamanig, D.: Structure-preserving signatures on equivalence classes and their application to anonymous credentials. In: Sarkar, P., Iwata, T. (eds.) ASIACRYPT 2014. LNCS, vol. 8873, pp. 491–511. Springer, Heidelberg (2014)
29. Baldimtsi, F., Lysyanskaya, A.: Anonymous credentials light. In: ACM CCS 13, pp. 1087–1098. ACM (2013)
30. Li, J., Au, M.H., Susilo, W., Xie, D., Ren, K.: Attribute-based signature and its applications. In: Feng, D., Basin, D.A., Liu, P. (eds.) ASIACCS 10, pp. 60–69. ACM (2010)

31. Maji, H.K., Prabhakaran, M., Rosulek, M.: Attribute-based signatures. In: Kiayias, A. (ed.) CT-RSA 2011. LNCS, vol. 6558, pp. 376–392. Springer, Heidelberg (2011)
32. Shahandashti, S.F., Safavi-Naini, R.: Threshold attribute-based signatures and their application to anonymous credential systems. In: Preneel, B. (ed.) AFRICACRYPT 2009. LNCS, vol. 5580, pp. 198–216. Springer, Heidelberg (2009)

Minimizing the Number of Bootstrappings in Fully Homomorphic Encryption

Marie Paindavoine[1,2](✉) and Bastien Vialla[3]

[1] Orange Labs, Applied Crypto Group, Caen, France
[2] Université Claude Bernard Lyon 1, LIP (CNRS/ENSL/INRIA/UCBL),
46 Allée d'Italie, 69364 Lyon Cedex 07, France
`marie.paindavoine@ens-lyon.fr`
[3] Université Montpellier, LIRMM, CNRS, 161 rue Ada, 34095 Montpellier, France
`bastien.vialla@lirmm.fr`

Abstract. There has been great progress regarding efficient implementations of fully homomorphic encryption schemes since the first construction by Gentry. However, evaluating complex circuits is still undermined by the necessary resort to the *bootstrapping* procedure. Minimizing the number of times such procedure is called is a simple yet very efficient way to critically improve performances of homomorphic evaluations. To tackle this problem, a first solution has been proposed in 2013 by Lepoint and Paillier, using boolean satisfiability. But their method cannot handle the versatility of fully homomorphic encryption schemes. In this paper, we go one step forward providing two main contributions. First, we prove that the problem of minimizing bootstrapping is NP-complete with a reduction from a graph problem. Second, we propose a *flexible* technique that permits to determine both such minimal number of bootstrappings and where to place them in the circuit. Our method is mainly based on linear programming. Our result can advantageously be applied to existing constructions. As an example, we show that for the Smart-Tillich AES circuit, published on the Internet in 2012, we find about 70 % less bootstrappings than naive methods.

Keywords: Fully homomorphic encryption · Bootstrapping · Complexity analysis · Mixed integer linear programming

1 Introduction

Homomorphic encryption extends traditional encryption in the sense that it becomes feasible to perform operations on ciphertexts, without the knowledge of the secret decryption key. As such, it enables someone to delegate heavy computations on his sensitive data to an untrusted third party, in a secure way. More

---

B. Vialla—This work is partially funded by the HPAC project and the CATREL project of the French Agence Nationale de la Recherche (ANR 11 BS02 013), (ANR 12 BS02 001).

© Springer International Publishing Switzerland 2016
O. Dunkelman and L. Keliher (Eds.): SAC 2015, LNCS 9566, pp. 25–43, 2016.
DOI: 10.1007/978-3-319-31301-6_2

precisely, with such a system, one user can encrypt his sensitive data such that the third party can evaluate a function on the encrypted data, without learning any information on the underlying plain data. Getting back the encrypted result, the user can use his secret key to decrypt it and obtain the result of the evaluation of the function on his sensitive plain data. For a cloud user, the applications are numerous, and reconcile both a rich user experience and a strong privacy protection.

Such a promising idea has first been proposed by Rivest, Adleman and Dertouzos in 1978 [20]. The first homomorphic cryptosystems were able to handle only additions (*e.g.* [14,19]), or only multiplications (*e.g.* [8]), or an arbitrary number of additions but only one multiplication [2]. The first fully homomorphic encryption (FHE) scheme, able to handle an arbitrary number of additions and multiplications on ciphertexts, has been proposed by Gentry in 2009 [10].

In homomorphic encryption schemes, the executed function is typically represented as an *arithmetic circuit*. In practice, any circuit can be described as a set of successive operation gates, each one being either a sum or a product performed over some ring. As we will see, the multiplication is the most important operation to be studied for efficiency optimization of a FHE schemes, and the *multiplicative depth* of a circuit, that is the maximum number of multiplications in a path, is an important parameter for FHE schemes.

In Gentry's construction, based on lattices, each ciphertext is associated with some noise, which grows at each operation (addition or multiplication) done throughout the evaluation of the function (procedure called HE.Eval in the sequel). When this noise reaches a certain limit, decryption is not possible anymore. To overcome this limitation, closely related to the number of operations that the HE.Eval procedure can handle, Gentry proposed in [10] a technique of noise refreshment called "bootstrapping".

The main idea behind this bootstrapping procedure is to homomorphically run the decryption procedure of the scheme on the ciphertext, using an encrypted version of the secret key. It comes along with a circular security assumption, as we have to feed the decryption circuit with an encryption of the secret key. This permits to get a "refreshed" ciphertext, which encrypts the same plaintext, but with less noise: the decryption is then always feasible. However, the counterpart is that its computational cost is quite heavy and it should be avoided as much as possible [16]. Ducas and Micciancio proposed a bootstrapping procedure in less than a second [7], but their procedure can only be applied to ciphertexts encrypting a single bit. HElib [15] procedure, on the other hand, takes roughly 6 min. However, the plaintext space is much larger, yielding an amortized cost per bit operation of the same order. In such a context, it is of great importance to determine the exact minimum number of bootstrappings needed to evaluate a given circuit. This way, the time execution for the evaluation of a function will be optimal for a given FHE scheme.

Noise Growth Model. Such a study requires a model to point out how noise grows operation after each operation. Following [17], we associate to each ciphertext c_i a discrete noise level l_i with $l_i = 1, 2, \ldots$. Level 1 corresponds to the noise

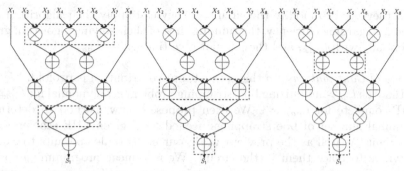

(a) Bootstrapping after (b) Bootstrapping before   (c) Optimal solution.
each multiplication.   each multiplication.

Fig. 1. In dashed rectangle, the bootstrapping positions given by the different heuristics in a FHE scheme with $l_{max} = 2$. (a) The first heuristic uses 5 bootstrappings. (b) The second heuristic uses 4 bootstrappings. (c) Whereas the optimal solution is 3 bootstrappings.

of encryption procedure output. The last level at which it is necessary to either stop the computation or to bootstrap the ciphertext is denoted l_{max}. The bootstrapping procedure does not reset the noise level of a ciphertext to 1 in general but to a level $1 \leq N < l_{max}$. As we will see later, FHE schemes can be divided into two categories depending on the effect of multiplication on noise level, the exponential ones and the linear ones.

Minimizing Bootstrapping. We introduce the l_{max}-*minimizing bootstrapping problem* as finding (one of) the minimal set of ciphertexts one has to bootstrap in order to correctly evaluate a given circuit. Naively, two heuristics can be used in order to avoid unnecessary bootstrappings.

Heuristic 1: One can bootstrap a ciphertext as soon as its noise level reaches l_{max}. It usually means to bootstrap a ciphertext just after a multiplicative gate.

Heuristic 2: When a ciphertext with noise level l_{max} is produced, one waits as long as possible before bootstrapping it. It usually means to bootstrap a ciphertext just before it is used as input into a multiplicative gate.

But, as shown in Fig. 1, these two heuristics most of the time fail to produce a solution to the l_{max}-minimizing bootstrapping problem. In this paper, our aim is then to provide a generic method to find such solution.

Previous Works. To the best of our knowledge, the only method to compute a minimal number of bootstrappings has been proposed by Lepoint and Paillier in [17]. It is based on the SAT problem, known to be NP-complete, and on the definition of some noise management rules. They focus on exponential schemes and proposed a method for any l_{max}. In order to handle linear schemes as well, they need to modify the circuit so they can apply their algorithm as a blackbox. Regarding efficiency, the running time of their solving algorithm grows exponentially with l_{max}, and they do not give timings for $l_{max} \geq 4$.

Outline and Contributions. In this context, our contribution is twofold. We first prove that the l_{max}-minimizing bootstrapping problem is polynomial for $l_{max} = 2$ and NP-complete for $l_{max} \geq 3$. We then propose a new method to determine the minimal number of bootstrappings needed for a given FHE scheme and a given circuit. As well as the previous work, our method also permits to exactly know where to place them in the circuit. We use linear programming to find the best outcome for our problem. The main advantage of our method over the previous one is that it is highly flexible and can be adapted for numerous types of homomorphic encryption schemes and circuits.

The paper is organized as follows. In the next section, we introduce the tools we need all along the paper. Section 3 provides our complexity analysis: the l_{max}-minimizing bootstrapping problem is polynomial for $l_{max} = 2$ and NP-complete for $l_{max} \geq 3$. Finally, Sect. 4 gives our new method for solving the l_{max}-minimizing bootstrapping problem.

2   Background

In this section, we first recall some technical details about graph theory, and in particular arithmetic circuits. We then describe noise growth model during homomorphic evaluation of an arithmetic circuit. Next, we introduce some basic notions of complexity theory. Finally, we present our main tool for solving the l_{max}-minimizing bootstrapping problem which is mixed integer linear programming.

2.1   Graph Theory

As sketched in the introduction, functions handled by homomorphic encryption are arithmetic circuits. They are a particular type of graph. This allows us to make use of complexity results over graph problems for our complexity analysis.

A *graph* G is a couple (V, E) where V is the set of *vertices* and E is the set of *edges*. An edge from a vertex u to a vertex v is noted (u, v). A *directed* graph is a graph where all the edges are oriented, meaning that $\forall u, v \in V$, $(u, v) \neq (v, u)$. For a directed edge (u, v), u is called the *tail* and v the *head*. A $(u_1 - u_{n+1})$-*path* of *length* n is a collection of n edges $((u_1, u_2), (u_2, u_3) \cdots, (u_n, u_{n+1}))$ and a *cycle* is a path where the first vertex equals the last. A directed graph is said *acyclic* if it does not contain any directed cycles. A directed acyclic graph is denoted DAG. The input degree of a vertex x is $|\{(u, x) \in E\}|$, and a vertex whose input degree is equal to 0 is called a *source*. The output degree of a vertex x is $|\{(x, u) \in E\}|$, and a vertex whose output degree is equal to 0 is called a *sink*.

Arithmetic Circuit. An *arithmetic circuit* $\mathcal{C} = (\mathcal{G}, \mathcal{W})$ is a DAG defined over a ring $\mathbb{R}$ and a set of n variables $X = \{X_1, X_2, \cdots, X_n\}$ as follows. The vertices $\mathcal{G}$ of $\mathcal{C}$ are called *gates*. The edges $\mathcal{W}$ of $\mathcal{C}$ are called *wires*. A gate of input degree 0 is an *input gate* and is labelled either by a variable from X or a ring element. Every other gate has an input degree 2, and is labelled either by $\times$ or $+$. We respectively call them *product gates* and *sum gates*. Every gate of output degree 0 is called an *output gate*. In the case of binary circuits defined over $\mathbb{F}_2$, we also have gates of input degree 1. They are labelled NOT, and are called *NOT gates*.

Let $\mathcal{P}$ be a path in $\mathcal{C}$. We call the *multiplicative length* of $\mathcal{P}$ the number of product gates in $\mathcal{P}$. Let us note that the multiplicative length of $\mathcal{P}$ is defined with respect to the number of product gates of $\mathcal{P}$, whereas its length is defined with respect to the number of edges. Therefore, for a path $\mathcal{P}$ that is only composed of k product gates (and no sum gates), its length is $k - 1$ and its multiplicative length is k.

2.2 Noise Growth Model

In existing homomorphic encryption schemes, each ciphertext has some noise attached to it. This noise grows throughout the HE.Eval procedure. In this section, we model how the noise grows operations-wise. As pointed out in the introduction, we use a discretized noise model.

Additions in homomorphic encryption are almost free. The noise growth induced by additions is indeed logarithmic with regard to the noise growth induced by multiplications. It can therefore be neglected most of the time. In this case, let c_1, c_2 be two ciphertexts of noise level l_1 and l_2, and let $c_3 = c_1 + c_2$. We have $l_3 = \max(l_1, l_2)$. However, this restriction is not necessary to apply our method for solving the l_{max}-minimizing bootstrapping problem, and the logarithmic noise induced by additions can be taken into account in our model.

The effect of a multiplication on noise levels divides FHE schemes into two categories. Let c_1, c_2 be two ciphertexts of noise level l_1, l_2 and $c_3 = c_1 \cdot c_2$ with noise level l_3.

- The *exponential* schemes [4,6,9,24]: in these schemes, we have $l_3 = l_1 + l_2$. Therefore, the evaluation of a circuit with a multiplicative depth D will require $l_{max} > 2^D$. This becomes quickly unacceptable and in practice l_{max} is set to 2.
- The *linear* schemes [3,11]: in these schemes, we have $l_3 = \max(l_1, l_2) + 1$. However, in those schemes, the user can set l_{max} to be greater than the multiplicative depth of the circuit to be evaluated. This comes at the cost of greater public parameters. When the multiplicative depth of the circuit is not known in advance, or is too important, one still has to resort to bootstrapping.

2.3 Complexity Theory

We recall the basic definitions of the classic complexity classes that we use in Sect. 3.

A *decision problem* is a yes-or-no question on an infinite set of inputs. A problem P is in the *NP* class if the verifying a feasible solution can be done in polynomial time. A problem P is *NP-hard* if P is at least as hard as the hardest problem in NP. In particular, a NP-hard problem is not necessary in NP.

To prove that a problem P is NP-hard we use a *reduction* that preserves the NP-hardness defined as follows:

Definition 1 (Reduction). *Let A and B be two decision problems, A NP-hard. Let x be an instance of A. A reduction is a pair of algorithms (f, g) such that:*

- *f is a polynomial algorithm transforming x into an instance $f(x)$ of B,*
- *g is a polynomial algorithm transforming a solution y of B in $f(x)$ into a solution $g(x, y)$ in x of A.*

A problem P is *NP-complete* if P is in NP and P is NP-hard.

2.4   Mixed Integer Linear Programming

To solve the l_{max}-minimizing bootstrapping problem, we use linear programming [21], and especially mixed integer linear programming (MILP). Linear programming is used to minimize a linear function whose variables are subject to linear constraints. An *integer linear programming problem* is expressed in the following form. Let A be a matrix in $\mathcal{M}_{m \times n}(\mathbb{R})$, $b \in \mathbb{R}^m$, $c \in \mathbb{R}^n$, $x, l, u \in \mathbb{Z}^n$. The program objective is:

$$
\begin{aligned}
\text{Minimize} \quad & c_1 x_1 + c_2 x_2 + \cdots + c_n x_n \\
\text{Subject To} \quad & a_{11} x_1 + a_{12} x_2 + \cdots + c_{1n} x_n \geq b_1 \\
& \vdots \\
& a_{n1} x_1 + a_{n2} x_2 + \cdots + c_{nn} x_n \geq b_n \\
& \forall x_i, l_i \leq x_i \leq u_i.
\end{aligned}
$$

We call $c^T x$ the *objective function*, x the *problem variables*, l the *lower bounds* on x, u the *upper bounds* on x and Ax the *linear constraints*. Constraints should not be defined with strict inequalities. If $x_i \in \{0, 1\}$, they are named *boolean variables*. The goal of this formulation is to find values for x that minimize the objective function without violating any constraints.

A *mixed integer linear programming problem* is an integer linear programming problem where some of the x_is (and the corresponding u_is and l_is) are allowed to be in $\mathbb{R}$.

Note that non-linear terms are not allowed in the model. Expressing constraints on the multiplication of variables or the maximum of variables is not straightforward, but is still possible with various techniques.

As for any optimization problem, a solution that satisfies all constraints is a *feasible solution*. An *optimal solution* is a feasible solution that achieves the best objective function value.

Theorem 1. *The decisional version of the mixed integer linear programming problem is NP-complete.*

Proof. See [21].   □

3   Complexity Analysis of the l_{max}-Minimizing Bootstrapping Problem

In this section, we first formally introduce the l_{max}-minimizing bootstrapping problem, before proving that it is polynomial for $l_{max} = 2$ and NP-complete for $l_{max} \geq 3$.

The l_{max}-minimizing bootstrapping problem is formally defined as a decision problem as follows.

Definition 2 (l_{max}-Minimizing Bootstrapping (l_{max}-MB)). *Let l_{max} be the desired maximum noise level and $C = (\mathcal{G}, \mathcal{W})$ be an arithmetic circuit. Is there a subset $S \subseteq \mathcal{G}$ of size ω such that each path $\mathcal{P} \subseteq C$ of multiplicative length l_{max} has at least one gate in S?*

3.1   A Polynomial Time Algorithm for $l_{max} = 2$

In order to prove that the l_{max}-minimizing bootstrapping problem is polynomial for $l_{max} = 2$, we design an algorithm that solves it using a *graph connectivity* algorithm as a blackbox.

In a DAG $G = (V, E)$, with a source s and a sink t we define a (s, t)-*separator*, that is, a subset $W \subseteq V$ such that each (s, t)-paths has at least one vertex in W. The *graph connectivity* problem consists in finding a minimal (s, t)-separator. This problem can be solved in $O(|V||E|\log(|V|^2/|E|))$ (see [1]).

In what follows, we describe the algorithm solving the 2-minimizing bootstrapping problem using the graph connectivity problem. Let C be a circuit and $G = (V, E)$ the underlying DAG. As only one level of product is allowed between each bootstrapping, the goal is to split G into subgraphs where each path has a multiplicative length of 1.

The first step is to delete every arc $(u, v) \in E$ where v is a product gate. The resulting graph is named G'. This step is depicted in Fig. 2a.

The connected components of G' are also directed acyclic graphs, but the underlying circuit has at most one level of multiplication. The second step is to add an edge from the source s to each product gate and one from each component's sinks to t. With this construction, each (s, t)-path passes through one and only one product gate. Therefore, in order to correctly evaluate the circuit C, we want to bootstrap each ciphertext once per path. In other words, we have to find the smallest subset of vertices $S \subseteq V$, for which each path has a gate in S. S is an (s, t)-separator of G.

In Fig. 3, we represent our algorithm which computes the minimal set of bootstrappings. A (toy) running example is depicted in Fig. 2.

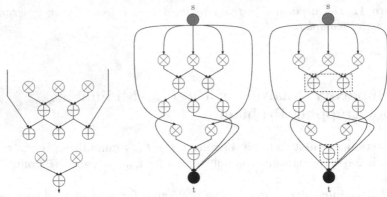

(a) Delete every edge entering a product gate.

(b) Add an edge from s to each product gate, and from every sink to t.

(c) Solve the graph connectivity problem on this instance.

Fig. 2. Algorithm for finding the optimal solution for $l_{max} = 2$ applied to the circuit from Fig. 1

Algorithm 1: Building the minimum set of bootstrappings for $l_{max} = 2$.

Data: $\mathcal{C}$ a circuit and $G = (V, E)$ the associated directed acyclic graph.
Result: The minimum set S of variables to bootstrap.
begin
    Delete every edge (u, v) where v is a product gate (figure 2a);
    Add two vertices s and t, s will be the source of G and t the sink;
    For each multiplication vertices v, add an edge (s, v) (figure 2b);
    For each edges (u, v) deleted in step 1, add an edge (u, t) (figure 2b);
    Compute the minimal (s, t)-vertex separator S (figure 2c);
    Return S;

Fig. 3. Algorithm to compute the minimal set of bootstrapping for exponential schemes.

Theorem 2. *The asymptotic complexity of Algorithm 1 is*

$$\mathcal{O}(|V||E| \log(|V|^2/|E|)).$$

Proof. The complexity of the first and third steps is $\mathcal{O}(|V|)$ and the complexity of the fourth step is $\mathcal{O}(|E|)$. The second step is executed in constant time. The complexity for computing a minimal (s, t)-separator is $\mathcal{O}(|V||E| \log(|V|^2/|E|))$, therefore, the general complexity of the algorithm is $\mathcal{O}(|V||E| \log(|V|^2/|E|))$. □

Thus, the 2-minimizing bootstrapping problem, which mostly corresponds to exponential schemes can be solved in polynomial time. Moreover, graph connectivity algorithms provide us with the optimal bootstrapping location in the circuit.

3.2 NP-Completeness of the l_{max}-Minimizing Bootstrapping Problem

In this section we prove that the l_{max}-minimizing bootstrapping problem is NP-complete for $l_{max} \geq 3$. We reduce the vertex cover problem known to be NP-complete to the l_{max}-MB problem. We need to introduce an intermediary problem: the k-path vertex cover problem.

Let us first recall the decison version of the vertex cover problem on a DAG [18].

Definition 3 (Vertex Cover in Directed Acyclic Graph (VCD)). *Let $G = (V, E)$ be a directed acyclic graph. Is there a subset $W \subseteq V$ of size ω, such that each edge in E admits a vertex of W as tail or head (or both)?*

Theorem 3. *The VCD problem is NP-complete.*

Proof. See [18]. □

Let us now extend the definition of VCD to a directed version of the k-path vertex cover problem from [5].

Definition 4 (k-Path Vertex Cover in Directed Acyclic Graph (k-PVCD)). *Let $G = (V, E)$ be a directed acyclic graph. Is there a subset $W \subset V$ of size ω, such that each path P in G of length k has a vertex in W, i.e., $P \cap W \neq \emptyset$?*

Problems	VCD		$k - $ PVCD
Instances	x	$\xmapsto{f}$	$f(x)$
Solutions	$g(x, Y)$ solution of x	$\xleftarrow{g}$	Y solution of $f(x)$

Fig. 4. Scheme of a reduction from the VCD problem to the k-PVCD problem.

Theorem 4. *The k-PVCD problem is NP-complete for $k \geq 2$.*

Proof. A scheme of the reduction is depicted Fig. 4.

Note that for $k = 2$, k-PVCD is the same as VCD which is NP-complete.

For $k > 2$ we show a reduction (f, g) from the VCD problem to the k-PVCD problem.

Let $G = (V, E)$ be an arbitrary directed acyclic graph. We transform G into a k-PVCD instance $f(G) = G'$. Let $G' = (V', E')$ be the graph obtained from G such that for all $x \in V$ we add a directed path of $\lfloor \frac{k}{2} \rfloor - 1$ new vertices where x is the head; and a path of size $\lceil \frac{k}{2} \rceil - 1$ new vertices where x is the tail. We call the vertices of G *original vertices*, and the others the *new vertices*. This transformation f has a linear complexity with respect to $|V|$. An example is depicted in Fig. 5.

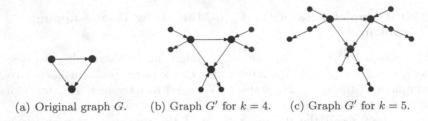

(a) Original graph G.   (b) Graph G' for $k = 4$.   (c) Graph G' for $k = 5$.

Fig. 5. Example of the G' construction.

We now have to transform back a k-PVCD feasible solution Y in G' into a VCD feasible solution $g(G, y)$ in G.

Let y be a k-path vertex cover in G'. Suppose that y contains a new vertex u that lies in one of the added path, *i.e.*, $\exists u \in Y,\ u \notin V$. Let $v \in V$ be the original vertex closest to u. Note that u only secures one path, hence we can swap u with v in Y. We can apply this procedure until all vertices of y are in V. Let us name g the algorithm just described. We claim that $g(G, Y) \subseteq V$ is a vertex cover in G.

Let us suppose otherwise. There is an edge $(u, v) \in E$ such that $u, v \notin g(G, Y)$. Depending on the orientation of the edge between u and v, consider the path P in G', composed of the path attached to x where u is the head (resp. the tail), of the edge (u, v) (resp. (v, u)), and of the path attached to v where v is the tail (resp. the head). Then P does not contain any vertex from y, and it has $\lfloor \frac{k}{2} \rfloor + \lceil \frac{k}{2} \rceil - 2 + 2 = k$ vertices, which is a contradiction. Hence, $g(G, Y)$ is a vertex cover in G.

The transformation g of a k-PVCD feasible solution in G' into a VCD feasible solution in G has a linear complexity with respect to $|V|$.

Conversely, we prove that a vertex cover X in G yields a k-path vertex cover in G'. Let us suppose otherwise. There is a path P of length k in G' such that $P \cap X = \emptyset$. By construction of G', at least one edge of P is in G, let $e = (u, v) \in P$ be this edge. So, $u, v \notin X$ which is a contradiction because X is a vertex cover. Hence, X is a k-path cover in G'.

Thus, there is reduction (f, g) from VCD to k-PVCD: k-PVCD is NP-hard.

We finally prove that an alleged solution of k-PVCD in a DAG $\tilde{G} = (\tilde{V}, \tilde{E})$ can be verified in polynomial time. Let Δ^- be the maximum output degree of $\tilde{G}$. The number of paths of size k in $\tilde{G}$ is at most $O(|\tilde{V}|\Delta^{-k})$, and the paths of length k can be computed using a truncated breadth first search on every vertex, with a complexity of $O(|\tilde{V}|(|\tilde{V}| + |\tilde{E}|))$. So a solution can be verified in polynomial time. k-PVCD lies in NP.

Hence k-PVCD is NP-hard and NP: it is NP-complete.   □

Now we can prove that l_{max}-MB is NP-complete by reducing the k-PVCD problem to the l_{max}-MB problem. A scheme of the reduction is depicted Fig. 7.

Theorem 5. l_{max}-*MB is NP-complete for* $l_{max} \geq 3$.

Proof. We show a reduction (f, g) from the k-path vertex cover problem to the l_{max}-minimizing bootstrapping, for $l_{max} \geq 3$ and $k = l_{max} - 1$.

Let $G = (V, E)$ be an arbitrary directed acyclic graph. We transform G into a l_{max}-MB instance $f(G) = \mathcal{C}$. $\mathcal{C}$ is not required evaluate any "interesting" function. For our reduction purpose, we only need that any path P of length k in G is transformed into a path $\mathcal{P}$ in $\mathcal{C}$ with multiplicative length l_{max}.

Let $\Delta^+(G)$ be the maximum input degree of G. In order to transform G into a circuit $\mathcal{C} = (\mathcal{G}, \mathcal{W})$, we distinguish three cases. When a vertex of G has input degree 2, it is directly transformed into a product gate. When a vertex of G has input degree 1, it is transformed into a product gate, the second input of the gate being a field constant. Finally, every vertex $x \in V$ with input degree at least 3 is transformed into a subcircuit only composed of sum gates (each of input degree 2) except for the last one that will be a product gate, see Fig. 6. Note that f is a bijection between the vertices of G and the product gates of $\mathcal{C}$, and that f has a linear complexity with regard to $|V|$.

We now have to transform a l_{max}-MB feasible solution y in $\mathcal{C}$ into a k-PVCD feasible solution $g(G, y)$ in G.

Let Y be a l_{max}-MB feasible solution in $\mathcal{C}$. The transformation g consists in moving every bootstrapping that is placed on a sum gate to the next product gate downwards. Every bootstrapping is now on a product gate. We claim that $g(x, Y)$ is a $(l_{max} - 1)$-path cover of G.

Let us suppose otherwise. There is a path $P \subseteq G$ of length $l_{max} - 1$ which is not covered by $g(x, Y)$. Let $\mathcal{P} \subseteq \mathcal{C}$ be the path obtained after the transformation of P. A path of length $l_{max} - 1$ is composed of l_{max} vertices. Each of these vertices is transformed into a subcircuit that contains exactly one multiplication. So the multiplicative length of $\mathcal{P}$ is equal to l_{max}. Therefore, there is a path in $\mathcal{C}$ of multiplicative length l_{max} that is not covered by y, which is a contradiction. Hence, $g(x, Y)$ is a $(l_{max} - 1)$-path cover of G. The transformation g between a l_{max}-MB feasible solution in $\mathcal{C}$ and a k-PVCD feasible solution in G has linear complexity with regard to $|V|$.

Conversely, using a similar reasoning, we can show that a $(l_{max} - 1)$-path vertex cover of in G yields a l_{max}-MB in $\mathcal{C}$.

Hence the l_{max}-MB problem is NP-hard. We have now to prove that the l_{max}-MB problem is NP. That is any alleged solution of l_{max}-MB in an arithmetic circuit $\tilde{\mathcal{C}} = (\tilde{\mathcal{G}}, \tilde{\mathcal{W}})$ can be verified in polynomial time. Let Δ^- be the maximum output degree of $\tilde{\mathcal{C}}$. The paths of multiplicative length l_{max} can be computed using a truncated breadth first search on every vertex, with a complexity of $O(|\tilde{\mathcal{G}}|(|\tilde{\mathcal{G}}| + |\tilde{\mathcal{W}}|))$. So a solution can be verified in polynomial time, so l_{max}-MB lies in NP.

Hence, l_{max}-MB is in NP and is NP-hard: it is NP-complete.   □

Thus, the l_{max}-minimizing bootstrapping problem, for $l_{max} \geq 3$, is NP-complete. In the following section we provide a constructive method to solve it.

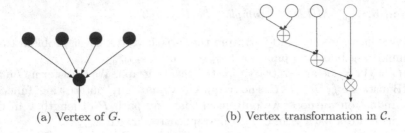

(a) Vertex of G.   (b) Vertex transformation in $\mathcal{C}$.

Fig. 6. Transformation a vertex of input degree greater than 2.

$$
\begin{array}{ccc}
\text{Problems} & k - \text{PVCD} & l_{max} - \text{MB} \\
\text{Instances} & x & \xmapsto{f} \quad f(x) \\
\text{Solutions } g(x, Y) \text{ solution of } x & \xleftarrow{g} Y \text{ solution of } f(x)
\end{array}
$$

Fig. 7. Scheme of a reduction from the k-PVCD problem to the l_{max}-MB problem.

4 Minimizing Bootstrappings with Mixed Integer Linear Programming

In this section we present a general and adaptable method based on mixed integer linear programming for solving the l_{max}-minimizing bootstrapping problem. We first introduce the model's variables and then we describe a general MILP model that can take into account many types of FHE operations. Moreover, one can choose the noise level at which the ciphertexts are refreshed after a bootstrapping.

4.1 Defining Variables and Objective Function of the Program

At each gate of the circuit, we attach a boolean variable which will take the value **true** if it is necessary to bootstrap after the node. The goal of our optimization program will be to minimize the sum of those bootstrapping variables.

For each gate $G^{(i)}$ of the circuit we denote by $G_1^{(i)}$ and $G_2^{(i)}$ the noise levels of the gate inputs. For each output wire of a gate, we add a fictive node corresponding to our bootstrapping boolean variable that we denote $B^{(i)}$. If $B^{(i)}$ equals to one, it means that a bootstrapping is necessary after the ith gate of the circuit. In order to keep the notations simple, $B^{(i)}$ will be used either for the boolean variable or for the fictive bootstrapping computation node. We consider that the $B^{(i)}$ node takes as input the noise level of the gate output it is attached to, which we denote $G_{in}^{(i)}$ and outputs a noise level variable $G_{out}^{(i)}$. These variables are depicted in Fig. 8.

Each of those variables admits 1 as lower bound and l_{max} as upper bound. Furthermore we require that the noise level of each circuit output is strictly less than l_{max} in order to have a correct decryption or to allow further computations.

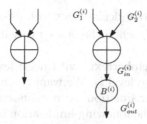

Fig. 8. Variables representing the noise level of a gate in the mixed integer linear programming problem.

Minimizing the number of bootstrappings is equivalent to minimizing the number of boolean variables set to true. Hence, the objective function to be minimized is:

$$\sum_i B^{(i)}.$$

4.2   Linear Constraints

We translate the relations between the noise levels of each gate into linear constraints. We describe them thoroughly for the main FHE operations: addition and multiplication. The model can easily be modified to include other kinds of FHE operations as long as the noise growth can be translated into linear constraints.

Bootstrapping. We first express the constraints that rule the noise growth after the bootstrapping gate added to each gate of the circuit. We recall that the scheme can handle l_{max} operations before the first bootstrapping and that each bootstrapping resets the noise level to N. If we do not bootstrap at a gate $B^{(i)}$, the noise level of the output of the gate is not affected, and we want $G_{in}^{(i)}$ to be equal to $G_{out}^{(i)}$. We can formulate these into a simple constraint:

$$G_{out}^{(i)} = G_{in}^{(i)} \cdot (1 - B^{(i)}) + N \cdot B^{(i)}. \tag{1}$$

This quadratic constraint can be written as a linear constraint using an auxiliary constant X such as $X \geq l_{max}$. The constraints system becomes:

$$\left\{ \begin{array}{lr} G_{out}^{(i)} \geq N \cdot B^{(i)} & (2) \\[4pt] G_{out}^{(i)} \leq N + (1 - B^{(i)}) \cdot X & (3) \\[4pt] G_{out}^{(i)} \geq G_{in}^{(i)} - X \cdot B^{(i)} & (4) \\[4pt] G_{out}^{(i)} \leq G_{in}^{(i)} + X \cdot B^{(i)}. & (5) \end{array} \right.$$

We can see that if the solver decides to bootstrap at gate i, both Eqs. (2) and (3) will force the equality $G_{out}^{(i)} = N$ while Eqs. (4) and (5) remain true.

On the other hand, if the solver decides not to bootstrap, Eqs. (4) and (5) will force the equality $G_{out}^{(i)} = G_{in}^{(i)}$ while the other two will remain true.

Addition. Let c_1, c_2 be two ciphertexts with noise levels l_1, l_2 respectively. We denote $c_3 = c_1 + c_2$ with noise level l_3. We want to ensure that $l_3 = \max(l_1, l_2)$. The maximum is not a linear function, so it cannot be directly used in a constraint. We prove later that the following implication is enough for our purposes:

$$A_{in}^{(i)} = \max(A_1^{(i)}, A_2^{(i)}) \implies \begin{cases} A_{in}^{(i)} \geq A_1^{(i)} \\ A_{in}^{(i)} \geq A_2^{(i)}. \end{cases}$$

These equations are linear so we can use them as constraints with the following bounds on the variables: $1 \leq A_{in}^{(i)} \leq l_{max}$ and $1 \leq A_j^{(i)} \leq l_{max}$.

Remark 1. If the proportion of sum gates in the circuit is overwhelming, our model can consider the logarithmic noise growth induce by additions. Let $\varepsilon \in [0,1]$ be the noise added by a sum gate normalized with respect to the noise added by a product gate. The noise level of a sum gate output is $l_3 = \max(l_1, l_2) + \varepsilon$. Working with mixed integer linear programming instead of integer linear programming allows to consider this noise using the following linear constraints:

$$\begin{cases} A_{in}^{(i)} \geq A_1^{(i)} + \varepsilon \\ A_{in}^{(i)} \geq A_2^{(i)} + \varepsilon, \end{cases}$$

with the same lower and upper bounds as for the addition case.

Multiplication. Let c_1, c_2 be two ciphertexts with noise levels l_1, l_2 respectively. We denote $c_3 = c_1 \cdot c_2$ with noise level l_3. We want to ensure have $l_3 = \max(l_1, l_2) + 1$. We have the following linear constraints:

$$l_3 = \max(l_1, l_2) + 1 \implies \begin{cases} M_{in}^{(i)} \geq M_1^{(i)} + 1 \\ M_{in}^{(i)} \geq M_2^{(i)} + 1, \end{cases}$$

with $1 \leq M_{in}^{(i)} \leq l_{max}$ and $1 \leq M_j^{(i)} \leq l_{max} - 1$ as upper and lower bounds for the linear program.

Other Operations. Other gates types can fit in our model as long as the noise growth rules can be expressed as linear constraints. For example, a multiplication by a constant roughly adds half a level [12] and therefore can be considered. In the GSW scheme [13], the authors used NAND gates. Our model can be applied to such a scheme as a NAND gate behaves with regard to noise growth exactly as a multiplicative gate.

Theorem 6. *The above MILP is equivalent to the l_{max}-minimizing bootstrapping problem.*

Proof. The constraints definition straightforwardly implies that every solution to the l_{max}-MB problem is a solution of the MILP.

Let us now show the converse. Let S be a MILP solution that is not a l_{max}-MB solution. There exists a path P in the circuit with multiplicative length l_{max} such that $P \cap S = \emptyset$. The noise level of a ciphertext along this path respects all the MILP constraints. In particular, it increases by at least 1 at each product gate. Its noise level at the end of the path is thus at least l_{max}. This is in contradiction with the noise variables constraints: each one of them is bounded by l_{max} and the circuit outputs has a noise level strictly less than l_{max}. Then S cannot be a MILP solution.   □

4.3 Practical Experimentations

In this section we discuss the practical results of our model on several circuits from [22], and on the AES circuit used in [12]. Circuits' characteristics are described in Table 1. We assume that the circuit's inputs noise level is equal to 1 and we require that the noise level of each circuit output is strictly less than l_{max}.

MILP Solvers. MILP solvers do not only solve the original program but also its dual. The transformation of a primal form of a MILP into its dual in our case is the following:

$$\min \left\{ c^T x \mid Ax \geq b,\ l \leq x \leq u \right\} \mapsto \max \left\{ b^T y \mid Ay \leq c,\ l \leq y \leq u \right\}.$$

A feasible solution of the dual problem gives a lower bound on the optimal solution [21]. The difference between a feasible solution of the linear program and a feasible solution of its dual is called the *gap*, until it reaches zero. It then means that the solution found is optimal. The gap gives a hint on how far the given solution is from the optimum in the worst case. As we will see in experimentations, the gap value is useful because it allows to get an approximate solution quickly.

Benchmarks. For the experimentation we ran both the Gurobi Optimizer 6[1] and IBM CPLEX 12.6[2] on an Apple MacBook Pro with 2.3 GHz Intel Core i7 and 16 GB of RAM. Each solver implements many different optimization routines, which makes difficult to predict the computation time. We tried both solvers on small circuits and choose the faster one to tackle the problem on bigger circuits. In our case, Gurobi performs better on all circuits. The results are displayed in Table 2. We tested two settings:

1. ($l_{max} = 2, N = 1$). For this setting, we found the same solutions as in [17].

---

[1] http://www.gurobi.com/.
[2] http://www-03.ibm.com/software/products/en/ibmilogcpleoptistud.

Table 1. Circuits' characteristics.

Circuits	Mult. gates	Add. gates	NOT gates	Mult. depth
Adder 32 bits	127	61	187	64
Adder 64 bits	265	115	379	128
Comparator 32 bits	150	0	150	23
Multiplier 32 × 32	5926	1069	5379	128
AES (expanded key)	5440	20325	1927	41
DES (expanded key)	18175	1351	10875	262
MD5	29084	14150	34627	2973
SHA256	90825	42029	103258	3977
Circuit	Mult. gates	Add. gates	Mult. cst gates	Mult. depth
AES [12]	30	220	230	40

2. ($l_{max} = 20, N = 9$) as more realistic parameters, similar to those used in [16], except for the AES from [12] where we chose the same parameters as the authors.

For the simplest circuits, such as **Adder** and **Comparator**, the heuristics find the optimal solution or a close one. For those circuits, computing the optimal solution is done in less than a second.

Table 2. Minimal number of bootstrappings.

Circuits	l_{max}	N	Solution heuristic 1	Solution heuristic 2	MILP solution
Adder 32 bits	2	1	127	127	**127**
Adder 32 bits	20	9	5	5	**4**
Adder 64 bits	2	1	265	267	**265**
Adder 64 bits	20	9	10	12	**10**
Comparator	20	9	1	1	**1**
Multiplier	2	1	6350	5926	**5924**
Multiplier	20	9	105	116	**69**
AES	2	1	4504	5440	**3040**
AES	20	9	736	1600	**220 ± 20**
AES [12]	23	11			**2**
DES	2	1	18399	18175	**18041**
DES	20	9	4435	4006	**440 ± 20**
MD5	2	1	29084	34496	**28896**
SHA256	2	1	90825	97009	**88178**

For bigger circuits, running time is difficult to predict. For l_{max} being small, as well as "close" to the circuit multiplicative depth the optimal solution is found in a couple of minutes. Between these settings, the solver can take hours to find the optimal solution. Nonetheless, the solver always finds a good approximation, better than both heuristics, in tens of minutes. But it can take a couple of hours to prove optimality. This is where the gap value is important: one can choose to stop the computation time when the gap reaches some desired threshold. For the DES circuit, we stopped the solver after 3.5 hours of computation, when the gap reached 5 % of error. In comparison with the more efficient heuristic, this spares 3566 bootstrappings.

Unlike circuits from [22], the AES circuit from [12] exploits all the possibilities offered by a FHE scheme. In particular they use SIMD [23], where ciphertexts are vectors of encrypted plaintexts, and operations are performed component-wise. These vectors are regularly permuted. This does not impact the noise level of ciphertexts. The plaintext space is also bigger than for the binary circuits from [22] which explains that much fewer bootstrappings are needed to correctly evaluate it. This circuit is described is Table 1.

5   Conclusion

While homomorphic encryption implementations are now available for anyone who wants to evaluate circuits on encrypted data, performances in the computation are largely undermined either by time taken by the bootstrapping step or by memory requirement when increasing l_{max}. In this paper we proposed an efficient and flexible technique to determine the minimal number of bootstrapping when evaluating circuits in homomorphic encryption. In [5], the authors give an upper bound on the size of the solution of the k-path vertex cover with respect to the vertices degree of the graph. It would be interesting to see if it is possible to adapt those formulas for the case of the l_{max}-minimizing bootstrapping problem, as that could give constraints on the design of arithmetic circuits. Also, it should be interesting to go further in the complexity analysis of the problem by finding a monadic second order logic formulation, which would allow to apply many meta-theorems giving better insights on the problem. A future work is to provide an automatic tool that, given a circuit and a FHE scheme, could generate a new circuit with optimal bootstrapping placement.

Acknowledgments. We thank Rémi Coletta, Tancrède Lepoint and Guillerme Duvillie for their insights and expertise, and Sébastien Canard, Pascal Giorgi, Laurent Imbert and Fabien Laguillaumie for discussion and comments that greatly improved the manuscript.

References

1. Berge, C.: Graphs. North-Holland Mathematical Library. North Holland Publishing Co., Amsterdam (1985)
2. Boneh, D., Goh, E.-J., Nissim, K.: Evaluating 2-DNF formulas on ciphertexts. In: Kilian, J. (ed.) TCC 2005. LNCS, vol. 3378, pp. 325–341. Springer, Heidelberg (2005)
3. Brakerski, Z., Gentry, C., Vaikuntanathan, V.: (Leveled) fully homomorphic encryption without bootstrapping. In: Innovations in Theoretical Computer Science, Cambridge, MA, USA, 8–10 January 2012, pp. 309–325 (2012)
4. Brakerski, Z., Vaikuntanathan, V.: Efficient fully homomorphic encryption from (standard) LWE. In: IEEE 52nd Annual Symposium on Foundations of Computer Science, FOCS, Palm Springs, CA, USA, 22–25 October 2011, pp. 97–106 (2011)
5. Bresar, B., Kardos, F., Katrenic, J., Semanisin, G.: Minimum k-path vertex cover. Discrete Appl. Math. **159**(12), 1189–1195 (2011)
6. Coron, J.-S., Mandal, A., Naccache, D., Tibouchi, M.: Fully homomorphic encryption over the integers with shorter public keys. In: Rogaway, P. (ed.) CRYPTO 2011. LNCS, vol. 6841, pp. 487–504. Springer, Heidelberg (2011)
7. Ducas, L., Micciancio, D.: FHEW: bootstrapping homomorphic encryption in less than a second. In: Oswald, E., Fischlin, M. (eds.) EUROCRYPT 2015. LNCS, vol. 9056, pp. 617–640. Springer, Heidelberg (2015)
8. El Gamal, T.: A public key cryptosystem and a signature scheme based on discrete logarithms. In: Blakely, G.R., Chaum, D. (eds.) CRYPTO 1984. LNCS, vol. 196, pp. 10–18. Springer, Heidelberg (1985)
9. Gentry, C.: A fully homomorphic encryption scheme. Ph.D. thesis, Stanford University (2009)
10. Gentry, C.: Fully homomorphic encryption using ideal lattices. In: Proceedings of the 41st Annual ACM Symposium on Theory of Computing, STOC, Bethesda, MD, USA, 31 May–2 June 2009, pp. 169–178 (2009)
11. Gentry, C., Halevi, S., Smart, N.P.: Fully homomorphic encryption with polylog overhead. In: Pointcheval, D., Johansson, T. (eds.) EUROCRYPT 2012. LNCS, vol. 7237, pp. 465–482. Springer, Heidelberg (2012)
12. Gentry, C., Halevi, S., Smart, N.P.: Homomorphic evaluation of the AES circuit. In: Safavi-Naini, R., Canetti, R. (eds.) CRYPTO 2012. LNCS, vol. 7417, pp. 850–867. Springer, Heidelberg (2012)
13. Gentry, C., Sahai, A., Waters, B.: Homomorphic encryption from learning with errors: conceptually-simpler, asymptotically-faster, attribute-based. In: Canetti, R., Garay, J.A. (eds.) CRYPTO 2013, Part I. LNCS, vol. 8042, pp. 75–92. Springer, Heidelberg (2013)
14. Goldwasser, S., Micali, S.: Probabilistic encryption and how to play mental poker keeping secret all partial information. In: Proceedings of the 14th Annual ACM Symposium on Theory of Computing, 5–7 May 1982, San Francisco, California, USA, pp. 365–377 (1982)
15. Shoup, V., Halevi, S.: Design and implementation of a homomorphic-encryption library
16. Shoup, V., Halevi, S.: Bootstrapping for helib. Cryptology ePrint Archive, Report 2014/873 (2014). http://eprint.iacr.org/2014/873
17. Lepoint, T., Paillier, P.: On the minimal number of bootstrappings in homomorphic circuits. In: Adams, A.A., Brenner, M., Smith, M. (eds.) FC 2013. LNCS, vol. 7862, pp. 189–200. Springer, Heidelberg (2013)

18. Naumann, U.: DAG reversal is NP-complete. J. Discrete Algorithms **7**(4), 402–410 (2009)
19. Pieprzyk, J.P., Harper, G., Menezes, A., Vanstone, S.A., Paillier, P.: Public-key cryptosystems based on composite degree residuosity classes. In: Stern, J. (ed.) EUROCRYPT 1999. LNCS, vol. 1592, pp. 223–238. Springer, Heidelberg (1999)
20. Rivest, R., Adleman, L., Dertouzos, M.: On data banks and privacy homomorphism. Found. Secur. Comput. **4**, 168–177 (1978)
21. Sierksma, G., Linear, I.P.: Theory and Practice. Advances in Applied Mathematics, 2nd edn. Taylor & Francis, London (2001)
22. Smart, N.P., Tillich, S.: Circuits of basic functions suitable for MPC and FHE. http://www.cs.bris.ac.uk/Research/CryptographySecurity/MPC/
23. Smart, N.P., Vercauteren, F.: Fully homomorphic SIMD operations. IACR Cryptology ePrint Archive 133 (2011)
24. van Dijk, M., Gentry, C., Halevi, S., Vaikuntanathan, V.: Fully homomorphic encryption over the integers. In: Gilbert, H. (ed.) EUROCRYPT 2010. LNCS, vol. 6110, pp. 24–43. Springer, Heidelberg (2010)

Privacy-Preserving Fingerprint Authentication Resistant to Hill-Climbing Attacks

Haruna Higo[1]([✉]), Toshiyuki Isshiki[1], Kengo Mori[1], and Satoshi Obana[2]

[1] NEC Corporation, Tokyo, Japan
h-higo@aj.jp.nec.com, {t-issiki,ke-mori}@bx.jp.nec.com
[2] Hosei University, Tokyo, Japan
obana@hosei.ac.jp

Abstract. This paper proposes a novel secure biometric authentication scheme that hides the biometric features and the distance between the enrolled and authenticated biometric features, and prevent impersonation. To confirm that the proposed scheme has such properties, we formally model secure biometric authentication schemes by generalizing the related and proposed schemes. As far as we know, the proposed scheme is the first one that has been proved to satisfy all the properties. In particular, the proposed scheme achieves security under the decisional Diffie-Hellman assumption.

Keywords: Biometric authentication · Fingerprint minutiae · Hill-climbing attack · Privacy-preserving technology

1 Introduction

Background. Traditionally, ID/passwords and tokens including cards are widely used as a means of authentication. However, they are at risk of being forgot or stolen. In contrast, biometric characteristics including fingerprint, face, palm veins, palm print, iris, and retina cannot be forgot or stolen. Therefore, biometric authentication has an advantage compared to traditional authentication means.

Biometric authentication makes use of the similarity of biometric features extracted from the same biometric characteristic. If a biometric feature presented by a client is similar enough to an enrolled biometric feature in some distance metrics, the client is successfully authenticated by the server. Since biometric features are unchangeable private information, it is required to prevent them from being leaked. Moreover, impersonation should also be prevented.

Related Works. For protecting biometric features from being leaked and preventing impersonation, many schemes have been proposed [16,18]. Some of them exploit secret information (i.e., helper data) that is remembered [6,12,13,15] or tokens (e.g., smart cards or devices) brought along [5,14] by the clients in addition to their biometric characteristics. However, to exploit the advantage of biometric characteristics that there is no risk of them being forgotten or stolen, it is preferable not to use other secret information that has such risks.

© Springer International Publishing Switzerland 2016
O. Dunkelman and L. Keliher (Eds.): SAC 2015, LNCS 9566, pp. 44–64, 2016.
DOI: 10.1007/978-3-319-31301-6_3

Without additional secrets, some schemes [1,3,4,10,11] employ a third entity called a decryptor that manages a secret key to protect biometric features from being leaked and prevent impersonation. In the schemes in [1,3,4,10,11], the decryptor computes the distance between the enrolled and authenticated biometric features and compares them with a predetermined threshold. However, it is known that the distances are useful for hill-climbing attacks [19] in which the attacker guesses the enrolled biometric feature by observing the change in the distance from multiple authentication trials.

In ACISP 2012, Shahandashti, Safavi-Naini, and Ogunbona [17] proposed a fingerprint matching scheme using minutiae. Minutiae are feature points in fingerprints and are widely used for fingerprint matching. Their scheme makes use of polynomials that are evaluated to be 0 or 1 in accordance with the correspondence of the two input fingerprints. Since the polynomials are evaluated to be binary values, the scheme prevents the distances from being leaked. However, the scheme requires the server to store the enrolled fingerprint itself (i.e., not an encrypted version) while the scheme hides the biometric features during the authentication due to leveraging homomorphic encryption.

Contributions of this Paper. The main contribution of this paper is proposing a secure biometric authentication scheme that uses fingerprint minutiae. The proposed scheme is designed for hiding information of enrolled and authenticated minutiae and the distance between them, and preventing impersonation. The comparison of the minutiae is done in accordance with their locations and orientations, which is the well-known method as in [14,17]. In the proposed scheme, enrolled minutiae are represented in the form of polynomials that are evaluated to be 1 or a random value in accordance with whether the input minutia is considered to be the same as the enrolled one. The scheme utilizes the modified Elgamal cryptosystem to evaluate the polynomials without leaking information of minutiae and distance [8,17]. Similar to the previous schemes [1,3,4,10,11], the decryptor is employed in addition to the server and clients, and it manages the secret key of the homomorphic encryption scheme. Since the operations handled by the decryptor are only decryption and comparison of plaintexts, the decryptor of our scheme can be implemented by hardware security modules (HSM). Therefore, it seems that utilization of the decryptor is realistic with respect to the proposed scheme.

To analyze the security of the proposed scheme, we formally model a secure biometric authentication scheme. The model is a generalization of the previous schemes [1,3,4,10,11] and the proposed one. That is, three types of entities, the server, clients, and decryptor, are employed. We formalize the following four security requirements: (a) hide biometric features from the server (which we call template protection against server), (b) hide biometric features from the decryptor (which we call template protection against decryptor), (c) prevent impersonation (which we call security for authentication), and (d) hide distances from the decryptor (which we call security against hill-climbing attacks). Requirements (a) and (c) are defined by generalizing the security definition provided by Hirano et al. [11] while we newly define requirement (b) and (d). We prove that the proposed scheme satisfies all requirements under standard cryptographic assumptions. In particular, we prove that the proposed scheme satisfies all requirements under the decisional Diffie-Hellman (DDH) assumption.

Table 1. Comparison with previous schemes.

Scheme	[17]	[11]	This paper
Representation of biometric features	Minutiae	Vector	Minutiae
Number of entities	2	3	3
BGN cryptosystem	Not used	Necessary	Not used
Template protection against server	No	Yes	Yes
Template protection against decryptor	–	No	Yes
Security for authentication	Yes	Yes	Yes
Security against hill-climbing attacks	Yes	No	Yes

We compare the related works and our work in Table 1. Shahandashti et al.'s scheme [17] is performed by a server and a client (a decryptor is not included) and hides authenticated biometric features. The template is the information of minutiae in the enrolled fingerprint itself and is not concealed. Therefore, it does not satisfy the requirements for template protection. On the other hand, the scheme satisfies the other two notions. Due to employing a third party, our scheme makes it possible to protect the information of both the enrolled and authenticated biometric features and the distance between them. As mentioned above, the decryptor in Hirano et al.'s scheme [11] does not hide the distance from the decryptor. Also, the scheme makes use of a special type of homomorphic encryption scheme with which evaluation of 2-DNF formulas is feasible when performed on ciphertext introduced by Boneh et al. [2].

2   Preliminaries

In this section, we describe preliminaries that are used in the proposed scheme.

2.1   Homomorphic Encryption Scheme

The homomorphic encryption scheme is a type of public key encryption scheme that has a special property. The property is that from ciphertexts, a new ciphertext corresponding to a result of some operation on the plaintexts can be generated without knowledge of the secret key. We focus on addition as the operation. That is, by using two ciphertexts $c_1 = \mathsf{Enc}(m_1)$ and $c_2 = \mathsf{Enc}(m_2)$, a ciphertext of $m_1 + m_2$ is computable. Such schemes are called additive homomorphic encryption schemes.

We utilize the modified (or lifted) Elgamal cryptosystem [7] in the proposed scheme. The modified Elgamal cryptosystem is an additive homomorphic encryption scheme where algorithms ($\mathsf{Gen}, \mathsf{Enc}, \mathsf{Dec}$) run as follows:

- $(pk, sk) := ((p, g, y), (g, x)) \leftarrow \mathsf{Gen}(1^\kappa)$ where G is a group of prime order p, g is a generator of G, $x \in \mathbb{Z}_p$, and $y := g^x$.
- $(c_1, c_2) := (g^r, y^r g^m) \leftarrow \mathsf{Enc}(pk, m)$ where message m is in $\mathbb{Z}_p$ and $r \in \mathbb{Z}_p$ is chosen randomly.
- $m' := \log_g c_2 / c_1^x = \mathsf{Dec}(sk, (c_1, c_2))$.

To divide Dec into two subalgorithms, we define two algorithms as $c_2/c_1^x = \mathsf{Dec}_1(x, (c_1, c_2))$ and $\log_g z = \mathsf{Dec}_2(g, z)$. Apparently $\mathsf{Dec}(sk, (c_1, c_2)) = \mathsf{Dec}_2(g, \mathsf{Dec}_1(x, (c_1, c_2)))$ holds. The first algorithm Dec_1 is computable for any ciphertext while the second Dec_2 is not always feasible since it requires computing of the discrete logarithm, which is assumed to be hard in general. However, in the proposed scheme, we just check if the plaintext is equal to 0, which is feasible by verifying the result of Dec_1 is 1 or not.

The modified Elgamal cryptosystem has been proved to be IND-CPA secure under the decisional Diffie-Hellman (DDH) assumption that states that solving the DDH problem is hard. It is easy to see that the modified Elgamal cryptosystem has the homomorphic property. From two ciphertexts $c = (c_1, c_2) \leftarrow \mathsf{Enc}(m)$ and $c' = (c_1', c_2') \leftarrow \mathsf{Enc}(m')$, it holds that $c \cdot c' := (c_1 c_1', c_2 c_2') = (g^{r+r'}, y^{r+r'} g^{m+m'})$ which is a ciphertext of $m + m'$.

This property is applicable in evaluating polynomials. An n-th degree polynomial $F(X)$ can be represented in the form of $F(X) = \sum_{i=0}^{n} a_i \cdot X^i$. As explained above, $\mathsf{Enc}(a_i \cdot x^i) = \mathsf{Enc}(a_i)^{x^i}$ holds for any x, i, and a_i. Therefore, from encrypted coefficients $\mathsf{Enc}(a_0), \cdots, \mathsf{Enc}(a_i)$ and any x, a ciphertext of the evaluated value $F(x)$ is computable by just multiplying $\mathsf{Enc}(a_i \cdot x^i)$ for every i, since it holds that $\mathsf{Enc}(F(x)) = \mathsf{Enc}(\sum_{i=0}^{n} a_i \cdot x^i) = \prod_{i=0}^{n} \mathsf{Enc}(a_i \cdot x^i)$.

2.2 Biometric Authentication and Fingerprint Minutiae

Biometric authentication is a technique that uses biometric characteristics such as fingerprints for authenticating individuals. Two biometric features extracted from the same biometric characteristic are, in most cases, different but close in some metric. Therefore, to verify if two biometric features are derived from the same individual, it is sufficient to check if they are close under that metric.

A client who would like to enroll himself extracts a biometric feature from his biometric characteristics using some devices such as sensors and cameras. The server stores a template that is generated from the biometric feature. To make the server authenticate a client, the client extracts a biometric feature again. Then, the server estimates the distance between the biometric feature to be authenticated and the biometric feature that has been generated and stored in a template to check if they have originated from the same biometric characteristic.

A fingerprint contains a number of ridges. Some of them abruptly end (called ridge endings), and others are divided into two ridges (called ridge bifurcations). Feature points such as ridge endings and ridge bifurcations are called *minutiae*. In general, a minutia is represented by its location (x, y) and orientation t. Different types (e.g., ridge endings or ridge bifurcations) are also used in some cases. We assume that the coordinate system is aligned every time biometric characteristics are captured. We refer the readers to [14] and its references for information on pre-alignment techniques.

In authenticating an individual with minutiae, a set of minutiae is extracted from the fingerprint. Two fingerprints are considered to match if they have more than a threshold number of pairs of corresponding minutiae.

Two minutiae are said to correspond if their locations and orientations are close enough. That is, two minutiae $((x, y), t)$ and $((x', y'), t')$ correspond if both $d_2((x, y), (x', y')) := \sqrt{(x - x')^2 + (y - y')^2} \leq \Delta_d$ and $d_1(t, t') := |t - t'| \leq \Delta_t$ hold where Δ_d and Δ_t are predetermined thresholds, and d_1 and d_2 stand for the Euclidean distance in one and two dimensions, respectively. In this paper, each location and orientation are assumed to be represented by integers.

3 Secure Biometric Authentication Schemes

We provide formal definitions of the secure biometric authentication scheme in this section. First, the components and procedures of the scheme are explained. After that, we define its security in accordance with Hirano et al.'s definition [11].

3.1 Algorithms and Procedures

There are three kinds of entities, a server, clients, and a decryptor, in the model of secure biometric authentication scheme. A client uses his own biometric feature to enroll or authenticate himself. Clients are not required to have any secret information other than their own biometric characteristics. The enrolled information is stored by the server. Authentication is performed with the aid of the decryptor who has the secret key. The server decides the authentication result in accordance with whether the enrolled and authenticated biometric features are considered to have originated from the same biometric characteristic.

The procedures of the secure biometric authentication scheme include three phases, setup, enrollment, and authentication. The setup phase is done only once, and afterward the enrollment phase and the authentication phase are executed repeatedly by the clients in an arbitrary order. We now describe the procedures of each phase in detail (Fig. 1).

In the setup phase, the decryptor executes the setup algorithm. It takes as input the security parameter and a tuple of parameters, which includes information about the metrics for evaluating distance and the thresholds of acceptance to generate a public parameter and a secret key. The public parameter is published while the secret key is kept secret from other entities.

In the enrollment phase, a client who would like to register himself on the system runs the pseudonymous identifier encoder (PIE). It generates a protected template from the client's biometric feature. The protected template is sent to the server. The server sets identification data for the client, and the client is informed of the identification data. The protected template with the identification data is stored by the server. Depending on the application, the identification data is decided by the client and the client notifies the client of it.

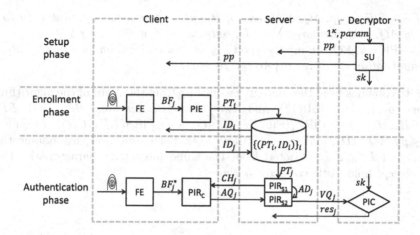

Fig. 1. Algorithms and procedures of secure biometric authentication scheme.

At the beginning of the authentication phase, a client who would like to be authenticated by the server shows his/her identification data to the server. The server selects the protected template that is associated with the identification data and interacts with the client through the pseudonymous identifier recoder (PIR) (for simplicity, we divide the PIR into three algorithms in the definition below). Finally, the server sends a verification query to the decryptor who runs the pseudonymous identifier comparator (PIC) with its secret key to determine the authentication result.

As mentioned in Sect. 2.2, the PIE and the PIR take as input a pre-aligned biometric feature extracted from a biometric characteristic. In Fig. 1, we denote the feature extraction algorithm by FE. This algorithm captures a biometric characteristic and outputs an appropriately aligned biometric feature. Since the pre-alignment technique is out of the range of the secure biometric authentication scheme, the feature extraction algorithm is not included in the tuple of the secure biometric authentication scheme.

The secure biometric authentication scheme is defined as follows. Note that the previous schemes [1,3,4,10,11] can be adapted to the formalization.

Definition 1. *A secure biometric authentication scheme is a tuple of six algorithms* $(\mathsf{SU}, \mathsf{PIE}, \mathsf{PIR}_{S,1}, \mathsf{PIR}_C, \mathsf{PIR}_{S,2}, \mathsf{PIC})$ *that satisfy the following:*

- $(pp, sk) \leftarrow \mathsf{SU}(1^\kappa, param)$ *on input security parameter κ and parameter param, outputs public parameter pp and secret key sk.*
- $PT \leftarrow \mathsf{PIE}(pp, BF)$ *on input public parameter pp and biometric feature BF, outputs protected template PT.*
- $(CH, AD) \leftarrow \mathsf{PIR}_{S,1}(pp, PT)$ *on input public parameter pp and protected template pp, outputs challenge CH and auxiliary data AD.*
- $AQ \leftarrow \mathsf{PIR}_C(pp, BF, CH)$ *on input public parameter pp, biometric feature BF, and challenge CH, outputs authentication query AQ.*

- $VQ \leftarrow \mathsf{PIR}_{S,2}(pp, AQ, AD)$ on input public parameter pp, authentication query AQ, and auxiliary data AD, outputs verification query VQ.
- $res \leftarrow \mathsf{PIC}(sk, VQ)$ on inputs secret key sk and verification query VQ, outputs authentication result $res \in \{Accept, Reject\}$.

When it is obvious from the context, we omit pp from the input of the algorithms.

For $(pp, sk) \leftarrow \mathsf{SU}(1^\kappa)$ and two biometric features BF_e and BF_a, let $(CH, AD) \leftarrow \mathsf{PIR}_{S,1}(\mathsf{PIE}(BF_e))$, $AQ \leftarrow \mathsf{PIR}_C(BF_a, CH)$, $(VQ) \leftarrow \mathsf{PIR}_{S,2}(pp, AQ, AD)$, and $res \leftarrow \mathsf{PIC}(sk, VQ)$. For correctness, we assume that if BF_e and BF_a are extracted from the same biometric characteristic, then $res = Accept$ holds; otherwise $res = Reject$.

3.2 Security

Adversarial clients may try to impersonate a legitimate user while an adversarial server and decryptor aim to obtain some information about the enrolled biometric features. We define these properties under the proposed framework described in the previous section. Note that the decryptor is assumed not to collude with any other entity. We formalize the following four security requirements: (a) hide biometric features from the server (which we call template protection against server), (b) hide biometric features from the decryptor (template protection against decryptor), (c) prevent impersonation (security for authentication), and (d) hide the distances between the enrolled and authenticated biometric features from the decryptor (security against hill-climbing attacks).

Hirano et al. [11] defined security that is specific to their scheme in the semi-honest model where the adversary is considered to corrupt some clients. We follow the definition of [11] about requirements (a) and (c) but we slightly modify the definitions to make them applicable to the proposed framework.

The major difference between the definitions in [11] and ours is to consider the security against an adversarial decryptor. Since the decryptor possesses the secret key, the decryptor is so powerful in the proposed framework that it can even obtain the biometric feature itself by mounting hill-climbing attack in some schemes (e.g., [11]). To capture such attacks by the decryptor, we newly define requirements (b) and (d). Here requirement (b) is defined in the semi-honest model similar to the definition of requirement (a). On the other hand, we newly define requirement (d) as the inability of the decryptor to obtain any information other than the authentication results.

Note that the definitions of requirements (a), (b), and (c) are in the semi-honest model similar to the definitions in [11] while we consider malicious adversaries in the definition of requirement (d). Below, let $(\mathsf{SU}, \mathsf{PIE}, \mathsf{PIR}_{S,1}, \mathsf{PIR}_C, \mathsf{PIR}_{S,2}, \mathsf{PIC})$ be a tuple that satisfies Definition 1.

Template protection against server. This security requirement captures an adversarial server that has templates and authentication queries from clients and try to obtain enrolled biometric features. We introduce a security game between challenger $\mathcal{C}$ and attacker $\mathcal{A}$ as follows (Fig. 2).

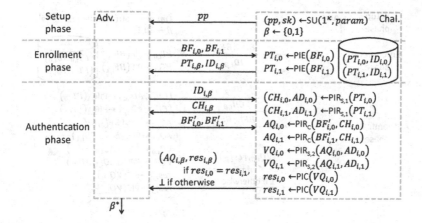

Fig. 2. Game for template protection against server.

Setup: $\mathcal{C}$ runs the setup algorithm to obtain (pp, sk) and chooses bit β randomly. pp is sent to $\mathcal{A}$.

Enrollment: As for the i-th query, $\mathcal{A}$ chooses and sends two biometric features $BF_{i,0}$ and $BF_{i,1}$ to $\mathcal{C}$. $\mathcal{C}$ runs $PT_{i,0} \leftarrow \mathsf{PIE}(BF_{i,0})$ and $PT_{i,1} \leftarrow \mathsf{PIE}(BF_{i,1})$ and selects two identification data $ID_{i,0}$ and $ID_{i,1}$ from $\mathcal{ID}$. $\mathcal{C}$ stores two pairs $(PT_{i,0}, ID_{i,0})$ and $(PT_{i,1}, ID_{i,1})$ and returns $(PT_{i,\beta}, ID_{i,\beta})$ to $\mathcal{A}$.

Authentication: For identification data ID_i from $\mathcal{A}$, $\mathcal{C}$ executes $(CH_{i,0}, AD_{i,0}) \leftarrow \mathsf{PIR}_{S,1}(PT_{i,0})$ and $(CH_{i,1}, AD_{i,1}) \leftarrow \mathsf{PIR}_{S,1}(PT_{i,1})$. Given $CH_{i,\beta}$ from $\mathcal{C}$, $\mathcal{A}$ chooses and sends to $\mathcal{C}$ two biometric features $BF'_{i,0}$ and $BF'_{i,1}$. Then, $\mathcal{C}$ runs $AQ_{i,0} \leftarrow \mathsf{PIR}_C(BF'_{i,0}, CH_i)$, $AQ_{i,1} \leftarrow \mathsf{PIR}_C(BF'_{i,1}, CH_i)$, $VQ_{i,0} \leftarrow \mathsf{PIR}_{S,2}(AQ_{i,0}, AD_i)$, $VQ_{i,1} \leftarrow \mathsf{PIR}_{S,2}(AQ_{i,1}, AD_i)$, $res_{i,0} \leftarrow \mathsf{PIC}(VQ_{i,0})$, and $res_{i,1} \leftarrow \mathsf{PIC}(VQ_{i,1})$, sequentially. $\mathcal{C}$ returns $(AQ_{i,\beta}, res_{i,\beta})$ to $\mathcal{A}$ if $res_{i,0} = res_{i,1}$ and $\perp$ otherwise.

Output: Finally, $\mathcal{A}$ outputs β^*.

Note that the enrollment and authentication phases can be repeated in an arbitrary order.

The advantage of $\mathcal{A}$ is defined as $\mathsf{Adv}_{\mathcal{A}}^{\mathsf{TP},\mathsf{S}}(\kappa) := \Pr[\beta = \beta^*] - 1/2$. With this advantage, the security property is defined as follows.

Definition 2. *We say that a biometric authentication scheme satisfies template protection against server if for any PPT $\mathcal{A}$, $\mathsf{Adv}_{\mathcal{A}}^{\mathsf{TP},\mathsf{S}}(\kappa) \leq negl(\kappa)$.*

Template protection against decryptor. As in Fig. 3, to capture an adversarial decryptor obtaining enrolled biometric features, we slightly modify the game for template protection against server.

Definition 3. *We say that a biometric authentication scheme satisfies template protection against decryptor if for any PPT $\mathcal{A}$, $\mathsf{Adv}_{\mathcal{A}}^{\mathsf{TP},\mathsf{D}}(\kappa) \leq negl(\kappa)$.*

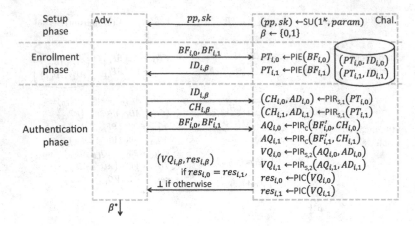

Fig. 3. Game for template protection against decryptor.

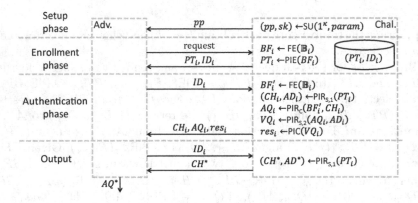

Fig. 4. Game for security for authentication.

Security for authentication. To capture an illegitimate client generating a valid authentication query, we define the following game (Fig. 4).

Setup: $\mathcal{C}$ runs the setup algorithm to obtain (pp, sk), and pp is sent to $\mathcal{A}$.

Enrollment: On the i-th request from $\mathcal{A}$, $\mathcal{C}$ chooses a biometric characteristic $\mathbb{B}_i$ and extracts a biometric feature $BF_i \leftarrow \mathbb{B}_i$. Also, $\mathcal{C}$ chooses identification data $ID_i \in \mathcal{ID}$. Then, $\mathcal{C}$ stores the pair (PT_i, ID_i) and also returns it to $\mathcal{A}$.

Authentication: For identification data ID_i from $\mathcal{A}$, $\mathcal{C}$ extracts a biometric feature from the i-th biometric characteristic $BF'_i \leftarrow \mathbb{B}_i$. With the protected template PT_i that is stored with ID_i, $\mathcal{C}$ executes $(CH_i, AD_i) \leftarrow \text{PIR}_{S,1}(PT_i)$, $AQ_i \leftarrow \text{PIR}_C(BF'_i, CH_i)$, $VQ_i \leftarrow \text{PIR}_{S,2}(AQ_i, AD_i)$, and $res_i \leftarrow \text{PIC}(VQ_i)$, sequentially, and returns (CH_i, AQ_i, res_i) to $\mathcal{A}$.

Output: For identification data ID_i from $\mathcal{A}$, $\mathcal{C}$ executes $(CH^*, AD^*) \leftarrow \text{PIR}_{S,1}(PT_i)$ and returns CH^*. Finally, $\mathcal{A}$ outputs AQ_i^*.

Note that the enrollment and authentication phases can be repeated in an arbitrary order.

The advantage of $\mathcal{A}$ is defined as $\mathsf{Adv}_{\mathcal{A}}^{\mathsf{Auth}}(\kappa) := \Pr[\mathsf{PIC}(\mathsf{PIR}_{S,2}(AQ^*, AD^*)) = Accept]$. With this advantage, security for authentication is defined as follows.

Definition 4. *We say that a biometric authentication scheme is secure in the sense of authentication if for any PPT $\mathcal{A}$, $\mathsf{Adv}_{\mathcal{A}}^{\mathsf{Auth}}(\kappa) \leq negl(\kappa)$.*

Security against hill-climbing attacks. It is preferable that the decryptor obtains as little information as possible. For example, distance is useful in guessing the enrolled biometric feature. Such guessing attacks are called hill-climbing attacks [19]. In the attacks, an attacker casts two queries and learns the distances between the queried biometric features and the enrolled one. From the distances, the attacker is able to learn which query is nearer to the enrolled one. Repeating this approach, he will successfully obtain some accepted queries. Therefore, it is preferable that the schemes do not to give out any information other than the authentication result (acceptance or rejection) to the decryptor. We define the following game (Fig. 5).

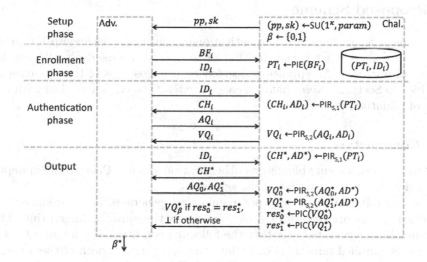

Fig. 5. Game for security against hill-climbing attacks.

Setup: $\mathcal{C}$ runs the setup algorithm to obtain (pp, sk) and chooses $\beta \in \{0, 1\}$ randomly. (pp, sk) is sent to $\mathcal{A}$.

Enrollment: On the i-th request from $\mathcal{A}$ for biometric feature BF_i, $\mathcal{C}$ runs $PT_i \leftarrow \mathsf{PIE}(pp, BF_i)$ and chooses index $ID_i \in \mathcal{ID}$. $\mathcal{C}$ stores pair (PT_i, ID_i) and returns ID_i to $\mathcal{A}$.

Authentication: For identification data ID_i from $\mathcal{A}$, $\mathcal{C}$ executes $(CH_i, AD_i) \leftarrow$ $\mathsf{PIR}_{S,1}(PT_i)$ where PT_i is the protected templates that are stored with ID_i. Given CH_i, $\mathcal{A}$ chooses and sends to $\mathcal{C}$ an authentication query AQ_i. Then, $\mathcal{C}$ runs $VQ_i \leftarrow \mathsf{PIR}_{S,2}(AQ_i, AD_i)$ and returns VQ_i to $\mathcal{A}$.

Output: For identification data ID_i and two biometric features BF_0^* and BF_1^* from $\mathcal{A}$, $\mathcal{C}$ executes $(CH^*, AD^*) \leftarrow \mathsf{PIR}_{S,1}(PT_i)$, $AQ_0^* \leftarrow$ $\mathsf{PIR}_C(BF_0^*, CH^*)$, $AQ_1^* \leftarrow \mathsf{PIR}_C(BF_1^*, CH^*)$, $VQ_0^* \leftarrow \mathsf{PIR}_{S,2}(AQ_0^*, AD^*)$, $VQ_1^* \leftarrow \mathsf{PIR}_{S,2}(AQ_1^*, AD^*)$, $res_0^* \leftarrow \mathsf{PIC}(VQ_0^*)$, and $res_1^* \leftarrow \mathsf{PIC}(VQ_1^*)$, sequentially. Then, $\mathcal{C}$ chooses $\beta \in \{0,1\}$ randomly and returns VQ_β^* to $\mathcal{A}$ if $res_0^* = res_1^*$ and returns $\perp$ otherwise. Finally, $\mathcal{A}$ outputs β^*.

Note that the enrollment and authentication phases can be repeated in an arbitrary order.

The advantage of $\mathcal{A}$ is defined as $\mathsf{Adv}_{\mathcal{A}}^{\mathsf{Dist}}(\kappa) := \Pr[\beta = \beta^*] - 1/2$. With this advantage, security against hill-climbing attacks is defined as follows.

Definition 5. *We say that a biometric authentication scheme is secure against hill-climbing attacks if for any PPT $\mathcal{A}$, $\mathsf{Adv}_{\mathcal{A}}^{\mathsf{Dist}}(\kappa) \leq negl(\kappa)$.*

4   Proposed Scheme

We propose a secure biometric authentication scheme that uses fingerprint minutiae. That is, the biometric features that the proposed scheme deals with consist of a set of minutiae. Acceptance is decided by the closeness of the minutiae as explained in Sect. 2.2. Note that it is easy to extend the scheme to deal with the types of minutiae.

4.1   Construction

Here, we propose a secure biometric authentication scheme. Prior to the in-depth description, we give an outline of the scheme.

In the enrollment phase, two polynomials are generated in accordance with the location and orientation of each minutia in the enrolled fingerprint. The polynomials are generated to satisfy the following condition: if a minutia that is close to the enrolled minutia is input into the polynomials, both of them result in 1. The template is a tuple of encrypted coefficients of the polynomials.

Later in the authentication phase, the polynomials are evaluated for every minutia of the fingerprint to be authenticated in the ciphertext domain. Then, the decryptor with the secret key checks to find out if the evaluated values of polynomials are 1 to determine if the authenticated minutia and the enrolled minutia are close enough to be corresponding minutiae.

As explained above, the proposed scheme proceeds in the same way for every pair of minutiae in the enrolled and authenticated fingerprints. Therefore, for simplicity, we explain the proposed scheme for matching a pair of minutiae. By just doing the same for every minutia pair, it is easy to extend the scheme to deal with fingerprints that consist of multiple minutiae.

Now we describe the algorithms of the proposed scheme $(\mathsf{SU}, \mathsf{PIE}, \mathsf{PIR}_{S,1},$ $\mathsf{PIR}_C, \mathsf{PIR}_{S,2}, \mathsf{PIC})$. In the following algorithms, $\mathsf{PKE} = (\mathsf{Gen}, \mathsf{Enc}, \mathsf{Dec})$ represents the modified Elgamal cryptosystem of which the plaintext domain is denoted by $\mathbb{Z}_p$. For simplicity, the homomorphic operations are described as $\mathsf{Enc}(m_1 + m_2) = \mathsf{Enc}(m_1)\mathsf{Enc}(m_2)$ and $\mathsf{Enc}(m)^i = \mathsf{Enc}(im)$. In addition, we define a ciphertext transformation function Trans by $\mathsf{Trans}(r; (c_1, c_2)) = (c_1^r, c_2)$ where $r \in \mathbb{Z}_p$ and (c_1, c_2) is a ciphertext. Informally, without r, the decryptor cannot decrypt the transformed ciphertexts. It is easy to see that the additive homomorphic property also holds for transformed ciphertexts.

$\mathsf{SU}(1^\kappa, param)$ where $param$ includes the thresholds Δ_d and Δ_t:

1. Generates a public/secret key pair of the homomorphic encryption scheme as $(pk, sk) \leftarrow \mathsf{Gen}(1^\kappa)$.
2. Outputs public parameter $pp := (pk, param)$ and the secret key sk.

$\mathsf{PIE}(BF = ((x, y), t))$:

1. Generates two polynomials

$$F(X, Y) = \sum_i \sum_j a_{i,j} \cdot X^i \cdot Y^j, \text{ and} \tag{1}$$

$$G(T) = \sum_k b_k \cdot T^k \tag{2}$$

that satisfy $F(X, Y) = 1$ if $d_2((X, Y), (x, y)) \leq \Delta_d$ and $G(T) = 1$ if $d_1(T, t) \leq \Delta_t$ (the details of the polynomials including the values i, j, and k are provided below).
2. Encrypts $A_{i,j} \leftarrow \mathsf{Enc}(a_{i,j})$ and $B_k \leftarrow \mathsf{Enc}(b_k)$, and outputs the protected template $PT := (\{A_{i,j}\}_{i,j}, \{B_k\}_k)$.

$\mathsf{PIR}_{S,1}(PT = (\{A_{i,j}\}_{i,j}, \{B_k\}_k))$:

1. Chooses r_F, r_G, and r randomly from $\mathbb{Z}_p$ and computes $A'_{i,j} := \mathsf{Trans}(r; A_{i,j}^{r_F})$, $B'_k := \mathsf{Trans}(r; B_k^{r_G})$ for every i, j, and k.
2. Encrypts $R \leftarrow \mathsf{Enc}(-r_F - r_G)$.
3. Lets $CH := (\{A'_{i,j}\}_{i,j}, \{B'_k\}_k)$ and $AD := (R, r)$ and outputs (CH, AD).

$\mathsf{PIR}_C(CH = (\{A'_{i,j}\}_{i,j}, \{B'_k\}_k), BF^* = ((x^*, y^*), t^*))$:

1. Outputs $AQ := \prod_i \prod_j \left(A'^{(x^*)^i (y^*)^j}_{i,j} \right) \cdot \prod_k B'^{(t^*)^k}_k$.

$\mathsf{PIR}_{S,2}(AQ, AD = (R, r))$:

1. Chooses $r' \in \mathbb{Z}_p$ randomly and outputs $VQ := (\mathsf{Trans}(1/r; AQ) \cdot R)^{r'}$.

$\mathsf{PIC}(VQ)$:

1. If $\mathsf{Dec}_1(VQ) = 1$ then outputs $Accept$; otherwise outputs $Reject$.

Before explaining the details of the construction of the polynomials, let us confirm the correctness of the proposed scheme. Since it holds that $A'_{i,j} = (\mathsf{Enc}(a_{i,j}))^{r_F} = \mathsf{Enc}(r_F \cdot a_{i,j})$ and $B'_k = (\mathsf{Enc}(b_k))^{r_G} = \mathsf{Enc}(r_G \cdot b_k)$ for any i, j, and k, the challenge from the server is a tuple of encrypted coefficients of randomized polynomials $F'(X, Y) = r_F \cdot F(X, Y) = \sum_i \sum_j (r_F \cdot a_{i,j}) \cdot X^i \cdot Y^j$

and $G'(T) = r_G \cdot G(T) = \sum_k (r_G \cdot b_k) \cdot T^k$. They satisfy $F'(X, Y) = r_F$ and $G'(T) = r_G$ if minutiae $((x, y), t)$ and $((X, Y), T)$ correspond. The authentication query is a ciphertext of the sum of the evaluated value of the polynomials for the authenticated minutia as

$$\mathsf{Trans}(1/r; AQ) = \prod_i \prod_j \mathsf{Enc}(r_F \cdot a_{i,j})^{(x^*)^i (y^*)^j} \cdot \prod_k \mathsf{Enc}(r_G \cdot b_k)^{(t^*)^k}$$
$$= \mathsf{Enc}\left(F'(x^*, y^*) + G'(t^*)\right).$$

Therefore, the verification query computed by the server is

$$VQ = \left(\mathsf{Enc}\left(F'(x^*, y^*) + G'(t^*)\right) \cdot \mathsf{Enc}\left(-r_F - r_G\right)\right)^{r'}$$
$$= \mathsf{Enc}\left(r'\left(r_F\left(F'(x^*, y^*) - 1\right) + r_G\left(G(t^*) - 1\right)\right)\right).$$

Thus, if the enrolled and authenticated minutiae correspond, VQ is a ciphertext of 0 that is $\mathsf{Dec}_1(VQ) = 1$ holds.

Description of polynomials. In the above scheme, polynomials $F(X, Y)$ and $G(T)$ that satisfy $F(X, Y) = 1$ if $d_2((X, Y), (x, y)) \leq \Delta_d$ and $G(T) = 1$ if $d_1(T, t) \leq \Delta_t$, respectively, are generated in accordance with enrolled minutia $\{((x, y), t)\}$. Such polynomials are constructible as

$$F(X, Y) = R_F \prod_{\ell \in L} \left\{(X - x)^2 + (Y - y)^2 - \ell\right\} + 1, \text{ and} \tag{3}$$

$$G(T) = R_G \prod_{\ell = -\Delta_t}^{\Delta_t} \left\{(T - t) - \ell\right\} + 1, \tag{4}$$

where R_F and R_G are randomly chosen and the set $L := \{d_2^2((X, Y), (X', Y')) \leq \Delta_d^2 \mid X, X', Y, Y' \in \mathbb{Z}\}$ consists of the possible values of the squared Euclidean distance that are smaller than the squared threshold Δ_d^2. In the above scheme, we describe the functions as an expanded form as $F(X, Y) = \sum_{i=0}^{2|L|} \sum_{j=0}^{2|L|} a_{i,j} \cdot X^i \cdot Y^j$ and $G(T) = \sum_{k=0}^{2\Delta_t + 1} b_k \cdot T^k$.

It is easy to see that they satisfy the required properties. $F(X, Y) - 1 = R_F \prod_{\ell \in L} \left\{(X - x)^2 + (Y - y)^2 - \ell\right\} = 0$ holds for all (X, Y) that satisfy $d_2^2((X, Y), (x, y)) = (X - x)^2 + (Y - y)^2 = \ell$ for some $\ell \in L$. $G(T) - 1 = R_G \prod_{\ell = -\Delta_t}^{\Delta_t} \left\{(T - t) - \ell\right\} = 0$ holds for all T that satisfy $d_1(T, t) = T - t = \ell$ for some $\ell \in \{-\Delta_t, \cdots, \Delta_t\}$.

Recall that all locations are integers, the set L is a subset of and is smaller than the set $\{0, 1, \cdots, \Delta_d^2\}$. For example, there exists no tuple (X, X', Y, Y') that satisfies $d_2^2((X, Y), (X', Y')) = 3$. Therefore, $3 \notin L$ holds. Set L is determined by Δ_d, for example, $L = \{0, 1, 2, 4, 5, 8, 9, 10, 13, 16, 17, 18, 20, 25\}$ when $\Delta_d = 5$. The number of terms in the derived polynomial F is computed to be $2|L|^2 + 3|L| + 1$.

4.2 Security

We show that the proposed scheme satisfies the security notions defined in Sect. 3.2. Detailed proofs are provided in Appendix A.

Template protection against server and template protection against decryptor. Under the IND-CPA security of the modified Elgamal cryptosystem, the proposed scheme can be proved to satisfy template protection against server and template protection against decryptor.

Theorem 1. *Under the IND-CPA security of the modified Elgamal cryptosystem, the proposed scheme satisfies template protection against server.*

Theorem 2. *Under the IND-CPA security of the modified Elgamal cryptosystem, the proposed scheme satisfies template protection against decryptor.*

Security for authentication. Under Assumption 1, we can prove Theorem 3.

Assumption 1. *For any PPT algorithm $\mathcal{A}$, it holds that*

$$\Pr\left[t = \mathsf{Dec}(t^*) \,\middle|\, \begin{matrix} (pk, sk) \leftarrow \mathsf{Gen}(1^\kappa); s, t \leftarrow \{0,1\}^\kappa; \\ t^* \leftarrow \mathcal{A}(pk, \mathsf{Enc}(s), \mathsf{Enc}(st)); \end{matrix}\right] \leq negl(\kappa).$$

Theorem 3. *Provided that the modified Elgamal cryptosystem satisfies Assumption 1, the proposed scheme satisfies security for authentication.*

Security against hill-climbing attacks. In the authentication phase, the decryptor checks if the decrypted value of the verification query is equal to 0. 0 implies that the distance does not exceed the threshold. Since non-0 values, which mean that the distance is greater than the threshold, are determined by the random values which the server chooses in every authentication, it looks random from the decryptor. Therefore, the decryptor can only know the authentication result but cannot guess the distance.

Theorem 4. *The proposed scheme satisfies security against hill-climbing attacks.*

5   Conclusion

We have defined the model of secure biometric authentication. A third party called a decryptor is employed in our model in addition to the normal entities in biometric authentication. Also, we have formally defined its security by adapting the security definition provided by Hirano et al. [11] to our model.

In the defined model, we have proposed a scheme that hides biometric features from the server and the decryptor. Moreover, no entity is able to obtain the distance between the enrolled and authenticated biometric features. Therefore, the proposed scheme is resistant to hill-climbing attacks [19]. Since the operations of the decryptor, key generation, and decryption are light enough, they can be implemented by hardware security modules (HSM). By utilizing the modified Elgamal cryptosystem [7], we have showed the security of the proposed scheme under the decisional Diffie-Hellman assumption.

Acknowledgment. We would like to appreciate Anja Lehmann and the anonymous reviewers for their valuable comments. The fourth author is supported by JSPS KAKENHI Grant Number 15K00193.

A Security of Proposed Scheme

In this section, we prove that the proposed scheme satisfies all security properties defined in Sect. 3.2. For simplicity, we show the proof of it where the adversary makes one enrollment query one authentication query. Also we prove three properties, template protection against server, security for authentication, and security against hill-climbing attacks for simpler but less secure version. The simpler version is without Trans. That is, it is different from the version in Sect. 4 in $\mathsf{PIR}_{S,1}$ and $\mathsf{PIR}_{S,2}$ as follows.

$\mathsf{PIR}_{S,1}(PT = (\{A_{i,j}\}_{i,j}, \{B_k\}_k))$:
1. Randomly generates r_F and r_G.
2. For every i, j, and k, computes $A'_{i,j} := A^{r_F}_{i,j}$, $B'_k := B^{r_G}_k$.
3. Encrypts $R \leftarrow \mathsf{Enc}(-r_F - r_G)$.
4. Lets $CH := (\{A'_{i,j}\}_{i,j}, \{B'_k\}_k)$ and $AD := R$ and outputs (CH, AD).

$\mathsf{PIR}_{S,2}(AQ, AD)$:
1. Randomly chooses r' and outputs $VQ := (AQ \cdot AD)^{r'}$.

Template protection against decryptor cannot be proved for the simpler version of the scheme. We explain the reason and that the scheme in Sect. 4 satisfies the property later.

A.1 Proof of Theorem 1 (Template Protection Against Server)

The modified Elgamal cryptosystem satisfies IND-CPA security under the DDH assumption.

Definition 6. *A public key encryption scheme* $\mathsf{PKE} = (\mathsf{Gen}, \mathsf{Enc}, \mathsf{Dec})$ *is indistinguishable against a chosen plaintext attack (CPA), or shortly IND-CPA-secure, if for any PPT algorithm* $\mathcal{A}$ *it holds that*

$$\mathsf{Adv}^{\mathsf{IND}}_{\mathcal{A}}(\kappa) := \Pr\left[\beta = \beta^* \,\middle|\, \begin{array}{l} (pk, sk) \leftarrow \mathsf{Gen}(1^\kappa); (m_0, m_1, \alpha) \leftarrow \mathcal{A}(pk); \\ \beta \leftarrow \{0,1\}; c \leftarrow \mathsf{Enc}(m_b); b^* \leftarrow \mathcal{A}(c, \alpha); \end{array}\right] \leq negl(\kappa).$$

The proof of Theorem 1 is provided using a sequence of games. We first present three games and after that discuss the relations among them.

Game 0. Game 0 is the game of security for template protection. We describe the game for the case of the proposed scheme as follows:

Setup: On input 1^κ and *param*, $\mathcal{C}$ runs $(pk, sk) \leftarrow \mathsf{Gen}(1^\kappa)$ and chooses $\beta \in \{0, 1\}$ at random. $(pk, param)$ is sent to $\mathcal{A}$.

Enrollment: $\mathcal{A}$ chooses and sends two minutiae $((x_0, y_0), t_0)$ and $((x_1, y_1), t_1)$ to $\mathcal{C}$. $\mathcal{C}$ generates two polynomials F_β and G_β according to $((x_\beta, y_\beta), t_\beta)$ as in Eqs. (3) and (4), and expands them as in Eqs. (1) and (2). $\mathcal{C}$ encrypts the coefficients of the polynomials as $A_{i,j} \leftarrow \mathsf{Enc}(a_{i,j})$ and $B_k \leftarrow \mathsf{Enc}(b_k)$ and sends the protected template $PT := (\{A_{i,j}\}_{i,j}, \{B_k\}_k)$ to $\mathcal{A}$.

Authentication: On a request from $\mathcal{A}$, $\mathcal{C}$ chooses r_F and r_G at random and let $A'_{i,j} \leftarrow \mathsf{Enc}(r_F a_{i,j})$ and $B'_k \leftarrow \mathsf{Enc}(r_G b_k)$. On receiving $CH = (\{A'_{i,j}\}_{i,j}, \{B'_k\}_k)$ from $\mathcal{C}$, $\mathcal{A}$ chooses two minutiae $((x'_0, y'_0), t'_0)$ and $((x'_1, y'_1), t'_1)$ and sends them to $\mathcal{C}$. If for two pairs of minutiae $(((x_0, y_0), t_0), ((x'_0, y'_0), t'_0))$ and $(((x_1, y_1), t_1), ((x'_1, y'_1), t'_1))$, one of them corresponds and the other is not, then $\mathcal{C}$ returns $\perp$ to $\mathcal{A}$. Otherwise, $\mathcal{C}$ sends $AQ := \mathsf{Enc}(r_F \cdot F_\beta(x'_\beta, y'_\beta) + r_G \cdot G_\beta(t'_\beta))$ to $\mathcal{A}$.

Output: Finally, $\mathcal{A}$ outputs β^*.

We define the advantage of the adversary in game 0 as $\mathsf{Adv}_{\mathcal{A}}^0(\kappa) := \Pr[\beta = \beta^*] - 1/2$. Apparently $\mathsf{Adv}_{\mathcal{A}}^0(\kappa)$ is the same as $\mathsf{Adv}_{\mathcal{A}}^{\mathsf{TP,S}}(\kappa)$ for the scheme.

Game 1. To make game 1, we only modify the way to generate AQ of game 0. The difference is as follows: Before generating AQ, $\mathcal{C}$ additionally chooses a random bit β' and let $AQ := \mathsf{Enc}(r_F \cdot F_\beta(x'_\beta, y'_\beta) + r_G \cdot G_\beta(t'_\beta))$ if $\beta' = 0$ and $AQ := \mathsf{Enc}(r_F \cdot F_{1-\beta}(x'_{1-\beta}, y'_{1-\beta}) + r_G \cdot G_{1-\beta}(t'_{1-\beta}))$ otherwise. $F_{1-\beta}$ and $G_{1-\beta}$ are the functions generated according to the minutia $((x_{1-\beta}, y_{1-\beta}), t_{1-\beta})$ which was obtained in the enrollment query through Eqs. (3) and (4) with the same random values R_F and R_G that are used in making F_β and G_β.

Obviously, if $\beta' = 0$, game 1 runs exactly the same way as the game 0. We define the advantage of the adversary in game 1 as $\mathsf{Adv}_{\mathcal{A}}^1 := \Pr[\beta = \beta^*] - 1/2$.

Game 2. We further modify the method of generating AQ to define game 2 as follows: Before generating AQ, $\mathcal{C}$ additionally chooses two random bit β' and let $AQ := \mathsf{Enc}(r_F \cdot F_\beta(x'_\beta, y'_\beta) + r_G \cdot G_\beta(t'_\beta))$ if $\beta' = 0$ and $AQ := \mathsf{Enc}(r_F \cdot F_{\beta''}(x'_{\beta''}, y'_{\beta''}) + r_G \cdot G_{\beta''}(t'_{\beta''}))$ otherwise. If $\beta' = 0$ or $\beta'' = \beta$, this game runs exactly the same way as the game 0. We define the advantage of the adversary in game 2 as $\mathsf{Adv}_{\mathcal{A}}^2 := \Pr[\beta'' = \beta^*] - 1/2$. We note that this probability is not taken over β but β''.

In order to prove Theorem 1, we show the following lemmas. The first lemma says that the advantage of game 2 is negligible under the IND-CPA security of the underlying cryptosystem.

Lemma 1. *Let* $\mathsf{PKE} = (\mathsf{Gen}, \mathsf{Enc}, \mathsf{Dec})$ *be an IND-CPA secure public key encryption scheme. Then for any PPT algorithm* $\mathcal{A}$ *it holds that* $\mathsf{Adv}_{\mathcal{A}}^2(\kappa) \leq negl(\kappa)$.

Proof. We assume that there exists an PPT algorithm $\mathcal{A}$ such that $\mathsf{Adv}_{\mathcal{A}}^2$ is non-negligible and show that PKE is not IND-CPA secure. In order to show it we construct an adversary $\mathcal{A}'$ of IND-CPA game as follows:

First, given an public key pk, $\mathcal{A}'$ input $(pk, param)$ into $\mathcal{A}$. When $\mathcal{A}$ queries two minutiae $((x_0, y_0), t_0)$ and $((x_1, y_1), t_1)$ to $\mathcal{C}$ for enrollment, $\mathcal{A}'$ chooses $\beta \in \{0, 1\}$ randomly, generates two polynomials F_β and G_β, and returns PT as in the description of the game 0.

On an authentication query from $\mathcal{A}$, $\mathcal{A}'$ chooses $\beta' \in \{0, 1\}$ and r_F and r_G randomly. If $\beta' = 0$, $\mathcal{A}'$ returns $AQ := \mathsf{Enc}(r_F \cdot F_\beta(x'_\beta, y'_\beta) + r_G \cdot G_\beta(t'_\beta))$ to $\mathcal{A}$. Otherwise $\mathcal{A}'$ outputs $m_0 := r_F \cdot F_0(x'_0, y'_0) + r_G \cdot G_0(t'_0))$ and $m_1 := r_F \cdot F_1(x'_1, y'_1) + r_G \cdot G_1(t'_1))$ as its challenge. Given a ciphertext c, $\mathcal{A}'$ returns it to $\mathcal{A}$. Finally, $\mathcal{A}'$ outputs $\beta^* \in \{0, 1\}$ that $\mathcal{A}$ outputs.

From the description of IND-CPA game, c is the ciphertext of $m_{\beta''}$ where β'' is a randomly chosen bit. By the assumption, the probability that the output β^* of $\mathcal{A}$ is equal to β'' is non-negligibly greater than $1/2$. Thus it is straightforward to see that $\mathsf{Adv}_{\mathcal{A}}^{\mathsf{IND}} \geq negl(\kappa)$.

Next we show that the advantage of game 1 is negligible from IND-CPA security of the underlying cryptosystem.

Lemma 2. *Let* $\mathsf{PKE} = (\mathsf{Gen}, \mathsf{Enc}, \mathsf{Dec})$ *be an IND-CPA secure public key encryption scheme. Then for any PPT algorithm* $\mathcal{A}$, *it holds that* $\mathsf{Adv}_{\mathcal{A}}^1(\kappa) \leq negl(\kappa)$.

To prove this lemma, we use the well-known result that for an IND-CPA secure public key encryption scheme sequences of ciphertexts are also indistinguishable.

Lemma 3. *Let* $\mathsf{PKE} = (\mathsf{Gen}, \mathsf{Enc}, \mathsf{Dec})$ *be a public key encryption scheme.* PKE *is* n-*IND-CPA secure if*

$$\mathsf{Adv}_{\mathcal{A}}^{n,\mathsf{IND}}(\kappa) := \Pr\left[\beta = \beta^* \left| \begin{array}{l} (pk, sk) \leftarrow \mathsf{Gen}(1^\kappa); \\ ((m_0^1, \ldots, m_0^n), (m_1^1, \ldots, m_1^n)), \alpha) \leftarrow \mathcal{A}(pk); \\ \beta \leftarrow \{0,1\}; \\ c^1 \leftarrow \mathsf{Enc}(m_b^1); \cdots; c^n \leftarrow \mathsf{Enc}(m_b^n); \\ b^* \leftarrow \mathcal{A}((c^1, \ldots, c^n), \alpha); \end{array} \right.\right] \leq negl(\kappa)$$

for any PPT algorithm $\mathcal{A}$ *and* $n \in \mathbb{Z}$. *Then* PKE *is IND-CPA secure if and only if* PKE *is* n-*IND-CPA secure.*

We refer to [9] for the formal proof. Instead of directly showing Lemma 2 from IND-CPA security of the underlying cryptosystem, we prove it from n-IND-CPA security as follows:

Proof. We assume that there exists an PPT algorithm $\mathcal{A}$ such that $\mathsf{Adv}_{\mathcal{A}}^2$ is non-negligible and show that PKE is not n-IND-CPA secure. In order to show it we construct an adversary $\mathcal{A}'$ of n-IND-CPA game as follows:

First, given an public key pk, $\mathcal{A}'$ input $(pk, param)$ into $\mathcal{A}$. When $\mathcal{A}$ queries two minutiae $((x_0, y_0), t_0)$ and $((x_1, y_1), t_1)$ to $\mathcal{C}$ for enrollment, $\mathcal{A}'$ generates four polynomials F_0, F_1, G_0, and G_1 by following Eqs. (3) and (4) where R_F is the same for F_0 and F_1 and R_G is the same for G_0 and G_1. Expanding them as in Eqs. (1) and (2), $\mathcal{A}'$ outputs two tuples of plaintexts where the first one consists of the coefficients of F_0 and G_0, and the second F_1 and G_1. when $\mathcal{A}'$ obtains the ciphertexts of one of the tuples, $\mathcal{A}'$ returns them to $\mathcal{A}$.

On a authentication query from $\mathcal{A}$, $\mathcal{A}'$ chooses $\alpha \in \{0,1\}$ and r_F and r_G randomly. $\mathcal{A}'$ returns $AQ := \mathsf{Enc}(r_F F_\alpha(x'_\alpha, y'_\alpha) + r_G G_\alpha(t'_\alpha))$ to $\mathcal{A}$. Finally, $\mathcal{A}'$ outputs $\beta^* \in \{0,1\}$ that $\mathcal{A}$ outputs.

Now let β be a bit that the challenger chooses. That is, the tuple that $\mathcal{A}'$ obtains are the ciphertexts of the coefficients of F_β and G_β. Then, if $\beta = \alpha$ $AQ := \mathsf{Enc}(r_F \cdot F_\beta(x'_\beta, y'_\beta) + r_G \cdot G_\beta(t'_\beta))$, otherwise $AQ := \mathsf{Enc}(r_F \cdot F_{1-\beta}(x'_{1-\beta}, y'_{1-\beta}) + r_G \cdot G_{1-\beta}(t'_{1-\beta}))$,

From the assumption, the probability that the output β^* of $\mathcal{A}$ is the same as β is non-negligibly greater than $1/2$. It implies that the cryptosystem does not satisfy n-IND-CPA security.

Next we show a relation among the games as Lemma 4. By combining Lemmas 1, 2, and 4, Theorem 1 is proved.

Lemma 4. *Let* $\mathsf{Adv}^2_{\mathcal{A}}(\kappa) \leq negl(\kappa)$ *then for any PPT algorithm* $\mathcal{A}'$ *it holds that* $|\mathsf{Adv}^0_{\mathcal{A}'}(\kappa) - \mathsf{Adv}^1_{\mathcal{A}'}(\kappa)| \leq negl(\kappa)$.

Proof. Recall that the differences among the games are only the ways to generate AQ. In game 2, if $\beta = \beta''$, $AQ = \mathsf{Enc}(r_F \cdot F_\beta(x'_\beta, y'_\beta) + r_G \cdot G_\beta(t'_\beta))$. In this case, the behavior of game 2 is the same as that of game 0. For the case where $\beta \neq \beta''$, $AQ = \mathsf{Enc}(r_F F_\beta(x'_\beta, y'_\beta) + r_G G_\beta(t'_\beta))$ if $\beta' = 0$ and $AQ = \mathsf{Enc}(r_F F_{1-\beta}(x'_{1-\beta}, y'_{1-\beta}) + r_G G_{1-\beta}(t'_{1-\beta}))$ otherwise. This is the same as game 1. Thus, if the advantage of adversary in game 2 is negligible, the difference between advantages of adversary in games 0 and 1 is also negligible.

A.2 Proof of Theorem 2 (Template Protection Against Decryptor)

In the simpler version of the scheme, the malicious decryptor who has the secret key can determine the bit β by decrypting the challenge. Therefore, we introduce the transform function Trans. Transformed ciphertexts cannot be decrypted by the secret key. This property is proved under the DDH assumption. The complete proof will be appeared in the final version of this manuscript.

A.3 Proof of Theorem 3 (Security for Authentication)

Here, we prove Theorem 3 under Assumption 1. Note that this assumption holds for the modified Elgamal cryptosystem under the computational Diffie-Hellman (CDH) assumption.

Lemma 5. *Under the CDH assumption, Assumption 1 holds for the modified Elgamal cryptosystem.*

Proof. Let $\mathcal{A}$ be a PPT algorithm that does not satisfy Assumption 1 for the modified Elgamal cryptosystem, and we show that there exists a PPT algorithm $\mathcal{A}'$ that breaks the CDH assumption. We construct $\mathcal{A}'$ as follows:

For tuple (G, p, g, g_1, g_2) where $g_1 = g^a$ and $g_2 = g^b$ for some $a, b \in \mathbb{Z}_p$ that is input into $\mathcal{A}'$, let $h_0 := g_1 = g^a$, $h_1 := g = h_0^{1/a}$ and $h_2 := g_2 = h_0^{b/a}$. $\mathcal{A}'$ chooses $x, r, r' \in \mathbb{Z}_p$, and let $y := h_0^x$. Then, $\mathcal{A}'$ inputs tuple $((p, g, y), (h_0^r, h_1 y^r), (h_0^{r'}, h_2 y^{r'}))$ into $\mathcal{A}$. Here, for the key pair $pk = (p, h_0, y)$ and $sk = x$, $(h_0^r, h_1 y^r)$ and $(h_0^{r'}, h_2 y^{r'})$ are ciphertexts of $1/a$ and b/a, respectively. Therefore, $\mathcal{A}$ outputs a ciphertext $(c_1, c_2) := (h_0^{r''}, h_0^b y^{r''})$ of b for some $r'' \in \mathbb{Z}_p$ with non-negligible probability. Now it is easy to see which $\mathcal{A}'$ outputs $c_2/c_1^x = h_0^b = g^{ab}$, which breaks the CDH assumption.

Now we prove Theorem 3.

Proof. We first describe the game of security for authentication for the proposed scheme between an adversary $\mathcal{A}$ and a challenger $\mathcal{C}$ as follows:

Setup: On input of 1^κ and *param*, $\mathcal{C}$ runs $(pk, sk) \leftarrow \mathsf{Gen}(1^\kappa)$ and chooses bit β at random. $(pk, param)$ is sent to $\mathcal{A}$.

Enrollment: On request from $\mathcal{A}$, $\mathcal{C}$ chooses a minutia $((x, y), t)$ and computes two polynomials F and G in accordance with it as in Eqs. (1), (2), (3), and (4). Then, $\mathcal{C}$ encrypts the coefficients of the polynomials as $A_{i,j} \leftarrow \mathsf{Enc}(a_{i,j})$ and $B_k \leftarrow \mathsf{Enc}(b_k)$ and returns the protected template $PT :=$ $(\{A_{i,j}\}_{i,j}, \{B_k\}_k)$ to $\mathcal{A}$.

Authentication: On request from $\mathcal{A}$, $\mathcal{C}$ chooses r_F and r_G at random and computes $A'_{i,j} \leftarrow \mathsf{Enc}(r_F \cdot a_{i,j})$ and $B'_k \leftarrow \mathsf{Enc}(r_G \cdot b_k)$. Then, $\mathcal{C}$ lets $CH := (\{A'_{i,j}\}_{i,j}, \{B'_k\}_k)$, $AQ := \mathsf{Enc}(-r_F - r_G)$ and sends back tuple $(CH, AQ, Accept)$ to $\mathcal{A}$.

Output: On request from $\mathcal{A}$, $\mathcal{C}$ chooses r'_F and r'_G randomly and sends back $CH^* := (\{A''_{i,j}\}_{i,j}, \{B''_k\}_k)$ to $\mathcal{A}$ where $A''_{i,j} \leftarrow \mathsf{Enc}(r'_F \cdot a_{i,j})$ and $B''_k \leftarrow \mathsf{Enc}(r'_G \cdot b_k)$. Finally, $\mathcal{A}$ outputs AQ^*.

The advantage of $\mathcal{A}$ that can be described as $\mathsf{Adv}_{\mathcal{A}}^{\mathsf{Auth}}(\kappa) := \Pr[\mathsf{Dec}(AQ^*) = r'_F + r'_G]$ is assumed to be non-negligible. We construct algorithm $\mathcal{A}'$ that breaks Assumption 1 for the underlying cryptosystem as follows:

On input of $(pk, \mathsf{Enc}(s), \mathsf{Enc}(st))$, $\mathcal{A}'$ chooses minutia $((x, y), t)$. For the location of the minutia, $\mathcal{A}'$ computes polynomial F as in Eqs. (1) and (3) and encrypts the coefficients of the polynomials as $A_{i,j} \leftarrow \mathsf{Enc}(a_{i,j})$. Also, $\mathcal{A}'$ generates the polynomial G as in Eqs. (2) and (4) where $R_G := s$. That is, if we write polynomial $\tilde{G}$ as $\tilde{G}(T) = \prod_{\ell=-\Delta_t}^{\Delta_t} \{(T-t) - \ell\} + 1 = \sum_k \tilde{b}_k \cdot T^k$, it holds that $b_k = s \cdot \tilde{b}_k$ for any k. Therefore, $\mathcal{A}'$ encrypts the coefficients as $B_k := \mathsf{Enc}(s)^{\tilde{b}_k} = \mathsf{Enc}(s \cdot \tilde{b}_k) = \mathsf{Enc}(b_k)$. Then, $\mathcal{A}'$ sends the protected template $PT := (\{A_{i,j}\}_{i,j}, \{B_k\}_k)$ to $\mathcal{A}$.

On authentication query from $\mathcal{A}$, $\mathcal{A}'$ chooses r_F and r_G at random and computes $A'_{i,j} \leftarrow \mathsf{Enc}(r_F \cdot a_{i,j})$ and $B'_k := \mathsf{Enc}(s)^{r_G \cdot \tilde{b}_k} = \mathsf{Enc}(r_G \cdot b_k)$. Then, $\mathcal{C}$ lets $CH := (\{A'_{i,j}\}_{i,j}, \{B'_k\}_k)$, $AQ := \mathsf{Enc}(-r_F - r_G)$ and sends back tuple $(CH, AQ, Accept)$ to $\mathcal{A}$.

On output query from $\mathcal{A}$, $\mathcal{A}'$ chooses r'_F randomly and computes $A''_{i,j} \leftarrow \mathsf{Enc}(r'_F \cdot a_{i,j})$ and $B''_k := \mathsf{Enc}(st)^{\tilde{b}_k} = \mathsf{Enc}(t \cdot b_k)$. Given $CH^* := (\{A''_{i,j}\}_{i,j}, \{B''_k\}_k)$, $\mathcal{A}$ outputs AQ^*. Finally, $\mathcal{A}'$ outputs $t^* := AQ^* \cdot \mathsf{Enc}(-r'_F)$.

Note that the random values r_G and r'_G that the challenger chooses in the game for authentication is set to be s and t, respectively, by $\mathcal{A}'$. Therefore, from the assumption, AQ^* is a ciphertext of $r'_F + r'_G = r'_F + t$ with non-negligible probability. This means that the output of $\mathcal{A}'$ is a ciphertext of t with non-negligible probability. That is, $\mathcal{A}'$ breaks Assumption 1.

A.4   Proof of Theorem 4 (Security Against Hill-Climbing Attacks)

Proof. We first describe the game of security against hill-climbing attacks for the proposed scheme between an adversary $\mathcal{A}$ and a challenger $\mathcal{C}$ as follows:

Setup: On input of 1^κ and *param*, $\mathcal{C}$ runs $(pk, sk) \leftarrow \mathsf{Gen}(1^\kappa)$ and chooses bit β at random. $((pk, param), sk)$ is sent to $\mathcal{A}$.

Enrollment: On request from $\mathcal{A}$ for a minutia $((x, y), t)$, $\mathcal{C}$ computes two polynomials F and G in accordance with it as in Eqs. (1)–(4). $\mathcal{C}$ encrypts the coefficients of the polynomials as $A_{i,j} \leftarrow \mathsf{Enc}(a_{i,j})$ and $B_k \leftarrow \mathsf{Enc}(b_k)$ and returns protected template $PT := (\{A_{i,j}\}_{i,j}, \{B_k\}_k)$ to $\mathcal{A}$.

Authentication: On request from $\mathcal{A}$, $\mathcal{C}$ chooses r_F and r_G at random and lets $A'_{i,j} \leftarrow \mathsf{Enc}(r_F \cdot a_{i,j})$ and $B'_k \leftarrow \mathsf{Enc}(r_G \cdot b_k)$. On receiving $CH = (\{A'_{i,j}\}_{i,j}, \{B'_k\}_k)$ from $\mathcal{C}$, $\mathcal{A}$ sends back AQ. Then, $\mathcal{C}$ chooses r at random and computes $VQ := (AQ \cdot \mathsf{Enc}(-r_F - r_G))^r$ and sends it to $\mathcal{A}$.

Output: $\mathcal{A}$ requests $\mathcal{C}$ for two minutiae $((x'_0, y'_0), t'_0)$ and $((x'_1, y'_1), t'_1)$. If for two pairs of minutiae $(((x, y), t), ((x'_0, y'_0), t'_0))$ and $(((x, y), t), ((x'_1, y'_1), t'_1))$, one of them corresponds and the other does not, then $\mathcal{C}$ returns $\bot$ to $\mathcal{A}$. Otherwise, $\mathcal{C}$ chooses bit β and r'_F, r'_G, and r' randomly and returns $VQ^* \leftarrow \mathsf{Enc}(r(r'_F \cdot F(x'_\beta, y'_\beta) + r'_G \cdot G(t'_\beta) - r'_F - r'_G))$ to $\mathcal{A}$. Finally, $\mathcal{A}$ outputs β^*.

The advantage of adversary $\mathcal{A}$ is defined as $\mathsf{Adv}^{\mathsf{Dist}}_{\mathcal{A}}(\kappa) := \Pr[\beta = \beta^*] - 1/2$. The adversary may obtain information related to β only in the output phase. If the correspondences of two pairs of minutiae are different, $\mathcal{A}$ does not obtain any information in the output phase. Also, if both pairs correspond, VQ^* that $\mathcal{A}$ obtains in the output phase is the ciphertext of 0 no matter which bit is chosen as β. In these two cases, we can say that $\mathcal{A}$ does not obtain any information on β. Thus, $\mathcal{A}$ has no chance to distinguish the bit β. The remaining case is where both pairs do not correspond. In this case, although $\mathcal{A}$ who has the secret key sk obtains VQ^*, which is related to β, it is randomized by a new random value r that is used only once. Therefore, $\mathcal{A}$ cannot guess β in this case as well.

References

1. Barbosa, M., Brouard, T., Cauchie, S., de Sousa, S.M.: Secure biometric authentication with improved accuracy. In: Mu, Y., Susilo, W., Seberry, J. (eds.) ACISP 2008. LNCS, vol. 5107, pp. 21–36. Springer, Heidelberg (2008)
2. Boneh, D., Goh, E.-J., Nissim, K.: Evaluating 2-DNF formulas on ciphertexts. In: Kilian, J. (ed.) TCC 2005. LNCS, vol. 3378, pp. 325–341. Springer, Heidelberg (2005)
3. Bringer, J., Chabanne, H.: An authentication protocol with encrypted biometric data. In: Vaudenay, S. (ed.) AFRICACRYPT 2008. LNCS, vol. 5023, pp. 109–124. Springer, Heidelberg (2008)
4. Bringer, J., Chabanne, H., Izabachène, M., Pointcheval, D., Tang, Q., Zimmer, S.: An application of the goldwasser-micali cryptosystem to biometric authentication. In: Pieprzyk, J., Ghodosi, H., Dawson, E. (eds.) ACISP 2007. LNCS, vol. 4586, pp. 96–106. Springer, Heidelberg (2007)
5. Campisi, P. (ed.): Security and Privacy in Biometrics. Springer, London (2013)
6. Dodis, Y., Reyzin, L., Smith, A.: Fuzzy extractors: how to generate strong keys from biometrics and other noisy data. In: Cachin, C., Camenisch, J.L. (eds.) EUROCRYPT 2004. LNCS, vol. 3027, pp. 523–540. Springer, Heidelberg (2004)
7. Elgamal, T.: A public key cryptosystem and a signature scheme based on discrete logarithms. IEEE Trans. Inf. Theor. **31**(4), 469–472 (1985). IEEE

8. Freedman, M.J., Nissim, K., Pinkas, B.: Efficient private matching and set intersection. In: Cachin, C., Camenisch, J.L. (eds.) EUROCRYPT 2004. LNCS, vol. 3027, pp. 1–19. Springer, Heidelberg (2004)
9. Goldreich, O.: The Foundations of Cryptography - Basic Applications, vol. 2. Cambridge University Press, Cambridge (2004)
10. Hattori, M., Matsuda, N., Ito, T., Shibata, Y., Takashima, K., Yoneda, T.: Provably-secure cancelable biometrics using 2-DNF evaluation. J. Inf. Process. **20**(2), 496–507 (2012). IPSJ
11. Hirano, T., Hattori, M., Ito, T., Matsuda, N.: Cryptographically-secure and efficient remote cancelable biometrics based on public-key homomorphic encryption. In: Sakiyama, K., Terada, M. (eds.) IWSEC 2013. LNCS, vol. 8231, pp. 183–200. Springer, Heidelberg (2013)
12. Juels, A., Sudan, M.: A fuzzy vault scheme. In: ISIT 2002, p. 408. IEEE (2002)
13. Juels, A., Wattenberg, M.: A fuzzy commitment scheme. In: CCS 1999, pp. 28–36. ACM (1999)
14. Maltoni, D., Maio, D., Jain, A.K., Prabhakar, S.: Handbook of Fingerprint Recognition, 2nd edn. Springer Publishing Company, Incorporated, London (2009)
15. Nandakumar, K., Jain, A.K., Pankanti, S.: Fingerprint-based fuzzy vault: implementation and performance. IEEE Trans. Inf. Forensics Secur. **2**(4), 744–757 (2007). IEEE
16. Rathgeb, C., Uhl, A.: A survey on biometric cryptosystems and cancelable biometrics. EURASIP J. Inf. Secur. **2011**(1), 3 (2011). Springer International Publishing AG
17. Shahandashti, S.F., Safavi-Naini, R., Ogunbona, P.: Private fingerprint matching. In: Susilo, W., Mu, Y., Seberry, J. (eds.) ACISP 2012. LNCS, vol. 7372, pp. 426–433. Springer, Heidelberg (2012)
18. Simoens, K., Bringer, J., Chabanne, H., Seys, S.: A framework for analyzing template security and privacy in biometric authentication systems. IEEE Trans. Inf. Forensics Secur. **7**(2), 833–841 (2012). IEEE
19. Uludag, U., Jain, A.K.: Attacks on biometric systems: a case study in fingerprints. In: Delp, E.J., Wong, P.W. (eds.) Proceedings of SPIE 5306, pp. 622–633. SPIE (2004)

Cryptanalysis of Symmetric-Key Primitives

Practical Cryptanalysis of Full Sprout with TMD Tradeoff Attacks

Muhammed F. Esgin[1,2]([✉]) and Orhun Kara[1]

[1] TÜBİTAK BİLGEM UEKAE, Gebze, Kocaeli, Turkey
{muhammed.esgin,orhun.kara}@tubitak.gov.tr
[2] Graduate School of Natural and Applied Sciences,
İstanbul Şehir University, İstanbul, Turkey

Abstract. The internal state size of a stream cipher is supposed to be at least twice the key length to provide resistance against the conventional Time-Memory-Data (TMD) tradeoff attacks. This well adopted security criterion seems to be one of the main obstacles in designing, particularly, ultra lightweight stream ciphers. At FSE 2015, Armknecht and Mikhalev proposed an elegant design philosophy for stream ciphers as fixing the key and dividing the internal states into equivalence classes where any two different keys always produce non-equivalent internal states.

The main concern in the design philosophy is to decrease the internal state size without compromising the security against TMD tradeoff attacks. If the number of equivalence classes is more than the cardinality of the key space, then the cipher is expected to be resistant against TMD tradeoff attacks even though the internal state (except the fixed key) is of fairly small length. Moreover, Armknecht and Mikhalev presented a new design, which they call Sprout, to embody their philosophy.

In this work, ironically, we mount a TMD tradeoff attack on Sprout within practical limits using 2^d output bits in 2^{71-d} encryptions of Sprout along with 2^d table lookups. The memory complexity is 2^{86-d} where $d \leq 40$. In one instance, it is possible to recover the key in 2^{31} encryptions and 2^{40} table lookups if we have 2^{40} bits of keystream output by using tables of 770 Terabytes in total. The offline phase of preparing the tables consists of solving roughly $2^{41.3}$ systems of linear equations with 20 unknowns and an effort of about 2^{35} encryptions. Furthermore, we mount a guess-and-determine attack having a complexity about 2^{68} encryptions with negligible data and memory. We have verified our attacks by conducting several experiments. Our results show that Sprout can be practically broken.

Keywords: Sprout · Stream cipher · Keystream generator · Time Memory Data tradeoff attacks · LFSR · NLFSR · Guess-and-determine · Divide-and-conquer

This author's work is based upon work from COST Action CRYPTACUS, supported by COST (European Cooperation in Science and Technology).

© Springer International Publishing Switzerland 2016
O. Dunkelman and L. Keliher (Eds.): SAC 2015, LNCS 9566, pp. 67–85, 2016.
DOI: 10.1007/978-3-319-31301-6_4

1   Introduction

One of the main design principles for stream ciphers is that internal state size should be at least twice the key size. Otherwise, the cipher will be vulnerable to generic Time-Memory-Data (TMD) tradeoff attacks [3,5–7,11,16]. This principle makes lightweight stream cipher design more challenging in comparison to designing block ciphers even though the same generic attacks are valid for block ciphers, particularly, when they are used in stream cipher modes such as OFB or CTR mode.

There have been several lightweight block cipher designs appeared in the literature such as Present [8], ITUBee [17], Lblock [21], Prince [9], Ktantan [10], Twine [20] and many more in the last decade. An example that does not follow the common tendency of designing lightweight block ciphers rather than lightweight stream ciphers is the new stream cipher design Sprout, presented at FSE 2015 [2] by Armknecht and Mikhalev. The cipher makes use of a variable internal state of only 80 bits and a fixed key of 80 bits as well. The authors show that a cipher design such as Sprout is resistant to classical TMD tradeoff attacks even though its (variable) internal state size is strictly less than twice the key size.

The property of having a smaller internal state size results from the uncommon design principle adopted by the designers of Sprout [2]. According to this principle, the key is also incorporated into the next state function of the cipher as a fixed vector and the internal states including the keys are divided into equivalence classes. Each key is in a different equivalence class. That is, it is impossible to obtain two equivalent internal states with different keys. Therefore, the conventional TMD tradeoff attacks are ineffective since the number of equivalence classes is not less than the cardinality of the key space and one needs to travel almost all the equivalence classes to recover one specific key, rendering the generic tradeoff attacks slower than the exhaustive search.

The design of Sprout is inspired by Grain 128a in [1]. The sizes of both the NLFSR and the LFSR of Grain family are reduced to half of their values and the functions are slightly changed. The design philosophy of output generation and feedback functions is almost conserved except adding a round key function to incorporate key bits for each clock. As a result, a stream cipher of area cost roughly 800 GE is developed whereas Grain needs 1162 GE for the same level of 80-bit security [2].

Sprout has immediately attracted the interest of the cryptology community intensively, indicating that the design philosophy itself introduces several open questions about the security of such ciphers. Even though it is a very recent cipher, there have been a couple of its analyses [4,12,18,19].

The first attack paper is by Lallemand and Naya-Plasencia, and they have shown that it is possible to recover the key with a time complexity equivalent to roughly 2^{70} Sprout encryptions by merging the sets of possible LFSRs and NLFSRs through a careful sieving [18]. Indeed, the actual workload is $2^{74.51}$ steps but since the cost of each step is considered to be $2^{5.64}$ times faster than one step of exhaustive search in [18], the overall time complexity is finalized as $2^{69.36}$. The memory complexity for leading the values for the registers is 2^{46}.

In another analysis, Maitra et al. show that when the whole (variable) internal state is known, the key can be found using a SAT solver [19]. The system of nonlinear equations generated from the output is quite easy to solve. The authors show that it is possible to solve the system with roughly 900 keystream bits in less than half a second on an ordinary PC. Moreover, the authors show that it is still possible to solve the system even though two thirds of LFSR bits are also unknown. However, solving the system takes around one minute this time. Hence, guessing the variables of whole NLFSR and one third of the variables of LFSR and then solving all the corresponding 2^{54} systems of equations, it is possible to recover the key. On the other hand, the authors do not compare the time complexity of solving 2^{54} equations with that of exhaustive search. Besides their algebraic attack, they give an example of a fault attack as well [19].

The guess-and-determine attack in [19] using a SAT solver is further improved in [4] by Banik. After guessing 50 bits of the state, the remaining bits including key bits can be solved in roughly half a minute on a standard PC by means of Cryptominisat 2.9.5 solver installed in SAGE 5.7. Moreover, Banik observed that the LFSR is expected to be initialized to the all zero vector in one of 2^{40} random synchronizations. When such an event occurs, the NLFSR is easily recovered with the key by guessing 33 bits of its state. The total time complexity is around 2^{67} encryptions with negligible memory and about 2^{42} bits of output. A distinguisher attack is also included by observing that it is possible to produce shifted versions of a keystream up to a factor of 80 with the same key by using different IVs. When the factor for shifting is limited as 2^{10}, it is possible to distinguish Sprout output with 2^{57} bits of memory and 2^{32} encryptions [4].

In [12], a related-key chosen-IV distinguisher is shown. However, the designers of the Sprout regard related-key attacks as out of scope since the key is assumed to be fixed.

Table 1. The time complexities are given in terms of number of encryptions except when specified as number of table lookups (TLs). The memory complexities are given in terms of number of rows. Additionally, we assume an effort of $1\,s = 2^{15}$ encryptions and that of $1\,min = 2^{21}$ encryptions on a standard PC.

	Time	Data	Memory
[18]	2^{70}	Negligible	2^{46}
[19]	2^{75}	Negligible	Negligible
Sect. 3 in [4]	2^{70}	Negligible	Negligible
Sect. 5 in [4]	$2^{66.7}$	2^{42} bits	Negligible
Sect. 3 in this work	2^{68}	Negligible	Negligible
Sect. 4.1 in this work	$2^{40}\ TLs + 2^{31}$	2^{40} bits	2^{46}
TMD Tradeoff in this work	$2^{d}\ TLs + 2^{71-d}$	2^{d} bits	2^{86-d}

In Table 1, we compare the complexities of known attacks on Sprout. Our straightforward C/C++ implementation of Sprout runs more than $2^{23.5}$ clocks in a second on a standard PC. Thus, we make the assumption that an effort of a second is equivalent to 2^{15} Sprout encryptions and that of a minute to 2^{21} Sprout encryptions. One may further improve these numbers by a more efficient and performance-oriented implementation. As we show in Appendix A, one clock of Sprout costs about $2^{-8.33}$ Sprout encryptions.

Our Contribution: None of the attacks in the literature so far has practical workloads. We introduce a new TMD tradeoff cryptanalysis of full Sprout within the practical bounds where all data, memory and time complexities can be upperbounded by 2^{45}. We first show that when the internal state is known (excluding the key), it is much easier to obtain the secret key when the cipher is run backwards compared to running the cipher forward. Later, guessing the internal states from the keystream bits is described based on a time-memory-data tradeoff approach. For the key-recovery part, our attack is a combination of both guess-and-determine and divide-and-conquer approaches. We show that the key recovery part itself can be transformed to a guess-and-determine attack of a complexity 2^{68} encryptions with negligible data and memory.

Our TMD tradeoff attack is based on a specific occasion where incorporation of key bits into the register can be discarded during the keystream production. So, Sprout behaves like Grain family since the round key function is bypassed. We can compute the internal states in advance for all the predefined occasions and store them with the produced keystream bits in a table. We can produce some keystream bits for these special internal states without knowing the key.

In general, one may expect that the special internal states to be stored are deduced by exhaustively searching them. However, we can deduce these states without a search-and-eliminate mechanism, reducing the time complexity of the precomputation phase dramatically. We show that computing the special internal states which cause to bypass the key bits is as easy as solving some systems of linear equations. Each of our guesses gives a valid solution. So, unlike generic TMD tradeoff attacks where the offline phase is generally as expensive as an exhaustive search, our offline phase of the attack is also in practical limits. To sum up, we explore and exploit a special kind of lack of sampling resistance in Sprout. The objective of this procedure is similar to the one of BSW sampling applied to A5/1 cipher [7].

In the online phase of the attack, we check if a special occasion occurs in the given keystream. The time complexity is 2^{79-d} Sprout clocks when we have 2^d bits of keystream (not necessarily from the same IV). The memory requirement is about 2^{86-d}. For example, when $d = 40$, we can recover the key in only 2^{31} encryptions and 2^{40} table lookups by using 2^{40} bits of keystream with 770 Terabytes of memory. The precomputation cost of preparing the tables is equivalent to solving about $2^{41.32}$ systems of linear equations with 20 unknowns and about 2^{35} encryptions. We have verified our results by conducting several experiments.

Organization of the Paper: Section 2 describes the high level structure of Sprout. In Sect. 3, we show how to efficiently recover the secret key when the (variable) internal state of Sprout is known and introduce a guess-and-determine attack. Section 4 introduces a TMD tradeoff approach to deal with filling the internal state, including how to construct systems of linear equations during the off-line phase of the attack. We conclude the paper with some remarks and propose a solution in Sect. 5. Some details of Sprout is given in Appendix A and we provide experimental results verifying our attack in Appendix B.

2   High Level Description of Sprout

Sprout [2] is a lightweight stream cipher inspired by Grain family [1,13–15]. The (variable) internal state of Sprout consist of an LFSR and an NLFSR, and there is also a fixed key that is used in the state update function. The sizes of LFSR and NLFSR are 40 bits each and the key length is 80 bits. An IV of size 70 bits is also incorporated during the initialization phase. The feedback functions of NLFSR and LFSR, and the nonlinear part of the output function are denoted by g, f and h, respectively (See Fig. 1 in the appendix). We follow the notations below throughout the paper.

- t - the clock-cycle number
- $\oplus$ - the XOR operation
- $L_t := (l_0^t, l_1^t, \ldots, l_{39}^t)$ - state of the LFSR at clock-cycle t
- $N_t := (n_0^t, n_1^t, \ldots, n_{39}^t)$ - state of the NLFSR at clock-cycle t
- $C_t := (c_0^t, c_1^t, \ldots, c_8^t)$ - state of the counter at clock-cycle t
- $K := (k_0, k_1, \ldots, k_{79})$ - the fixed key
- $IV := (iv_0, iv_1, \ldots, iv_{69})$ - the initialization vector
- k_t^* - the round key bit generated during clock t
- n_t - the output bit of NLFSR during clock t
- l_t - the output bit of LFSR during clock t
- z_t - the keystream bit generated during clock t

A 9-bit counter is used in the algorithm to count the number of rounds for the initialization phase (which has 320 rounds). After initialization, its first seven bits run cyclically from 0 to 79 and determine the index of the key bit selected at the current time. Moreover, the fourth bit of the counter c_4^t is involved in the NLFSR feedback.

The linear relation of the LFSR is $l_{39}^{t+1} = f(L_t) = l_0^t \oplus l_5^t \oplus l_{15}^t \oplus l_{20}^t \oplus l_{25}^t \oplus l_{34}^t$. The function g is the nonlinear feedback function for the NLFSR. Its output is XORed with the round key bit k_t^*, the counter bit c_4^t and the output of the LFSR as $n_{39}^{t+1} = g(N_t) \oplus k_t^* \oplus l_0^t \oplus c_4^t$ where $k_t^* = k_{t \bmod 80} \cdot \delta_t$ with $\delta_t := l_4^t \oplus l_{21}^t \oplus l_{37}^t \oplus n_9^t \oplus n_{20}^t \oplus n_{29}^t$. Remark that one clock of Sprout is equivalent to $2^{-8.33}$ encryption of an exhaustive key search. We give the necessary details in the analysis sections. Also, one can refer to [2] or Appendix A for details.

3   A Key Recovery Attack

It is easy to see that Sprout's next state function is invertible. So, once we obtain the whole state (including the key), we can clock the internal states of the cipher forward and backwards, as well. So, it is straightforward to recover the internal state at clock-cycle t from the internal state at clock-cycle $t+1$. We first decrease the counter. Then, for the LFSR feedback, we have

$$l_0^t = l_{39}^{t+1} \oplus l_4^{t+1} \oplus l_{14}^{t+1} \oplus l_{19}^{t+1} \oplus l_{24}^{t+1} \oplus l_{33}^{t+1} \tag{1}$$

and $l_{i+1}^t = l_i^{t+1}$ for $0 \le i \le 38$. For the NLFSR feedback, we have

$$
\begin{aligned}
n_0^t = {}& k_t^* \oplus c_4^t \oplus l_{39}^{t+1} \oplus l_4^{t+1} \oplus l_{14}^{t+1} \oplus l_{19}^{t+1} \oplus l_{24}^{t+1} \oplus l_{33}^{t+1} \\
& \oplus n_{39}^{t+1} \oplus n_{12}^{t+1} \oplus n_{18}^{t+1} \oplus n_{34}^{t+1} \oplus n_{38}^{t+1} \oplus n_1^{t+1} n_{24}^{t+1} \\
& \oplus n_2^{t+1} n_4^{t+1} \oplus n_6^{t+1} n_7^{t+1} \oplus n_{13}^{t+1} n_{20}^{t+1} \oplus n_{15}^{t+1} n_{17}^{t+1} \oplus n_{21}^{t+1} n_{23}^{t+1} \oplus n_{25}^{t+1} n_{31}^{t+1} \\
& \oplus n_{32}^{t+1} n_{35}^{t+1} n_{36}^{t+1} n_{37}^{t+1} \oplus n_9^{t+1} n_{10}^{t+1} n_{11}^{t+1} \oplus n_{26}^{t+1} n_{29}^{t+1} n_{30}^{t+1}
\end{aligned}
\tag{2}
$$

where

$$
\begin{aligned}
k_t^* &= k_t, \quad 0 \le t \le 79 \\
k_t^* &= k_{t \bmod 80} \cdot (l_3^{t+1} \oplus l_{20}^{t+1} \oplus l_{36}^{t+1} \oplus n_8^{t+1} \oplus n_{19}^{t+1} \oplus n_{28}^{t+1})
\end{aligned}
$$

and $n_{i+1}^t = n_i^{t+1}$ for $0 \le i \le 38$ (see [2] or Appendix A). Now, the keystream z_t can be generated while the index t is decreasing.

Maitra et al. has shown in their recent paper that it is possible to recover the key once the (variable) internal state is known by solving a system of nonlinear equations by a SAT Solver in less than half second on a single PC, using roughly 900 bits of keystream sequence [19]. We have a similar problem indeed: We make a guess for the internal state and then, we do not just want to determine the key from the internal state but also we would like to check if our guess is correct simultaneously without recovering the whole set of the key bits. The simple observation below gives us a much faster key recovery and internal state checking mechanism. Indeed, we do not need to solve a system of nonlinear equations. The following property suggests that recovering key from the internal state and output is much easier if we trace backwards through the registers.

Proposition 1. *Assume that at time $t + 1$, we know the internal states of both registers NLFSR and LFSR, but the whole key is unknown and that $\delta_t = 1$. While clocking the registers backwards, when a key bit appears in the keystream for the first time, it will appear as a single unknown inside n_1^{t-1} in the keystream bit z_{t-1}. This happens before the key bit is incorporated into the feedback of NLFSR through the g function.*

The proof of Proposition 1 is straightforward. Assume that while the cipher is run backwards, at some clock-cycle $t + 1$, we guess the value of the whole internal state (excluding the key) and $\delta_t = 1$. One clock later, n_0^t becomes a term of the form $k_i \oplus a$ where a is a known value obtained from the NLFSR feedback, the LFSR output and the counter. Now, at time t, n_0^t is not incorporated into the NLFSR feedback function. Thus, $k_i \oplus a$ shifts to the position n_1^{t-1} and n_0^{t-1} does not depend on k_i (it may depend on another key bit but that is not important). Now, at time $t-1$, we know all the register values except for n_0^{t-1} and $n_1^{t-1} = k_i \oplus a$. But, n_0^{t-1} is not involved in the output function. So, we can easily determine k_i as $k_i = z_{t-1} \oplus a \oplus a'$ where a' is a known value coming from the tap points of the output function.

As a result, when we make a guess for the registers, at each clock, we will either have opportunity to check if the keystream bit we generate matches the corresponding output bit, or a key bit will appear as a single unknown and we will determine the key bit from the output. Hence, if the state candidate does not yield a contradiction, we again end up with registers that are completely known except maybe for the first bit n_0^t of NLFSR. However, if a key bit is involved in that term, it will be determined one clock later before going into the feedback. To sum up, continuing the procedure recursively, either we recover a single key bit or we have a check bit for each clock. Let us illustrate this with a simple example.

Example 1. Assume that at time $t+1$ we know the whole internal state but the secret key and let $\delta_t = \delta_{t-1} = \delta_{t-2} = 1$ and $\delta_{t-3} = 0$. Let k_0, k_1, k_2 and k_3 be the key bits selected in given order. In Table 1, we show how the values of NLFSR bits and keystream bits proceed.

Table 2. ✓ denotes a known value, ? a value that is either known or unknown, $D(k_i)$ determining the value of k_i and C is a check if the keystream bit generated matches the actual one.

Clock-cycle	δ_i	n_0^i	n_1^i	n_2^i	n_3^i	$\cdots$	n_{39}^i	z_i	D/C
$i = t+1$	-	✓	✓	✓	✓	✓	✓	✓	C
$i = t$	1	$k_0 \oplus$ ✓	✓	✓	✓	✓	✓	✓	C
$i = t-1$	1	$k_1 \oplus$ ✓	$k_0 \oplus$ ✓	✓	✓	✓	✓	$k_0 \oplus$ ✓	$D(k_0)$
$i = t-2$	1	$k_2 \oplus$ ✓	$k_1 \oplus$ ✓	✓	✓	✓	✓	$k_1 \oplus$ ✓	$D(k_1)$
$i = t-3$	0	✓	$k_2 \oplus$ ✓	✓	✓	✓	✓	$k_2 \oplus$ ✓	$D(k_2)$
$i = t-4$	-	?	✓	✓	✓	✓	✓	✓	C

The probability that a key bit does not appear in the output for p blocks of length 80 bits is 2^{-p}. Hence, after roughly 160 clocks, 60 different key bits will appear in the output and will thus be determined. The time complexity of recovering roughly 60 bits of the key for a correct guess of internal state is almost 160 clocks of Sprout. The remaining 20 bits can be recovered by searching exhaustively. On the other hand, the probability that a guess for an internal state survives for $2r$ clocks is 2^{-r}. On average for each 2 clocks, half of the possible guesses will be eliminated. So, the average number of clocks for each elimination among 2^s possible guesses is

$$\sum_{i=0}^{s} \frac{2 \cdot 2^{s-i}}{2^s} = \sum_{i=0}^{s} \frac{1}{2^{i-1}} \approx 4 \quad \text{for } 21 \le s \le 40.$$

We see that we can check if a given state is correct in 4 clocks on average. Moreover, it is possible to mount a guess-and-determine attack by using Proposition 1. We can guess 77 bits of the internal state and determine the three remaining bits that appear as XOR in the output (such as $l_{31}^t, l_{30}^t, l_{29}^t$) since three bits of keystream can be produced without knowing the key. Observe that the bits l_i^t for $28 \ge i \ge 25$ are incorporated into the output as XOR at time $t+i-30$ for the first time after clock t during the backward clocking. So, the guess-and-determine attack can be improved by further determining l_i^t for $28 \ge i \ge 25$, if a key bit is not involved in the output along with l_i^t (that is, $\delta_{t+i-29} = 0$). If $\delta_{t+i-29} = 1$, then guess l_i^t as well and determine the related key bit. So, we guess at least 73 and at most 77 bits according to the values of δ_{t+i-29}. The overall complexity is around 2^{70} encryptions. It can be further improved by assuming that $\delta_{t+i-29} = 0$ for $28 \ge i \ge 25$. In this case, we guess 69 bits and determine 11 bits ($l_{36}^t, l_{35}^t, l_{34}^t, l_{33}^t$ from $\delta_{t+i-29} = 0$ and $l_{31}^t, \ldots, l_{25}^t$ from keystream bits). The cipher is clocked 9 times on average to come up with a contradiction (5 clocks during determining $l_{31}^t, \ldots, l_{25}^t$ and 4 clocks for the key recovery and checking). Repeat the attack 16 times by using shifted keystreams to fulfill the assumption. Hence, the average complexity is $2^4 \cdot 2^{69} \cdot 2^{3.17} = 2^{76.17}$ clocks and thus $2^{76.17} \cdot 2^{-8.33} = 2^{67.84}$ encryptions of Sprout. The data and memory complexities are negligible.

4 A Time-Memory-Data Tradeoff Attack

Recall that, treating the bits of the registers as the terms of a sequence, we denote $l_{t+i+j} := l_i^{t+j}$ and $n_{t+i+j} := n_i^{t+j}$. We mount an attack on Sprout with 2^d data and a time complexity of 2^{79-d} clocks where $d \le 40$ by enhancing the idea of the guess-and-determine attack given in Sect. 3. We make use of memory also, having roughly 2^{86-d} entries.

The attack scenario is simple. Assume that δ_t is zero for consecutive d clocks. That is, the key bits are not incorporated into the NLFSR during d consecutive clocks: $t-9, t-8, \ldots, t+d-10$. Then we can make a guess to the internal state at time t and then check if the guess is correct and the condition is satisfied since we can produce d bits of outputs without knowing any key bit.

Assuming that $\delta_{t-9} = \delta_{t-8} = \cdots = \delta_{t+d-10} = 0$, we get the following d linear equations of the internal state bits.

$$l_{t-5} \oplus l_{t+12} \oplus l_{t+28} \oplus n_t \oplus n_{t+11} \oplus n_{t+20} = 0$$
$$l_{t-4} \oplus l_{t+13} \oplus l_{t+29} \oplus n_{t+1} \oplus n_{t+12} \oplus n_{t+21} = 0$$

$$\vdots$$

$$l_{t+4} \oplus l_{t+21} \oplus l_{t+37} \oplus n_{t+9} \oplus n_{t+20} \oplus n_{t+29} = 0$$
$$l_{t+5} \oplus l_{t+22} \oplus l_{t+38} \oplus n_{t+10} \oplus n_{t+21} \oplus n_{t+30} = 0$$

$$\vdots$$

$$l_{t+d-6} \oplus l_{t+d+11} \oplus l_{t+d+27} \oplus n_{t+d-1} \oplus n_{t+d+10} \oplus n_{t+d+19} = 0$$

If $d \leq 20$, we have d linear equations with at most 80 unknowns since there will be no feedback for NLFSR. Note that we can write an LFSR feedback as a linear equation without adding new unknowns. So, we simply exclude both the linear equations and new unknowns coming from LFSR feedback. However, if $d > 20$, the new unknowns from the feedback of the NLFSR will appear with some nonlinear equations. The new equations coming from the feedback are

$$c_4^t \oplus l_t \oplus n_{t+40} \oplus g(N_t) = 0$$
$$c_4^{t+1} \oplus l_{t+1} \oplus n_{t+41} \oplus g(N_{t+1}) = 0$$

$$\vdots$$

$$c_4^{t+d-21} \oplus l_{t+d-21} \oplus n_{t+d+19} \oplus g(N_{t+d-21}) = 0,$$

which are adding $d - 20$ more equations with $d - 20$ new unknowns $n_{t+40}, \ldots, n_{t+d+19}$. We have $2d - 20$ equations with $60 + d$ unknowns. We see that by carefully choosing $80 - d$ unknowns to be guessed, we mostly come up with $2d-20$ linear equations with $2d-20$ unknowns. Solving the linear system for each counter set, we can determine $2d - 20$ unknown bits. That is, we can determine the whole internal state. Then we can produce the output up to $d + 3$ bits for all possible counter combinations. See Sect. 4.1 for solving the system of equations in the most extreme case.

Let us store all the guessed internal states where $\delta_t = 0$ for d consecutive clocks with their outputs up to $d + 3$ bits for each counter, sorted according to the outputs. Note that we can generate keystream bits $z_{t-10}, z_t, \ldots, z_{t+d-8}$ due to the fact that the output is not affected by the most and the least significant taps of the NLFSR.

We need at most 80 tables having 2^{80-d} rows each to produce a table for each counter. During the online phase of the attack, any $d + 3$ clock output x at a certain time (so counter is known) is searched in the table with the related counter. There are 2^{77-2d} internal states producing the output x. Check if any of the internal state is correct. Repeat this procedure 2^d times since the probability that $\delta_t = 0$ for d consecutive clocks is 2^{-d}. We expect this event to occur once. Then, we can recover the internal state from the tables and then recover the key easily once the internal state is known. Recovering the key from a known internal state is explained in Sect. 3.

For each output of $d + 3$ bits, we have on average 2^{77-2d} internal states producing the output. We both check the validity of the internal state and recover the key bits for each candidate. On average, clocking 4 times is enough for the checking. Hence, the time complexity is $4 \cdot 2^{77-d} = 2^{79-d}$ clocks which is equivalent to 2^{71-d} encryptions of Sprout along with 2^d table lookups.

4.1 Detailed Workload for $d = 40$

We focus on the extreme case $d = 40$ (i.e., $\delta_t = 0$ for 40 consecutive t values) and give the workloads in detail for this case. We need to solve the following systems of equations: a linear system $\mathcal{LS}$ and a nonlinear system $\mathcal{NS}$.

$$\mathcal{LS} := \begin{cases} l_{t-5} \oplus l_{t+12} \oplus l_{t+28} \oplus n_t \quad\ \oplus n_{t+11} \oplus n_{t+20} = 0 \\ l_{t-4} \oplus l_{t+13} \oplus l_{t+29} \oplus n_{t+1} \quad \oplus n_{t+12} \oplus n_{t+21} = 0 \\ \quad\quad\quad\quad\quad \vdots \\ l_{t+4} \oplus l_{t+21} \oplus l_{t+37} \oplus n_{t+9} \quad \oplus n_{t+20} \oplus n_{t+29} = 0 \\ l_{t+5} \oplus l_{t+22} \oplus l_{t+38} \oplus n_{t+10} \oplus n_{t+21} \oplus n_{t+30} = 0 \\ \quad\quad\quad\quad\quad \vdots \\ l_{t+33} \oplus l_{t+50} \oplus l_{t+66} \oplus n_{t+38} \oplus n_{t+49} \oplus n_{t+58} = 0 \\ l_{t+34} \oplus l_{t+51} \oplus l_{t+67} \oplus n_{t+39} \oplus n_{t+50} \oplus n_{t+59} = 0 \end{cases}$$

$$\mathcal{NS} := \begin{cases} c_4^t \quad\ \oplus l_t \quad\ \oplus n_{t+40} \oplus g(N_t) \quad = 0 \\ c_4^{t+1} \oplus l_{t+1} \oplus n_{t+41} \oplus g(N_{t+1}) = 0 \\ \quad\quad\quad\quad \vdots \\ c_4^{t+18} \oplus l_{t+18} \oplus n_{t+58} \oplus g(N_{t+18}) = 0 \\ c_4^{t+19} \oplus l_{t+19} \oplus n_{t+59} \oplus g(N_{t+19}) = 0 \end{cases}$$

First of all, we can easily write all l_i's in $\mathcal{LS}$ as linear combinations of l_j's for $t \leq j \leq t + 39$. Let $\mathcal{LS}'$ be the new system of equations where all l_i's for $t - 5 \leq i \leq t - 1$ and $t + 40 \leq i \leq t + 67$ are replaced with l_j's for $t \leq j \leq t + 39$ in accordance with the LFSR feedback function. Now denoting $\mathcal{L} := (l_t, l_{t+1}, \ldots, l_{t+39})^T$ and

$$\mathcal{B} := (n_t \oplus n_{t+11} \oplus n_{t+20}, n_{t+1} \oplus n_{t+12} \oplus n_{t+21}, \ldots, n_{t+39} \oplus n_{t+50} \oplus n_{t+59})^T,$$

we can write $\mathcal{LS}'$ as $\mathcal{M} \cdot \mathcal{L} = \mathcal{B}$ where $\mathcal{M}$ is the 40×40 coefficient matrix of l_j's and ^{T} is the transpose operation.

The sequence $l_{t-5} \oplus l_{t+12} \oplus l_{t+28}$ can be also produced by the LFSR of Sprout. Since its characteristic polynomial is primitive, the coefficient matrix $\mathcal{M}$ is a power of the next state matrix and hence it is invertible. We also have verified on a computer that $\mathcal{M}$ is invertible. Hence, $\mathcal{L} = \mathcal{M}^{-1}\mathcal{B}$ implying we can equate each l_j, $t \leq j \leq t + 39$, to linear combinations of some n_i's for $t \leq i \leq t + 59$. Plugging in the values of l_j's for $t \leq j \leq t+19$ in $\mathcal{NS}$, we end up with a system, denoted by $\mathcal{NS}'$, of 20 nonlinear equations in 60 variables (ignoring the counter values for the moment). As a result, the main goal is to find all the solutions of $\mathcal{NS}'$ and store them in a table with their outputs.

It's expected that there exist 2^{40} solutions for the system. One approach for solving it may be to use a SAT solver. However, this approach is quite inefficient compared to solving a system of linear equations. Having a more detailed look at the equations in $\mathcal{NS}'$, we see that by carefully guessing 40 n_i values, the system becomes almost linear. To do that, one possible choice for selected n_i values, $SEC\{n_i\}$ is as follows:

$$
\begin{array}{cccccccccc}
n_{t+5} & n_{t+6} & n_{t+8} & n_{t+9} & n_{t+11} & n_{t+12} & n_{t+14} & n_{t+15} & n_{t+17} & n_{t+18} \\
n_{t+19} & n_{t+21} & n_{t+22} & n_{t+23} & n_{t+25} & n_{t+26} & n_{t+27} & n_{t+29} & n_{t+30} & n_{t+31} \\
n_{t+32} & n_{t+33} & n_{t+34} & n_{t+35} & n_{t+36} & n_{t+38} & n_{t+39} & n_{t+40} & n_{t+41} & n_{t+42} \\
n_{t+43} & n_{t+45} & n_{t+46} & n_{t+47} & n_{t+48} & n_{t+49} & n_{t+50} & n_{t+52} & n_{t+54} & n_{t+55}
\end{array}
$$

Selection of these n_i values mostly (but not always) follows the rule that n_{i+1} and n_{i+2} should be guessed to make an n_i value appear as a linear term. Fixing the values in $SEC\{n_i\}$, there still exists (at most 3) nonlinear terms $n_{t+51}n_{t+53}$, $n_{t+51}n_{t+56}$ and $n_{t+56}n_{t+57}$ with probability $\frac{1}{2}$. We see some nonlinear terms when $(n_{t+48}, n_{t+52}, n_{t+54}, n_{t+55}) \in S$ where

$$
\begin{aligned}
S := \{ & (1,1,1,1), (0,1,1,1), (1,0,1,1), (1,1,0,1), \\
& (1,1,1,0), (0,0,1,1), (0,1,0,1), (1,1,0,0) \}.
\end{aligned}
$$

It is expected that each 40-bit guess in both cases,

$$
(n_{t+48}, n_{t+52}, n_{t+54}, n_{t+55}) \in S \text{ and } (n_{t+48}, n_{t+52}, n_{t+54}, n_{t+55}) \notin S
$$

yields 1 solution on average, which we have verified experimentally (see Appendix B). Hence, by solving all the cases where the system is linear, we can obtain around 2^{39} solutions. As a result, we can obtain half of the solutions with an effort of solving $\frac{1}{2} \cdot 2^{40} = 2^{39}$ systems of linear equations. If one wishes to obtain all the solutions, then in order to make the system linear, 2 more n_i values (e.g., adding n_{t+51} and n_{t+57} to $SEC\{n_i\}$ and forming $EXSEC\{n_i\}$) need to be guessed. Hence, the effort of finding all the solutions is equivalent to solving $\frac{1}{2} \cdot 2^{42} + 2^{39} \approx 2^{41.32}$ systems of linear equations. The nonlinear cases can be solved using a SAT solver as well. However, this is still less efficient than guessing 2 more bits and solving a linear system.

We can repeat the computations for each candidate of the counter values. For $d = 40$, there are 39 different values of $(c_4^t, c_4^{t+1}, \ldots, c_4^{t+d-21})$, which we will refer as a *counter array*. Observe that the counter values are added to the system $\mathcal{NS}'$ linearly. Therefore, we can do row reduction operation once and find solutions for different counter arrays. However, we still need to store separate tables for

---

Algorithm 1. Creating Tables

for each choice of 40 n_i values in $SEC\{n_i\}$ with $(n_{t+48}, n_{t+52}, n_{t+54}, n_{t+55}) \in S$ do

   $N_t \leftarrow Solve(NS')$

   for each C_i where C_i's denote possible counter arrays do

      Find $N_t^{C_i}$ from N_t by plugging in the values in C_i

      $L_t^{C_i} \leftarrow \mathcal{M}^{-1} \cdot \mathcal{B}$

      $Z_{t-10}^{t+32} \leftarrow (z_{t-10}, \ldots, z_{t+32})$ // generate keystream for 43 clocks.

      Store $(Z_{t-10}^{t+32}, N_{t-9}^{C_i}, L_{t-9}^{C_i})$ in a table $T_Z^{C_i}$ sorted by Z_{t-10}^{t+32}

   end for

end for

for each choice of 42 n_i values in $EXSEC\{n_i\}$ with $(n_{t+48}, n_{t+52}, n_{t+54}, n_{t+55}) \notin S$ do

   $N_t \leftarrow Solve(NS')$

   for each C_i where C_i's denote possible counter arrays do

      Find $N_t^{C_i}$ from N_t by plugging in the values in C_i

      $L_t^{C_i} \leftarrow \mathcal{M}^{-1} \cdot \mathcal{B}$

      $Z_{t-10}^{t+32} \leftarrow (z_{t-10}, \ldots, z_{t+32})$ // generate keystream for 43 clocks.

      Store $(Z_{t-10}^{t+32}, N_{t-9}^{C_i}, L_{t-9}^{C_i})$ in a table $T_Z^{C_i}$ sorted by Z_{t-10}^{t+32}

   end for

end for

---

each counter array to produce 43 bits of output for each internal state. Since there are 74 possible counter arrays to produce 43 output bits, we need to store 74 separate tables. The solutions give internal states at the t-th clock. Store the internal states at time $(t - 9)$ just to avoid clocking backwards 9 times before mounting the key recovery attack in Sect. 3.

Algorithms 1 and 2 summarizes the full attack. Algorithm 1 is the off-line phase of the attack, related to preparing tables for each counter value. Algorithm 2 gives the set of instructions of how to recover key or come to a contradiction in a formal way for each trial of 43-bit keystream segment. We verified by producing some test values that Algorithm 1 generates internal states $((N_t, L_t)$ pairs) for which $(\delta_{t-9}, \delta_{t-8}, \ldots, \delta_{t+30}) = (0, 0, \ldots, 0)$ and that Algorithm 2 finds the correct key when our assumption about δ_t is fulfilled during keystream generation (see Appendix B).

The data complexity is given as $D = 2^{40}$ bits of output, not necessarily produced by just one IV, and we have 74 tables of each having 2^{40} rows. Each row contains 80-bit internal state and 3 output bits, indexed by the remaining 40 bits of the output. Hence, the memory requirement M is roughly 770 Terabytes[1] (this can be reduced by storing some of the tables in cost of increased data complexity). The precomputation for creating the tables is $2^{41.32}$ row reduction operations of at most 20 by 20 matrices along with producing 43 bit outputs for each solution. Since 60 n_i values and all the l_i values are known, 33-bit

---

[1] One may choose to store only 60 n_i values to reduce the memory requirement. In this case about 580TB of memory is needed. But, in the online phase, the values of LFSR need to be computed.

Algorithm 2. Online Phase of the Attack

Take 2^{40} keystream bits (not necessarily generated using the same IV)
for each 43-bit keystream block **do**
    // C_j := corresponding counter array for the current clock-cycle
    if Keystream block exists in $T_Z^{C_j}$ **then**
        Fill NLFSR and LFSR according to values in $T_Z^{C_j}$
        while Internal state does not produce a contradiction **do**
            Clock cipher backwards
            if keystream is in $\{0, 1\}$ **then**
                Check state!
            else
                Determine key bit value involved
            end if
        end while
    end if
end for

---

output may be generated by substituting the appropriate values in the output function without needing to clock the whole cipher. In addition, we need to clock backwards 9 times to produce the remaining 10 output bits, which brings an additional effort of about $9 \cdot 2^{40} \cdot 2^{-8.33} \approx 2^{34.84}$ encryptions. The workload of adding an output with its internal state to the appropriate table in a sorted way is negligible. The time complexity during the online phase is 2^{40} table lookups along with $2^{-3} \cdot 2^{40} \cdot 5 \cdot 2^{-8.33} \approx 2^{31}$ encryptions. Recall that only $1/8$ of the keystreams are in one of the tables. And, if a keystream is found in a table, the corresponding internal state obtained is for time $t - 9$, but we already make use of z_{t-10}. Hence, a candidate is checked in $1 + 4 = 5$ clocks on average.

4.2 Reducing the Data Complexity

Let $\Delta_t^d := (\delta_t, \delta_{t+1}, \ldots, \delta_{t+d-1})$. We focused on the case when $\Delta_t^{40} = (0, 0, \ldots, 0)$ for some t. However, suppose Δ_t^{40} contains a single 1 at i-th index and k_i is incorporated into the NLFSR feedback at clock-cycle $t + i$. In that case, we can create the tables for $k_i = 0$ and $k_i = 1$ separately (which would double the table size) and apply the same attack. If we do this for each possible index i, M increases by a factor of $2 \cdot 40 = 80$ and D decreases by a factor of 40. Generalizing this idea, we can create tables for each possible Δ_t^d. If there are n 1s in Δ_t^d, M increases by a factor of $2^n \cdot \binom{d}{n}$ while D decreases by a factor of $\binom{d}{n}$.

5   Conclusion and Discussion

We illustrated Time-Memory-Data tradeoff attacks and a guess-and-determine attack mounted on full Sprout. The TMD attacks combine both guess-and-determine and divide-and-conquer techniques. We have guessed some taps of the internal state satisfying a certain property and then determined the remaining taps by solving a specific system of linear equations. After storing all such internal states in tables with their outputs (we can produce some output bits without knowing the key bits, thanks to the special property that the internal states satisfy), we mount a divide-and-conquer attack to recover the key from the given output keystream bits. The complexities indicate that the attack is highly feasible. We have verified our statements by conducting several experiments. The detailed experimental results including an implementation of the attack with a small-scale table are given in Appendix B.

Designing ultra lightweight stream ciphers allocating less than 1K GE in hardware with a moderate security level such as 80 bits, is a new challenge, initiated by Armknecht and Mikhalev [2]. Even though their first design attempt does not achieve the required security level, we believe that several other designs will likely appear in the literature soon and some of them will probably be secure enough to be used in the industry.

We claim that producing output during each clocking of the registers is not a convenient design philosophy for ciphers whose internal state sizes are not large enough. Otherwise, they may be prone to guess-and-determine or divide-and-conquer attacks. We think that adopting the design philosophy of Grain family, particularly in bitwise operations and clockwise output generation, has brought the security failure to Sprout. The research question is about the optimization of the rate of the output generation over the register clocking in terms of security versus throughput. This can be considered as a generic question relating to all ultra lightweight stream ciphers. One straightforward countermeasure against all the attacks on Sprout may be decreasing the throughput of Sprout and giving only one bit output in, for instance, 16 clockings of Sprout registers. That output may be the sum of all the 16 bit outputs of the original Sprout.

Acknowledgments. We would like to thank anonymous reviewers for their helpful comments, especially the second reviewer who makes a comprehensive analysis of the paper and a lot of useful suggestions. We would also like to thank Ferhat Karakoç and Güven Yücetürk for commenting on the paper.

References

1. Ågren, M., Hell, M., Johansson, T., Meier, W.: Grain-128a: a new version of Grain-128 with optional authentication. Int. J. Wire. Mob. Comput. 5(1), 48–59 (2011)
2. Armknecht, F., Mikhalev, V.: On lightweight stream ciphers with shorter internal states. In: Leander, G. (ed.) FSE 2015. LNCS, vol. 9054, pp. 451–470. Springer, Heidelberg (2015)
3. Babbage, S.: Improved exhaustive search attacks on stream ciphers. Security and Detection, European Convention IET (1995)
4. Banik, S.: Some results on Sprout. Cryptology ePrint Archive, Report 2015/327 (2015). http://eprint.iacr.org/
5. Barkan, E., Biham, E., Shamir, A.: Rigorous bounds on cryptanalytic time/memory tradeoffs. In: Dwork, C. (ed.) CRYPTO 2006. LNCS, vol. 4117, pp. 1–21. Springer, Heidelberg (2006)

6. Biryukov, A., Shamir, A.: Cryptanalytic time/memory/data tradeoffs for stream ciphers. In: Okamoto, T. (ed.) ASIACRYPT 2000. LNCS, vol. 1976, pp. 1–13. Springer, Heidelberg (2000)
7. Biryukov, A., Shamir, A., Wagner, D.: Real time cryptanalysis of A5/1 on a PC. In: Schneier, B. (ed.) FSE 2000. LNCS, vol. 1978, pp. 1–18. Springer, Heidelberg (2001)
8. Bogdanov, A.A., Knudsen, L.R., Leander, G., Paar, C., Poschmann, A., Robshaw, M., Seurin, Y., Vikkelsoe, C.: PRESENT: an ultra-lightweight block cipher. In: Paillier, P., Verbauwhede, I. (eds.) CHES 2007. LNCS, vol. 4727, pp. 450–466. Springer, Heidelberg (2007)
9. Borghoff, J., Canteaut, A., Güneysu, T., Kavun, E.B., Knezevic, M., Knudsen, L.R., Leander, G., Nikov, V., Paar, C., Rechberger, C., Rombouts, P., Thomsen, S.S., Yalçın, T.: PRINCE – A Low-Latency Block Cipher for Pervasive Computing Applications. In: Wang, X., Sako, K. (eds.) ASIACRYPT 2012. LNCS, vol. 7658, pp. 208–225. Springer, Heidelberg (2012)
10. De Cannière, C., Dunkelman, O., Knežević, M.: KATAN and KTANTAN — a family of small and efficient hardware-oriented block ciphers. In: Clavier, C., Gaj, K. (eds.) CHES 2009. LNCS, vol. 5747, pp. 272–288. Springer, Heidelberg (2009)
11. Golić, J.D.: Cryptanalysis of alleged A5 stream cipher. In: Fumy, W. (ed.) EURO-CRYPT 1997. LNCS, vol. 1233, pp. 239–255. Springer, Heidelberg (1997)
12. Hao, Y.: A related-key chosen-iv distinguishing attack on full Sprout stream cipher. Cryptology ePrint Archive, Report 2015/231 (2015). http://eprint.iacr.org/
13. Hell, M., Johansson, T., Maximov, A., Meier, W.: A stream cipher proposal: Grain-128. In: 2006 IEEE International Symposium on Information Theory, pp. 1614–1618, July 2006
14. Hell, M., Johansson, T., Maximov, A., Meier, W.: The Grain family of stream ciphers. In: Robshaw, M., Billet, O. (eds.) New Stream Cipher Designs. LNCS, vol. 4986, pp. 179–190. Springer, Heidelberg (2008)
15. Hell, M., Johansson, T., Meier, W.: Grain: a stream cipher for constrained environments. Int. J. Wire. Mob. Comput. **2**(1), 86–93 (2007)
16. Hellman, M.E.: A cryptanalytic time-memory trade-off. IEEE Trans. Inf. Theor. **26**(4), 401–406 (1980)
17. Karakoç, F., Demirci, H., Harmancı, A.E.: ITUbee: a software oriented lightweight block cipher. In: Avoine, G., Kara, O. (eds.) LightSec 2013. LNCS, vol. 8162, pp. 16–27. Springer, Heidelberg (2013)
18. Lallemand, V., Naya-Plasencia, M.: Cryptanalysis of full Sprout. In: Gennaro, R., Robshaw, M. (eds.) Advances in Cryptology - CRYPTO 2015. LNCS, vol. 9215, pp. 663–682. Springer, Heidelberg (2015)
19. Maitra, S., Sarkar, S., Baksi, A., Dey, P.: Key recovery from state information of Sprout: application to cryptanalysis and fault attack. Cryptology ePrint Archive, Report 2015/236 (2015). http://eprint.iacr.org/
20. Suzaki, T., Minematsu, K., Morioka, S., Kobayashi, E.: TWINE: a lightweight block cipher for multiple platforms. In: Knudsen, L.R., Wu, H. (eds.) SAC 2012. LNCS, vol. 7707, pp. 339–354. Springer, Heidelberg (2013)
21. Wu, W., Zhang, L.: LBlock: a lightweight block cipher. In: Lopez, J., Tsudik, G. (eds.) ACNS 2011. LNCS, vol. 6715, pp. 327–344. Springer, Heidelberg (2011)

A Details of Sprout

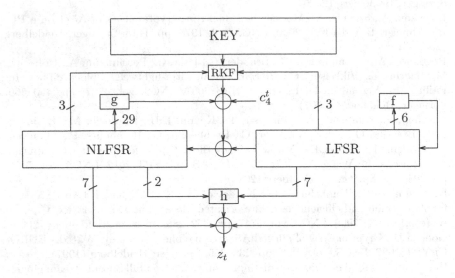

Fig. 1. The high level structure of Sprout

RKF in Fig. 1 represents the round key function. The LFSR is clocked as $l_{39}^{t+1} = f(L_t) = l_0^t \oplus l_5^t \oplus l_{15}^t \oplus l_{20}^t \oplus l_{25}^t \oplus l_{34}^t$ and $l_i^{t+1} = l_{i+1}^t$ for $0 \leq i \leq 38$. The function g is the nonlinear feedback function for the NLFSR. Its output is XORed with the round key bit k_t^*, the counter bit c_4^t and the output of the LFSR $l_t = l_0^t$. So, the feedback of the NLFSR n_{39}^{t+1} is given as follows:

$$
\begin{aligned}
n_{39}^{t+1} &= g(N_t) \oplus k_t^* \oplus l_0^t \oplus c_4^t \\
&= k_t^* \oplus l_0^t \oplus c_4^t \oplus n_0^t \oplus n_{13}^t \oplus n_{19}^t \oplus n_{35}^t \oplus n_{39}^t \oplus n_2^t n_{25}^t \\
&\quad \oplus n_3^t n_5^t \oplus n_7^t n_8^t \oplus n_{14}^t n_{21}^t \oplus n_{16}^t n_{18}^t \oplus n_{22}^t n_{24}^t \oplus n_{26}^t n_{32}^t \\
&\quad \oplus n_{33}^t n_{36}^t n_{37}^t n_{38}^t \oplus n_{10}^t n_{11}^t n_{12}^t \oplus n_{27}^t n_{30}^t n_{31}^t
\end{aligned}
$$

where

$$
\begin{aligned}
k_t^* &= k_t, \quad 0 \leq t \leq 79 \\
k_t^* &= k_{t \bmod 80} \cdot (l_4^t \oplus l_{21}^t \oplus l_{37}^t \oplus n_9^t \oplus n_{20}^t \oplus n_{29}^t)
\end{aligned}
$$

with $\delta_t := l_4^t \oplus l_{21}^t \oplus l_{37}^t \oplus n_9^t \oplus n_{20}^t \oplus n_{29}^t$. Once the internal state is determined for clock-cycle t, the keystream bit z_t is generated as follows:

$$
\begin{aligned}
z_t &= n_4^t l_6^t \oplus l_8^t l_{10}^t \oplus l_{32}^t l_{17}^t \oplus l_{19}^t l_{23}^t \oplus n_4^t l_{32}^t n_{38}^t \\
&\quad \oplus l_{30}^t \oplus n_1^t \oplus n_6^t \oplus n_{15}^t \oplus n_{17}^t \oplus n_{23}^t \oplus n_{28}^t \oplus n_{34}^t
\end{aligned}
$$

Initialization Phase: The feedback registers are initialized as follows:

$$n_i = iv_i, \quad \text{for } 0 \leq i \leq 39$$
$$l_i = iv_{40+i}, \text{for } 0 \leq i \leq 29$$
$$l_i = 1, \qquad \text{for } 30 \leq i \leq 38$$
$$l_{39} = 0$$

After filling the registers, the cipher is run 320 clocks without producing keystream while the output z_t is fed back into both feedback registers such that $l_{39}^{t+1} = z_t \oplus f(L_t)$ and $n_{39}^{t+1} = z_t \oplus k_t^* \oplus l_0^t \oplus c_4^t \oplus g(N_t)$. The keystream generator starts generating the keystream after the initialization phase.

The designers of Sprout suggest to generate up to 2^{40} keystream bits with one (key, IV) pair.

Workload of Exhaustive Search: To exhaustively search a key, one has to run the initialization phase first (320 clocks), and then generate 80 bits of keystream for a unique match. However, since each keystream bit generated matches the corresponding actual keystream bit one with probability $\frac{1}{2}$, 2^{80} keys are tried for 1 clock and roughly half of them are eliminated, 2^{79} for one more clock and half of the remaining keys are eliminated, and so on. Hence, the average number of clocks per one trial after the initialization step among 2^{80} keys is

$$\sum_{i=0}^{79} \frac{2^{80-i}}{2^{80}} = \sum_{i=0}^{79} \frac{1}{2^i} \approx 2$$

As a result, we will assume that clocking the registers once will cost roughly $\frac{1}{322} \approx 2^{-8.33}$ encryptions.

B Experiments

We did not implement the whole attack due to our memory shortage. Instead, we have conducted several experiments to verify our attack. We have solved millions of the systems of linear equations to collect some of the special internal states where $\delta_t = 0$ for consecutive 40 values of t and stored the solutions in tables. Then, we have accomplished to recover an internal state in the table with the key used in the given keystream sequence.

First of all, we performed a run test on δ_t values to check if the probability that 40 consecutive δ_t values vanish simultaneously is around 2^{-40} for the Sprout internals. We chose 30 random (IV, K) pairs, and run the cipher 2^{40} clocks for each selection. We depict the average number of runs having length i in Table 3. The values for $10 \leq i \leq 21$ are not given in the table since they are too big to display. However, we can simply say that the empirical result for any $10 \leq i \leq 21$ does not deviate from the corresponding expected value more than 0.0678 percent.

Moreover, we have implemented the attack with a small-scale table in Sage 6.5. We found more than 5 million solutions for the system $\mathcal{NS}'$ and generated tables for different counter values. The total size of all the tables is about 215

Table 3. Comparison between the empirical results and the expected number of runs.

	i		$i+1$		$i+2$	
	Empirical	Expected	Empirical	Expected	Empirical	Expected
$i = 22$	65596.66	65536	32802.60	32768	16398.17	16384
$i = 25$	8202.73	8192	4114.03	4096	2042.5	2048
$i = 28$	1025.9	1024	509.9	512	259.4	256
$i = 31$	127.8	128	64.2	64	31.33	32
$i = 34$	17.63	16	7.7	8	4.13	4
$i = 37$	1.83	2	1.2	1	0.400	0.500
$i = 40$	0.333	0.250	0.133	0.125	0	0.062
$i = 43$	0	0.031	0.067	0.016	0.033	0.008

Table 4. Number of state candidates eliminated at each clock i.

Clock-cycle number	# of states eliminated at i-th clock
$i = 1$	250071
$i = 2$	187325
$i = 3$	140862
$i = 4$	105620
$i = 5$	79061
$i = 6$	59067
$i = 7$	44714
$i = 8$	33287
$i = 9$	25044
$i = 10$	18590
Average # of clocks run	$3.999 \cdots$

MB. For a randomly chosen state in the tables, we have found a valid (IV, K) pair generating the state. Then, we mounted the attack on the corresponding keystream sequence and successfully recovered the key. The (IV, K) pair was

$$IV = 110101011010000011101001100101110111101001101110011011001100000100101 0$$
$$0110111001101100110000001001010$$
$$K = 100010010001100111111111100000011000111110$$
$$0011001110100100010110100010000100101001$$

where the left-most bit represents the value for index 0. At time $t = 25916$ (after initialization), $(\delta_{t-9}, \delta_{t-8}, \ldots, \delta_{t+30}) = (0, 0, \ldots, 0)$ is satisfied. The target

Table 5. Number of solutions for $\mathcal{NS}'$ for each cases when $(n_{t+48}, n_{t+52}, n_{t+54}, n_{t+55})$ $\in S$ or $(n_{t+48}, n_{t+52}, n_{t+54}, n_{t+55}) \notin S$.

# of solutions	# of eqns with i solns when $(n_{t+48}, n_{t+52}, n_{t+54}, n_{t+55}) \in S$	# of eqns with i solns when $(n_{t+48}, n_{t+52}, n_{t+54}, n_{t+55}) \notin S$
$i = 0$	6049	6478
$i = 1$	5198	4376
$i = 2$	3704	4637
$i = 3$	671	0
$i = 4$	344	501
$i = 5$	9	0
$i = 6$	20	0
$i = 7$	1	0
$i = 8$	4	8
Average	1.012	0.982

internal state at time t was

$$L_t = 01001111111100100000101111110111001011111$$
$$N_t = 11110110010111111110111110010001011100000$$

which is in one of the tables. Running the key recovery attack, we have found 62 key bits in 160 clocks, verifying our statements. The remaining bits can be recovered by an exhaustive search.

We also verified the average number of clocks iterated before arriving at a contradiction during the internal-state-check mechanism. We chose 1 million random internal states and ran the key recovery attack for each of them. Each state was clocked backwards once before starting the test since the output at this clock is definitely a check bit. Table 4 shows how many states were eliminated at each clock i. The average number of clocks run for a candidate was about 3.999, as expected (See Sect. 3). Another experiment was about verifying the assumption that there exists approximately 2^{40} solutions for the system $\mathcal{NS}'$ in total. We have solved the system 16000 times for each case when

$$(n_{t+48}, n_{t+52}, n_{t+54}, n_{t+55}) \in S \text{ and } (n_{t+48}, n_{t+52}, n_{t+54}, n_{t+55}) \notin S.$$

Table 5 summarizes the results of the experiment, supporting our assumption about the number of solutions.

Related-Key Attack on Full-Round PICARO

Anne Canteaut[✉], Virginie Lallemand, and María Naya-Plasencia

Inria, project-team SECRET, Rocquencourt, France
{Anne.Canteaut,Virginie.Lallemand,Maria.Naya_Plasencia}@inria.fr

Abstract. Side-channel cryptanalysis is a very efficient class of attacks that recover secret information by exploiting the physical leakage of a device executing a cryptographic computation. To address this type of attacks, many countermeasures have been proposed, and some papers addressed the question of constructing an efficient masking scheme for existing ciphers. In their work, G. Piret, T. Roche and C. Carlet took the problem the other way around and specifically designed a cipher that would be easy to mask. Their careful analysis, that started with the design of an adapted Sbox, leads to the construction of a 12-round Feistel cipher named PICARO. In this paper, we present the first full-round cryptanalysis of this cipher and show how to recover the key in the related-key model. Our analysis takes advantage of the low diffusion of the key schedule together with the non-bijectivity of PICARO Sbox. Our best trade-off has a time complexity equivalent to $2^{107.4}$ encryptions, a data complexity of 2^{99} plaintexts and requires to store 2^{17} (plaintext, ciphertext) pairs.

Keywords: Related-key attack · Differential cryptanalysis · PICARO

1 Introduction

While performance and side-channel attacks resistance are most of the time considered separately and as distinct problems — new design papers focus on performance figures while countermeasure papers focus on how to implement a specific protection in order to reduce performance overheads —, some new cipher proposals tackle the two problems together by designing new primitives that fit given protections. Examples of such constructions are PICARO [8], Zorro [6], and the family of LS-design [7] (including Robin and Fantomas as concrete instantiations).

The countermeasure studied in these three designs is the *masking scheme* [10] for which the heavier parts to protect are the operations which are not linear with respect to the group operation used to share the sensitive variables. For these three ciphers as for most ciphers, these non-linear operations are concentrated in the Sboxes, and then the straightforward way to limit the masking cost is to

---

Partially supported by the French Agence Nationale de la Recherche through the BLOC project under Contract ANR-11-INS-011.

© Springer International Publishing Switzerland 2016
O. Dunkelman and L. Keliher (Eds.): SAC 2015, LNCS 9566, pp. 86–101, 2016.
DOI: 10.1007/978-3-319-31301-6_5

reduce the number of Sbox applications[1] as well as to choose Sboxes that are masking-friendly, for instance by reducing the number of field multiplications processed[2]. This later direction is the one followed by Piret, Roche and Carlet while devising PICARO [8,9]: their design relies on a non-bijective Sbox defined by two bivariate polynomials over $GF(2^4)$, reducing the number of non-linear multiplications to four. This Sbox is integrated to the round function of a Feistel scheme, composed of four operations which are an expansion, a key addition, an Sbox-layer and a compression.

In this paper we show that despite the fact that the authors of PICARO aim at resisting to related-key attacks, as pointed out in [9][3], we have been able to mount related-key attacks on the full cipher. Our attacks exploit some weaknesses in the key-schedule, the non-bijective properties of the Sbox and some properties of the linear layer. They provide different trade-offs between time and data complexities.

The paper is organized as follows. In Sect. 2, we give a brief description of the PICARO block cipher and of its design choices, in Sect. 3 we give some definitions and in Sect. 4, we present an analysis of PICARO key-schedule that leads to the related-key attack described in Sect. 5. The paper ends with a conclusion.

2 Description of PICARO

2.1 Round Function

One of the main objectives of G. Piret, T. Roche and C. Carlet when designing PICARO was to propose a 128-bit block cipher that would get an advantage over other ciphers regarding the ease to protect against *side-channel attacks*. To achieve this, they started from Rivain and Prouff's masking scheme [10] and determined which operations are difficult to mask to derive some new design criteria. Their analysis brought to light that efforts have to focus on the Sboxes since the masking scheme implies heavy overheads for the non-linear operations.

This condition must be added to the usual criteria coming from the (non-physical) usual attacks (including the prominent linear, differential, algebraic cryptanalyses and their variants) that require the Sbox be highly non-linear, have a high algebraic degree and a low differential-uniformity.

Their deep analysis resulted in the selection of an Sbox defined as the concatenation of two polynomials:

$$S : GF(2^4) \times GF(2^4) \to GF(2^4) \times GF(2^4)$$
$$(x, y) \mapsto (xy, (x^3 + \text{0x02})(y^3 + \text{0x04}))$$

where 0x02 and 0x04 represent elements of $GF(2^4)$ defined as $GF(2)[x]/(X^4 + X^3 + 1)$. Its full look-up table is given in Appendix A.

---

[1] This direction has been followed in the design of Zorro.

[2] When considering binary masking, this criterion is equivalent to limiting the number of AND processed (see for instance the LS-design [7]).

[3] Section 7.2 of [9]: "We want our scheme to resist known attacks on a key schedule algorithm, in particular related-key attacks...".

This non-bijective Sbox has already been studied in [5] and possesses the following desirable characteristics: a non-linearity equal to 94, an algebraic degree equal to 4, a maximal differential probability equal to 4/256. Furthermore, to fit well the masking protection, it can be implemented with only 4 non-linear operations; moreover all these non-linear operations are defined in the small field $GF(2^4)$.

An obvious but quite important remark that has to be made is that since the Sbox is not bijective, there exist sets of values that all have the same image through the Sbox. In other words, it is possible to cancel a difference entering the Sbox: a byte that is active before applying the Sbox can lead to an inactive output. Since the cube function is 3-to-1 over $GF(2^4)^*$, we deduce from the definition of the Sbox that it is a 3-to-1 function over $GF(2^4)^2 \backslash \{(0,0)\}$. Moreover, it is very easy to prove that, for any non-zero input difference Δ, the transition $\Delta \to 0$ holds with probability 2^{-7}. In other words, the equation $S(x + \Delta) + S(x) = 0$ has exactly two solutions x.

The choice of a non-bijective Sbox was motivated by the fact that finding a good Sbox is easier in this case. However, it requires to find a way to include it in a construction that makes the cipher invertible, and also to take the resulting differential properties into account. Therefore, this non-bijective Sbox is used within the Feistel construction. But, as noticed by the designers, the use of a basic Sbox layer as an inner function would make the cipher vulnerable to a quite simple but very efficient differential attack with only one active Sbox every 2 rounds.

To thwart this sort of attacks, PICARO designers choose to use an expansion and a compression function that have good diffusion, more precisely that ensure that a minimum of 7 Sboxes are active in each active round. This is achieved by using linear operations deduced from linear codes with minimum distance 7. The expansion G takes as input a 64-bit word and outputs an extended word of 112 bits. The corresponding matrix is depicted below (its entries are elements in $GF(2^8)$ defined as $GF(2)[X]/(1 + X^2 + X^3 + X^4 + X^8)$). Then, the inner state must be compressed back at the end of the round function. To this end, another linear operation is used, defined by the matrix $H = G^T$, which converts 112-bit words into 64-bit ones.

$$
G = \begin{pmatrix}
01 & 00 & 00 & 00 & 00 & 00 & 00 & 00 & 01 & 01 & 0A & 01 & 09 & 0C \\
00 & 01 & 00 & 00 & 00 & 00 & 00 & 00 & 05 & 01 & 01 & 0A & 01 & 09 \\
00 & 00 & 01 & 00 & 00 & 00 & 00 & 00 & 06 & 05 & 01 & 01 & 0A & 01 \\
00 & 00 & 00 & 01 & 00 & 00 & 00 & 00 & 0C & 06 & 05 & 01 & 01 & 0A \\
00 & 00 & 00 & 00 & 01 & 00 & 00 & 00 & 09 & 0C & 06 & 05 & 01 & 01 \\
00 & 00 & 00 & 00 & 00 & 01 & 00 & 00 & 01 & 09 & 0C & 06 & 05 & 01 \\
00 & 00 & 00 & 00 & 00 & 00 & 01 & 00 & 0A & 01 & 09 & 0C & 06 & 05 \\
00 & 00 & 00 & 00 & 00 & 00 & 00 & 01 & 01 & 0A & 01 & 09 & 0C & 06
\end{pmatrix}
$$

To sum up, the round function is made of four operations, as depicted on Fig. 1: first, the 64-bit left part is extended by the expansion function defined by G, then the round-key is added. This is followed by the Sbox application (14 parallel applications of S) and finally the state is compressed back to 64 bits, and is XORed to the right part of the internal state.

PICARO encryption routine is made of 12 iterations of this round function.

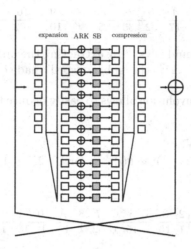

Fig. 1. One round of PICARO block cipher.

2.2   Key-Schedule

The previous algorithm requires twelve 112-bit subkeys that are derived from the 128-bit master key K. The designers wanted a simple and efficient key-schedule, together with resistance to the two main attacks exploiting the key-schedule, namely related-key attacks [1] and slide attacks [3]. In addition to that, Piret *et al.* looked for round-keys that could be computed on the fly, i.e. for which the computation of the subkey of round i can be done from the knowledge of the subkey at round $(i-1)$ (or from the knowledge of the subkey at round $(i+1)$, in decryption mode).

Their analysis has resulted in a linear key-schedule composed of rotations, bitwise additions and truncations. Namely, K denotes the 128-bit master key and $K^{(1)}, K^{(2)}, K^{(3)}$ and $K^{(4)}$ are the four 32-bit chunks composing K: $K = (K^{(1)}, K^{(2)}, K^{(3)}, K^{(4)})$. Let $T : (GF(2)^{32})^4 \rightarrow (GF(2)^{32})^4$ be the linear transformation defined as follows:

$$\begin{pmatrix} T(K)^{(1)} \\ T(K)^{(2)} \\ T(K)^{(3)} \\ T(K)^{(4)} \end{pmatrix} = \begin{pmatrix} 0\ 1\ 1\ 1 \\ 1\ 0\ 1\ 1 \\ 1\ 1\ 0\ 1 \\ 1\ 1\ 1\ 0 \end{pmatrix} \times \begin{pmatrix} K^{(1)} \\ K^{(2)} \\ K^{(3)} \\ K^{(4)} \end{pmatrix}$$

Then, to compute the subkeys from the master key K, we first compute the extended key $(\kappa^1, \kappa^2, \cdots \kappa^{12})$ (where each κ^i is 128-bit long) with the following formulas:

$$\begin{cases} \kappa^1 = K \\ \kappa^i = T(K) \ggg \Theta(i) & \text{for } i = 2, 4, 6, 8, 10, 12 \\ \kappa^i = K \ggg \Theta(i) & \text{for } i = 3, 5, 7, 9, 11 \end{cases}$$

where $\Theta(i)$ is:

i	2	3	4	5	6	7	8	9	10	11	12
$\Theta(i)$	1	16	17	32	33	85	86	101	102	117	118

and then we obtain the round keys k^i by extracting the first 112 bits from κ^i. We denote by $skw = (k^1||k^2||...k^{12})$ the 'subkey word' made of the concatenation of the k^i.

The fact that T is an involution allows to easily deduce the 'on-the-fly' expressions, which are:

$$\begin{cases} \kappa^1 = K \\ \kappa^i = T(\kappa^{i-1}) \ggg \theta(i) \end{cases} \qquad \text{for } i = 2, \cdots, 12$$

where θ is defined by:

i	2	3	4	5	6	7	8	9	10	11	12
$\theta(i)$	1	15	1	15	1	52	1	15	1	15	1

Since we do not need it here, we refer to the design document [8] for the formulas of the round-key derivation in decryption mode.

3   Definitions and Notation

In most block cipher constructions, the Sboxes are bijective, implying that in a differential attack a non-zero difference at the input of an Sbox is equivalent to a non-zero difference at its output. Here, the situation is different since the Sboxes are not bijective. Therefore, in order to avoid any ambiguity, we give a more precise definition of the notion of *active Sbox*:

Definition 1 (Active Sbox). *An active Sbox is an Sbox with a nonzero input difference (and a possibly zero output difference).*

Definition 2 (Data, Time and Memory Complexities).

- *The data complexity is defined as the number of plaintext/ciphertext pairs necessary to conduct the attack;*
- *The time complexity, which is expressed as a number of full 12-round encryptions, incorporates all the operations performed by the attacker to recover the key, and includes the encryptions needed to compute the necessary data;*
- *The memory complexity, which is expressed as a number of 128-bit blocks, measures the memory needed during the attack.*

Definition 3 (nR-attack [2]). *A nR-attack is a differential attack that covers R rounds with a differential characteristic and attacks $R+n$ rounds in total. We extend this definition to the related-key setting and call the additional rounds the key-recovery rounds.*

4 Key-Schedule Analysis

In this section we focus on PICARO key-schedule. We first focus on keys under which a given plaintext leads to the same ciphertext and then extend our analysis to a related-key attack.

4.1 Keys Leading to Colliding Ciphertexts

In this section we are interested in sets of keys that with high probability encrypt a given plaintext into the same ciphertext after the 12 rounds of PICARO. Note that in the ideal case, if one plaintext is fixed and encrypted under different keys, assuming that the resulting function is random, a ciphertext collision is expected to occur with probability 2^{-128}.

In the case of PICARO, we can remark that the round structure enables us to cancel a key difference quite easily: indeed, since the key addition is immediately followed by the Sbox layer, composed of non-bijective Sboxes, a key difference can be immediately canceled by going through the Sbox, resulting in an internal-state collision at the input of the compression function.

Obtaining colliding ciphertexts becomes then possible if we can construct keys differing in as few positions as possible, in order to make the probability of the event "all the subkey differences are canceled by the Sboxes" higher than 2^{-128}. This means that we can cancel a maximum of s Sboxes, with s satisfying $2^{-7 \times s} > 2^{-128}$, so we have to find keys such that the corresponding subkey words differ in at most 18 bytes.

To find such keys, we first remark that the key-schedule algorithm is linear over $GF(2)$. This implies that the words corresponding to the concatenation of the subkeys (skw) belong to a linear code C of dimension $k = 128$ and length $n = 112 \times 12 = 1344$. Our search then boils down to the search for codewords with a low Hamming weight. We first focus on the Hamming weights of the codewords over $GF(2)$ and we will then move to the Hamming weight over $GF(2^8)$, i.e., the number of non-zero bytes, which is the relevant parameter in our attack.

To determine if it is possible to obtain keys differing in less than 18 bytes, i.e., at the inputs of less than 18 Sboxes, a first idea is to compute the minimum distance of the binary code, hereafter denoted by d. A straightforward computation of d would be too complicated due to the large size of the code. Some algorithms for finding low-weight codewords, e.g [4], could be used to determine d. But we now show that it can be easily deduced from the structure of the code with very simple arguments. We first make some observations coming from the structure of the generator matrix depicted on Fig. 2.

- If we consider all possible master keys of weight 1, the minimum weight among all corresponding codewords is 20, implying that $d \leq 20$
- According to the key-schedule description, 6 subkeys, namely k^1, k^3, k^5, k^7, k^9 and k^{11} consist of a selection of bits from the master key K. Following this, if we consider a master key of weight 1, then the word made by the

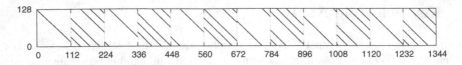

Fig. 2. Graphical representation of the generator matrix of linear code corresponding to the key-schedule.

concatenation of the subkeys skw contains at least 4 ones. Indeed, we have a 1 for sure in κ^1, κ^3, κ^5, κ^7, κ^9 and κ^{11} but after the truncation, a maximum of two of these ones may disappear. Accordingly, every time we add a one in the master key, the weights of the odd subkey words increase by at least 4.

All in one, those remarks show that, if they exist, the codewords of weight strictly less than 20 are obtained from master keys of weight at most 4. An exhaustive search over all these master keys shows that the minimum distance of the code is $d = 18$. The active positions of all master keys that reach this minimum are given in Table 1.

Table 1. Positions of the active bits in the round-keys that correspond to a minimum-weight subkey word (the bits entering the first Sbox are indexed from 0 to 7).

set	K	k^1	k^2	k^3	k^4	k^5	k^6	k^7	k^8	k^9	k^{10}	k^{11}	k^{12}
1	27, 123	27	28	11, 43	12, 44	27, 59	28, 60	80	81	0, 96	1, 97	16	17
2	28, 124	28	29	12, 44	13, 45	28, 60	29, 61	81	82	1, 97	2, 98	17	18
3	29, 125	29	30	13, 45	14, 46	29, 61	30, 62	82	83	2, 98	3, 99	18	19
4	30, 126	30	31	14, 46	15, 47	30, 62	31, 63	83	84	3, 99	4, 100	19	20
5	91, 123	91	92	11, 107	12, 108	27	28	48, 80	49, 81	64, 96	65, 97	80	81
6	92, 124	92	93	12, 108	13, 109	28	29	49, 81	50, 82	65, 97	66, 98	81	82
7	93, 125	93	94	13, 109	14, 110	29	30	50, 82	51, 83	66, 98	67, 99	82	83
8	94, 126	94	95	14, 110	15, 111	30	31	51, 83	52, 84	67, 99	68, 100	83	84

Note that these configurations might have been expected: the best scenarios are the ones for which the differences do not propagate a lot, so for keys differing in more than 2 bits we are looking for differences that cancel each other by the T map, i.e. that are at the same relative position in the 32-bit chunks of the master key. Moreover, the best configurations correspond to differences that end up as often as possible in the truncated part of the subkeys.

We note that the previous analysis determines the subkey words having a low Hamming weight counted in bits, while the relevant quantity is the number of nonzero bytes since it corresponds to the number of active Sboxes. However, the structure of the previous result indicates that the words with minimal weight over $GF(2^8)$ are derived from master keys having at most two nonzero bytes. An exhaustive search over this set enables us to determine a set W of 30 minimum-weight codewords over $GF(2^8)$, corresponding to the $(2^4 - 1)$ non-zero linear combinations of the first four rows in Table 1 and to the $(2^4 - 1)$ non-zero linear combinations of the last four rows. This analysis exhibits the following bias in PICARO:

The probability that any given plaintext is encrypted to the same ciphertext under two keys whose difference belongs to the set W is $(2^{-7})^{18} = 2^{-126}$.

5   Related-Key Attack on the Full-Round PICARO

We have shown in the previous section that we can ensure that the differences introduced by the two keys are integrally canceled with probability greater than 2^{-128}. In this section, we show how to use this property in order to build a distinguisher and recover the encryption key with a related-key attack. We consider a scenario in which the attacker is given the right to ask for the encryption of plaintexts under the secret encryption key and under a second key which is related to the first one by a fixed difference.

In the following, we denote by a_i, $i = 1 \cdots 12$ the number of active Sboxes in each round (including the rounds not covered by the characteristic) and by $a_{i \to j}$ the number of active Sboxes in rounds i to j. We provide in Fig. 3 a concrete example providing the best trade-off.

5.1   A First 2R-Attack

We describe here a simple related-key attack on the full cipher and give some optimizations in the following sections. Our attack is based on a differential characteristic (in which all the differences are canceled by the Sboxes) that covers the first 10 rounds and ends with a 2-round key-recovery.

The two points that make the 2-round key-recovery holds are the following: first, the characteristic - that cancels all the differences - allows us to obtain a good filter on the ciphertexts. Second, a property of the compression function allows us to invert the compression step with a limited complexity.

Ciphertext Filter. For a plaintext and a pair of keys following the characteristic, the state entering round 11 is free of differences, and we know the value of the round-key differences entering the Sboxes at round 11, so we can determine a set of possible differences at the output of round 11. In the worst case scenario, there are 2^7 possible output differences for each active Sbox, so $2^{7 \times a_{11}}$ differences are possible out of 2^{64} for the corresponding half of the ciphertext. This remark implies that we can filter out some pairs that for sure do not follow the characteristic.

Property of the Compression Function. The following proposition will be extensively used in the attack. It shows that the knowledge of the output of the compression function and of any 6 bytes of the input uniquely determines all input bytes.

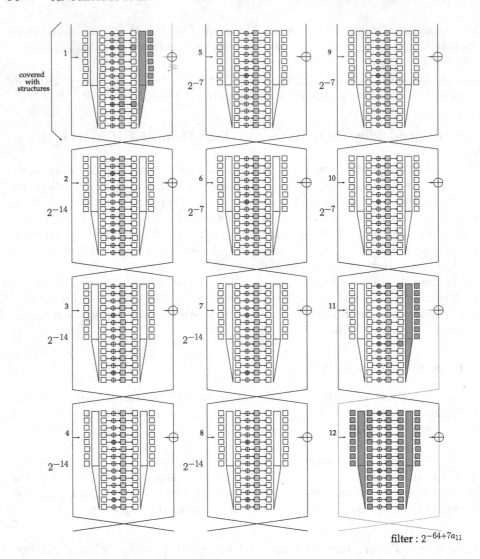

Fig. 3. Related-key attack on full-round PICARO. This figure represents the variant providing the best data complexity: 2^{99}, with a time complexity of $2^{107.4}$. The master key differences are positioned in bits 16 and 80.

Proposition 1. *Two distinct 14-byte words having the same image under the compression function coincide on at most five bytes.*

Proof. The compression function is defined by

$$\mathsf{Compr} : GF(2^8)^{14} \rightarrow GF(2^8)^8$$
$$x \mapsto xH$$

where H is a 14×8 matrix over $GF(2^8)$ defined in Sect. 2.1. Then, the set of all inputs which have the same image under Compr is an affine subspace of $GF(2^8)^{14}$ of dimension 6, which is a coset of the kernel

$$\mathcal{C} = \ker(\mathsf{Compr}) = \{x : xH = 0\}.$$

Therefore, any two elements x and x' having the same output under Comp are such that $(x + x')$ belongs to $\mathcal{C}$. Moreover, $\mathcal{C}$ can also be seen as the linear code of length 14 and dimension 6 over $GF(2^8)$ with parity-check matrix H^T. Since H^T has been chosen by the designers of PICARO as the generator matrix of an MDS code, it also corresponds to the parity-check matrix of an MDS code, implying that the minimum distance of $\mathcal{C}$ is $14 - 6 + 1 = 9$. Now, assume that there exist two distinct inputs x and x' which have the same output under Compr and which coincide on 6 input bytes. Then, $(x + x')$ is a non-zero element of $\mathcal{C}$ and it has at most $14 - 6 = 8$ nonzero bytes, which contradicts the fact that the minimum distance of $\mathcal{C}$ equals 9. $\square$

Moreover, for any fixed output y of Compr and any set $I \subset \{1, \ldots, 14\}$ of 6 input positions, the unique 14-byte word equal to a given α on I can be determined by elementary algebra. We first observe that the matrix H defining the compression function is equal to

$$\begin{pmatrix} \mathsf{Id}_8 \\ Z^T \end{pmatrix}$$

where Z denotes the 8×6-right part of matrix G defined in Sect. 2.1. Then, the set of all words whose image equals y is equal to $\mathcal{C} + \widetilde{y}$ where $\widetilde{y}$ is the 14-byte word equal to y on the first 8 bytes and which vanishes on the other 6 bytes, and $\mathcal{C} = \ker(\mathsf{Compr})$ is the linear space spanned by the rows of matrix $M = (Z, \mathsf{Id}_6)$. Let M_I (resp. $M_{\overline{I}}$) denote the columns of M corresponding to I (resp. to $\overline{I} = \{1, \ldots, 14\} \backslash I$). The previous proposition shows that M_I is non-singular. Therefore, the unique element x in $\mathsf{Compr}^{-1}(y)$ which is equal to α on I corresponds to the sum of $\widetilde{y}$ and of the element in $\mathcal{C}$ which is equal to $\alpha + \widetilde{y}_{|I}$ on I. This implies that the value of x on $\overline{I}$ is equal to $\widetilde{y}_{|\overline{I}} + (\alpha + \widetilde{y}_{|I}) M_I^{-1} M_{\overline{I}}$.

Obtaining Suggestions for the Values of the Subkey at Round 12.

1. Choose a master-key difference Δ with a minimum number of active Sboxes in the first 10 rounds and ask for $2^{7 \times a_{1 \rightarrow 10}}$ triples (P_i, C_i, C_i') generated from different plaintexts P_i, where C_i and C_i' are respectively the corresponding ciphertext under the secret master key K and the corresponding ciphertext under the key $K' = K \oplus \Delta$.
2. Filter out the triples by looking at the ciphertext differences: the number of remaining triples is $2^{7 \times a_{1 \rightarrow 10} - (64 - 7 \times a_{11})}$.
3. Since the internal state entering round 11 is free of difference, the left part of the ciphertext difference is equal to the difference at the output of the compression function at round 12. Given that, guess 6 bytes of the difference

entering the compression function and deduce the value of the full 112-bit difference entering the compression function at round 12 (this value is uniquely determined as shown by Proposition 1). There are 2^{48} possibilities for each triple.

4. Starting from the right half of the ciphertext difference, compute the expansion function and add the difference coming from the key difference to deduce the value of the difference entering the Sboxes at round 12.

5. With the help of the difference distribution table, check the difference transitions of the 14 Sboxes at round 12. If all the transitions are valid, compute the possible intermediate state values that permit these transitions. In order to estimate the average number of states returned by this procedure, we use that, for any Sbox, the product between the probability that a transition is valid and the average number of values following this transition is equal to 1. For PICARO Sbox S (note that the reasoning would be similar for any Sbox), we consider the number $\delta(a, b)$ of inputs x such that $S(x + a) + S(x) = b$, and the number τ of valid transitions, i.e., τ is the number of pairs (a, b) such that $\delta(a, b) > 0$. The sum of all (non-zero) entries in the difference table of S equals $2^{2 \times 8}$, i.e., the sum over all valid (a, b) of $\delta(a, b)$ equals $2^{2 \times 8}$. It then follows that

$$\left(\frac{1}{\tau} \sum_{\text{valid } (a,b)} \delta(a, b) \right) \times \left(\frac{\tau}{2^{2 \times 8}} \right) = 1.$$

Therefore, for each Sbox transition considered in the attack, we obtain in average one intermediate state which satisfies this transition. So, we get a total of

$$2^{7 \times a_{1 \to 10} - (64 - 7 \times a_{11})} \times 2^{48} = 2^{7 \times a_{1 \to 11} - 16} \text{ candidates.}$$

6. From the ciphertext value, deduce the intermediate state value before the key addition in round 12. From that, we obtain $2^{7 \times a_{1 \to 11} - 16}$ candidate values for the subkey of round 12.

This algorithm requires $2^{7 \times a_{1 \to 10} + 1}$ full encryptions for obtaining the initial triples, then $2^{7 \times a_{1 \to 11} - 16 + 1} \times \frac{1}{12}$ encryptions (since only round 12 is computed) to obtain candidates for k^{12}, which leads to an overall time complexity corresponding to the cost of

$$2^{7 \times a_{1 \to 10} + 1} + 2^{7 \times a_{1 \to 11} - 18.58} \text{ encryptions.}$$

5.2   Optimizations

Reducing the Initial Number of Encryptions. A potential bottleneck in the previous algorithm comes from the data complexity and then from the cost of the generation of the initial data. In this section, we detail a technique (close to the notion of structures which is commonly used in differential attacks) that reduces it by a factor of $2^{7 \times a_1}$. The idea here is to let the first round-key difference spread freely and to cancel this difference by introducing a difference in another plaintext.

For each plaintext P_i encrypted under the master key K, we ask for the encryption of a set of plaintexts P_i^{set} encrypted under the second key. These plaintexts are made by adding to P_i all the possible differences coming from the subkey difference. To compute these differences, we exhaust all the $(2^{8 \times a_1})$ possibilities for the difference at the output of the active Sboxes at round 1 and apply the (linear) compression function.

Since this set contains all the possible differences, one of them corresponds to the actual difference and will cancel the first round difference. Now, we extend that remark by asking the oracle for the encryption of the same set P_i^{set} under the master key. We then possesses $2^{8 \times a_1 + 1}$ plaintexts and each plaintext in the first set is such that the difference introduced by the first round-key is canceled by exactly one plaintext in the second set. To sum up, we have exactly $2^{8 \times a_1}$ pairs that lead to a zero difference at the end of round 1 from $2^{8 \times a_1 + 1}$ encryptions only. The other parts in the algorithm remain the same. So to obtain one pair that leads to a zero difference at round 10 we need

$$2^{7 \times a_{2 \to 10} - 8 \times a_1 + 8 \times a_1 + 1} = 2^{7 \times a_{2 \to 10} + 1} \text{ encryptions.}$$

If we apply this technique to the previous algorithm, the number of initial encryptions is reduced from $2^{7 \times a_{1 \to 10} + 1}$ to $2^{7 \times a_{2 \to 10} + 1}$.

Speeding up the Master-Key Recovery. The previous algorithm returns candidates for the 112 bits of the subkey at round 12. From that point, we can naively perform an exhaustive search for the remaining 16 bits of the master key, which would lead to an attack with data complexity $2^{7 \times a_{2 \to 10} + 1}$, time complexity $2^{7 \times a_{2 \to 10} + 1} + 2^{7 \times a_{1 \to 11} - 16} \times 2^{16}$ and with memory complexity $2^{8 \times a_1 + 1}$ (necessary to store the initial 'structures').

We can further reduce the second term in the time complexity by using the previous rounds to filter out wrong candidates faster than with a trial encryption. Indeed, we can check if the candidate key gives the right Sbox transitions round after round, and discard a candidate as soon as the Sbox transition is wrong.

The optimal configuration would be the one for which we can deduce from the candidate value for k^{12} the key bits of k^1 and k^{11} that enter the active Sboxes at rounds 1 and 11. In such a case, the attacker can directly (i.e. without any additional key guess) check if the candidate value for k^{12} leads to the right Sbox transitions at rounds 1 and 11. This corresponds to a filter of about 2^{-7a_1} and $2^{-7a_{11}}$. The remaining operations (checking the Sbox transitions at the other rounds and guessing the remaining bits) are comparatively of negligible time complexity.

In case only a few key bits are known, the attacker can also reduce the time complexity by doing this gradual check. For the $2^{7a_{1 \to 11} - 16}$ pairs and candidate values for k^{12}, she can compute the expansion function at the round in which the largest number of information bits is known. At this round, she needs to guess the unknown bits in order to check the active Sbox transitions and then, for the right guesses, she can compute the entire round function in order to access another round. The full round function is then computed only for the guesses

Table 2. List of all master-key differences (of weight less than 4) minimizing $a_{2\to10}$ ($a_{2\to10} = 14$) with the smallest value for $a_{1\to11}$ ($a_{1\to11} = 18$). The last 2 columns indicate the number of key bytes (and bits) deduced from k^{12} at the positions corresponding to the active Sboxes at rounds 1 and 11.

Diff. positions		a_1	$a_{2\to10}$	a_{11}	$a_{1\to11}$	Known bytes (bits) in k_1	Known bytes (bits) in k_{11}
11	107	2	14	2	18	1 (14)	2 (16)
12	108	2	14	2	18	1 (14)	2 (16)
13	109	2	14	2	18	1 (14)	2 (16)
14	110	2	14	2	18	1 (14)	2 (16)
15	111	2	14	2	18	1 (14)	2 (16)
16	80	2	14	2	18	2 (16)	2 (16)
17	81	2	14	2	18	2 (16)	2 (16)
18	82	2	14	2	18	2 (16)	2 (16)
19	83	2	14	2	18	2 (16)	0 (14)
20	84	2	14	2	18	2 (16)	0 (14)
21	85	2	14	2	18	2 (16)	0 (14)
22	86	2	14	2	18	2 (16)	0 (14)
23	87	2	14	2	18	2 (16)	0 (14)
24	88	2	14	2	18	0 (4)	0 (14)
25	89	2	14	2	18	0 (4)	0 (14)
26	90	2	14	2	18	0 (4)	0 (14)
27	91	2	14	2	18	0 (4)	0 (0)
28	92	2	14	2	18	0 (4)	0 (0)
29	93	2	14	2	18	0 (4)	0 (0)
30	94	2	14	2	18	0 (4)	0 (0)
31	95	2	14	2	18	0 (4)	0 (0)
32	96	2	14	2	18	0 (0)	0 (0)
33	97	2	14	2	18	0 (0)	0 (0)
34	98	2	14	2	18	0 (0)	0 (0)
35	99	2	14	2	18	0 (0)	0 (2)
36	100	2	14	2	18	0 (0)	0 (2)
37	101	2	14	2	18	0 (0)	0 (2)
38	102	2	14	2	18	0 (0)	0 (2)
39	103	2	14	2	18	0 (0)	0 (2)
40	104	2	14	2	18	0 (12)	0 (2)
41	105	2	14	2	18	0 (12)	0 (2)

leading to the valid Sbox transitions. This occurs with probability 2^{-7} while at most 8 bits must be guessed. Therefore, we can consider that this step has negligible time complexity compared to $2^{7 \times a_{1\to11} - 18.58}$.

Hence, the total data complexity is $2^{7 \times a_{2\to10}+1}$, the time complexity is reduced to $2^{7 \times a_{2\to10}+1} + 2^{7 \times a_{1\to11} - 18.58}$, and the memory is unchanged ($2^{8 \times a_1+1}$).

Choosing $a_{2\to10}$ and $a_{1\to11}$. According to the previous analysis the bottleneck in the time complexity is:

$$2^{7 \times a_{2\to10}+1} + 2^{7 \times a_{1\to11} - 18.58}.$$

It is then parametrized by $a_{2\to10}$ (the amount of active Sboxes from round 2 to 10) and by $a_{1\to11}$ (the amount of active Sboxes from round 1 to 11).

A simple search finds that the minimum for $a_{2\to10}$ is 14 and that the minimum for $a_{1\to11}$ is 17. Unfortunately, the two minima cannot be reached simultaneously, which gives raise to the following two possible options.

Variant 1: minimizing $a_{2\to10}$. We consider here situations for which $a_{2\to10} = 14$, the minimum for $a_{1\to11}$ in these cases is 18 (see Table 2 for the difference positions that reach these minima and Fig. 3 for an example of a related-key attack using this variant).

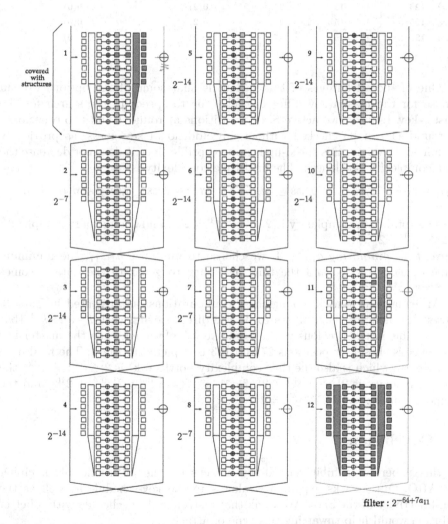

Fig. 4. Related-key attack on full-round PICARO . This figure represents the attack with the best time complexity: 2^{106}, with a data complexity of 2^{106}. The master-key differences are located at bits 27 and 123.

Table 3. List of all master-key differences (of weight less than 4) minimizing $a_{1\to11}$ ($a_{1\to11} = 17$) with the smallest value for $a_{2\to10}$ ($a_{2\to10} = 15$). The last 2 columns indicate the number of key bytes (and bits) deduced from k^{12} at the positions corresponding to the active Sboxes at rounds 1 and 11.

diff. positions		a_1	$a_{2\to10}$	a_{11}	$a_{1\to11}$	known bytes (bits) in k_1	known bytes (bits) in k_{11}
27	123	1	15	1	17	0(2)	0(0)
28	124	1	15	1	17	0(2)	0(0)
29	125	1	15	1	17	0(2)	0(0)
30	126	1	15	1	17	0(2)	0(0)
91	123	1	15	1	17	0(2)	0(0)
92	124	1	15	1	17	0(2)	0(0)
93	125	1	15	1	17	0(2)	0(0)
94	126	1	15	1	17	0(2)	0(0)

One of the advantages of this variant is that some of the options we can choose for the master-key difference allow us to speed up the search for the master-key. Indeed, two active Sbox transitions at round 1 and two other ones at round 11 can be checked without any additional key guess, as previously explained. However, this speed-up is imperceptible since it does not decrease the time complexity bottleneck. The final time complexity is

$$2^{7\times a_{2\to10}+1} + 2^{7\times a_{1\to11}-18.58} = 2^{7\times14+1} + 2^{7\times18-18.58} = 2^{99} + 2^{107.4} = 2^{107.4}.$$

The total data complexity is $2^{7\times a_{2\to10}+1} = 2^{99}$, and the memory complexity is $2^{8\times a_1+1} = 2^{17}$.

Variant 2: minimizing $a_{1\to11}$. If we choose to minimize $a_{1\to11}$, the minimum value of $a_{2\to10}$ is 15, and the time necessary to generate the data becomes $2^{7\times a_{2\to10}+1} = 2^{106}$.

An exhaustive search among these configurations, as presented in Table 3, shows that none of them allows to do the direct sieving with k^1 and k^{11} that was possible in the previous variant. The attack obtained when the master-key difference is chosen at positions 27 and 123 is depicted in Fig. 4. The final data complexity (which is also the time complexity bottleneck) is $2^{7\times15+1} = 2^{106}$, the time complexity is $2^{7\times a_{2\to10}+1} + 2^{7\times a_{1\to11}-18.58} = 2^{106} + 2^{100.4} = 2^{106}$ and the memory complexity is $2^{8\times a_1+1} = 2^9$.

6 Conclusion

In this paper we exhibit related-key attacks on the full-round block cipher PICARO. Our attacks exploit a weakness in the key-schedule as well as the non-bijectivity of the Sbox. We think that a stronger key-schedule with a better diffusion would help thwarting this type of attack.

Despite the fact that our related-key attacks do not represent a threat to the cipher in other more realistic scenarios, the authors aimed at providing related-key resistance, and we believe that such claim should be revised.

A PICARO Sbox

	00	01	02	03	04	05	06	07	08	09	0a	0b	0c	0d	0e	0f
00	08	0c	03	06	06	04	05	06	05	04	0c	0c	04	03	05	03
10	0a	1f	29	3b	4b	55	62	7b	82	95	af	bf	c5	d9	e2	f9
20	01	2d	45	6a	8a	ac	cf	ea	9f	bc	dd	fd	1c	35	5f	75
30	0f	34	61	52	c2	fb	a3	92	13	2b	74	44	db	e1	b3	81
40	0f	44	81	c2	92	db	13	52	b3	fb	34	74	2b	61	a3	e1
50	0e	59	a4	f8	d8	87	7c	28	3c	67	99	c9	e7	b4	4c	14
60	02	63	ca	ad	1d	71	d7	bd	27	41	e3	83	31	5a	f7	9a
70	0f	74	e1	92	52	2b	b3	c2	a3	db	44	34	fb	81	13	61
80	02	83	9a	1d	bd	31	27	ad	f7	71	63	e3	41	ca	d7	5a
90	0e	99	b4	28	f8	67	4c	d8	7c	e7	c9	59	87	14	3c	a4
a0	0a	af	d9	7b	3b	95	e2	4b	62	c5	bf	1f	55	f9	82	29
b0	0a	bf	f9	4b	7b	c5	82	3b	e2	55	1f	af	95	29	62	d9
c0	0e	c9	14	d8	28	e7	3c	f8	4c	87	59	99	67	a4	7c	b4
d0	01	dd	35	ea	6a	bc	5f	8a	cf	1c	fd	2d	ac	75	9f	45
e0	02	e3	5a	bd	ad	41	f7	1d	d7	31	83	63	71	9a	27	ca
f0	01	fd	75	8a	ea	1c	9f	6a	5f	ac	2d	dd	bc	45	cf	35

References

1. Biham, E.: New Types of Cryptanalytic Attacks Using Related Keys. J. Cryptology **7**(4), 229–246 (1994)
2. Biham, E., Shamir, A.: Differential Cryptanalysis of DES-like Cryptosystems. J. Cryptology **4**(1), 3–72 (1991)
3. Biryukov, A., Wagner, D.: Slide Attacks. In: Knudsen, L.R. (ed.) FSE 1999. LNCS, vol. 1636, pp. 245–259. Springer, Heidelberg (1999)
4. Canteaut, A., Chabaud, F.: A New algorithm for Finding Minimum-Weight Words in a Linear Code: Application to McEliece's Cryptosystem and to Narrow-Sense BCH Codes of Length 511. IEEE Trans. Inf. Theory **44**(1), 367–378 (1998)
5. Carlet, C.: Relating Three Nonlinearity Parameters of Vectorial Functions and Building APN Functions from Bent Functions. Des. Codes Crypt. **59**(1–3), 89–109 (2011)
6. Gérard, B., Grosso, V., Naya-Plasencia, M., Standaert, F.-X.: Block Ciphers That Are Easier to Mask: How Far Can We Go? In: Bertoni, G., Coron, J.-S. (eds.) CHES 2013. LNCS, vol. 8086, pp. 383–399. Springer, Heidelberg (2013)
7. Grosso, V., Leurent, G., Standaert, F.-X., Varici, K.: LS-Designs: Bitslice Encryption for Efficient Masked Software Implementations. In: Cid, C., Rechberger, C. (eds.) FSE 2014. LNCS, vol. 8540, pp. 18–37. Springer, Heidelberg (2015)
8. Piret, G., Roche, T., Carlet, C.: PICARO – A Block Cipher Allowing Efficient Higher-Order Side-Channel Resistance. In: Bao, F., Samarati, P., Zhou, J. (eds.) ACNS 2012. LNCS, vol. 7341, pp. 311–328. Springer, Heidelberg (2012)
9. Piret, G., Roche, T., Carlet, C.: PICARO - A Block Cipher Allowing Efficient Higher-Order Side-Channel Resistance, extended version, IACR Cryptology ePrint Archive 2012, 358 (2012). http://eprint.iacr.org/2012/358
10. Rivain, M., Prouff, E.: Provably Secure Higher-Order Masking of AES. In: Mangard, S., Standaert, F.-X. (eds.) CHES 2010. LNCS, vol. 6225, pp. 413–427. Springer, Heidelberg (2010)

Cryptanalysis of Feistel Networks with Secret Round Functions

Alex Biryukov[1], Gaëtan Leurent[2], and Léo Perrin[3(✉)]

[1] University of Luxembourg, Luxembourg City, Luxembourg
alex.biryukov@uni.lu
[2] Inria, project-team SECRET, Rocquencourt, France
gaetan.leurent@inria.fr
[3] SnT, University of Luxembourg, Luxembourg City, Luxembourg
leo.perrin@uni.lu

Abstract. Generic distinguishers against Feistel Network with up to 5 rounds exist in the regular setting and up to 6 rounds in a multi-key setting. We present new cryptanalyses against Feistel Networks with 5, 6 and 7 rounds which are not simply distinguishers but actually recover completely the unknown Feistel functions.

When an exclusive-or is used to combine the output of the round function with the other branch, we use the so-called *yoyo game* which we improved using a heuristic based on particular cycle structures. The complexity of a complete recovery is equivalent to $O(2^{2n})$ encryptions where n is the branch size. This attack can be used against 6- and 7-round Feistel Networks in time respectively $O(2^{n2^{n-1}+2n})$ and $O(2^{n2^{n}+2n})$. However when modular addition is used, this attack does not work. In this case, we use an optimized guess-and-determine strategy to attack 5 rounds with complexity $O(2^{n2^{3n/4}})$.

Our results are, to the best of our knowledge, the first recovery attacks against generic 5-, 6- and 7-round Feistel Networks.

Keywords: Feistel Network · Yoyo · Generic attack · Guess-and-determine

1 Introduction

The design of block ciphers is a well researched area. An overwhelming majority of modern block ciphers fall in one of two categories: Substitution-Permutation Networks (SPN) and Feistel Networks (FN). Examples of those two structures are the block ciphers standardized by the American National Institute for Standards and Technology, respectively the AES [1] and the DES [2]. However, since block ciphers are simply keyed permutations, the same design strategies can be applied to the design of so-called S-Boxes.

S-Boxes are "small" functions operating usually on at most 8-bits of data which are used to ensure a high non-linearity. For instance, both the DES and

© Springer International Publishing Switzerland 2016
O. Dunkelman and L. Keliher (Eds.): SAC 2015, LNCS 9566, pp. 102–121, 2016.
DOI: 10.1007/978-3-319-31301-6_6

the AES use S-Boxes, respectively mapping 6 bits to 4 and permuting 8 bits. If a bijective S-Box is needed, it can be built like a small unkeyed block cipher. For example, the S-Box of Khazad [3] is a 3-round SPN and the S-Box of Zorro [4] is a 3-round FN. These 8×8 bits S-Boxes are built from smaller 4×4 ones to diminish memory requirements.

Keeping the design process of an S-Box secret might be necessary in the context of white-box cryptography, as described e.g. in [5]. In this paper, Biryukov *et al.* describe a memory-hard white-box encryption scheme relying on a SPN with large S-Boxes built like smaller SPN. Their security claim needs the fact that an adversary cannot decompose these S-Boxes into their different linear and non-linear layers. Such memory-hard white-box implementation may also be used to design proofs-of-work such that one party has an advantage over the others. Knowing the decomposition of a memory-hard function would effectively allow a party to bypass this memory-hardness. Such functions can have many use cases including password hashing and crypto-currency design.

Decomposing SPNs into their components is possible for up to 3 S-Box layers when the S-Boxes are small using the multi-set attack on SASAS described in [6]. A more general strategy for reverse-engineering of unknown S-Boxes was proposed recently in [7]. Our work pursues the same line of research but targets FN specifically: what is the complexity of recovering all Feistel functions of a R-round FN? Our results are different depending on whether the Feistel Network attacked uses an exclusive-or ($\oplus$) or a modular addition ($\boxplus$). Thus, we refer to a Feistel Network using XOR as a $\oplus$-Feistel and to one based on modular addition as a $\boxplus$-Feistel.

This work also has implications for the analysis of format-preserving encryption schemes, which are designed to encrypt data with a small plaintext set, for instance credit-card numbers (a 16 decimal digits number). In particular, the BPS [8] and FFX [9] constructions are Feistel schemes with small blocks; BPS uses 8 rounds, while FFX uses between 12 and 36 rounds (with more rounds for smaller domains). When these schemes are instantiated with small block sizes, recovering the full round functions might be easier than recovering the master key and provides an equivalent key.

Previous Work. Lampe *et al.* [10], followed by Dinur *et al.* [11], studied Feistel Networks where the Feistel function at round i consists in $x \mapsto F_i(x \oplus k_i)$, with F_i being public but k_i being kept secret. If the subkeys are independent then it is possible to recover all of them for a 5-round (respectively 7-round) FN in time $O(2^{2n})$ (resp. $O(2^{3n})$) using only 4 known plaintexts with the optimised Meet-in-the-Middle attack described in [11]. However, we consider the much more complex case where the Feistel functions are completely unknown.

A first theoretical analysis of the Feistel structure and the first generic attacks were proposed in the seminal paper by Luby and Rackoff [12]. Since then, several cryptanalyses have been identified with the aim to either distinguish a Feistel Network from a random permutation or to recover the Feistel functions. Differential distinguishers against up to 5 rounds in the usual setting and 6 rounds in a multi-key setting are presented in [13], although they assume that the Feistel functions are random functions and thus have inner-collisions. Conversely, an impossible differential covering 5 rounds in the case where the Feistel functions are permutations is described in [14] and used to attack DEAL, a block cipher based on a 6-round Feistel Network. Finally, a method relying on a SAT-solver was recently shown in [7]. It is capable of decomposing Feistel Networks with up to $n = 7$ in at most a couple of hours. How this time scales for larger n is unclear. These attacks, their limitations and their efficiency are summarized in Table 1. A short description of some of them is given in Sect. 2 for the sake of completeness.

Table 1. Generic attacks against Feistel Networks.

R	Type	Power	Restrictions	Time	Data	Ref.
4	Differential	Distinguisher	Non bij. round func.	$2^{n/2}$	$2^{n/2}$	[13]
	Guess & Det.	Full recovery	–	$2^{3n/2}$	$2^{3n/2}$	Sect. 5.2
5	Differential	Distinguisher	Non bij. round func.	2^{n}	2^{n}	[13]
	Imp. diff	Distinguisher	Bij. round func.	2^{2n}	2^{n}	[14]
	SAT-based	Full recovery	$n \leq 7$	Practical	2^{2n}	[7]
	Yoyo	Full Recovery	Only for $\oplus$-Feistel	2^{2n}	2^{2n}	Sect. 4.3
	Integral	Full recovery	S_1 or S_3 bij.	$2^{2.81n}$	2^{2n}	Sect. 6
	Guess & Det.	Full recovery	–	$2^{n2^{3n/4}}$	2^{2n}	Sect. 5.3
6	Differential	Distinguisher	Multi-key setting	2^{2n}	2^{2n}	[13]
	Yoyo	Full recovery	Only for $\oplus$-Feistel	$2^{n2^{n-1}+2n}$	2^{2n}	Sect. 4.4
7	Yoyo	Full recovery	Only for $\oplus$-Feistel	$2^{n2^{n}+2n}$	2^{2n}	Sect. 4.4

Our Contribution. We present attacks against generic 5-round Feistel Networks which recover all Feistel functions efficiently instead of only distinguishing them from random. Furthermore, unlike distinguishers from the litterature, our attacks do not make any assumptions about whether the Feistel functions are bijective or not. Our attack against $\oplus$-Feistel uses the *yoyo game*, a tool introduced in [15] which we improve by providing a more general theoretical framework for it and leveraging particular cycle structures to diminish its cost. The principle of the yoyo game is introduced in Sect. 3 and how to use cycles to improve it is described in Sect. 4. We also present an optimized guess-and-determine attack which, unlike yoyo cryptanalysis, works against $\boxplus$-Feistel. It exploits a boomerang-like property related to the one used in our yoyo game to quickly explore the implications of the guess of an S-Box entry, see Sect. 5. Finally, an integral attack is given in Sect. 6.

We note that several of our attack have a double exponential complexity, and can only be used in practice for small values of n, as used for 8-bit S-Boxes ($n = 4$) or format-preserving encryption with a small domain.

Notation. We introduce some notation for the different states during encryption (see Fig. 1). Each of the values is assigned a letter, e.g. the left side of the input is in position "A". When we look at 5-round Feistel Networks, the input is fed in positions A and B and the output is read in G, F. For 6 rounds, the input is the same but the output is read in H, G with $H = S_5(G) + F$. If we study a ⊞-Feistel then "+" denotes modular addition (⊞); it denotes exclusive-or (⊕) if we attack a ⊕-Feistel. Concatenation is denoted "||" and encryption is denoted $\mathcal{E}$ (the number of rounds being clear from the context). For example, $\mathcal{E}(a||b) = g||f$ for a 5-round Feistel Network. The bit-length of a branch of the Feistel Network is equal to n.

In addition we remark that, for an R-round Feistel, we can fix one entry of the last $R-2$ Feistel functions (or the first $R-2$ ones) arbitrarily. For example, the output of the 5-round Feistel Network described in Fig. 2 does not depend on α_0, α_1 or α_2.

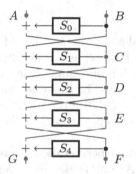

Fig. 1. Internal state notation.

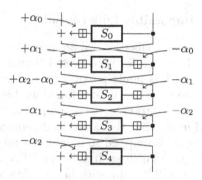

Fig. 2. Equivalent Feistel Networks.

2   Previous Attacks Against 5- and 6-Round Feistel Networks

2.1   Differential Distinguishers

In [13], Patarin shows a differential distinguisher against 5-round Feistel Networks. However, it only works if the Feistel functions have inner-collisions. It is based on the following observation. Let $(g_i||f_i)$ be the image of $(a_i||b_i)$ by a permutation and let b_i be constant. Then for $i \neq j$, such that $f_i = f_j$, count how many times $a_i \oplus a_j = g_i \oplus g_j$. This number is roughly twice as high for a 5-round Feistel Network than for a random permutation.

In the same paper, Patarin suggests two distinguishers against 6-round ⊕-Feistel Networks. However, these do not target a permutation but a generator of permutation. This can be interpreted as a multi-key attack: the attacker has a black-box access to several permutations and either none or all of which are 6-round $\oplus$-Feistel Networks. The first attack uses that the signature of a $\oplus$-Feistel Network is always even. The second attack exploits a statistical bias too weak to be reliably observable using one codebook but usable when several permutations are available. It works by counting all quadruples of encryptions $(a_i||b_i) \to (g_i||h_i)$, $i = 1..4$ satisfying this system:

$$\begin{cases} b_1 = b_3, \ b_2 = b_4 \\ g_1 = g_2, \ g_3 = g_4 \\ a_1 \oplus a_3 = a_2 \oplus a_4 = g_1 \oplus g_3 \\ h_1 \oplus h_2 = h_3 \oplus h_4 = b_1 \oplus b_2. \end{cases}$$

If there are λ black-boxes to distinguish and if m queries are performed for each then we expect to find about $\lambda m^4 2^{-8n}$ solutions for a random permutation and $2\lambda m^4 2^{-8n}$ for 6-round Feistel Networks, i.e. twice as much.

2.2 Impossible Differential

Knudsen described in [14] an impossible differential attack against his AES proposal, DEAL, a 6-round Feistel Network using the DES [2] as a round function. This attack is made possible by the existence of a 5-round impossible differential caused by the Feistel functions being permutations. In this case, an input difference $(\alpha||0)$ cannot be mapped to a difference of $(\alpha||0)$ after 5 rounds. This would imply that the non-zero difference which has to appear in D as the image of α by S_2 is mapped to 0, which is impossible.

To distinguish such a 5-round FN from a random permutation we need to generate $\lambda \cdot 2^{2n}$ pairs with input difference $(\Delta||0)$. Among those, about λ should have an output difference equal to $(\Delta||0)$ if the permutation is a random permutation while it is impossible to observe if for a 5-round FN with bijective Feistel functions. Note that while the time complexity is $O(2^{2n})$, the data complexity can be brought down to $O(2^n)$ using structures.

An attack on 6 rounds uses this property by identifying pairs of encryptions with difference $(\alpha||0)$ in the input and $(\alpha||\Delta)$ for the output for any $\Delta \neq 0$. A pair as a correct output difference with probability $2^{-n}(1 - 2^{-n})$ since α is fixed and Δ can take any value except 0. We repeat this process for the whole codebook and all $\alpha \neq 0$ to obtain $2^{n+(2n-1)} \cdot 2^{-n}(1 - 2^{-n}) = 2^{2n-1} - 2^{n-1}$ pairs. Each of them gives an impossible equation for S_5: if $\{(a||b) \to (g||h), (a \oplus \alpha||b) \to (g \oplus \alpha||h \oplus \Delta)\}$ is a pair of encryptions then it is impossible that $S_5(g) \oplus S_5(g \oplus \alpha) = \Delta$ as it would imply the impossible differential. In the end, we have a system of about $2^{2n-1} - 2^{n-1}$ impossible equations, a random Feistel function satisfying an impossible equation with probability $(1 - 2^{-n})$. Thus, this attack filters out all but the following fraction of candidates for S_5:

$$\text{Impossible differential filter} = \left(1 - 2^{-n}\right)^{2^{2n-1} - 2^{n-1}} \approx 2^{0.72 - 1.443 \cdot 2^{n-1}}.$$

3   Yoyo Game and Cryptanalysis

3.1   The Original Yoyo Game

Several cryptanalyses have been proposed in the literature that rely on encrypting a plaintext, performing an operation on the ciphertext and then decrypting the result. For example, the "double-swiping" used against newDES [16] in the related-key setting relies on encrypting a pair of plaintexts using two related-keys and decrypting the result using two different related-keys. Another example is the boomerang attack introduced by Wagner [17] in the single-key setting. A pair with input difference δ is encrypted. Then, a difference Δ is added to the ciphertexts and the results are decrypted, hopefully yielding two plaintexts with a difference of δ.

The yoyo game was introduced by Biham $et\ al.$ in [15] where it was used to attack the 16 center rounds of Skipjack [18], a block cipher operating on four 16-bits words. We describe this attack using slightly different notation and terminology to be coherent with the rest of our paper. In this paragraph, $\mathcal{E}_k$ denotes an encryption using round-reduced Skipjack under key k.

It was noticed that if the difference between two encryptions at round 5 is $(0, \Delta, 0, 0)$ where $\Delta \neq 0$ then the other three words have difference 0 between rounds 5 and 12. Two encryptions satisfying this truncated differential are said to be $connected$. The key observation is the following. Consider two plaintexts $x = (x_0, x_1, x_2, x_3)$ and $x' = (x'_0, x'_1, x_2, x'_3)$ where x_2 is constant. If they are connected, then the pair $\phi(x, x') = ((x_0, x'_1, x_2, x_3), (x'_0, x_1, x_2, x'_3))$ is connected as well (see [15] for a detailed explanation on why it is the case). Furthermore, let $y = (y_0, y_1, y_2, y_3) = \mathcal{E}_k(x)$ and $y' = (y'_0, y'_1, y'_2, y'_3) = \mathcal{E}_k(x')$. We can form two new ciphertexts by swapping their first words to obtain $z = (y'_0, y_1, y_2, y_3)$ and $z' = (y_0, y'_1, y'_2, y'_3)$. If we decrypt them to obtain $(u, u') = (\mathcal{E}_k^{-1}(z), \mathcal{E}_k^{-1}(z'))$, then u and u' are connected. If we denote $\psi(x, x')$ the function which encrypts x and x', swaps the first words of the ciphertexts obtained and decrypts the result then ψ preserves connection, just like ϕ. It is thus possible to iterate ϕ and ψ to obtain many connected pairs, this process being called the $yoyo\ game$.

In this section, we present other definitions of the connection and of the functions ϕ and ψ which allow us to play a similar yoyo game on 5-round Feistel Networks.

3.2   Theoretical Framework for the Yoyo Game

Consider two plaintexts $a||b$ and $a'||b'$ such that the difference between their encryptions in positions (C, D) is equal to $(\gamma, 0)$ with $\gamma \neq 0$. Then the difference in position E is equal to γ. Conversely, the difference in (E, D) being $(\gamma, 0)$ implies that the difference in C is γ. When this is the case, the two encryptions satisfy the systems of equations and the trail described in Fig. 3.

Definition 1. *If the encryptions of $a||b$ and $a'||b'$ follow the trail in Fig. 3 then they are said to be* connected in γ.

Top equations

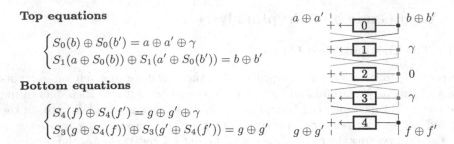

$$\begin{cases} S_0(b) \oplus S_0(b') = a \oplus a' \oplus \gamma \\ S_1(a \oplus S_0(b)) \oplus S_1(a' \oplus S_0(b')) = b \oplus b' \end{cases}$$

Bottom equations

$$\begin{cases} S_4(f) \oplus S_4(f') = g \oplus g' \oplus \gamma \\ S_3(g \oplus S_4(f)) \oplus S_3(g' \oplus S_4(f')) = g \oplus g' \end{cases}$$

Fig. 3. The equations defining connection in γ and the corresponding differential trail.

This *connection* is an "exclusive" relation: if $(a||b)$ and $(a'||b')$ are connected, then neither $(a||b)$ nor $(a'||b')$ can be connected to anything else. Furthermore, we can replace (a, a') by $(a \oplus \gamma, a' \oplus \gamma)$ in the top equations and still have them being true. Indeed, the two γ cancel each other in the first one. In the second, the values input to each call to S_1 are simply swapped as a consequence of the first equation. Similarly, we can replace (g, g') by $(g \oplus \gamma, g' \oplus \gamma)$ in the bottom equations.[1] As consequence of these observation, we state the following lemma.

Lemma 1. *We define the following two involutions*

$$\phi_\gamma(a||b) = (a \oplus \gamma)||b, \ \psi_\gamma = \mathcal{E}^{-1} \circ \phi_\gamma \circ \mathcal{E}.$$

If $a||b$ and $a'||b'$ are connected then, with probability 1:

- *$\phi_\gamma(a||b)$ and $\phi_\gamma(a'||b')$ are connected,*
- *$\psi_\gamma(a||b)$ and $\psi_\gamma(a'||b')$ are connected.*

By repeatedly applying ϕ_γ and ψ_γ component-wise on a pair of plaintexts (x, x'), we can play a yoyo game which preserves connection in γ. This process is defined formally below.

Definition 2. *Let $\big(x_0 = (a_0||b_0), x_0' = (a_0'||b_0')\big)$ be a pair of inputs. The yoyo game in γ starting in (x_0, x_0') is defined recursively as follows:*

$$\big(x_{i+1}, x_{i+1}'\big) = \begin{cases} \big(\phi_\gamma(x_i), \phi_\gamma(x_i')\big) & \text{if } i \text{ is even,} \\ \big(\psi_\gamma(x_i), \psi_\gamma(x_i')\big) & \text{if } i \text{ is odd} \end{cases}$$

Lemma 2. *If (x_0, x_0') is connected in γ then all pairs in the game starting in (x_0, x_0') are connected in γ. In other words, either all pairs within the game played using ϕ_γ and ψ_γ are connected in γ or none of them are.*

---

[1] However, such a yoyo game cannot be played against a ⊞-Feistel, as explained in Sect. 3.4. It only works in characteristic 2.

3.3 The Yoyo Cryptanalysis Against 5-Round ⊕-Feistel Networks

Given a yoyo game connected in γ, it is easy to recover Feistel functions S_0 and S_4 provided that the yoyo game is long enough, i.e. that it contains enough connected pairs to be able to recover all 2^n entries of both S-Boxes. If the yoyo game is not connected in γ then *yoyo cryptanalysis* (Algorithm 1) identifies it as such very efficiently.

It is a differential cryptanalysis using that all pairs in the game are (supposed to be) right pairs for the differential trail defining connection in γ. If it is not the case, S_0 or S_4 will end up requiring contradictory entries, e.g. $S_0(0) = 0$ and $S_0(0) = 1$. In this case, the game is not connected in γ and must be discarded. Yoyo cryptanalysis is described in Algorithm 1[2]. It only takes as inputs a (possible) yoyo game and the value of γ. Algorithm 2 describes AddEntry, a subroutine handling some linear equations. Note that one entry can be set arbitrarily (here, $S_0(0) = 0$) as summarized in Fig. 2.

Let $\mathcal{Y}$ be a (supposed) yoyo game of size $|\mathcal{Y}|$. For each pair in it, either an equation is added to the list, FAIL is returned or AddEntry is called. While the recursive calls to AddEntry may lead to a worst time complexity quadratic in $|\mathcal{Y}|$ if naively implemented, this problem can be mitigated by using a hashtable indexed by the Feistel functions inputs instead of a list. Furthermore, since already solved equations are removed, the total time complexity is $O(|\mathcal{Y}|)$.

Algorithm 1. Yoyo cryptanalysis against a 5-round ⊕-Feistel Network
Inputs supposed yoyo game $\left(a_i\|b_i, a_i'\|b_i'\right)$; difference γ | **Output** S_0 or FAIL

$L_e \leftarrow []$ ▷ List of equations
$S_0 \leftarrow$ empty S-Box
$\delta_0 \leftarrow a_0 \oplus a_0' \oplus \gamma$
$S_0(b_0) \leftarrow 0,\ S_0(b_0') \leftarrow \delta_0$
for all $i \geq 1$ **do**
    $\delta_i \leftarrow a_i \oplus a_i' \oplus \gamma$
    if $S_0(b_i)$ and $S_0(b_i')$ are already known and $S_0(b_i) \oplus S_0(b_i') \neq \delta_i$ **then**
        return FAIL
    else if $S_0(b_i)$ is known but not $S_0(b_i')$ **then**
        AddEntry$\left(S_0, b_i', S_0(b_i) \oplus \delta_i, L_e\right)$; **if** it fails **then return** FAIL
    else if $S_0(b_i')$ is known but not $S_0(b_i)$ **then**
        AddEntry$\left(S_0, b_i, S_0(b_i') \oplus \delta_i, L_e\right)$; **if** it fails **then return** FAIL
    else
        add "$S_0(b_i') \oplus S_0(b_i) = \delta_i$" to L_e.
    end if
end for
return S_0

---

[2] It can also recover S_4 in an identical fashion but this part is omitted for the sake of clarity.

Algorithm 2. Adding new entry to S_0 (AddEntry)
Inputs S-Box S_0 ; input x ; output y ; List of equations L_e | **Output** SUCCESS
or FAIL

---

    if $S_0(x)$ already set and $S_0(x) = y$ **then**
        return SUCCESS   ▷ No new information
    else if $S_0(x)$ already set and $S_0(x) \neq y$ **then**
        return FAIL   ▷ Contradiction identified
    else
        $S_0(x) \leftarrow y$
        for all Equation $S_0(x_i) \oplus S_0(x_i') = \Delta_i$ in L_e **do**
            if $S_0(x_i)$ and $S_0(x_i')$ are set **then**
                if $S_0(x_i) \oplus S_0(x_i') \neq \Delta_i$ **then return** FAIL ; **else** Remove eq. from L_e
                                                        ▷ Eq. satisfied
            else if $S_0(x_i)$ is set but not $S_0(x_i')$ **then**
                AddEntry$\big(S_0, x_i', S_0(x_i) \oplus \Delta_i, L_e\big)$; **if** it fails **then return** FAIL
                                                     ▷ Eq. gives new entry
            else if $S_0(x_i')$ is set but not $S_0(x_i)$ **then**
                AddEntry$\big(S_0, x_i, S_0(x_i') \oplus \Delta_i, L_e\big)$; **if** it fails **then return** FAIL
                                                     ▷ Eq. gives new entry
        end if
        end for
    end if
    return SUCCESS

3.4   On the Infeaseability of Our Yoyo Game Against an ⊞-Feistel

Assume that the following equations holds:

$$\begin{cases} \big(S_0(b) + a\big) - \big(S_0(b') + a'\big) = \gamma \\ \big(S_1(S_0(b) + a) + b\big) - \big(S_1(S_0(b') + a') + b'\big) = 0. \end{cases} \tag{1}$$

In order to be able to play a yoyo game against the corresponding ⊞-Feistel, we need to be able to replace a by $a+\gamma$ and a' by $a'+\gamma$ in System (1) and still have it hold. In other words, we need that Eq. (1) holding implies that the following equations hold as well:

$$\begin{cases} \big(S_0(b) + a + \gamma\big) - \big(S_0(b') + a' + \gamma\big) = \gamma \\ \big(S_1(S_0(b) + a + \gamma) + b\big) - \big(S_1(S_0(b') + a' + \gamma) + b'\big) = 0. \end{cases} \tag{2}$$

The first one trivially does. Using it, we note that $S_0(b)+a+\gamma = S_0(b')+a'+2\gamma$. Let $X = S_0(b')+a'$. Then the left-hand side of the second equation in System (2) can be re-written as $S_1(X + 2\gamma) - S_1(X + \gamma) + b - b'$. Furthermore, the second equation in System (1), which is assumed to hold, implies that $S_1(X + \gamma) - S_1(X) = b' - b$. Thus, the left-hand side of the second equation in System (2) is equal to

$$S_1(X + 2\gamma) - (b' - b + S_1(X)) + b - b' = S_1(X + 2\gamma) - S_1(X) - 2(b' - b).$$

The term $S_1(X+2\gamma)-S_1(X)$ has an unknown value unless $\gamma = 2^{n-1}$. Nevertheless, in this case, we would need $2(b'-b)=0$ which does not have a probability equal to 1. However both $S_1(X+2\gamma)-S_1(X)$ and $2(b'-b)$ are always equal to 0 in characteristic 2 which is why our yoyo game can always be played against a $\oplus$-Feistel.

4   An Improvement: Using Cycles

4.1   Cycles and Yoyo Cryptanalysis

A yoyo game is a cycle of ψ_γ and ϕ_γ applied iteratively component-wise on a pair of elements. Thus, it can be decomposed into two cycles, one for each "side" of the game: $(x_0, x_1, x_2, ...)$ and $(x'_0, x'_1, x'_2, ...)$. This means that both cycles must have the same length, otherwise the game would imply that x_0 is connected to x'_j for $j \neq 0$, which is impossible. Since both ϕ_γ and ψ_γ are involutions, the cycle can be iterated through in both directions. Therefore, finding one cycle gives us two directed cycles.

In order to exploit yoyo games, we could generate pairs (x_0, x'_0) at random, generate the yoyo game starting at this pair and then try and recover S_0 and S_4 but this endeavour would only work with probability 2^{-2n} (the probability for two random points to be connected). Instead, we can use the link between cycles, yoyo games and connection in γ as is described in this section. Note that the use of cycles in cryptography is not new; in fact it was used in the first cryptanalyses against ENIGMA. More recently, particular distribution of cycle sizes were used to distinguish rounds-reduced PRINCE-core [19] from random [20] and to attack involutional ciphers [21].

4.2   Different Types of Cycles

Let $\mathcal{C} = (x_i)_{i=0}^{\ell-1}$ be a cycle of length ℓ of ψ_γ and ϕ_γ, with $x_{2i} = \psi_\gamma(x_{2i-1})$ and $x_{2i+1} = \phi_\gamma(x_{2i})$. We denote the point connected to x_i as y_i, where all indices are taken modulo ℓ. Since x_i and y_i are connected, and the connection relation is one-to-one, we also have $y_{2i} = \psi_\gamma(y_{2i-1})$ and $y_{2i+1} = \phi_\gamma(y_{2i})$. Therefore, $\mathcal{C}' = (y_i)_{i=0}^{\ell-1}$ is also a cycle of length ℓ.

We now classify the cycles according to the relationship between $\mathcal{C}$ and $\mathcal{C}'$.

- If $\mathcal{C}$ and $\mathcal{C}'$ are **Distincts**, $\mathcal{C}$ is a **Type-D** cycle. A representation is given in Fig. 4a. Otherwise, there exists k such that $y_0 = x_k$.
- If k is even, we have $x_{k+1} = \phi_\gamma(x_k)$. Since $x_k = y_0$ is connected to x_0, $x_{k+1} = \phi_\gamma(x_k)$ is connected to $\phi_\gamma(x_0) = x_1$, i.e. $y_1 = x_{k+1}$. Further, $x_{k+2} = \psi_\gamma(x_{k+1})$ is connected to $\psi_\gamma(x_1) = x_2$, i.e. $y_2 = x_{k+2}$. By induction, we have $y_i = x_{k+i}$. Therefore x_0 is connected to x_k and x_k is connected to x_{2k}. Since the connection relation is one-to-one, this implies that $2k = \ell$.

We denote this setting as a **Type-S** cycle. Each element x_i is connected to $x_{i+\ell/2}$. Thus, if we represent the cycle as a circle, the connections between the elements would all cross in its center, just like **S**pokes, as can be seen in Fig. 4b.

– If k is odd, we have $x_{k-1} = \phi_\gamma(x_k)$. Since $x_k = y_0$ is connected to x_0, $x_{k-1} = \phi_\gamma(x_k)$ is connected to $\phi_\gamma(x_0) = x_1$, i.e. $y_1 = x_{k-1}$. Further, $x_{k-2} = \psi_\gamma(x_{k-1})$ is connected to $\psi_\gamma(x_1) = x_2$, i.e. $y_2 = x_{k-2}$. By induction, we have $y_i = x_{k-i}$. We denote this setting as a **Type-P** cycle. If we represent the cycle as a circle, the connections between the elements would all be **P**arallel to each other as can be seen in Fig. 4c.

In particular, there at exactly two pairs (x_i, x_{i+1}) such that x_i and x_{i+1} are connected. Indeed, we have $x_{i+1} = y_i$ if and only if $i + 1 \equiv k - i \bmod \ell$ i.e. $i \equiv (k-1)/2 \bmod \ell/2$. As a consequence, the existence of w connected pairs (x, x') with $x' = \phi_\gamma(x)$ or $x' = \psi_\gamma(x)$ implies the existence of $w/2$ Type-P cycles.

In addition, Type-P cycles can only exist if either S_1 or S_3 are not bijections. Indeed, if $(a||b)$ and $(a \oplus \gamma||b)$ are connected then the difference in position D cannot be zero unless S_1 can map a difference of γ to zero. If it is a permutation, this is impossible. The situation is identical for S_3. Furthermore, each value c such that $S_1(c) = S_1(c \oplus \gamma)$ implies the existence of 2^n values $(a||b)$ connected to $\phi_\gamma(a||b)$ as b can be chosen arbitrarily and a computed from b and c. Again, the situation is identical for S_3. Thus, if $S_1(x) = S_1(x \oplus \gamma)$ has w_1 solutions and if $S_3(x) = S_3(x \oplus \gamma)$ has w_3 solutions then there are $(w_1 + w_3) \cdot 2^{n-2}$ Type-P cycles.

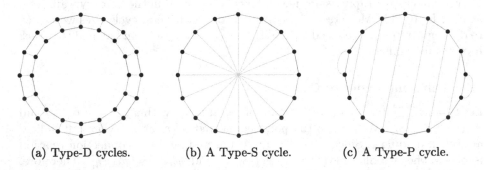

| (a) Type-D cycles. | (b) A Type-S cycle. | (c) A Type-P cycle. |

Fig. 4. All the types of cycles that can be encountered. ϕ_γ is a blue line, ψ_γ is a red one and connection is a green one (remember that ϕ_γ and ψ_γ are involutions) (Color figure online).

4.3   The Cycle-Based Yoyo Cryptanalysis

Exploiting a Type-S cycle is a lot easier than exploiting a Type-P or a pair of Type-D cycles. Indeed, the connected pairs $(x_i, x_{i+\ell/2})$ can be immediately derived from the length ℓ of the cycle, while we have to guess a shift amount for connected pairs in a Type-P cycle, or between two type D cycles. Thus, it makes sense to target those specifically, for instance by implementing Algorithm 3.

Algorithm 3. The cycle based yoyo cryptanalysis.

for all $\gamma \in \{0,1\}^{2n} \setminus \{0\}$ do
  for all $s \in \{0,1\}^{2n}$ do
    if s was not encountered before for this γ then
      $\mathcal{C} \leftarrow$ empty list
      $x \leftarrow s$
      repeat
        $x \leftarrow \phi(x)$; append x to $\mathcal{C}$
        $x \leftarrow \psi(x)$; append x to $\mathcal{C}$
      until $x = s$
      if $|\mathcal{C}| \geq 2^{n+2}$ then
        Build yoyo game $\mathcal{Y} = \big(\mathcal{C}[0,..,\ell-1], \mathcal{C}[\ell,..,2\ell-1]\big)$ with $\ell = \frac{|\mathcal{C}|}{2}$
        Run yoyo cryptanalysis (Algorithm 1) against $\mathcal{Y}$
        if yoyo cryptanalaysis is a success then
          return S_0, S_4
      end if
    end if
  end for
end for

Let $q_S(n)$ be the probability that a Type-S cycle exists for the chosen γ for a 5-round Feistel Network built out of bijective Feistel functions. When averaged over all such Feistel Networks, this probability does not depend on γ. The full version of this paper [22] contains a more detailed discussion of this probability in Sect. 3.4, and examples of functional graphs in the Appendix.

This attacks requires $O(2^{2n}/n)$ blocks of memory to store which plaintexts were visited and $O(2^{2n})$ time. Indeed, at most all elements of the codebook will be evaluated and inspected a second time when attempting a yoyo cryptanalysis on each cycle large enough. Even though the attack must be repeated about $1/q_S(n)$ times to be able to obtain a large enough Type-S cycle, $q_S(n)$ increases with n so that $1/q_S(n)$ can be upper-bounded by a constant independent of n.[3] Note also that special points can be used to obtain a time-memory tradeoff: instead of storing whether all plaintexts were visited or not, we only do so for those with, say, the first $\mathcal{B}$ bits equal to 0. In this case, the time complexity becomes $O(\mathcal{B} \cdot 2^{2n})$ and the memory complexity $O\big(2^{2n}/(n \cdot \mathcal{B})\big)$. Access to the hash table storing whether an element has been visited or not is a bottle-neck in practice so special points actually give a "free" memory improvement in the sense that memory complexity is decreased without increasing time. In fact, wall clock time may actually decrease. An attack against a $\oplus$-Feistel with $n = 14$ on a regular desktop computer[4] takes about 1 hour to recover both S_0 and S_4.

---

[3] We experimentally found that a lower bound of 0.1 is more than sufficient even for n as small as 4.

[4] CPU: Intel core i7-3770 (3.40 GHz); 8 Gb of RAM. The program was compiled with g++ and optimization flag -O3.

4.4   Attacking 6 and 7 Rounds

An Attack on 6 Rounds. A naive approach could consist in guessing all of the entries of S_5 and, for each guess, try running a cycle-based yoyo cryptanalysis. If it fails then the guess is discarded. Such an attack would run in time $O(2^{n2^n + 2n})$. However, it is possible to run such an attack at a cost similar to that of guessing only half of the entries of S_5, namely $O(2^{n2^{n-1} + 2n})$ which corresponds to a gain of $2^{n2^{n-1}}$.

Instead of guessing all the entries, this attack requires guessing the values of $\Delta_5(x, \gamma) = S_5(x) \oplus S_5(x \oplus \gamma)$. Once these are know, we simply need to replace ψ_γ by ψ'_γ with

$$(\mathcal{E} \circ \psi'_\gamma \circ \mathcal{E}^{-1})(g||h) = (g \oplus \gamma \ || \ h \oplus \Delta_5(x, \gamma)).$$

The cycle-based yoyo cryptanalysis can then be run as previously because, again, both ϕ_γ and ψ'_γ preserve connection in γ. Once it succeeds, the top S-Box is known which means that it can be peeled of. The regular attack is then performed on the remaining 5 rounds. Note that if the yoyo cryptanalysis fails because of inner collisions in S_1 or S_3 then we can still validate a correct guess by noticing that there are $O(2^n)$ cycles instead of $O(2n)$ as would be expected[5].

In this algorithm, 2^{n-1} values of $[0, 2^n - 1]$ must be guessed and for each of those an attack with running time $O(2^{2n})$ must be run. Hence, the total running time is $O(2^{n2^{n-1} + 2n})$. The time necessary to recover the remainder of the Feistel functions is negligible.

An Attack on 7 Rounds. A $\oplus$-Feistel with 7 rounds can be attacked in a similar fashion by guessing both $\Delta_0(x, \gamma)$ and $\Delta_6(x, \gamma)$ for all x. These guesses allow the definition of ϕ''_γ and ψ''_γ, as follows:

$$\phi''_\gamma(a||b) = (a \oplus \Delta_0(x, \gamma) \ || \ b \oplus \gamma)$$
$$(\mathcal{E} \circ \psi''_\gamma \circ \mathcal{E}^{-1})(g||h) = (h \oplus \Delta_6(x, \gamma) \ || \ g \oplus \gamma).$$

For each complete guess $((\Delta_0(x, \gamma_0), \forall x), (\Delta_6(x, \gamma_0), \forall x))$, we run a yoyo cryptanalysis. If it succeeds, we repeat the attack for a new difference γ_1. In this second step, we don't need to guess 2^{n-1} values for each $\Delta_0(x, \gamma_1)$ and $\Delta_6(x, \gamma_1)$ but only 2^{n-2} as $\Delta_i(x \oplus \gamma_0, \gamma_1) = \Delta_i(x, \gamma_0) \oplus \Delta(x \oplus \gamma_1, \gamma_0) \oplus \Delta_i(x, \gamma_1)$. We run again a cycle-based yoyo cryptanalysis to validate our guesses. The process is repeated $n - 1$ times in total so as to have $\sum_{k=0}^{n-1} 2^k = 2^n$ independent linear equations connecting the entries of S_0 and another 2^n for the entries of S_6. Solving those equations gives the two outer Feistel functions, meaning that they can be peeled off. We then run a regular yoyo-cryptanalysis on the 5 inner rounds to recover the remainder of the structure.

Since $\sum_{k=0}^{n-1} 2^{n2^k + 2n} = O(2^{n2^n + 2n})$, the total time complexity of this attack is $O(2^{n2^n + 2n})$, which is roughly the complexity of a naive 6-round attack based on guessing a complete Feistel function and running a cycle-based yoyo cryptanalysis on the remainder.

---

[5] A random permutation of a space of size N is expected to have about $\log_e(N)$ cycles.

5   Guess and Determine Attack

Since the yoyo game is only applicable to an $\oplus$-Feistel, we now describe a different attack that works for any group operation $+$. This guess and determine attack is based on a well-known boomerang-like distinguisher for 3-round Feistel Networks, initially described by Luby and Rackoff [12]. This is used to attack Feistel Networks with 4 or 5 rounds using a guess and determine approach: we guess entries of S_3 and S_4 in order to perform partial encryption/decryption for some values, and we use the distinguisher on the first three rounds in order to verify the consistency of the guesses, and to recover more values of S_3 and S_4.

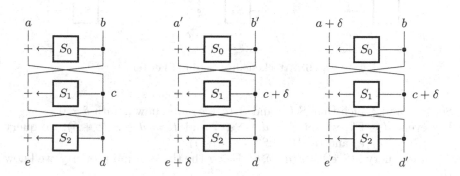

Fig. 5. Distinguisher for a 3-round Feistel

5.1   Three-Round Property

The distinguisher is illustrated by Fig. 5, and works as follows:

- Select arbitrary values a, b, δ ($\delta \neq 0$);
- Query $(e, d) = \mathcal{E}(a, b)$ and $(e', d') = \mathcal{E}(a + \delta, b)$;
- Query $(a', b') = \mathcal{E}^{-1}(e + \delta, d)$;
- If $\mathcal{E}$ is a three-round Feistel, then $d - b' = d' - b$.

The final equation is always true for a 3-round Feistel Network because the input to the third Feistel function is $c + \delta$ for both queries $\mathcal{E}(a + \delta, b)$ and $\mathcal{E}^{-1}(e + \delta, d)$. Therefore the output of S_1 is the same in both cases. On the other hand, the relation only holds with probability 2^{-n} for a random permutation.

5.2   Four-Round Attack

We now explain how to use this property to decompose a four-round Feistel network. We first fix $S_3(0) = 0$, and guess the value $S_3(1)$. Then we use known values of S_3 to iteratively learn new values as follows (see Fig. 6):

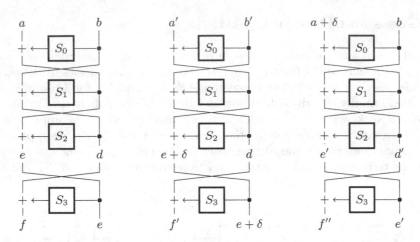

Fig. 6. Attack against 4-round Feistel

- Select e and δ such that $S_3(e)$ and $S_3(e+\delta)$ are known, with $\delta \neq 0$.
- For every $d \in \mathbb{F}_2^n$, we set $f = d + S_3(e)$ and $f' = d + S_3(e+\delta)$; we query $(a,b) = \mathcal{E}^{-1}(f,e)$ and $(a',b') = \mathcal{E}^{-1}(f',e+\delta)$
- Then we query $(f'',e') = \mathcal{E}(a+\delta,b)$. Using the three-round property, we know that:
$$d - b' = d' - b, \quad \text{where } d' = f'' - S_3(e').$$
This gives the value of $S_3(e')$ as $f'' - d + b' - b$.

We iterate the deduction algorithm until we either detect a contradiction (if the guess of $S_3(1)$ is wrong), or we recover the full S_3. Initially, we select $e = 0, \delta = 1$, or $e = 1, \delta = -1$, with 2^n choices of d: this allows 2^{n+1} deductions. If the guess of $S_3(1)$ is wrong, we expect to find a contradiction after about $2^{n/2}$ deductions. If the guess is correct, almost all entries of S_3 will be deduced with a single choice of e and δ, and we will have many options for further deduction. Therefore, the complexity of this attack is about $2^{3n/2}$.

5.3   Five-Round Attack

The extension from 4 rounds to 5 rounds is similar to the extension from a three-round distinguisher to a four-round attack. First, we guess some entries of the last S-Box, so that we can invert the last round for a subset of the outputs. Then, we use those pairs to perform an attack on a reduced version so as to test whether the guess was valid. However, we need to guess a lot more entries in this context. The deductions are performed as follows (see Fig. 7):

- Select d, e and δ such that $S_3(e)$, $S_3(e+\delta)$, $S_4(d+S_3(e))$ and $S_4(d+S_3(e+\delta))$ are known.
- Let $(f,g) = (d+S_3(e), e+S_4(f))$ and $(f',g') = (d+S_3(e+\delta), e+\delta+S_4(f'))$, then query $(a,b) = \mathcal{E}^{-1}(g,f)$ and $(a',b') = \mathcal{E}^{-1}(g',f')$

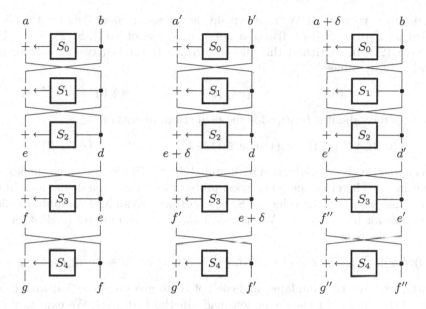

Fig. 7. Attack against 5-round Feistel

- Finally, query $(g'', f'') = \mathcal{E}(a + \delta, b)$. Assuming that $S_4(f'')$ is known, we can use the three-round property and deduce:

$$d - b' = d' - b, \quad \text{where } d' = f'' - S_3(g'' - S_4(f''))$$

This gives the value of $S_3(g'' - S_4(f''))$ as $f'' - d + b' - b$.

Guessing Strategy. The order in which we guess entries of S_3 and S_4 is very important in order to obtain a low complexity attack. We first guess the values of $S_3(i)$ and $S_4(S_3(i))$ for $i < \ell$, with $\ell > 2^{n/2}$. This allows to try deductions with $d = 0$ and any $e, e + \delta \leq \ell$, i.e. ℓ^2 attempts. Since ℓ entries of S_4 are known, each attempt succeeds with probability $\ell 2^{-n}$, and we expect to guess about $\ell^3 2^{-n}$ new values of S_3. With $\ell > 2^{n/2}$, this will introduce a contradiction with high probability.

When an initial guess is non-contradictory, we select x such that $S_3(x)$ has been deduced earlier, we guess the corresponding value $S_4(S_3(x))$, and run again the deduction. The new guess allows to make ℓ new deduction attempts with $d = 0$, $e < \ell$ and $e' = x$. We expect about $\ell^2 2^{-n}$ successful new deductions. With $\ell = 2^{3n/4+\varepsilon}$ with a small $\varepsilon > 0$, the probability of finding a contradiction is higher than 2^{-n}, and the size of the search tree decreases.

The attack will also work if we start with $2^{n/4}$ entries in S_3 and $2^{3n/4}$ entries in S_4: the first step will deduce $2^{3n/4}$ values in S_3. Therefore, we have to make only $2^{n/4} + 2^{3n/4} \approx 2^{3n/4}$ guesses, and the total complexity is about $2^{n 2^{3n/4}}$.

Application to $n = 4$. We now explain the attack in more detail with $n = 4$. We first set $S_4(0) = S_3(0) = 0$ and we guess the values of $S_3(1), S_3(2), S_4(S_3(1))$, and $S_4(S_3(2))$. In particular this allows to compute the last two rounds for the following (e, d) values:

$$(0,0) \qquad\qquad (1,0) \qquad\qquad (2,0)$$

This gives 6 candidates $(e, d), \delta$ for the deduction algorithm:

$$(0,0), \delta \in \{1,2\} \qquad (1,0), \delta \in \{1,-1\} \qquad (2,0), \delta \in \{-1,-2\}$$

Each candidate gives a deduction with probability $3/16$ because three entries are known in S_4. Therefore there is a good probability to get one deduction $S_4(x)$. In this case, we guess the value $S_3(S_4(x))$, so that we can also compute the last two rounds for $(d, e) = (x, 0)$. We have at least 6 new candidates $(e, d), \delta$ for the deduction algorithm:

$$(x,0), \delta \in \{-x, 1-x, 2-x\} \quad (0,0), \delta = x \quad (1,0), \delta = x-1 \quad (2,0), \delta = x-2$$

In total, we have 12 candidates, and each of them gives a deduction with probability $4/16$, including the deduction made in the first step. We expect about 3 deductions in total, which leads to 7 known values in S_3. Since $7 > 2^{n/2}$, there is already a good chance to detect a contradiction. For the remaining cases, we have to make further guesses of S_4 entries, and repeat the deduction procedure.

Since we had to make five guesses for most branches of the guess and determine algorithm, the complexity is about 2^{20}. In practice, this attack takes less than one second on a single core with $n = 4$ (on a 3.4 GHz Haswell CPU).

6   Integral attack

Finally, we present an integral attack against 5-round Feistels, that was shown to us by one of the anonymous reviewers. This attack has a complexity of 2^{3n}, and works for any group operation $+$, but it requires S_1 or S_3 to be a permutation.

The attack is based on an integral property, as introduced in the cryptanalysis of SQUARE [23,24]. In the following, we assume that S_1 is a permutation; if S_3 is a permutation instead, the attack is performed against the decryption oracle. An attacker uses a set of 2^n plaintexts (a_i, b) where a_i takes all possible values, and b is fixed to a constant value. She can then trace the evolution of this set of plaintext through the Feistel structure:

- $c_i = a_i + S_0(b)$ takes all possible values once;
- $d_i = b + S_1(c_i)$ takes all possible values once, since S_1 is a permutation;
- $e_i = c_i + S_2(d_i)$ has a fixed sum:

$$\sum_i e_i = \sum_{x \in \{0...2^n - 1\}} x + \sum_{x \in \{0...2^n - 1\}} S_2(x) = S.$$

The first term is 0 for a $\oplus$-Feistel, and 2^{n-1} for a $\boxplus$-Feistel, while the second term is equal to the first if S_2 is a permutation, but otherwise unknown.

After collecting the 2^n ciphertexts (g_i, f_i) corresponding to the set of plaintexts, she can express $e_i = g_i - S_4(f_i)$. The fixed sum $\sum e_i = S$ gives a linear equation between the values $S_4(f_i)$ and S. This can be repeated with 2^n different sets of plaintexts, in order to build 2^n linear equations. Solving the equations recovers the values $S_4(f_i)$, i.e. the full S_4 box.

When S_2 is a permutation, S is known, and the system has a single solution with high probability. However, when S is unknown, the system has n equations and $n + 1$ unknowns; with high probability it has rank n and 2^n solutions. Therefore, an attacker has to explore the set of solutions, and to use a 4-round distinguisher to verify the guess. Using the attack of Sect. 5.2, this has complexity $2^{5n/2}$.

For a $\oplus$-Feistel, the cost of solving the linear system is 2^{3n} with Gaussian elimination, but can be improved to $O(2^{2.81n})$ with Strassen's Algorithm (the currently best known algorithm [25] has complexity only $O(2^{2.3729n})$ but is probably more expensive for practical values of n).

For a $\boxplus$-Feistel, solving a linear system over $\mathbb{Z}/2^n\mathbb{Z}$ is harder. However, we can solve the system bit-by-bit using linear algebra over $\mathbb{F}_2$. We first consider the equations modulo 2, and recover the least significant bit of S_4. Next, we consider the equations modulo 4. Since the least significant bits are known, this also turns into linear equations over $\mathbb{F}_2$, with the second bits as unknowns. We repeat this technique from the least significant bit to the most significant. At each step, we might have to consider a few different candidates if the system is not full rank.

In total, this attack has a time complexity about $2^{2.81n}$.

7 Conclusion

We presented new generic attacks against Feistel Networks capable of recovering the full description of the Feistel functions without making any assumptions regarding whether they are bijective or not. To achieve this, we have improved the yoyo game proposed in [15] using cycles and found an efficient guessing strategy to be used if modular addition is used instead of exclusive-or. We implemented our attacks to check our claims. We finally described an integral attack suggested by an anonymous reviewer.

Our attacks allow an efficient recovery of the Feistel functions for 5-round Feistel Networks and cycle-based yoyo cryptanalysis can be pushed to attack 6-round and(respectively 7-round) $\oplus$-Feistel at a cost similar to guessing half (resp. all) of the entries of a Feistel function.

Our results differ significantly between $\oplus$-Feistel and $\boxplus$-Feistel. It remains an open problem to find a more efficient attack against $\boxplus$-Feistel or to theoretically explain such a difference.

Acknowledgment. We would like to thank the anonymous reviewers for their valuable comments, and in particular for showing us the attack of Sect. 6. The work of Léo Perrin is supported by the CORE ACRYPT project (ID C12-15-4009992) funded by the *Fonds National de la Recherche* (Luxembourg).

References

1. Daemen, J., Rijmen, V.: The Design of Rijndael: AES-The Advanced Encryption Standard. Springer, Heidelberg (2002)
2. U.S. Department of commerce, National Institute of Standards and Technology: Data encryption standard. Federal Information Processing Standards Publication (1999)
3. Barreto, P., Rijmen, V.: The Khazad legacy-level block cipher. Primitive submitted to NESSIE, vol. 97 (2000)
4. Gérard, B., Grosso, V., Naya-Plasencia, M., Standaert, F.-X.: Block ciphers that are easier to mask: how far can we go? In: Bertoni, G., Coron, J.-S. (eds.) CHES 2013. LNCS, vol. 8086, pp. 383–399. Springer, Heidelberg (2013)
5. Biryukov, A., Bouillaguet, C., Khovratovich, D.: Cryptographic schemes based on the ASASA structure: black-box, white-box, and public-key (extended abstract). In: Sarkar, P., Iwata, T. (eds.) ASIACRYPT 2014. LNCS, vol. 8873, pp. 63–84. Springer, Heidelberg (2014)
6. Biryukov, A., Shamir, A.: Structural cryptanalysis of SASAS. In: Pfitzmann, B. (ed.) EUROCRYPT 2001. LNCS, vol. 2045, pp. 394–405. Springer, Heidelberg (2001)
7. Biryukov, A., Perrin, L.: On reverse-engineering S-Boxes with hidden design criteria or structure. In: Gennaro, R., Robshaw, M. (eds.) CRYPTO 2015. LNCS, vol. 9215, pp. 116–140. Springer, Heidelberg (2015)
8. Brier, E., Peyrin, T., Stern, J.: BPS: a format-preserving encryption proposal. Submission to NIST (2010). http://csrc.nist.gov/groups/ST/toolkit/BCM/modes_development.html
9. Bellare, M., Rogaway, P., Spies, T.: The FFX mode of operation for format-preserving encryption. Submission to NIST (2010). http://csrc.nist.gov/groups/ST/toolkit/BCM/modes_development.html
10. Lampe, R., Seurin, Y.: Security analysis of key-alternating Feistel ciphers. In: Cid, C., Rechberger, C. (eds.) FSE 2014. LNCS, vol. 8540, pp. 243–264. Springer, Heidelberg (2015)
11. Dinur, I., Dunkelman, O., Keller, N., Shamir, A.: New attacks on Feistel structures with improved memory complexities. In: Gennaro, R., Robshaw, M. (eds.) CRYPTO 2015. LNCS, vol. 9215, pp. 433–454. Springer, Heidelberg (2015)
12. Luby, M., Rackoff, C.: How to construct pseudorandom permutations from pseudorandom functions. SIAM J. Comput. **17**(2), 373–386 (1988)
13. Patarin, J.: Generic attacks on Feistel schemes. Cryptology ePrint Archive, Report 2008/036 (2008). http://eprint.iacr.org/
14. Knudsen, L.R.: DEAL - a 128-bit block cipher, AES submission (1998)
15. Biham, E., Biryukov, A., Dunkelman, O., Richardson, E., Shamir, A.: Initial observations on Skipjack: cryptanalysis of Skipjack-3XOR. In: Tavares, S., Meijer, H. (eds.) SAC 1998. LNCS, vol. 1556, pp. 362–375. Springer, Heidelberg (1999)
16. Kelsey, J., Schneier, B., Wagner, D.: Related-key cryptanalysis of 3-way, Biham-DES, CAST, DES-X, newDES, RC2, and TEA. In: Proceedings of the First International Conference on Information and Communication Security, ICICS 1997, pp. 233–246. Springer, London (1997). ISBN: 3-540-63696-X. http://dl.acm.org/citation.cfm?id=646277.687180
17. Wagner, D.: The boomerang attack. In: Knudsen, L.R. (ed.) FSE 1999. LNCS, vol. 1636, pp. 156–170. Springer, Heidelberg (1999)

18. National Security Agency, N.S.A.: SKIPJACK and KEA Algorithm Specifications (1998)
19. Borghoff, J., et al.: PRINCE - a low-latency block cipher for pervasive computing applications. In: Wang, X., Sako, K. (eds.) ASIACRYPT 2012. LNCS, vol. 7658, pp. 208–225. Springer, Heidelberg (2012)
20. Derbez, P., Perrin, L.: Meet-in-the-middle attacks and structural analysis of round-reduced PRINCE. In: Leander, G. (ed.) FSE 2015. LNCS, vol. 9054, pp. 190–216. Springer, Heidelberg (2015)
21. Biryukov, A.: Analysis of involutional ciphers: Khazad and Anubis. In: Johansson, T. (ed.) FSE 2003. LNCS, vol. 2887, pp. 45–53. Springer, Heidelberg (2003)
22. Biryukov, A., Leurent, G., Perrin, L.: Cryptanalysis of Feistel Networks with Secret Round Functions. IACR eprint report 2015/723, July 2015
23. Daemen, J., Knudsen, L.R., Rijmen, V.: The block cipher SQUARE. In: Biham, E. (ed.) FSE 1997. LNCS, vol. 1267, pp. 149–165. Springer, Heidelberg (1997)
24. Knudsen, L.R., Wagner, D.: Integral cryptanalysis. In: Daemen, J., Rijmen, V. (eds.) FSE 2002. LNCS, vol. 2365, pp. 112–127. Springer, Heidelberg (2002)
25. Gall, F.L.: Powers of tensors and fast matrix multiplication. In: Nabeshima, K., Nagasaka, K., Winkler, F., Szántó, Á. (eds.) International Symposium on Symbolic and Algebraic Computation, ISSAC 2014, Kobe, Japan, 23–25 July 2014, pp. 296–303. ACM (2014)

Improved Meet-in-the-Middle Distinguisher on Feistel Schemes

Li Lin[1,2,3]($\boxtimes$), Wenling Wu[1,2,3], and Yafei Zheng[1,2,3]

[1] TCA Laboratory, SKLCS, Institute of Software,
Chinese Academy of Sciences, Beijing, China
[2] State Key Laboratory of Cryptology, P.O. Box 5159, Beijing 100878, China
[3] University of Chinese Academy of Science, Beijing, China
{linli,wwl,zhengyafei}@tca.iscas.ac.cn

Abstract. Improved meet-in-the-middle cryptanalysis with efficient tabulation technique has been shown to be a very powerful form of cryptanalysis against SPN block ciphers, especially AES. However, few results have been proposed on Balanced-Feistel-Networks (BFN) and Generalized-Feistel-Networks (GFN). This is due to the stagger of affected trail and special truncated differential trail in the precomputation phase, i.e. these two trails differ a lot from each other for BFN and GFN ciphers. In this paper, we describe an efficient and generic algorithm to search for an optimal improved meet-in-the-middle distinguisher with efficient tabulation technique on word-oriented BFN and GFN block ciphers. It is based on recursive algorithm and greedy algorithm. To demonstrate the usefulness of our approach, we show key recovery attacks on 14/16-round CLEFIA-192/256 which are the best attacks. We also give key recovery attacks on 13/15-round Camellia-192/256 (without FL/FL^{-1}).

Keywords: Block ciphers · Improved meet-in-the-middle attack · Efficient tabulation technique · Automatic search algorithm · CLEFIA · Camellia

1 Introduction

Meet-in-the-middle attack is first proposed by Diffie and Hellman to attack DES [8]. In recent years, it is widely researched due to its effectiveness against block cipher AES [4]. For AES, Gilbert and Minier show in [12] some collision attacks on 7-round AES. At FSE 2008, Demirci and Selçuk improve the Gilbert and Minier attacks using meet-in-the-middle technique instead of collision idea. More specifically, they show that the value of each byte of 4-round AES ciphertext can be described by a function of the δ-*set*, i.e. a set of 256 plaintexts where a byte (called active byte) can take all values and the other 15 bytes are constant, parameterized by 25 [5] and 24 [6] 8-bit parameters. The last improvement is due to storing differences instead of values. This function is used to build a distinguisher in the offline phase, i.e. they build a lookup table containing all the

© Springer International Publishing Switzerland 2016
O. Dunkelman and L. Keliher (Eds.): SAC 2015, LNCS 9566, pp. 122–142, 2016.
DOI: 10.1007/978-3-319-31301-6_7

possible sequences constructed from a δ-set. In the online phase, they identify a δ-set, and then partially decrypt the δ-set through some rounds and check whether it belongs to the table. At ASIACRYPT 2010, Dunkelman, Keller and Shamir develop many new ideas to solve the memory problems of the Demirci and Selçuk attacks [10]. First of all, they only store *multiset*, i.e. an unordered sequence with multiplicity, rather than the ordered sequence. The second and main idea is the differential enumeration technique which uses a special property on a truncated differential characteristic to reduce the number of parameters that describes the set of functions from 24 to 16. Furthermore, Derbez, Fouque and Jean present a significant improvement to the Dunkelman et al. attacks at EUROCRYPT 2013 [7], called efficient tabulation technique. Using this rebound-like idea, they show that many values in the precomputation table are not reached at all under the constraint of a special truncated differential trail. Actually, the size of the precomputation table is determined by 10 byte-parameters only. At FSE 2014, Li et al. give an attack on 9-round AES-192 using some relations among subkeys [17].

In [18], Lin et al. summarize former works of improved MITM distinguisher, and then define T-δ-set which is a special δ-set of T active cells and S-multiset which is a *multiset* of S cells. With these definitions, they get affected trail which is a function connecting a T-δ-set to an S-multiset with the minimal number of active cells after R-round encryption. After that, they introduce a general algorithm to search for the affected trail from a T-δ-set to an S-multiset, and find that building a better distinguisher is equivalent to a positive integer optimization problem.

Although new results on Substitution-Permutation Networks(SPN) block ciphers using improved meet-in-the-middle attack with efficient tabulation technique are given in many literatures [7,16–18], few results have been proposed on Balanced-Feistel-Networks (BFN) [14] and Generalized-Feistel-Networks (GFN) [23][1]. This is due to the stagger of affected trail and special truncated differential trail in the precomputation phase, i.e. these two trails differ a lot from each other for BFN and GFN ciphers. However, they are almost the same for SPN block ciphers.

Our Contribution. In this paper, we describe an efficient and generic algorithm to search for an optimal improved meet-in-the-middle distinguisher with efficient tabulation technique on word-oriented BFN and GFN block ciphers. It is based on recursive algorithm and greedy algorithm. Given an affected trail $\mathcal{D}$ connecting a T-δ-set to an S-multiset by the algorithm of [18], our algorithm can get an optimal truncated differential trail. The algorithm is made up of two phases: table construction phase and searching phase.

The table construction phase is based on the precomputation phase proposed by Fouque et al. in [11]. In this phase, we build a 2-equipartite directed acyclic graph G containing all the possible one-round transitions.

The searching phase is based on the algorithm proposed by Matsui to find the best differential characteristics for DES [22]. Our algorithm works by recursion

---

[1] [13] was published after the accomplishment of this paper.

and can be seen as a tree traversal in a depth-first manner. One truncated differential trail is a path in this tree, and its weight equals the product of all traversed edges. We are looking for the path with the minimum number of guessed-cells in this tree under certain transition probability. The knowledge of the previous best truncated differential trail allows pruning during the procedure. To speed up this algorithm, We also use greedy algorithm to divide the search process into 2 parts: one starts from the beginning, the other one starts from the end.

To demonstrate the usefulness of our approach, we apply our algorithm to CLEFIA-192/256 [24] and Camellia-192/256 (without FL/FL^{-1})[2] [1]. For CLEFIA-256, we give a 10-round distinguisher, and then give a 16-round key recovery attack with data complexity of 2^{113} chosen-plaintexts, time complexity of 2^{219} encryptions and memory complexity of 2^{210} 128-bit blocks. To the best of our knowledge, this is currently the best attack with respect to the number of attacked rounds. For CLEFIA-192, we give a 9-round distinguisher, and then give a 14-round key recovery attack with data complexity of 2^{113} chosen-plaintexts, time complexity of 2^{155} encryptions and memory complexity of 2^{146} 128-bit blocks. To the best of our knowledge, this is currently the best attack with respect to the complexity.

We also give an 8-round distinguisher on Camellia*-192, and then give a 13-round key recovery attack with data complexity of 2^{113} chosen-plaintexts, time complexity of 2^{180} encryptions and memory complexity of 2^{130} 128-bit blocks. For Camellia*-256, we give a 9-round distinguisher, and then give a 15-round key recovery attack with data complexity of 2^{113} chosen-plaintexts, time complexity of 2^{244} encryptions and memory complexity of 2^{194} 128-bit blocks. Although Lu et al. proposed a 14-round attack on Camellia*-192 and a 16-round attack on Camellia*-256 [21], they didn't consider the whitening operations. So we think our works on Camellia* have certain significance.

We present here a summary of our attack results on CLEFIA and Camellia*, and compare them to the best attacks known for them. This summary is given in Table 1.

Organization of the paper. The rest of this paper is organized as follows. Section 2 provides notations and definitions used throughout this paper, and then gives a brief review of the previous improved MITM distinguishers. Section 2 also gives the general attack scheme, and discusses distinguisher on Feistel schemes with efficient tabulation technique. We provide the automatic search algorithm to search for an optimal improved meet-in-the-middle distinguisher with efficient tabulation technique on Feistel schemes in Sect. 3. Our attacks on CLEFIA-192/256 and Camellia*-192/256 are described in Sect. 4. Finally, we conclude this paper in Sect. 5.

2   Preliminaries

In this section, we give notations used throughout this paper, and then briefly recall the previous improved MITM distinguishers, and after that the attack

---

[2] We call it Camellia* in this paper.

Table 1. Summary of the best attacks on CLEFIA-192/256, Camellia*-192/256.

Cipher	Attack type	Rounds	Data	Memory (Bytes)	Time (Euc)	Source
CLEFIA-192	Improbable	14	$2^{127.0}$ CPs	$2^{127.0}$	$2^{183.2}$	[25]
	Multidim. ZC	14	$2^{127.5}$ KPs	2^{115}	$2^{180.2}$	[2]
	Improved MITM	14	2^{113} CPs	2^{150}	2^{155}	Sect. 4.1
CLEFIA-256	Improbable	15	$2^{127.4}$ CPs	$2^{127.4}$	$2^{247.5}$	[25]
	Multidim. ZC	15	$2^{127.5}$ KPs	2^{115}	$2^{244.2}$	[2]
	Improved MITM	16	2^{113} CPs	2^{214}	2^{219}	Sect. 4.1
Camellia*-192	Impossible Diff	12	2^{119} CPs	2^{124}	$2^{147.3}$	[19]
	Impossible Diff	14^{ww}	2^{117} CPs	$2^{122.1}$	$2^{182.2}$	[20]
	HO MITM	14^{ww}	2^{118} CPs	2^{166}	$2^{164.6}$	[21]
	Improved MITM	13	2^{113} CPs	2^{134}	2^{180}	Sect. 4.2.1
Camellia*-256	Impossible Diff	15^{ww}	$2^{122.5}$ KPs	2^{233}	$2^{236.1}$	[3]
	Impossible Diff	16^{ww}	2^{123} KPs	2^{129}	2^{249}	[20]
	HO MITM	16^{ww}	2^{120} CPs	2^{230}	2^{252}	[21]
	Improved MITM	15	2^{113} CPs	2^{198}	2^{244}	Sect. 4.2.2

KPs: Known-Plaintexts. CPs: Chosen-Plaintexts. ww: Without Whiten Operation

scheme is given. Finally, we discuss the improved meet-in-the-middle distinguisher on Feistel schemes with efficient tabulation technique.

2.1 Notation

The following notations will be used throughout this paper (the sizes are counted in number of cells):

- b: number of cells in a state.
- k: the size of the master key.
- o: the size of one branch.
- r: number of rounds.
- c: number of bits in a cell.
- n: number of branches that will get through F-function in a state.
- $|X|$: number of active cells in a state X.
- x_i: the input state of round-i.
- y_i: the state after the key addition layer of round-i.
- z_i: the state after the S-box layer of round-i.
- w_i: the state after the linear transformation layer of round-i.
- $s[i]$: the ith branch of a state.
- $s[i][j]$: the jth cell in the ith branch of a state.
- $K_i[j]$: the jth cell of the ith round key.

2.2 Reviews of Former Works

In this section, we present a unified view of the previously known MITM distinguishers on AES in [5,7,10] and the algorithm to search for the affected trail given by Lin et al. in [18]. Let's start with particular structures of messages captured by Definitions 1 and 2.

Definition 1 (δ-set, [4]). *Let a δ-set be a set of 256 AES-states that are all different in one state byte (active byte) and all equal in the other state bytes (inactive bytes).*

Definition 2 (Multisets of bytes, [7]). *A multiset generalizes the set concept by allowing elements to appear more than once. Here, a multiset of 256 bytes can take as many as $\binom{2^8+2^8-1}{2^8} \approx 2^{506.17}$ different values.*

Proposition 1 (Differential Property of S, [7]). *Given Δ_i and Δ_0 two non-zero differences, the equation of S-box*

$$S(x) \oplus S(x \oplus \Delta_i) = \Delta_0, \tag{1}$$

has one solution in average.

Demirci and Selçuk distinguisher. Consider the set of functions

$$f : \{0,1\}^8 \longrightarrow \{0,1\}^8$$

that maps a byte of a δ-set to another byte of the state after four AES rounds. A convenient way is to view f as an ordered byte sequence $(f(0),\ldots,f(255))$ so that it can be represented by 256 bytes. The crucial observation made by the generalizing Gilbert and Minier attacks [12] is that this set is tiny since it can be described by 25 byte-parameters ($2^{25\cdot 8} = 2^{200}$) compared with the set of all functions of this type which counts as may as $2^{8\cdot 2^8} = 2^{2048}$ elements [5]. Considering the differences $(f(0) - f(0), f(1) - f(0), \ldots, f(255) - f(0))$ rather than values, the set of functions can be described by 24 parameters [6]. The 24 byte-parameters which map $x_1[0]$ to $\Delta x_5[0]$ are presented as gray cells in Fig. 1.

Dunkelman et al. Distinguisher and Derbez et al. Distinguisher. In [10], Dunkelman et al. introduced two new improvements to further reduce the memory complexity of [6]. The first uses *multiset* which is an unordered sequence

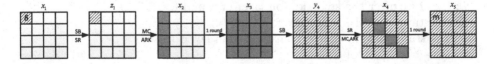

Fig. 1. The 4-round AES distinguisher used in [6]. The gray cells represent 24 byte-parameters, δ represents the δ-set and m represents the differential sequence to be stored.

with multiplicity to replace ordered sequence in the offline phase, since there is enough information so that the attack succeeds. The second improvement uses a novel idea named differential enumeration technique. The main idea of this technique is to use a special 4-round property on a truncated differential characteristic to reduce the number of parameters which describes the set of functions from 24 to 16.

In [7], Derbez et al. presented an improvement to Dunkelman et al.'s differential enumeration technique, called efficient tabulation technique (Proposition 2). Combining with the rebound-like idea, many values in the precomputation table are not reached at all under the constraint of a special truncated differential trail.

Proposition 2 (Efficient Tabulation Technique, [7]). *If a message of δ-set belongs to a pair conforming to the 4-round truncated differential trail outlined in Fig. 2, the values of multiset are only determined by 10 byte-parameters of intermediate state $\Delta z_1[0]||x_2[0,1,2,3]||\Delta x_5[0]||z_4[0,1,2,3]$ presented as gray cells in this figure.*

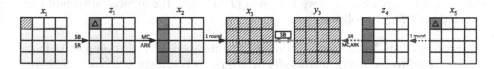

Fig. 2. The truncated differential trail of 4-round AES used in [5], the gray cells represent 10 byte-parameters, Δ represents difference.

The main idea of their works is that suppose one get a pair of messages conforming to this truncated differential trail, the differences Δx_3 and Δy_3 can be determined by these 10 byte-parameters. By Proposition 1, part of the 24 byte-parameters in the Demirci and Selçuk distinguisher, i.e. x_3, can be determined.

Lin et al. Algorithm. In [18], Lin et al. summarized former works of improved MITM distinguisher and gave a general model for this kind of distinguisher on SPN block ciphers. In their works, they define T-δ-set which is a special δ-set of T active cells and S-multiset which is a multiset of S cells to extend the definitions of δ-set and multiset. With these definitions, they get **affected trail** which is a function connecting a T-δ-set to an S-multiset with the minimal number of active cells[3] after r-round encryption. We call these active cells **guessed-cells** in this paper. As shown in Fig. 1, $x_1[0]$ is an 1-δ-set and $\Delta x_5[0]$ is an 1-multiset. $(x_4[0,5,10,15]||x_3||x_2[0,1,2,3])$ is the affected trail connecting $x_1[0]$ to $x_5[0]$, i.e. the value of Δx_5 can be determined by $x_1[0]$ and these bytes. After that, they introduce a general algorithm to search for the best affected trail from a T-δ-set

---

[3] These active cells are before the S-box layer, so we need to guess their values to get the S-multiset.

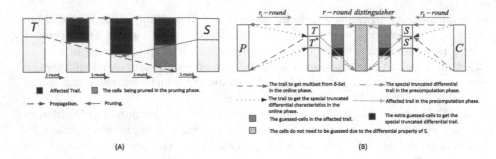

Fig. 3. (A) 4-round example of propagation-then-pruning algorithm; (B) General scheme of the improved meet-in-the-middle attack with efficient tabulation technique.

to an S-*multiset*, and find that building a better affected trail is equivalent to a positive integer optimization problem. In this paper, T represents a T-δ-*set* and S represents an S-*multiset*. With these two definitions in mind, we can choose appropriate values of $|T|$ and $|S|$ according to the success probability [18].

The search algorithm is based on the propagation-then-pruning algorithm as shown in Fig. 3(A). Suppose we have one (T, S) pair, the algorithm to build an affected trail $\mathcal{D}$ is as follows:

1. **Propagation.** In the forward direction, differences of T can propagate from one round to the next. Active cells before the S-box layer need to be guessed, and then S can be got from this trail.
2. **Pruning.** In this trail, some guessed-cells have nothing to do with the building of S. These cells are pruned.

Using this algorithm, we can get an affected trail $\mathcal{D}$ for (T, S).

2.3   Attack Scheme

In this subsection, we present our new unified view of the improved meet-in-the-middle attack, where R rounds of block cipher can be split into three consecutive parts: r_1, r, and r_2.

The general attack scheme is shown in Fig. 3(B), here T represents a T-δ-*set* and S represents an S-*multiset*, T^* represents input difference of the truncated differential trail and S^* represents the output difference. For SPN block ciphers, (T, S) and (T^*, S^*) are almost the same. But for Feistel block ciphers, they are always different. The reason will be explained in Subsect. 2.4.

The general attack scheme uses two successive phases:

Precomputation phase
1. In the precomputation phase, we build an affected trail from T to S by guessing some cells.
2. Use efficient tabulation technique to build a special truncated differential trail from T^* to one middle round, and then build a truncated differential trail from S^* to another middle round in the reverse direction. This step needs to guess some more cells. Using Proposition 1, we can prune some guessed-cells made in step 1.
3. Build a lookup table L containing all the possible sequences constructed from T such that one message verifies the truncated differential trail.

Online phase
1. In the online phase, we need to identify a T-δ-set containing a message m verifying the desired property. This is done by using a large number of plaintexts and ciphertexts, and expecting that for each key candidate, there is one pair of plaintexts satisfying the truncated differential trail from $P \to T^*$ and $C \to S^*$. m is one member of this plaintext pair. Then use m to build a T-δ-set.
2. Finally, we partially decrypt the associated T-δ-set through the last r_2 rounds and check whether it belongs to L.

2.4 Feistel Ciphers with Efficient Tabulation Technique

In this paper, we focus on the application of efficient tabulation technique on word-oriented BFN and GFN. The round function F is made up of 3 layers: key addition layer, S-box layer and linear transformation layer. This is true for most BFN and GFN ciphers.

The advantage of improved meet-in-the-middle attack with efficient tabulation technique on Feistel ciphers is that it can use Proposition 1 in $\lfloor \frac{b}{n \times o} \rfloor$ rounds, i.e. less cells are guessed in these rounds. We take CLEFIA [24] as an example. CLEFIA is a 4-branch type-2 GFN cipher with $b = 16$, $n = 2$ and $o = 4$.

As shown in Fig. 4(A), $\Delta y_i[0] = \Delta x_i[0]$, $\Delta z_i[0] = M_0^{-1}(\Delta x_i[1] \oplus \Delta x_{i+2}[3])$. If Δx_i and Δx_{i+2} are known, using Proposition 1, the values of active bytes in $y_i[0]$ can be got. Using the same method, if Δx_i and Δx_{i+2} are known, the values of active bytes in $y_i[2]$, $y_{i+1}[0]$ and $y_{i+1}[2]$ can be got as well. So if the special truncated differential trail is got, some cells need not to be guessed.

Although affected trail and truncated differential trail are almost the same for SPN ciphers, they differ a lot from each other for BFN and GFN ciphers. We also take CLEFIA as an example. As shown in Fig. 4(B), the guessed-cells of affected trail and truncated differential trail are totally different in the backward direction. For affected trail, if we want to get $\Delta x_{i+2}[1][0]$ from Δx_i, $\Delta x_{i+2}[1][0] = \Delta x_{i+1}[1][0] = \Delta w_i[2][0] \oplus \Delta x_i[3][0]$, only $y_i[2]$ need to be guessed. For truncated differential trail, if $\Delta x_{i+2}[1][0]$ is active and we know its value, by guessing $y_{i+1}[2][0]$, $\Delta x_{i+1}[2][0]$ and $\Delta x_{i+1}[3]$ can be got. By guessing $y_i[0]$, Δx_i can be got. So $y_{i+1}[2][0]$ and $y_i[0]$ need to be guessed in addition. But for another

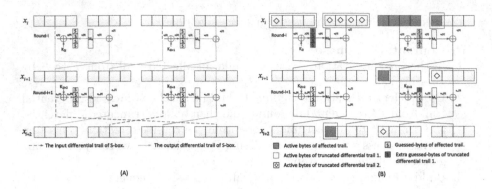

Fig. 4. (A) The truncated differential trail using Proposition 1; (B) The stagger of an affected trail and a truncated differential trail (Color figure online).

truncated differential trail, if $\Delta x_{i+2}[2][0]$ is active, by guessing $y_i[0][0]$, Δx_i can be got. So only one additional cell need to be guessed.

In order to apply this technique to Feistel schemes, first of all, we should get an affected trail; and then we can use a truncated differential trail to restrain values of guessed-cells in the middle $\lfloor \frac{b}{n \times o} \rfloor$ rounds by guessing some extra cells. These two trails may differ a lot from each other, so we need to find a method to minimize the total number of guessed-cells.

To solve this problem, we present an automatic search algorithm in Sect. 3 to search for an optimal improved meet-in-the-middle distinguisher with efficient tabulation technique on BFN and GFN ciphers.

3   Automatic Search Algorithm Using Efficient Tabulation Technique

In this section, we present a practical algorithm to derive an optimal improved meet-in-the-middle distinguisher on Feistel schemes with efficient tabulation technique, which combines the precomputation phase of [11] and the search procedure of [22].

Suppose we get the affected trail [18] gives, our algorithm works by recursion and consists of 2 phases: table construction phase and searching phase.

3.1   Table Construction Phase

As shown in [11], the table construction phase builds a 2-equipartite directed acyclic graph G which contains all the possible one-round transitions. This graph can be built and stored efficiently by observing its inner structure: the block cipher internal state output depends only on the block cipher internal state input. Unlike [11], since we don't consider the key schedule, the graph is small and can be stored in truncated differential characteristic[4]. A toy example of G is shown in Fig. 5.

---

[4] The memory cost of CLEFIA is less than 20 MB.

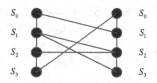

Fig. 5. Example of graph product to build G, with 4 possible internal states s_0, s_1, s_2 and s_3. There is an edge from s_i to s_j if and only if $s_i \rightarrow s_j$ after one round encryption.

A 2-equipartite directed acyclic graph G^{-1} that contains all the possible one-round transitions in the backward direction should also be built as well.

3.2 Searching Phase

As the algorithm in [22], our searching phase finds an optimal n-round improved meet-in-the-middle distinguisher. The algorithm works by recursion and can be seen as a tree traversal in a depth-first manner, where the tree represents all the possible truncated differential trail in the cipher layered by round. The nodes represent the truncated differential characteristics and the edges the possible transitions between them, and are labeled by their numbers of guessed-cells and transition probabilities. One truncated differential trail is a path in this tree, and its weight equals the product of all traversed edges. We are looking for the path with the minimum number of guessed-cells in this tree under certain transition probability. The knowledge of the previous best truncated differential trail allows pruning during the procedure. Since we only consider truncated differential characteristic and the pruning is very efficient, the running time will not be too long.

3.2.1 Trail Probability

Almost all the truncated differential trails used in the distinguishers on SPN ciphers are with probability 1. In our algorithm, we consider the truncated differential trail with probability less than 1, i.e. operations such as XOR can output inactive cells with certain probability. Suppose there is one less active cell in the backward direction (with probability 2^{-c}), it may cause less extra cells to be guessed. Getting a trail with probability 2^{-c} means that the online phase should be repeated 2^c times. So our algorithm require an "initial value" for the minimum trail probability, which is presented as $\overline{P}$. This value can be determined by analyzing the online phase.

3.2.2 Comparing Trails

Next we introduce a (quasi-)order relation for two truncated differential trail T_1 and T_2 as follows:

Definition 3 ($\succ$). $\mathcal{T}_1$ *is better than* $\mathcal{T}_2$ *if and only if it has less guessed-cells or its probability is higher with the same number of guessed-cells, i.e.*

$$\mathcal{T}_1 \succ \mathcal{T}_2 \Leftrightarrow \begin{cases} \mathcal{G}(\mathcal{T}_1) < \mathcal{G}(\mathcal{T}_2) \\ or \\ \mathcal{G}(\mathcal{T}_1) = \mathcal{G}(\mathcal{T}_2) \ and \ \mathcal{P}(\mathcal{T}_1) > \mathcal{P}(\mathcal{T}_2) \end{cases} \tag{2}$$

Also $\mathcal{T}_1 \equiv \mathcal{T}_2 \Leftrightarrow \mathcal{G}(\mathcal{T}_1) = \mathcal{G}(\mathcal{T}_2)$ and $\mathcal{P}(\mathcal{T}_1) = \mathcal{P}(\mathcal{T}_2)$[5].

3.2.3 Ending Condition

Given key length k, probability lower bound $\overline{P}$ and the best trail $\mathcal{T}_{best}$ we found so far, we define ending condition $\mathcal{E}$ as follows:

$$\mathcal{T} \in \mathcal{E} \Leftrightarrow \begin{cases} \mathcal{P}(\mathcal{T}) \leq \overline{P} \\ or \ \mathcal{G}(\mathcal{T}) \geq k \\ or \ \mathcal{T}_{best} \succ \mathcal{T} \end{cases} \tag{3}$$

If a trail belongs to $\mathcal{E}$, we should stop the search procedure and try another trail.

3.2.4 Finding an Optimal Trail

Although we could test all the truncated differential trail under the probability lower bound $\overline{P}$ to find the best one, we have a more efficient way using the greedy algorithm. Since the affected trail is unique for each (S, T) pair, we can find out $\lfloor \frac{b}{n \times o} \rfloor$ successive rounds which need to guess more cells than others, i.e. there are more active cells in these rounds. This means that more cells need not to be guessed using Proposition 1. If the number of these successive rounds is more than one, the beginning of these rounds can be represented as a set, called SR-set.

With SR-set in mind, the search procedure can be divided into 2 parts: one starts at the first round in the forward direction, the other starts at the last round in the backward direction. This algorithm will divide an r-round truncated differential trail search process into 2 parts: r_1 and r_2, where $r = r_1 + \lfloor \frac{b}{n \times o} \rfloor + r_2$.

At the beginning of this algorithm, we should decrease $\mathcal{G}(\mathcal{D})$ by the number of guessed-cells in the middle $\lfloor \frac{b}{n \times o} \rfloor$ rounds. We may meet the situation that not all the guessed-cells can be pruned in the middle $\lfloor \frac{b}{n \times o} \rfloor$ rounds, i.e. there are inactive cells in the truncated differential trail. However, since this also restrains the values of guessed-cells in the truncated differential trail, the total number of guessed-cells remains unchanged.

Although the first and last truncated differential characteristics in the trail can take any values, we put some limitations on them by some observations. Since there is little difference between an affected trail and a truncated differential trail in the first r_1-round, we fix the first truncated differential characteristic T^* to T.

---

[5] $\mathcal{G}$: total number of guessed-cells including the affected trail. $\mathcal{P}$: trail probability.

And by the propagation of differences, we constrain values of the last truncated differential characteristics S^* with $|S^*| \leq |S|$.

The framework of our algorithm for the first r_1 and the last r_2 rounds is now established by Algorithms 1 and 2 including essentially recursive calls.

The inputs of Algorithms 1 and 2 are affected trail $\mathcal{D}$, input truncated differential trail $\mathcal{T}$, best truncated differential trail $\mathcal{T}_{best}$, probability lower bound $\overline{P}$ and graph G/G^{-1}.

Algorithm 1. Search the first r_1-round

1: **function** $Procedure\ Round_i^{begin}(\mathcal{D}, \mathcal{T}, \mathcal{T}_{best}, \overline{P})$
2:   Find all truncated values this trail can lead to in graph G
3:   Sort these truncated differential characteristics
4:   **for all** truncated differential characteristics **do**
5:     Add this characteristic to $\mathcal{T}$
6:     **if** $\mathcal{T} \notin \mathcal{E}$ **then**
7:       **if** $i < r_1 - 1$ **then**
8:         Call $Procedure\ Round_{i+1}^{begin}$
9:       **else**
10:        **for all** truncated differences S^* satisfying $|S^*| \leq |S|$ **do**
11:          Call $Procedure\ Round_0^{end}$ with S^*
12:        **end for**
13:      **end if**
14:    **end if**
15: **end for**
16: **end function**

The sorting algorithm of line 3 in Algorithms 1 and 2 is according to $\succ$.

Line 10 of Algorithm 2 means $s_{r_1} \xrightarrow{\lfloor \frac{b}{n \times o} \rfloor rounds} s_{r_2}$, where s_{r_1} and s_{r_2} are truncated differential characteristics of round-r_1 and round-$(r_1 + \lfloor \frac{b}{n \times o} \rfloor)$ in the trail, respectively. Take CLEFIA as an example, since the branch number of M_0 and M_1 is 5, if the total number of active cells before and after M_0 or M_1 is less than 5, we should stop the search procedure and try another trail.

Algorithm 3 presents the search algorithm for a (T, S) pair.

We can loop through all possible(T, S) pairs to find an optimal r-round distinguisher under $\overline{P}$, and then find $r_{max} = max\{r | \mathcal{P}(\mathcal{T}_{best}) > \overline{P}$ and $\mathcal{G}(\mathcal{T}_{best}) < k\}$.

4 Applications

In this section, we give our attacks on CLEFIA-192/256 and Camellia*-192/256.

4.1 Applications to CLEFIA-192/256

CLEFIA is a lightweight 128-bit block cipher designed by Shirai et al. in 2007 [24] and based on a 4-branch type-2 GFN. It is adopted as an international ISO/IEC 29192 standard in lightweight cryptography. We refer to [24] for a detailed description.

Algorithm 2. Search the last r_2-round

1: **function** *Procedure* $Round_i^{end}(\mathcal{D}, \mathcal{T}, \mathcal{T}_{best}, \overline{P})$
2:    Find all truncated values this trail can lead to in graph G^{-1}
3:    Sort these truncated differential characteristics
4:    **for all** truncated differential characteristics **do**
5:       Add this characteristic to $\mathcal{T}$
6:       **if** $\mathcal{T} \notin \mathcal{E}$ **then**
7:          **if** $i < r_2 - 1$ **then**
8:             Call *Procedure* $Round_{i+1}^{end}$
9:          **else**
10:             **if** Combining 2 parts of $\mathcal{T}$ leads to a trail and $\mathcal{T} \notin \mathcal{E}$ **then**
11:                $\mathcal{T}_{best} \leftarrow \mathcal{T}$
12:             **end if**
13:          **end if**
14:       **end if**
15:    **end for**
16: **end function**

Algorithm 3. Search for an optimal trail for (T, S)

1: **function** *Procedure* $SerachingTrail(\mathcal{D}, \overline{P}, T, S, \lfloor \frac{b}{n \times o} \rfloor)$
2:    Initial $\mathcal{T}_{best}$ with k and $\overline{P}$
3:    Get the SR-set from $\mathcal{D}$
4:    **for all** r_1 in SR-set **do**
5:       Decrease $\mathcal{G}(\mathcal{D})$ by the number of guessed-cells in these $\lfloor \frac{b}{n \times o} \rfloor$ rounds
6:       Call $Round_0^{begin}$ with r_1 and T
7:    **end for**
8: **return** $\mathcal{T}_{best}$
9: **end function**

4.1.1 9/10-Round Distinguisher on CLEFIA

First, we use our search algorithm to find an optimal 10-round distinguisher on CLEFIA-256 and 9-round distinguisher on CLEFIA-192. They are shown in Figs. 6 and 7.

In the attack of CLEFIA, we apply an equivalent transformation to the 10-round and 9-round distinguishers, as shown in Figs. 6 and 7. Namely, the right linear transformations of round $i + 8$ and $i + 7$ are removed from these rounds, and these linear transformations are added to three different positions in order to obtain distinguishers that are computationally equivalent to the original one.

The 10-round distinguisher on CLEFIA-256 is based on the proposition below.

Proposition 3. *Considering to encrypt 2^8 values of the (1-)δ-set after 10-round CLEFIA-256 starting from round-i, where $x_i[1][0]$ is the active byte, in the case of that a message of the δ-set belongs to a pair which conforms to the truncated differential trail outlined in Fig. 6, then the corresponding (1-)multiset of $\Delta x_{i+10}[1][0]$ only contains about 2^{208} values.*

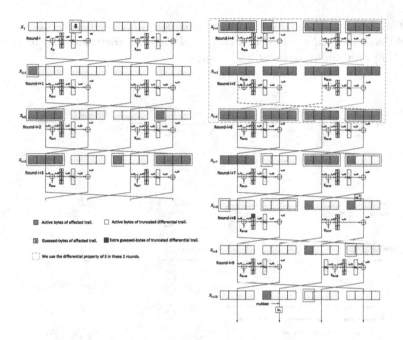

Fig. 6. 10-Round Distinguisher on CLEFIA-256

Proof. As shown in Fig. 6, we first consider the affected trail from $x_i[1][0]$ to $x_{i+10}[1][0]$. This affected trail is determined by 39 byte-parameters: $y_{i+1}[0][0]$ $||y_{i+2}[0]||y_{i+3}[0]||y_{i+3}[2][0]||y_{i+4}[0]||y_{i+4}[2]||y_{i+5}[0]||y_{i+5}[2]||y_{i+6}[0]||y_{i+6}[2]||y_{i+7}$ $[2]||y_{i+8}[2][0]$.

This can be easily seen from the figure.

Furthermore, if there exists a message of the $(1\text{-})\delta$-set belongs to a pair which conforms the truncated differential trail as in Fig. 6, the 35 byte-parameters $y_{i+1}[0][0]||y_{i+2}[0]||y_{i+3}[0]||y_{i+3}[2][0]||y_{i+4}[0]||y_{i+4}[2]||y_{i+5}[0]||y_{i+5}[2]$ $||y_{i+6}[0]||y_{i+6}[2][0]||y_{i+7}[2]$ is determined by 22 byte-parameters: $\Delta x_i[1][0]||y_{i+1}[0$ $][0]||y_{i+2}[0]||y_{i+3}[0]||y_{i+3}[2][0]||y_{i+6}[0]||y_{i+6}[2][0]||y_{i+7}[2]||y_{i+8}[0][0]||\Delta x_{i+10}[2][0]$.

Using 11 byte-parameters $\Delta x_i[1][0]||y_{i+1}[0][0]||y_{i+2}[0]||y_{i+3}[0]||y_{i+3}[2][0]$, we can deduce Δx_{i+4}. In the backward direction, using $\Delta x_{i+10}[2][0]||y_{i+8}[0][0]||y_{i+7}$ $[2]||y_{i+6}[0]||y_{i+6}[2][0]$,we can deduce Δx_{i+6}. By Proposition 1, this can deduce $y_{i+4}[0]$, $y_{i+4}[2]$, $y_{i+5}[0]$ and $y_{i+5}[2]$.

In conclusion, the corresponding *multiset* of byte $\Delta x_{i+10}[1][0]$ only contains about 2^{208} values with the truncated differential trail. $\square$

The 9-round distinguisher on CLEFIA-192 is based on the proposition below, and we will meet the situation Sect. 3.2.4 gives.

Proposition 4. *Considering to encrypt 2^8 values of the $(1\text{-})\delta$-set after 9-round CLEFIA-192 starting from round-i, where $x_i[1][0]$ is the active byte, in the case*

of that a message of the δ-set belongs to a pair which conforms to the trun-
cated differential trail outlined in Fig. 7, and then the corresponding multiset of
$\Delta x_{i+9}[1][0]$ *only contains about* 2^{144} *values.*

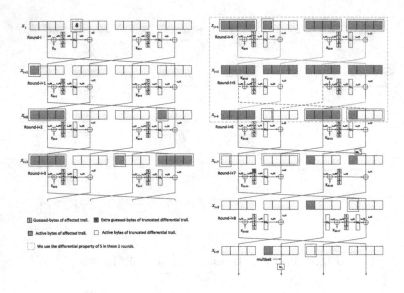

Fig. 7. 9-Round Distinguisher on CLEFIA-192

The proof of this proposition is the same as before and is shown in Fig. 7.

The online phase of 16/14-round attack on CLEFIA-256/192 is shown in Appendix A. For CLEFIA-256, we give a 16-round key recovery attack with data complexity of 2^{113} chosen-plaintexts, time complexity of 2^{219} encryptions and memory complexity of 2^{210} 128-bit blocks. For CLEFIA-192, we give a 14-round key recovery attack with data complexity of 2^{113} chosen-plaintexts, time complexity of 2^{155} encryptions and memory complexity of 2^{146} 128-bit blocks.

4.2   Applications to Camellia*-192/256

Camellia is a 128-bit block cipher designed by Aoki et al. in 2000 [1]. It is a Feistel-like construction where two key-dependent layer FL and FL^{-1} are applied every 6 rounds to each branch. In this paper, we analyze Camellia without FL and FL^{-1}, and call it Camellia* here. We refer to [1] for the detailed description of Camellia.

4.2.1   Attack on Camellia*-192

The 8-round distinguisher of Camellia*-192 with 16 guessed-bytes is shown in Fig. 8. We apply an equivalent transformation to the distinguisher. Namely, the linear transformation of round $i + 6$ is removed, and this linear transformation is added to three different positions in order to obtain distinguishers that are computationally equivalent to the original one. This idea is inspired by [9, 15].

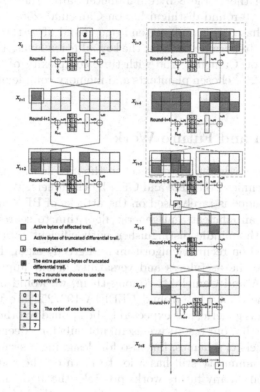

Fig. 8. 8-Round Distinguisher on Camellia*-192

By guessing $y_{i+1}[0][0]$, $y_{i+2}[0][0, 1, 2, 4, 7]$, $y_{i+3}[0]$, $y_{i+4}[0]$, $y_{i+5}[0][1, 2, 4, 6, 7]$ and $y_{i+6}[0][5]$, we can get *multiset* of $\Delta x_{i+8}[1][5]$. If there is a truncated differential trail from $x_i[1][0]$ to $x_{i+8}[0][0]$ as the figure shows, then $y_{i+1}[0][0]$, $y_{i+2}[0][0, 1, 2, 4, 7]$, $y_{i+3}[0]$, $y_{i+4}[0]$ and $y_{i+5}[0][1, 2, 4, 7]$ can be determined by $\Delta x_i[1][0]$, $y_{i+1}[0][0]$, $y_{i+2}[0][0, 1, 2, 4, 7]$, $\Delta x_{i+8}[0][0]$, $y_{i+6}[0][0]$, $y_{i+5}[0][0, 1, 2, 4, 7]$.

In a word, the *multiset* of byte $\Delta x_{i+8}[1][5]$ can be determined by 16 byte-parameters.

The online phase of this attack is the same as the 12-round attack on Camellia-192 in [15]. So we can extend 2 rounds on the top and 3 rounds on the bottom to build a 13-round attack on Camellia*-192 with time complexity of 2^{180} encryptions, data complexity of 2^{113} chosen plaintexts and memory complexity of 2^{130} 128-bit blocks.

4.2.2   Attack on Camellia*-256

For the distinguisher of Camellia*-256, we simply extend one round after round-$(i + 3)$ by guessing the whole 8 byte-parameters after the key addition layer. Then we can get a 10-round distinguisher on Camellia*-256.

In the online phase, we simply extend one round after the distinguisher by guessing all the 8 byte-parameters after the key addition layer. Then we can build a 15-round attack on Camellia*-256 with time complexity of 2^{244} encryptions, data complexity of 2^{113} chosen plaintexts and memory complexity of 2^{194} 128-bit blocks.

5   Conclusion and Future Work

This paper has shown the improved meet-in-the-middle distinguisher with efficient tabulation technique on BFN and GFN block ciphers. We discuss the problem why this technique is rarely used on the attacks of BFN and GFN ciphers, and then describe an efficient and generic algorithm to search for an optimal improved meet-in-the-middle distinguisher with efficient tabulation technique on them. It is based on recursive algorithm and greedy algorithm.

To demonstrate the usefulness and versatility of our approaches, we show attacks on CLEFIA and Camellia*. Among them, we would like to stress that the presented attacks on 14/16-round CLEFIA-192/256 are the best attacks. Since our approach is generic, it is expected to be applied to other BFN and GFN ciphers. We believe that our results are useful not only for a deeper understanding of the security of Feistel schemes, but also for designing a secure block cipher.

The research community still has a lot to learn on the way to build better attacks and there are many future works possible: the algorithm combining the precomputation phase and online phase together, and the link between this kind of attack with other kinds of attacks, such as truncated differential attack and impossible differential attack.

Acknowledgements. We would like to thank the anonymous reviewers for providing valuable comments. The research presented in this paper is supported by the National Basic Research Program of China (No. 2013CB338002) and National Natural Science Foundation of China (No. 61272476, No.61232009 and No. 61202420).

References

1. Aoki, K., Ichikawa, T., Kanda, M., Matsui, M., Moriai, S., Nakajima, J., Tokita, T.: *Camellia*: a 128-bit block cipher suitable for multiple platforms - design and analysis. In: Stinson, D.R., Tavares, S. (eds.) SAC 2000. LNCS, vol. 2012, pp. 39–56. Springer, Heidelberg (2001)
2. Bogdanov, A., Geng, H., Wang, M., Wen, L., Collard, B.: Zero-correlation linear cryptanalysis with FFT and improved attacks on ISO standards Camellia and CLEFIA. In: Lange, T., Lauter, K., Lisoněk, P. (eds.) SAC 2013. LNCS, vol. 8282, pp. 306–323. Springer, Heidelberg (2014)

3. Chen, J., Jia, K., Yu, H., Wang, X.: New impossible differential attacks of reduced-round Camellia-192 and Camellia-256. In: Parampalli, U., Hawkes, P. (eds.) ACISP 2011. LNCS, vol. 6812, pp. 16–33. Springer, Heidelberg (2011)
4. Daemen, J., Rijmen, V.: The Design of Rijndael: AES-the Advanced Encryption Standard. Springer, Heidelberg (2002)
5. Demirci, H., Selçuk, A.A.: A meet-in-the-middle attack on 8-round AES. In: Nyberg, K. (ed.) FSE 2008. LNCS, vol. 5086, pp. 116–126. Springer, Heidelberg (2008)
6. Demirci, H., Taşkın, I., Çoban, M., Baysal, A.: Improved meet-in-the-middle attacks on AES. In: Roy, B., Sendrier, N. (eds.) INDOCRYPT 2009. LNCS, vol. 5922, pp. 144–156. Springer, Heidelberg (2009)
7. Derbez, P., Fouque, P.-A., Jean, J.: Improved key recovery attacks on reduced-round AES in the single-key setting. In: Johansson, T., Nguyen, P.Q. (eds.) EURO-CRYPT 2013. LNCS, vol. 7881, pp. 371–387. Springer, Heidelberg (2013)
8. Diffie, W., Hellman, M.E.: Special feature exhaustive cryptanalysis of the NBS data encryption standard. Computer 10(6), 74–84 (1977)
9. Dong, X., Li, L., Jia, K., Wang, X.: Improved attacks on reduced-round Camellia-128/192/256. In: Nyberg, K. (ed.) CT-RSA 2015. LNCS, vol. 9048, pp. 59–83. Springer, Heidelberg (2015)
10. Dunkelman, O., Keller, N., Shamir, A.: Improved single-key attacks on 8-round AES-192 and AES-256. In: Abe, M. (ed.) ASIACRYPT 2010. LNCS, vol. 6477, pp. 158–176. Springer, Heidelberg (2010)
11. Fouque, P.-A., Jean, J., Peyrin, T.: Structural evaluation of AES and Chosen-Key Distinguisher of 9-Round AES-128. In: Canetti, R., Garay, J.A. (eds.) CRYPTO 2013, Part I. LNCS, vol. 8042, pp. 183–203. Springer, Heidelberg (2013)
12. Gilbert, H., Minier, M.: A collisions attack on the 7-rounds Rijndael. In: AES Candidate Conference, Citeseer (2000)
13. Guo, J., Jean, J., Nikolić, I., Sasaki, Y.: Meet-in-the-middle attacks on generic Feistel constructions. In: Sarkar, P., Iwata, T. (eds.) ASIACRYPT 2014. LNCS, vol. 8873, pp. 458–477. Springer, Heidelberg (2014)
14. Isobe, T., Shibutani, K.: All subkeys recovery attack on block ciphers: extending meet-in-the-middle approach. In: Knudsen, L.R., Wu, H. (eds.) SAC 2012. LNCS, vol. 7707, pp. 202–221. Springer, Heidelberg (2013)
15. Li, L., Jia, K.: Improved meet-in-the-middle attacks on reduced-round Camellia-192/256. IACR Cryptology ePrint Archive 2014, 292 (2014)
16. Li, L., Jia, K., Wang, X.: Improved meet-in-the-middle attacks on AES-192 and PRINCE. IACR Cryptology ePrint Archive 2013, 573 (2013)
17. Li, L., Jia, K., Wang, X.: Improved single-key attacks on 9-round AES-192/256. In: Cid, C., Rechberger, C. (eds.) FSE 2014. LNCS, vol. 8540, pp. 127–146. Springer, Heidelberg (2015)
18. Lin, L., Wu, W., Wang, Y., Zhang, L.: General model of the single-key meet-in-the-middle distinguisher on the word-oriented block cipher. In: Lee, H.-S., Han, D.-G. (eds.) ICISC 2013. LNCS, vol. 8565, pp. 203–223. Springer, Heidelberg (2014)
19. Jiqiang, L.: Cryptanalysis of block ciphers. Ph.D. thesis. University of London, UK (2008)
20. Jiqiang, L., Wei, Y., Fouque, P.-A., Kim, J.: Cryptanalysis of reduced versions of the Camellia block cipher. IET Inf. Secur. 6(3), 228–238 (2012)
21. Jiqiang, L., Wei, Y., Kim, J., Pasalic, E.: The higher-order meet-in-the-middle attack and its application to the Camellia block cipher. Theoret. Comput. Sci. 527, 102–122 (2014)

22. Matsui, M.: On correlation between the order of S-Boxes and the strength of DES. In: De Santis, A. (ed.) EUROCRYPT 1994. LNCS, vol. 950, pp. 366–375. Springer, Heidelberg (1995)
23. Nyberg, K.: Generalized Feistel networks. In: Kim, K., Matsumoto, T. (eds.) ASIACRYPT 1996. LNCS, vol. 1163, pp. 91–104. Springer, Heidelberg (1996)
24. Shirai, T., Shibutani, K., Akishita, T., Moriai, S., Iwata, T.: The 128-bit blockcipher CLEFIA (extended abstract). In: Biryukov, A. (ed.) FSE 2007. LNCS, vol. 4593, pp. 181–195. Springer, Heidelberg (2007)
25. Tezcan, C.: The improbable differential attack: cryptanalysis of reduced round CLEFIA. In: Gong, G., Gupta, K.C. (eds.) INDOCRYPT 2010. LNCS, vol. 6498, pp. 197–209. Springer, Heidelberg (2010)

Appendix A: 16/14-Round Attack on CLEIFA-256/192

Based on the 10-round distinguisher, we extend 3 rounds on the top and 3 rounds on the bottom to present a 16-round improved meet-in-the-middle attack on CLEFIA-256, and based on the 9-round distinguisher, we extend 3 rounds on the top and 2 rounds on the bottom to present a 14-round improved meet-in-the-middle attack on CLEFIA-192. The procedure of the attack on CLEFIA-256 is shown in Fig. 9.

The following proposition is important in finding the special truncated differential trail.

Proposition 5. If $M_i(\Delta s_0) = (a_0, 0, 0, 0)$ and $M_i(\Delta s_1) = (a_1, 0, 0, 0)$, then $M_i(\Delta s_0 \oplus \Delta s_1) = (a_2, 0, 0, 0)$, where a_0, a_1, a_2 are any bytes[6].

This is because of the linearity of M_i. The set of 2^8 differences that results in $(a, 0, 0, 0)$ after the linear transformation layer is called α-set, and it is marked by red triangle in Fig. 9.

The detailed attack is shown below:

1. **Precomputation phase.** In the precomputation phase of the attack, we build the lookup table L that contains the multiset of size 2^{208} for difference $\Delta x_{13}[1][0]$ by following the method of Proposition 4.

2. **Online Phase.**

   (a) **Detecting the Right Pair:**

      i. We prepare a structure of 2^{80} plaintexts where $\Delta P[0][0]$, $\Delta P[2]$ and $\Delta P[3]$ take all the 2^{72} values and $\Delta P[1]$ takes all the values of the α-set. Hence, we can generate $2^{80} \times (2^{80} - 1)/2 \approx 2^{159}$ pairs satisfying the plaintext differences. Choose 2^{33} structures and get the corresponding ciphertexts. Among the $2^{159+33} = 2^{192}$ corresponding ciphertext pairs, we expect $2^{192} \times 2^{-48} = 2^{144}$ pairs to verify the truncated difference pattern where $\Delta C[0][0]$, $\Delta C[2]$, $\Delta C[3]$ have differences, and $\Delta C[1]$ has difference in α-set. Store the 2^{144} remaining pairs in a hash table. This step require 2^{113} plaintext and ciphertext pairs.

---

[6] M_i denotes the linear transformation layer.

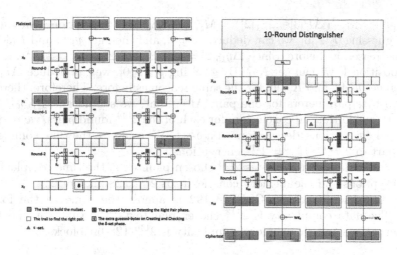

Fig. 9. 16-Round Attack on CLEFIA-256

ii. Guess the values of $K_0[0]$, K_1, $y_1[0]$ and $y_2[2][0]$, using the guessed values to encrypt the remaining pairs to x_3. We choose the pairs that have difference only in byte $x_3[1][0]$, there are $2^{144-72} = 2^{72}$ pairs left.

iii. Guess the values of $K_{30}[0]$, K_{31}, $y_{14}[2]$ and $y_{13}[2][0]$, using the guessed values to decrypt the remaining pairs to x_{13}. We choose the pairs that have differences only in byte $x_{13}[2][0]$, there are $2^{72-72} = 1$ pair left.

(b) **Creating and Checking the Multiset:**

i. For each guess of the 20 bytes made in Phase (a), decrypt the 2^8 possible differences in Δx_3 to Δx_0 using the knowledge of $K_0[0]$, K_1, $y_1[0]$ and $y_2[2][0]$. Then XOR it with one plaintext P_0 of the pair.

ii. Using P_0 as the standard plaintext, denote the other 255 plaintexts as P_1 to P_{255}, and the corresponding ciphertexts as C_0 to C_{255}. Partially decrypt the ciphertexts to x_{13} (here $x'_{13}[1] = M_1(x_{13}[1])$), and then get the *multiset* of $\Delta x_{13}[1][0]$ by guessing $K_{30}[1, 2, 3]$, $y_{13}[0]$, and using the knowledge of $K_{30}[0]$, K_{31}, $y_{14}[2]$.

iii. Checking whether the *multiset* exists in L. If not, discard the key guess. The probability for a wrong guess to pass this test is smaller than $2^{208}2^{-506.17} = 2^{-306.17}$.

(c) **Searching the Rest of Key:** For each remaining key guess, find the remaining key bytes by exhaustive search.

Complexity. The look up table of the 2^{208} possible values requires about 2^{210} 128-bit blocks to be stored [7]. To construct the table, we have to perform 2^{208} partial encryptions on 256 messages, which we estimate to be equivalent to $2^{208+8-4} = 2^{212}$ encryptions.

We take another look at online phase to evaluate the complexity. For each of the 2^{144} remaining pairs after the first part of Phase (a), since $\Delta y_0[2] = \Delta P[2]$

and $\Delta z_0[2] = M_1^{-1}(\Delta P[3] \oplus \Delta x_1[2]) = M_1^{-1}(\Delta P[3])$, by Proposition 1, we can get K_1. By guessing $y_2[2][0]$, we can deduce $\Delta x_2[3]$, and then get $y_1[0]$ and $K_0[0]$ for the same reason as before. Since $\Delta y_{15}[2] = \Delta C[2]$ and $\Delta z_{15}[2] = M_1^{-1}(\Delta C[3])$, by Proposition 1, we can get K_{31}. By guessing $y_{14}[2][0]$, we can deduce $\Delta x_{14}[2]$, and then get $y_{14}[2]$ and $K_{30}[0]$ for the same reason as before. Therefore, there are 2^{16} of 20 byte-parameters for one pair. After that, we should guess $K_{30}[1, 2, 3]$ and $y_{13}[0]$ in addition. In conclusion, for each of the 2^{144} found pairs, we perform 2^{72} partial encryptions/decryptions of a δ-set. We evaluate the time complexity of this part to $2^{144+72+8-5} = 2^{219}$ encryptions.

Hence, the data complexity is 2^{113} chosen-plaintexts, the time complexity is 2^{219} encryptions and the memory complexity is 2^{210} 128-bit blocks.

The 14-round attack on CLEFIA-192 is almost the same as the former attack. The data complexity is 2^{113} chosen-plaintexts, the time complexity is 2^{155} encryptions and the memory complexity is 2^{146} 128-bit blocks.

Implementation of Cryptographic Schemes

Sandy2x: New Curve25519 Speed Records

Tung Chou[✉]

Department of Mathematics and Computer Science,
Technische Universiteit Eindhoven,
P.O. Box 513, 5600 MB Eindhoven, The Netherlands
blueprint@crypto.tw

Abstract. This paper sets speed records on well-known Intel chips for the Curve25519 elliptic-curve Diffie-Hellman scheme and the Ed25519 digital signature scheme. In particular, it takes only 159 128 Sandy Bridge cycles or 156 995 Ivy Bridge cycles to compute a Diffie-Hellman shared secret, while the previous records are 194 036 Sandy Bridge cycles or 182 708 Ivy Bridge cycles.

There have been many papers analyzing elliptic-curve speeds on Intel chips, and they all use Intel's serial $64 \times 64 \rightarrow$ 128-bit multiplier for field arithmetic. These papers have ignored the 2-way vectorized $32 \times 32 \rightarrow$ 64-bit multiplier on Sandy Bridge and Ivy Bridge: it seems obvious that the serial multiplier is faster. However, this paper uses the vectorized multiplier. This is the first speed record set for elliptic-curve cryptography using a vectorized multiplier on Sandy Bridge and Ivy Bridge. Our work suggests that the vectorized multiplier might be a better choice for elliptic-curve computation, or even other types of computation that involve prime-field arithmetic, even in the case where the computation does not exhibit very nice internal parallelism.

Keywords: Elliptic curves · Diffie-Hellman · Signatures · Speed · Constant time · Curve25519 · Ed25519 · Vectorization

1 Introduction

In 2006, Bernstein proposed Curve25519, which uses a fast Montgomery curve for Diffie-Hellman (DH) key exchange. In 2011, Bernstein, Duif, Schwabe, Lange and Yang proposed the Ed25519 digital signature scheme, which uses a fast twisted Edwards curve that is birationally equivalent to the same Montgomery curve. Both schemes feature a conservative 128-bit security level, very small key sizes, and consistent fast speeds on various CPUs (cf. [1,8]), as well as microprocessors such as ARM ([3,16]), Cell ([2]), etc.

Curve25519 and Ed25519 have gained public acceptance and are used in many applications. The IANIX site [17] has lists for Curve25519 and Ed25519

This work was supported the Netherlands Organisation for Scientific Research (NWO) under grant 639.073.005. Permanent ID of this document: 33050f87509019 320b8192d4887bc053. Date: 2015.09.30.

© Springer International Publishing Switzerland 2016
O. Dunkelman and L. Keliher (Eds.): SAC 2015, LNCS 9566, pp. 145–160, 2016.
DOI: 10.1007/978-3-319-31301-6_8

deployment, which include the Tor anonymity network, the QUIC transport layer network protocol developed by Google, openSSH, and many more.

This paper presents Sandy2x, a new software which sets speed records for Curve25519 and Ed25519 on the Intel Sandy Bridge and Ivy Bridge microarchitectures. Previous softwares set speed records for these CPUs using the serial multiplier. Sandy2x, instead, uses of a vectorized multiplier. Our results show that previous elliptic-curve cryptography (ECC) papers using the serial multiplier might have made a suboptimal choice.

A part of our software (the code for Curve25519 shared-secret computation) has been submitted to the SUPERCOP benchmarking toolkit, but the speeds have not been included in the eBACS [8] site yet. We plan to submit the whole software soon for public use.

1.1   Serial Multipliers Versus Vectorized Multipliers

Prime field elements are usually represented as big integers in softwares. The integers are usually divided into several small chunks called *limbs*, so that field operations can be carried out as sequences of operations on limbs. Algorithms involving field arithmetic are usually bottlenecked by multiplications, which are composed of limb multiplications. On Intel CPUs, each core has a powerful $64 \times 64 \rightarrow 128$-bit serial multiplier, which is convenient for limb multiplications. There have been many ECC papers that use the serial multiplier for field arithmetic. For example, [1] uses the serial multipliers on Nehalem/Westmere; [6] uses the serial multipliers on Sandy Bridge; [5] uses the serial multipliers on Ivy Bridge.

On some other chips, it is better to use a vectorized multiplier. The Cell Broadband Engine has 7 Synergistic Processor Units (SPUs) which are specialized for vectorized instructions; the primary processor has no chance to compete with them. ARM has a 2-way vectorized $32 \times 32 \rightarrow 64$-bit multiplier, which is clearly stronger than the $32 \times 32 \rightarrow 64$ serial multiplier. A few ECC papers exploit the vectorized multipliers, including [3] for ARM and [2] for Cell. In 2014, there is finally one effort for using a vectorized multiplier on Intel chips, namely [4]. The paper uses vectorized multipliers to carry out hyperelliptic-curve cryptography (HECC) formulas that provide a natural 4-way parallelism. ECC formulas do not exhibit such nice internal parallelism, so vectorization is expected to induce much more overhead than HECC.

Our speed records rely on using a 2-way vectorized multipliers on Sandy Bridge and Ivy Bridge. The vectorized multiplier carries out only a pair of $32 \times 32 \rightarrow 64$-bit multiplication in one instruction, which does not seem to have any chance to compete with the $64 \times 64 \rightarrow 128$-bit serial multiplier, which is used to set speed records in previous Curve25519/Ed25519 implementations. In this paper we investigate how serial multipliers and vectorized multipliers work (Sect. 2), and give arguments on why the vectorized multiplier can compete.

Our work is similar to [4] in the sense that we both use vectorized multipliers on recent Intel microarchitectures. The difference is that our algorithm does not have very nice internal parallelism, especially for verification. Our work is also

Table 1. Performance results for Curve25519 and Ed25519 of this paper, the CHES 2011 paper [1], and the implementation by Andrew Moon "floodyberry" [7]. All implementations are benchmarked on the Sandy Bridge machine "h6sandy" and the Ivy Bridge machine "h9ivy" (abbreviated as SB and IB in the table), of which the details can be found on the eBACS website [8]. Each cycle count listed is the measurement result of running the software on one CPU core, with Turbo Boost disabled. The table sizes (in bytes) are given in two parts: read-only memory size + writable memory size.

	SB cycles	IB cycles	Table size	Reference	Implementation
Curve25519 public-key generation	54 346	52 169	30720 + 0	(new) this paper	
	61 828	57 612	24576 + 0	[7]	
	194 165	182 876	0 + 0	[1] CHES 2011	amd64-51
Curve25519 shared secret computation	159 128	156 995	0 + 0	(new) this paper	
	194 036	182 708	0 + 0	[1] CHES 2011	amd64-51
Ed25519 public-key generation	57 164	54 901	30720 + 0	(new) this paper	
	63 712	59 332	24576 + 0	[7]	
	64 015	61 099	30720 + 0	[1] CHES 2011	amd64-51-30k
Ed25519 sign	63 526	59 949	30720 + 0	(new) this paper	
	67 692	62 624	24576 + 0	[7]	
	72 444	67 284	30720 + 0	[1] CHES 2011	amd64-51-30k
Ed25519 verification	205 741	198 406	10240 + 1920	(new) this paper	
	227 628	204 376	5120 +960	[7]	
	222 564	209 060	5120 +960	[1] CHES 2011	amd64-51-30k

similar to [3] in the sense that the vectorized multipliers have the same input and output size. We stress that the low-level optimization required on ARM is different to Sandy/Ivy Bridge, and it is certainly harder to beat the serial multiplier on Sandy/Ivy Bridge.

1.2 Performance Results

The performance results for our software are summarized in Table 1, along with the results for [1,7]. [1] is chosen because it holds the speed records on the eBACS site for publicly verifiable benchmarks [7,8] is chosen because it is the fastest constant-time public implementation for Ed25519 (and Curve25519 public-key generation) to our knowledge. The speeds of our software (as [1,7]) are fully protected against simple timing attacks, cache-timing attacks, branch-prediction attacks, etc.: all load addresses, all store addresses, and all branch conditions are public.

For comparison, Longa reported ≈ 298 000 Sandy Bridge cycles for the "ECDHE" operation, which is essentially 1 public-key generation plus 1 secret-key computation, using Microsoft's 256-bit NUMS curve [19]. OpenSSL 1.0.2, after heavy optimization work from Intel, compute a NIST P-256 scalar multiplication in 311 434 Sandy Bridge cycles or 277 994 Ivy Bridge cycles.

For Curve25519 public-key generation, [7] and our implementation gain much better results than [1] by performing the fixed-base scalar multiplications on the twisted Edwards curve used in Ed25519 instead of the Montgomery curve; see Sect. 3.2. Our implementation strategy for Ed25519 public-key generation and signing is the same as Curve25519 public-key generation. Also see Sect. 3.1 for Curve25519 shared-secret computation, and Sect. 4 for Ed25519 verification.

We also include the tables sizes of [1, 7] and Sandy2x in Table 1. Note that our current code uses the same window sizes as [1, 7] but larger tables for Ed25519 verification. This is because we use a data format that is not compact but more convenient for vectorization. Also note that [1] has two implementations for Ed25519: amd64-51-30k and amd64-64-24k. The tables sizes for amd64-64-24k are 20 % smaller than those of amd64-51-30k, but the speed records on eBACS are set by amd64-51-30k.

1.3   Other Fast Diffie-Hellman and Signature Schemes

On the eBACS website [8] there are a few DH schemes that achieve fewer Sandy/Ivy Bridge cycles for shared-secret computation than our software: gls254prot from [12] uses a GLS curve over a binary field; gls254 is a non-constant-time version of gls254prot; kummer from [4] is a HECC scheme; kumfp127g from [13] implements the same scheme as [4] but uses an obsolete approach to perform scalar multiplication on hyperelliptic curves as explained in [4].

GLS curves are patented, making them much less attractive for deployment, and papers such as [14,15] make binary-field ECC less confidence-inspiring. There are algorithms that are better than the Rho method for high-genus curves; see, for example, [20]. Compared to these schemes, Curve25519, using an elliptic curve over a prime field, seems to be a more conservative (and patent-free) choice for deployment.

The eBACS website also lists some signature schemes which achieve better signing and/or verification speeds than our work. Compared to these schemes, Ed25519 has the smallest public-key size (32 bytes), fast signing speed (super-seded only by *multivariate* schemes with much larger key sizes), reasonably fast verification speed (can be much better if batched verification is considered, as shown in [1]), and a high security level (128-bit).

2   Arithmetic in $\mathbb{F}_{2^{255}-19}$

A radix-2^r representation represents an element f in a b-bit prime field as $(f_0, f_1, \ldots, f_{\lceil b/r \rceil - 1})$, such that

$$f = \sum_{i=0}^{\lceil b/r \rceil - 1} f_i 2^{\lceil ir \rceil}.$$

This is called a radix-2^r representation. Field arithmetic can then be carried out using operations on limbs; as a trivial example, a field addition can be carried out by adding corresponding limbs of the operands.

Since the choice of radix is often platform-dependent, several radices have been used in existing software implementations of Curve25519 and Ed25519.

This section describes and compares the radix-2^{51} representation (used by [1]) with the radix-$2^{25.5}$ representation (used by [3] and this paper), and explains how a small-radix implementation can beat a large-radix one on Sandy Bridge and Ivy Bridge, even though the vectorized multiplier seems to be slower. The radix-2^{64} representation by [1] appears to be slower than the radix-2^{51} representation for Curve25519 shared-secret computation, so only the latter is discussed in this section.

2.1 The Radix-2^{51} Representation

[1] represents an integer f modulo $2^{255} - 19$ as

$$f_0 + 2^{51}f_1 + 2^{102}f_2 + 2^{153}f_3 + 2^{204}f_4$$

As the result, the product of $f_0 + 2^{51}f_1 + 2^{102}f_2 + 2^{153}f_3 + 2^{204}f_4$ and $g_0 + 2^{51}g_1 + 2^{102}g_2 + 2^{153}g_3 + 2^{204}g_4$ is $h_0 + 2^{51}h_1 + 2^{102}h_2 + 2^{153}h_3 + 2^{204}h_4$ modulo $2^{255} - 19$ where

$$
\begin{aligned}
h_0 &= f_0 g_0 + 19 f_1 g_4 + 19 f_2 g_3 + 19 f_3 g_2 + 19 f_4 g_1, \\
h_1 &= f_0 g_1 + f_1 g_0 + 19 f_2 g_4 + 19 f_3 g_3 + 19 f_4 g_2, \\
h_2 &= f_0 g_2 + f_1 g_1 + f_2 g_0 + 19 f_3 g_4 + 19 f_4 g_3, \\
h_3 &= f_0 g_3 + f_1 g_2 + f_2 g_1 + f_3 g_0 + 19 f_4 g_4, \\
h_4 &= f_0 g_4 + f_1 g_3 + f_2 g_2 + f_3 g_1 + f_4 g_0.
\end{aligned}
$$

One can replace g by f to derive similar equations for squaring.

The radix-2^{51} representation is designed to fit the $64 \times 64 \to$ 128-bit serial multiplier, which can be accessed using the mul instruction. The usage of the mul is as follows: given a 64-bit integer (either in memory or a register) as operand, the instruction computes the 128-bit product of the integer and rax, and stores the higher 64 bits of in rdx and lower 64 bits in rax.

The field multiplication function begins with computing $f_0 g_0, f_0 g_1, \ldots, f_0 g_4$. For each g_j, f_0 is first loaded into rax, and then a mul instruction is used to compute the product; some mov instructions are required to move the rdx and rax to the registers where h_j is stored. Each monomial involving f_i where $i > 0$ also takes a mul instruction, and an addition (add) and an addition with carry (adc) are required to accumulate the result into h_k. Multiplications by 19 can be handled by the imul instruction. In total, it takes 25 mul, 4 imul, 20 add, and 20 adc instructions to compute $h_0, h_1, \ldots, h_4$[1]. Note that some *carries* are required to bring the h_k back to around 51 bits. We denote such a radix-51 field multiplication including carries as **m**; **m**$^-$ represents **m** without carries.

2.2 The Radix-$2^{25.5}$ Representation

[3] represents an integer f modulo $2^{255} - 19$ as

$$f_0 + 2^{26}f_1 + 2^{51}f_2 + 2^{77}f_3 + 2^{102}f_4 + 2^{128}f_5 + 2^{153}f_6 + 2^{179}f_7 + 2^{204}f_8 + 2^{230}f_9.$$

---

[1] [1] uses one more imul; perhaps this is for reducing memory access.

As the result, the product of $f_0 + 2^{26}f_1 + 2^{51}f_2 + \cdots$ and $g_0 + 2^{26}g_1 + 2^{51}g_2 + \cdots$ is $h_0 + 2^{26}h_1 + 2^{51}h_2 + \cdots$ modulo $2^{255} - 19$ where

$$
\begin{aligned}
h_0 &= f_0g_0 + 38f_1g_9 + 19f_2g_8 + 38f_3g_7 + 19f_4g_6 + 38f_5g_5 + 19f_6g_4 + 38f_7g_3 + 19f_8g_2 + 38f_9g_1, \\
h_1 &= f_0g_1 + f_1g_0 + 19f_2g_9 + 19f_3g_8 + 19f_4g_7 + 19f_5g_6 + 19f_6g_5 + 19f_7g_4 + 19f_8g_3 + 19f_9g_2, \\
h_2 &= f_0g_2 + 2f_1g_1 + f_2g_0 + 38f_3g_9 + 19f_4g_8 + 38f_5g_7 + 19f_6g_6 + 38f_7g_5 + 19f_8g_4 + 38f_9g_3, \\
h_3 &= f_0g_3 + f_1g_2 + f_2g_1 + f_3g_0 + 19f_4g_9 + 19f_5g_8 + 19f_6g_7 + 19f_7g_6 + 19f_8g_5 + 19f_9g_4, \\
h_4 &= f_0g_4 + 2f_1g_3 + f_2g_2 + 2f_3g_1 + f_4g_0 + 38f_5g_9 + 19f_6g_8 + 38f_7g_7 + 19f_8g_6 + 38f_9g_5, \\
h_5 &= f_0g_5 + f_1g_4 + f_2g_3 + f_3g_2 + f_4g_1 + f_5g_0 + 19f_6g_9 + 19f_7g_8 + 19f_8g_7 + 19f_9g_6, \\
h_6 &= f_0g_6 + 2f_1g_5 + f_2g_4 + 2f_3g_3 + f_4g_2 + 2f_5g_1 + f_6g_0 + 38f_7g_9 + 19f_8g_8 + 38f_9g_7, \\
h_7 &= f_0g_7 + f_1g_6 + f_2g_5 + f_3g_4 + f_4g_3 + f_5g_2 + f_6g_1 + f_7g_0 + 19f_8g_9 + 19f_9g_8, \\
h_8 &= f_0g_8 + 2f_1g_7 + f_2g_6 + 2f_3g_5 + f_4g_4 + 2f_5g_3 + f_6g_2 + 2f_7g_1 + f_8g_0 + 38f_9g_9, \\
h_9 &= f_0g_9 + f_1g_8 + f_2g_7 + f_3g_6 + f_4g_5 + f_5g_4 + f_6g_3 + f_7g_2 + f_8g_1 + f_9g_0.
\end{aligned}
$$

One can replace g by the f to derive similar equations for squaring.

The representation is designed to fit the vector multiplier on Cortex-A8, which performs a pair of $32 \times 32 \to 64$-bit multiplications in one instruction. On Sandy Bridge and Ivy Bridge a similar vectorized multiplier can be accessed using the vpmuludq[2] instruction. The AT&T syntax of the vpmuludq instruction is as follows:

$$\text{vpmuludq src2, src1, dest}$$

where src1 and dest are 128-bit registers, and src2 can be either a 128-bit register or (the address of) an aligned 32-byte memory block. The instruction multiplies the lower 32 bits of the lower 64-bit words of src1 and src2, multiplies the lower 32 bits of the higher 64-bit words of src1 and src2, and stores the 64 bits products in 64-bit words of dest.

To compute $h = fg$ and $h' = f'g'$ at the same time, we follow the strategy of [3] but replace the vectorized addition and multiplication instructions by corresponding ones on Sandy/Ivy Bridge. Given $(f_0, f'_0), \ldots (f_9, f'_9)$ and $(g_0, g'_0), \ldots (g_9, g'_9)$, first prepare 9 vectors $(19g_1, 19g'_1), \ldots, (19g_9, 19g'_9)$ with 10 vpmuludq instructions and $(2f_1, 2f'_1), (2f_3, 2f'_3), \ldots, (2f_9, 2f'_9)$ with 5 vectorized addition instructions vpaddq. Note that the reason to use vpaddq instead of vpmuludq is to balance the loads of different execution units on the CPU core; see analysis in Sect. 2.3. Each $(f_0g_j, f'_0g'_j)$ then takes 1 vpmuludq, while each $(f_ig_j, f'_ig'_j)$ where $i > 0$ takes 1 vpmuludq and 1 vpaddq. In total, it takes 109 vpmuludq and 95 vpaddq to compute $(h_0, h'_0), (h_1, h'_1), \ldots, (h_9, h'_9)$. We denote such a vector of two field multiplications as $\mathbf{M}^2$, including the carries that bring h_k (and also h'_k) back to $26 - (k \bmod 2)$ bits; $\mathbf{M}^{2-}$ represents $\mathbf{M}^2$ without carries. Similarly, we use $\mathbf{S}^2$ and $\mathbf{S}^{2-}$ for squarings.

We perform a carry from h_k to h_{k+1} (the indices work modulo 10), which is denoted by $h_k \to h_{k+1}$, in 3 steps:

- Perform a logical right shift for the 64-bit words in h_k using a vpsrlq instruction. The shift amount is $26 - (k \bmod 2)$.

---

[2] The starting 'v' indicate that the instruction is the VEX extension of the pmuludq instruction. The benefit of using vpmuludq is that it is a 3-operand instruction. In this paper we show vector instructions in their VEX extension form, even though vector instructions are sometimes used without the VEX extension.

- Add the result of the first step into h_{k+1} using a vpaddq instruction.
- Mask out the most significant $38 + (k \bmod 2)$ bits of h_k using a vpand instruction.

For $h_9 \to h_0$ the result of the shift has to be multiplied by 19 before being added to h_0. Note that the usage of vpsrlq suggests that we are using *unsigned* limbs; there is no vectorized arithmetic shift instruction on Sandy Bridge and Ivy Bridge.

To reduce number of bits in all of $h_0, h_1, \ldots, h_9$, the simplest way is to perform the carry chain

$$h_0 \to h_1 \to h_2 \to h_3 \to h_4 \to h_5 \to h_6 \to h_7 \to h_8 \to h_9 \to h_0 \to h_1.$$

The problem of the simple carry chain is that it suffers severely from the instruction latencies. To mitigate the problem, we instead interleave the 2 carry chains

$$h_0 \to h_1 \to h_2 \to h_3 \to h_4 \to h_5 \to h_6,$$

$$h_5 \to h_6 \to h_7 \to h_8 \to h_9 \to h_0 \to h_1.$$

It is not always the case that there are two multiplications that can be paired with each other in an elliptic-curve operation; sometimes there is a need to vectorize a field multiplication internally. We use a similar approach to [3] to compute $h_0, h_1, \ldots, h_9$ in this case; the difference is that we compute vectors $(h_0, h_1), \ldots, (h_8, h_9)$ as result. The strategy for performing the expensive carries on $h_0, h_1, \ldots, h_9$ is the same as [3]. Such an internally-vectorized field multiplication is denoted as **M**.

2.3 Why Is Smaller Radix Better?

m takes 29 multiplication instructions (mul and imul), while $\mathbf{M}^2$ takes $109/2 = 54.5$ multiplication instructions (vpmuludq) per field multiplication. How can our software, (which is based on $\mathbf{M}^2$) be faster than [1] (which is based on **m**) using almost twice as many multiplication instructions?

On Intel microarchitechtures, an instruction is decoded and decomposed into some *micro-operations* (μops). Each μop is then stored in a pool, waiting to be executed by one of the *ports* (when the operands are ready). On each Sandy Bridge and Ivy Bridge core there are 6 ports. In particular, Port 0, 1, 5 are responsible for arithmetic. The remaining ports are responsible for memory access, which is beyond the scope of this paper (Table 2).

The arithmetic ports are not identical. For example, vpmuludq is decomposed into 1 μop, which is handled by Port 0 each cycle with latency 5. vpaddq is decomposed into 1 μop, which is handled by Port 1 or 5 each cycle with latency 1. Therefore, an $\mathbf{M}^{2-}$ would take at least 109 cycles. Our experiment shows that $\mathbf{M}^{2-}$ takes around 112 Sandy Bridge cycles, which translates to 56 cycles per multiplication.

Table 2. Instructions field arithmetic used in [1] and this paper. The data is mainly based on the well-known survey by Fog [10]. The survey does not specify the port utilization for `mul`, so we figure this out using the performance counter (accessed using `perf-stat`). Throughputs are per-cycle. Latencies are given in cycles .

Instruction	Port	Throughput	Latency
vpmuludq	0	1	5
vpaddq	either 1 or 5	2	1
vpsubq	either 1 or 5	2	1
mul	0 and 1	1	3
imul	1	1	3
add	either 0, 1, or 5	3	1
adc	either two of 0,1,5	1	2

The situation for $\mathbf{m}$ is more complicated: `mul` is decomposed into 2 μops, which are handled by Port 0 and 1 each cycle with latency 3. `imul` is decomposed into 1 μop, which is handled by Port 1 each cycle with latency 3. `add` is decomposed into 1 μop, which is handled by one of Port 0,1,5 each cycle with latency 1. `adc` is decomposed into 2 μops, which are handled by two of Port 0,1,5 each cycle with latency 2. In total it takes 25 `mul`, 4 `imul`, 20 `add`, and 20 `adc`, accounting for at least $(25 \cdot 2 + 4 + 20 + 20 \cdot 2)/3 = 38$ cycles. Our experiment shows that $\mathbf{m}^-$ takes 52 Sandy Bridge cycles. The `mov` instructions explain a few cycles out of the $52 - 38 = 14$ cycles. Also, the performance counter shows that the core fails to distribute μops equally over the ports.

Of course, by just looking at these cycle counts it seems that $\mathbf{M^2}$ is still a bit slower, but at least we have shown that the serial multiplier is not as powerful as it seems to be. Here are some more arguments in favor of $\mathbf{M^2}$:

- $\mathbf{m}$ spends more cycles on carries than $\mathbf{M^2}$ does: $\mathbf{m}$ takes 68 Sandy Bridge cycles, while $\mathbf{M^2}$ takes 69.5 Sandy Bridge cycles per multiplication.
- The algorithm built upon $\mathbf{M^2}$ might have additions/subtractions. Some speedup can be gained by interleaving the code; see Sect. 2.5.
- The computation might have some non-field-arithmetic part which can be improved using vector unit; see Sect. 3.2.

2.4 Importance of Using a Small Constant

For the ease of reduction, the prime fields used in ECC and HECC are often a big power of 2 subtracted by a small constant c. It might seem that as long as c is not too big, the speed of field arithmetic would remain the same. However, in the following example, we show that using the slightly larger $c = 31$ ($2^{255} - 31$ is the large prime before $2^{255} - 19$) might already cause some overhead.

Consider two field elements f, g which are the results of two field multiplications. Because the limbs are reduced, the upper bound of f_0 would be close to 2^{26}, and the upper bound of f_1 would be close to 2^{25}, and so on; the same bounds apply for g. Now suppose we need to compute $(f - g)^2$, which is batched with another squaring to form an S^2. To avoid possible underflow, we compute the limbs of $h = f - g$ as $h_i = (f_i + 2 \cdot q_i) - g_i$ instead of $h_i = f_i - g_i$, where q_i is the corresponding limb of $2^{255} - 19$. As the result, the upper bound of h_6 is around $3 \cdot 2^{26}$. To perform the squaring $c \cdot h_6^2$ is required. When $c = 19$ we can simply multiply h_6 by 19 using 1 vpmuludq, and then multiply the product by h_6 using another vpmuludq. Unfortunately the same instructions do not work for $c = 31$, since $31 \cdot h_6$ can take more than 32 bits.

To overcome such problem, an easy solution is to use a smaller radix so that each (reduced) limb takes fewer bits. This method would increase number of limbs and thus increase number of vpmuludq required. A better solution is to delay the multiplication by c: instead of computing $31 f_{i_1} g_{j_1} + 31 f_{i_2} g_{j_2} + \cdots$ by first computing $31 g_{j_1}, 31 g_{j_2}, \ldots$, compute $f_{i_1} g_{j_1} + f_{i_2} g_{j_2} + \cdots$ and then multiply the sum by 31. The sum can take more than 32 bits (and vpmuludq takes only 32-bit inputs), so the multiplication by 31 cannot be handled by vpmuludq. Let $s = f_{i_1} g_{j_1} + f_{i_2} g_{j_2} + \cdots$, one way to handle the multiplication by 19 is to compute $32s$ with one shift instruction vpsllq and then compute $32s - s = 31s$ with one subtraction instruction vpsubq. This solution does not make Port 0 busier as vpsllq also takes only one cycle in Port 0 as vpmuludq, but it does make Port 1 and 5 busier (because of vpsubq), which can potentially increase the cost for S^{2-} by a few cycles.

It is easy to imagine for some c's the multiplication can not be handled in such a cheap way as 31. In addition, delaying multiplication cannot handle as many c's as using a smaller radix; as a trivial example, it does not work if $c f_{i_1} g_{j_1} + c f_{i_2} g_{j_2} + \cdots$ takes more than 64 bits. We note that the computation pattern in the example is actually a part of elliptic-curve operation (see lines 6–9 in Algorithm 1), meaning a bigger constant c actually can slow down elliptic-curve operations.

We comment that usage of a larger c has bigger impact on constrained devices. If c is too big for efficient vectorization, at least one can go for the $64 \times 64 \rightarrow 128$-bit serial multiplier, which can handle a wide range of c without increasing number of limbs. However, on ARM processors where the serial multiplier can only perform $32 \times 32 \rightarrow 64$-bit multiplications, even the serial multiplier would be sensitive to the size of c. For even smaller devices the situation is expected to be worse.

2.5 Instruction Scheduling for Vectorized Field Arithmetic

The fact that μops are stored in a pool before being handled by a port allows the CPU to achieve so called *out-of-order execution*: a μop can be executed before another μop which is from an earlier instruction. This feature is sometimes viewed as the CPU core being able to "look ahead" and execute a later instruction whose operands are ready. However, the ability of out-of-order execution is limited: the core is not able to look too far away. It is thus better to arrange the code so that each code block contains instructions for each port.

While Port 0 is quite busy in $\mathbf{M}^2$, Port 1 and 5 are often idle. In an elliptic-curve operation (see the following sections) an $\mathbf{M}^2$ is often preceded by a few field additions/subtractions. Since vpaddq and the vectorized subtraction instruction vpsubq can only be handled by either Port 1 and Port 5, we try to interleave the multiplication code with the addition/subtraction code to reduce the chance of having an idle port. Experiment results show that the optimization brings a small yet visible speedup. It seems more difficult for an algorithm built upon $\mathbf{m}$ to use the same optimization.

3   The Curve25519 Elliptic-Curve-Diffie-Hellman Scheme

[11] defines Curve25519 as a function that maps two 32-byte input strings to a 32-byte output string. The function can be viewed as an x-coordinate-only scalar multiplication on the curve

$$E_M : y^2 = x^3 + 486662x^2 + x$$

over $\mathbb{F}_{2^{255}-19}$. The curve points are denoted as $E_M(\mathbb{F}_{2^{255}-19})$. The first input string is interpreted as an integer scalar s, while the second input string is interpreted as a 32-byte encoding of x_P, the x-coordinate of a point $P \in E_M(\mathbb{F}_{2^{255}-19})$; the output is the 32-byte encoding of x_{sP}.

Given a 32-byte secret key and the 32-byte encoding of a standard base point defined in [11], the function outputs the corresponding public key. Similarly, given a 32-byte secret key and a 32-byte public key, the function outputs the corresponding shared secret. Although the same routine can be used for generating both public keys and shared secrets, the public-key generation can be done much faster by performing the scalar multiplication on an equivalent curve. The rest of this section describes how we implement the Curve25519 function for shared-secret computation and public-key generation.

3.1   Shared-Secret Computation

The best known algorithm for x-coordinate-only variable-base-point scalar multiplication on Montgomery curves is the *Montgomery ladder*. [1,3] and our software all use the Montgomery ladder for Curve25519 shared secret computation. Similar to the double-and-add algorithm, the algorithm also iterates through each bit of the scalar, from the most significant to the least significant one. For each bit of the scalar the ladder performs a differential addition and a doubling. The differential addition and the doubling together are called a *ladder step*. Since the ladder step can be carried out by a fixed sequence of field operations, the Montgomery ladder is almost intrinsically constant-time. We summarize the ladder step for Curve25519 in Algorithm 1. Note that Montgomery uses projective coordinates.

In order to make the best use of the vector unit (see Sect. 2), multiplications and squarings are handled in pairs whenever convenient. The way we pair multiplications is shown in the comments of Algorithm 1. It is not specified in [3]

Algorithm 1. The Montgomery ladder step for Curve25519

1: **function** LADDERSTEP(x_2, z_2, x_3, z_3, x_P)
2:   $\quad t_0 \leftarrow x_3 - z_3$
3:   $\quad t_1 \leftarrow x_2 - z_2$
4:   $\quad x_2 \leftarrow x_2 + z_2$
5:   $\quad z_2 \leftarrow x_3 + z_3$
6:   $\quad z_3 \leftarrow t_0 \cdot x_2; \; z_2 \leftarrow z_2 \cdot t_1$ ▷ batched multiplications
7:   $\quad x_3 \leftarrow z_3 + z_2$
8:   $\quad z_2 \leftarrow z_3 - z_2$
9:   $\quad x_3 \leftarrow x_3^2; \; z_2 \leftarrow z_2^2$ ▷ batched squarings
10: $\quad z_3 \leftarrow x_P \cdot z_2;$
11: $\quad t_0 \leftarrow t_1^2; \; t_1 \leftarrow x_2^2$ ▷ batched squarings
12: $\quad x_2 \leftarrow t_1 - t_0$
13: $\quad z_3 \leftarrow x_2 \cdot 121666$
14: $\quad z_2 \leftarrow t_0 + z_3$
15: $\quad z_2 \leftarrow x_2 \cdot z_2; \; x_2 \leftarrow t_1 \cdot t_0$ ▷ batched multiplications
16: $\quad$ **return** (x_2, z_2, x_3, z_3)
17: **end function**

whether they pair multiplications and squarings in the same way, but this seems to be the most natural way. Note that the multiplication by 121666 (line 13) and the multiplication by x_1 (line 10) are not paired with other multiplications. We deal with the two multiplications as follows:

- Compute multiplications by 121666 without carries using 5 vpmuludq.
- Compute multiplications by x_1 without carries. This can be completed in 50 vpmuludq since we precompute the products of small constants (namely, 2, 19, and 38) and limbs in x_1 before the ladder begins.
- Perform batched carries for the two multiplications.

This uses far fewer cycles than handling the carries for the two multiplications separately.

Note that we often have to "transpose" data in the ladder step. More specifically, after an $\mathbf{M^2}$ which computes $(h_0, h'_0), \ldots, (h_9, h'_9)$, we might need to compute $h + h'$ and $h - h'$; see lines 6–8 of Algorithm 1. In this case, we compute $(h_i, h_{i+1}), (h'_i, h'_{i+1})$ from $(h_i, h'_i), (h_{i+1}, h'_{i+1})$ for $i \in \{0, 2, 4, 6, 8\}$, and then perform additions and subtractions on the vectors. The transpositions can be carried out using the "unpack" instructions vpunpcklqdq and vpunpckhqdq. Similarly, to obtain the operands for $\mathbf{M^2}$ some transpositions are also required. Unpack instructions are the same as vpaddq and vpsubq in terms of port utilization, so we also try to interleave them with $\mathbf{M^2}$ or $\mathbf{S^2}$ as described in Sect. 2.5.

3.2 Public-Key Generation

Instead of performing a fixed-base scalar multiplication directly on the Montgomery curve, we follow [7] to perform a fixed-base scalar multiplication on the twisted Edwards curve

$$E_T : -x^2 + y^2 = 1 - 121665/121666x^2y^2$$

over $\mathbb{F}_{2^{255}-19}$ and convert the result back to the Mongomery curve with one inversion. The curve points are denoted as $E_T(\mathbb{F}_{2^{255}-19})$. There is an efficiently computable birational equivalence between E_T and E_M, which means the curves share the same group structure and ECDLP difficulty. Unlike Mongomery curves, there are complete formulas for point addition and doubling on twisted Edwards curves; we follow [1] to use the formulas for the extended coordinates proposed in [9]. The complete formulas allow utilization of a table of many precomputed points to accelerate the scalar multiplication, which is the reason fixed-base multiplications (on both E_M and E_T) can be carried out much faster than variable-base scalar multiplications.

In [1] a fixed-base scalar multiplication sB where $s \in \mathbb{Z}$ and $B \in E_T(\mathbb{F}_{2^{255}-19})$ (B corresponds to the standard base point in $E_M(\mathbb{F}_{2^{255}-19})$) is performed as follows: write s (modulo the order of B) as $\sum_{i=0}^{15} 16^i s_i$ where $s_i \in \{-8, -7, \ldots, 7\}$ and obtain sB by computing the summation of $s_0B, s_1 16B, \ldots, s_{15}16^{15}B$. To obtain $s_i 16^i B$, the strategy is to precompute several multiples of $16^i B$ and store them in a table, and then perform a constant-time table lookup using s_i as index on demand. [1] also shows how to reduce the size of the table by dividing the sum into two parts:

$$P_0 = s_0 B + s_2 16^2 B + \cdots + s_{14} 16^{14} B$$

and

$$P_1 = s_1 B + s_3 16^2 B + \cdots + s_{15} 16^{14} B.$$

$sB = P_0 + 16P_1$ is then obtained with 4 point doublings and 1 point addition. In this way, the table contains only multiples of $B, 16^2 B, \ldots, 16^{14}B$.

We do better by vectorizing between computations of P_0 and P_1: all the data related to P_0 and P_1 are loaded into the lower half and upper half of the 128-bit registers, respectively. This type of computation pattern is very friendly for vectorization since there no need to "transpose" the data as in the case of Sect. 3.1.

While parallel point additions can be carried out easily, an important issue is how to perform parallel constant-time table lookups in an efficient way. In [1], suppose there is a need to lookup $s_i P$, the strategy is to precompute a table containing $P, 2P, \ldots, 8P$, and then the lookup is carried out in two steps:

- Load $|s_i|P$ in constant time, which is the main bottleneck of the table lookup.
- Negate $|s_i|P$ if s_i is negative.

For the first step it is convenient to use the conditional move instruction (cmov): To obtain each limb (of each coordinate) of $|s_i|P$, first initialize a 64-bit register to the corresponding limb of ∞, then for each of $P, 2P, \ldots, 8P$, conditionally move the corresponding limb into the register. Computation of the conditions and conditional negation are relatively cheap compared to the cmov instructions.

[1] uses a 3-coordinate system for precomputed points, so the table-lookup function takes $3 \cdot 8 \cdot 5 = 120$ cmov instructions. The function takes 159 Sandy Bridge cycles or 158 Ivy Bridge cycles.

We could use the same routine twice for parallel table lookups, but we do better by using vector instructions. Here is a part of the inner loop of our qhasm ([18]) code.

```
v0 = mem64[input_1 + 0] x2
v1 = mem64[input_1 + 40] x2
v2 = mem64[input_1 + 80] x2
v0 &= mask1
v1 &= mask1
v2 &= mask1
t0 |= v0
t1 |= v1
t2 |= v2
```

The first line `v0 = mem64[input_1 + 0] x2` loads the first and second limb (each taking 32 bits) of the first coordinate of P and broadcasts the value to the lower half and upper half of the 128-bit register v0 using the movddup instruction. The line `v0 &= mask1` performs a bitwise AND of v0 and a mask; the value in the mask depends on whether $s_i = 1$. Finally, v0 is ORed into t0, which is initialized in a similar way as in the cmov-based approach. Similarly, the rest of the lines are for the second and third coordinates. Similar code blocks are repeated 7 more times for $2P, 3P, \ldots, 8P$, and all the code blocks are surrounded by a loop which iterates through all the limbs. In total it takes $3 \cdot 8 \cdot 5 \cdot 2 = 240$ logic instructions. The parallel table-lookup function (inlined in our implementation) takes less than 160 Sandy/Ivy Bridge cycles, which is almost twice as fast as the cmov-based table lookup function.

4   Vectorizing the Ed25519 Signature Scheme

This section describes how the Ed25519 verification is implemented with focus on the challenge of vectorization. Since the public-key generation and signing process, as the Curve25519 public-key generation, is bottlenecked by a fixed-base scalar multiplication on E_T, the reader can check Sect. 3.2 for the implementation strategy.

4.1   Ed25519 Verification

[1] verifies a message by computing the double-scalar multiplication of the form $s_1 P_1 + s_2 P_2$. The double-scalar multiplication is implemented using a generalization of the sliding-window method such that $s_1 P_1$ and $s_2 P_2$ share the doublings. With the same window sizes, we do better by vectorizing the point doubling and point addition functions.

Algorithm 2. The doubling function for twisted Edwards curves

1: **function** GE_DBL_P2(X, Y, Z)
2:     $A \leftarrow X^2; B \leftarrow Y^2$ ▷ batched squarings
3:     $G \leftarrow A - B$
4:     $H \leftarrow A + B$
5:     $C \leftarrow 2Z^2; D = (X + Y)^2$ ▷ batched squarings
6:     $E \leftarrow H - D$
7:     $I \leftarrow G + C$
8:     $X' \leftarrow E \cdot I; Y' \leftarrow G \cdot H$ ▷ batched multiplications
9:     $Z' \leftarrow G \cdot I$
10:    **return** (X', Y', Z')
11: **end function**

On average each verification takes about 252 point doublings, accounting for more than 110000 cycles. There are two doubling functions in our implementation; ge_dbl_p2, which is adapted from the "$\mathcal{E} \leftarrow 2\mathcal{E}$" doubling described in [9], is the most frequently used one; see [9] for the reason to use different doubling and addition functions. On average ge_dbl_p2 is called 182 times per verification, accounting for more than 74000 cycles. The function is summarized in Algorithm 2. Given $(X : Y : Z)$ representing $(X/Z, Y/Z) \in E_T$, the function returns $(X' : Y' : Z') = (X : Y : Z) + (X : Y : Z)$. As in Sect. 3.1, squarings and multiplications are paired whenever convenient. However it is not always possible to do so, as the multiplication in line 9 can not be paired with other operations. The single multiplication slows down the function, and the same problem also appears in addition functions.

Another problem is harder to see. $E = X^2 + Y^2 - (X + Y)^2$ has limbs with upper bound around $4 \cdot 2^{26}$, and $I = X^2 - Y^2 + 2Z^2$ has limbs with upper bound around $5 \cdot 2^{26}$. For the multiplication $E \cdot I$, limbs of either E or I have to be multiplied by 19 (see Sect. 2.2), which can be more than 32 bits. This problem is solved by performing extra carries on limbs in E before the multiplication. The same problem appears in the other doubling function.

In general the computation pattern for verification is not so friendly for vectorization. However, even in this case our software still gains non-negligible speedup over [1,7]. We conclude that the power of vector unit on recent Intel microarchitectures might have been seriously underestimated, and implementors for ECC software should consider trying vectorized multipliers instead of serial multipliers.

References

1. Bernstein, D.J., Duif, N., Lange, T., Schwabe, P., Yang, B.-Y.: High-speed high-security signatures. In: Preneel, B., Takagi, T. (eds.) CHES 2011. LNCS, vol. 6917, pp. 124–142. Springer, Heidelberg (2011). Citations in this document: §1, §1.1, §1.2, §1.2, §1.2, §1.2, §1.2, §1, §1, §1.2, §1.2, §1.2, §1.2, §1.2, §1.2, §1.2, §1.3, §2, §2, §2.1, §1, §2, §2, §2.3, §3.1, §3.2, §3.2, §3.2, §3.2, §3.2, §4.1, §4.1

2. Costigan, N., Schwabe, P.: Fast elliptic-curve cryptography on the cell broad- band engine. In: AFRICACRYPT 2009, pp. 368–385 (2009). Citations in this document: §1, §1.1

3. Bernstein, D.J., Schwabe, P.: NEON crypto. In: Prouff, E., Schaumont, P. (eds.) CHES 2012. LNCS, vol. 7428, pp. 320–339. Springer, Heidelberg (2012). Citations in this document: §1, §1.1, §1.1, §2, §2.2, §2.2, §2.2, §2.2, §3.1, §3.1

4. Bernstein, D.J., Chuengsatiansup, C., Lange, T., Schwabe, P.: Kummer strikes back: new DH speed records. In: EUROCRYPT 2015, pp. 317–337 (2014). Citations in this document: §1.1, §1.1, §1.3, §1.3, §1.3

5. Costello, C., Hisil, H., Smith, B.: Faster compact Diffie-Hellman: endomorphisms on the x-line. In: EUROCRYPT 2014, pp. 183–200 (2014). Citations in this document: §1.1

6. Longa, P., Sica, F., Smith, B.: Four-dimensional Gallant-Lambert-Vanstone scalar multiplication. In: Asiacrypt 2012, pp. 718–739 (2012). Citations in this document: §1.1

7. Andrew, M.: "Floodyberry", Implementations of a fast Elliptic-curve Digital Signature Algorithm (2013). https://github.com/floodyberry/ed25519-donna. Citations in this document: §1.2, §1.2, §1.2, §1.2, §1, §1, §1.2, §1.2, §1.2, §1.2, §1.2, §1.2, §3.2, §4.1

8. Bernstein, D.J., Lange, T. (eds.): eBACS: ECRYPT Benchmarking of Cryptographic Systems (2014). http://bench.cr.yp.to. Citations in this document: §1, §1, §1, §1, §1.2, §1.3

9. Hisil, H., Wong, KK.-H., Carter, G., Dawson, Ed.: Twisted Ed- wards curves revisited. In: Asiacrypt 2008, pp. 326–343 (2008). Citations in this document: §3.2, §4.1, §4.1

10. Fog, A.: Instruction tables (2014). http://www.agner.org/optimize/instruction. tables.pdf. Citations in this document: §2, §2

11. Bernstein, D.J.: Curve25519: new Diffie-Hellman speed records. In: PKC 2006, pp. 207–228 (2006). Citations in this document: §3, §3

12. Oliveira, T., López, J., Aranha, D.F., Rodríguez-Henríquez, F.: Lambda coordinates for binary elliptic curves. In: Bertoni, G., Coron, J.-S. (eds.) CHES 2013. LNCS, vol. 8086, pp. 311 330. Springer, Heidelberg (2013). Citations in this document: §1.3

13. Bos, J.W., Costello, C., Hisil, H., Lauter, K.: Fast cryptography in genus 2. In: Johansson, T., Nguyen, P.Q. (eds.) EUROCRYPT 2013. LNCS, vol. 7881, pp. 194–210. Springer, Heidelberg (2013). Citations in this document: §1.3

14. Petit, C., Quisquater, J.-J.: On polynomial systems arising from a Weil descent. In: ASIACRYPT 2012, pp. 451–466 (2012). Citations in this document: §1.3

15. Semaev, I.: New algorithm for the discrete logarithm problem on elliptic curves (2015). https://eprint.iacr.org/2015/310.pdf. Citations in this document: §1.3

16. Düll, M., Haase, B., Hinterwälder, G., Hutter, M., Paar, C., Sánchez, A.H., Schwabe, P.: High-speed Curve25519 on 8-bit, 16-bit, and 32-bit microcontrollers. Des. Codes Crypt. 77(2), 493–514 (2015). http://cryptojedi.org/papers/ \#mu25519. Citations in this document: §1

17. IANIX. www.ianix.com. Citations in this document: §1

18. Bernstein, D.J.: qhasm sofware package (2007). http://cr.yp.to/qhasm.html. Citations in this document: §3.2

19. Longa, P.: NUMS Elliptic Curves and their Implementation. http://patricklonga. webs.com/NUMS_Elliptic_Curves_and_their_Implementation-UoWashington.pdf. Citations in this document: §1.2
20. Thériault, N.: Index calculus attack for hyperelliptic curves of small genus. In: Asiacrypt 2003, pp. 75–92 (2003). Citations in this document: §1.3

ECC on Your Fingertips: A Single Instruction Approach for Lightweight ECC Design in $GF(p)$

Debapriya Basu Roy$^{(\boxtimes)}$, Poulami Das, and Debdeep Mukhopadhyay

Secured Embedded Architecture Laboratory (SEAL),
Department of Computer Science and Engineering,
Indian Institute of Technology Kharagpur, Kharagpur, India
{deb.basu.roy,debdeep}@cse.iitkgp.ernet.in, poulamidas22@gmail.com

Abstract. Lightweight implementation of Elliptic Curve Cryptography on FPGA has been a popular research topic due to the boom of ubiquitous computing. In this paper we propose a novel single instruction based ultra-light ECC crypto-processor coupled with dedicated hard-IPs of the FPGAs. We show that by using the proposed single instruction framework and using the available block RAMs and DSPs of FPGAs, we can design an ECC crypto-processor for NIST curve *P-256*, requiring only 81 and 72 logic slices on Virtes-5 and Spartan-6 devices respectively. To the best of our knowledge, this is the first implementation of ECC which requires less than 100 slices on any FPGA device family.

Keywords: Elliptic curve · Single instruction · URISC · SBN · FPGA · Hard-IPs

1 Introduction

With the recent boom in ubiquitous computing, specially in Internet-of-Things (IoT), the need of lightweight crypto-algorithms, either at algorithmic or implementation level, has increased significantly. Though the researchers have proposed various lightweight symmetric ciphers, the most popular options for public key cryptography are RSA and Elliptic Curve Cryptography (ECC). ECC based crypto-system is being preferred over its counterpart RSA because of its wonderful property of increased security level per key bit over RSA. Any ECC based protocol or algorithm is based on underlying elliptic curve scalar multiplication whose computation is based on a number of field operations, making it computationally extensive. Software implementations of ECC, running on smart cards or AVR are slow and can become performance bottleneck for many applications. As an alternative, dedicated ECC-crypto processors are being built on hardware platforms like ASICs (Application Specific Integrated Circuits) and FPGAs (Field Programmable Gate Arrays).

D. Mukhopadhyay—This work was partially supported by project from Defence Research and Development Organization (DRDO), India [Sanction No: ERIP/ER/1100420/M/01/1517].

O. Dunkelman and L. Keliher (Eds.): SAC 2015, LNCS 9566, pp. 161–177, 2016.
DOI: 10.1007/978-3-319-31301-6_9

Although ASIC implementations are faster than those based on FPGAs, FPGAs are sometimes preferred over ASIC for cryptographic applications due to its inherent properties of reconfigurability, short time to market and in house security. The entire design cycle of an FPGA based system can be completed inside a single lab unlike ASIC based systems where several different parties are involved in the design cycle. Moreover, modern FPGAs with various device families provide interesting design choices to the designer. Additionally, these FPGAs are now equipped with dedicated hard IPs like DSP blocks, Block RAMs, which when properly utilized results in efficient design of dedicated ECC-based crypto-processors in $GF(p)$ with improved timing performance and reduced area overhead.

There have been many works in the literature which focus on efficient implementation of ECC crypto-processor in $GF(p)$ on FPGAs. An overview of such implementations can be found in [1]. A lightweight ASIC design was reported in [2]. Considerably high speed designs for FPGAs can be found in [3] which is significantly faster than previous designs reported in [4,5]. But, though the proposed design requires much less area compared to the previous designs (1715 logic slices on Virtex-4 platform for NIST P-256), it is still considerably large for lightweight applications. A fast pipelined modular multiplier for ECC field multiplication was proposed in [6], whereas optimized tiling methodology targeting rectangular DSP blocks of Virtex-5 FPGA was proposed in [7]. However, both of them have considerable area overhead, hence can not be applied in lightweight applications.

A lightweight ECC algorithm for RFID tags was presented in [8] and authentication and ID transfer protocols based on lightweight ECC was introduced in [9]. On implementation level, authors have proposed a lightweight architecture, known as *Micro-ECC*, in [10]. The proposed design methodology shows significant improvement in terms of *area-time product* compared to the previous implementation [11–13]. However, *Micro-ECC* was implemented on Virtex-II platform which is no longer a recommended design platform by Xilinx [14]. Moreover, unlike [11–13], *Micro-ECC* architecture does not support generalized ECC scalar multiplication on any prime field. Nevertheless, for fixed P-256 and P-224 curve, the performance of *Micro-ECC* outperforms other by big margin. Lightweight implementation of IPsec protocols comprising implementation of lightweight block cipher PRESENT, lightweight hash function PHOTON and ECC crypto-processor (P-160 and P-256) was presented in [15]. The ECC implementation requires 670 logic slices on Spartan-6 platform for NIST P-256 curve. Consequently, a lightweight architecture supporting both RSA and ECC along with some side channel countermeasure was proposed in [16]. The slice consumption of the proposed design is 1914 logic slices on Virtex-5 platform which is quite low considering dual support of RSA and ECC, provided by the design. As an alternative of standard NIST specified curves, many researchers have recommended use of Edward curve and hyper elliptic curve (HECC). Efficient lightweight implementation of ECC scalar multiplication on such curves can be found in [17,18].

In this paper, we want to propose an alternative single instruction approach for designing lightweight ECC scalar multiplier which has not been adopted in the previous works. It is well known that using a single instruction like *SBN* *(subtract and branch if negative)*, *SUBLEQ (subtract and branch if the answer is* *negative or equal to zero)*, we can construct a Turing complete computer processor. However, though single instruction processor can execute any arithmetical or logical operation, the execution time of some operations become so large that it can not be used in practical scenarios. Hence, a stand alone URISC processor can not be used to design computationally intensive ECC applications. However, in this paper we will show that using the dedicated hard-IPs of FPGA, and with some simple modification of a URISC processor, it is possible to design an immensely lightweight and yet practical ECC architecture.

This architecture is extremely lightweight and to our best of knowledge this is the first implementation of ECC scalar multiplication which requires less than 100 slices on Virtex-5 and Spartan-6 platform. This significant reduction in slice consumption has been achieved by the lightweight architecture of single instruction processor along with intensive usage of hard-IPs of the modern FPGAs. ECC scalar multiplication execution requires to compute and store multiple temporary variables along with the inputs and outputs. This contributes to significant number of register usage and hence increases the slice consumption. In this paper, we will show an alternative design approach where we intensively use the block RAMs and reduce the slice consumption significantly. Further reduction is obtained by replacing the LUT logics with high speed DSP blocks whenever possible. The strategy of using block RAMs to reduce the slice consumption has already been applied for lightweight block ciphers like PRESENT [19], where the authors have shown that block RAM based block cipher design can be extremely lightweight resulting in more slices left for other applications.

Thus the contribution of the present paper can be listed as below:

- We propose a single instruction ECC crypto-processor for NIST P-256 curve, and analyze various challenges along with their solutions that a designer will face while applying single instruction approach in the context of lightweight implementation of ECC designs.
- We show that single instruction based ECC crypto-processor, coupled with intensive usage of block RAMs and DSP blocks, can yield extremely lightweight design for ECC scalar multiplication execution. The proposed processor requires less than 100 slices on both Virtex-5 and Spartan-6 family and involves thorough usage of FPGA hard-IPs.

The rest of the paper is structured as below: Sect. 2 gives a very brief introduction of ECC and single instruction processor. Section 3 gives a detailed description of single instruction processor along with the modifications required for efficient ECC scalar multiplication. Consequently, Sect. 4 focuses on the architecture of the proposed ECC crypto-processor. Next, in Sect. 6, we discuss the timing and area performance of our design followed by conclusion in Sect. 7.

2 Preliminaries

In this section, we will give a brief summary of ECC and single instruction processors.

2.1 Elliptic Curve Cryptography

As we have previously mentioned, elliptic curve cryptography (ECC) is a public key cryptography based on elliptic curves and finite field. Security of ECC depends upon the mathematical intractability of discrete logarithm of a point in elliptic curve with respect to a known base point.

ECC in finite field $GF(p)$ is defined by the following equation

$$y^2 = x^3 + ax + b; a, b \in GF(p), b \neq 0. \tag{1}$$

Scalar multiplication is the most important operation in ECC for performing key agreement or digital signature schemes. Given a point P on an elliptic curve and a scalar k, scalar multiplication is computed by adding the point P, k times. The basic algorithm used for scalar multiplication is *Double-and-Add* algorithm, defined in Algorithm 1 in Appendix A, which shows that scalar multiplication is executed by a repeated sequence of point doubling and point additions. It is advantageous to use standard projective coordinates [20] for ECC scalar multiplication as it requires less number of field inversion operations compared to affine coordinate system. In this paper, we have used standard projective coordinates during implementation of ECC scalar multiplication.

Now, each point addition and point doubling operation involves multiple field multiplication operation, making it most critical operation for efficient scalar multiplication execution. NIST specified curves are efficient for hardware implementation as modular reduction operation in those curves are simple as it involves a combination of few addition and subtraction. The fast modular reduction algorithm for NIST P-256 is shown in Appendix A.

In our proposed design, we have concentrated on the NIST P-256 curve. Nevertheless, our approach can be extended to other NIST certified curves also.

2.2 Single Instruction Processor

The concept of single instruction computer or one instruction set computer (OISC) was first proposed in [21]. It has been shown in [22] that using just a single instruction it is possible to create a Turing complete machine. The idea of applying URISC on cryptographic applications was proposed in [23]. In the similar direction, application of one instruction set computer on encrypted data computation was analyzed in [24], but in that paper the authors have investigated OISC in the context of homomorphic encryption and have not considered elliptic curves, which is the precise objective of the present paper.

A standard single instruction processor can be designed by instruction like

1. ADDLEQ (Add the operands and branch if the answer is less than or equal to zero)
2. SUBLEQ (Subtract the operands and branch if the answer is less than or equal to zero)
3. SBN (Subtract the operands and branch if the answer is less than zero)
4. RSSB (Reverse subtract and skip if borrow)
5. SBNZ (Subtract the operands and branch if the answer is non-zero)

The main advantage of OISC is that we don't need any instruction decoding mechanism, which makes the processor architecture exceptionally simple and lightweight. The instruction format of a standard OISC is shown in Fig. 1.

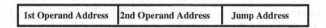

| 1st Operand Address | 2nd Operand Address | Jump Address |

Fig. 1. Instruction format of OISC

For the present work, we have chosen SBN as the single instruction. However, the proposed design strategy can be tweaked to adopt any of the above described instructions. The operation of SBN instruction is described in Table 1 (code 1.1):

Table 1. SBN and addition using SBN

Code 1.1 SBN: Subtract and Branch if negative	Code 1.2 Addition using SBN

SBN A,B,C
D.Mem[A]=D.Mem[A]-D.Mem[B]
if(D.Mem[A]<0)//D.Mem=Data Memory
jump to C
else
jump to next instruction
	ADD C,A,B //D.Mem[C]=D.Mem[A]+D.Mem[B]
	1. SBN X,X,2 // D.Mem[X]=0
	2. SBN X,A,3 // D.Mem[X]=-D.Mem[A]
	3. SBN X,B,4 // D.Mem[X]=-D.Mem[A]-D.Mem[B]
	4. SBN C,C,5 // D.Mem[C]=0
	5. SBN C,X,6 // D.Mem[C]=D.Mem[A]+D.mem[B]

Using this instruction, we can execute any mathematical, logical, flow-control, memory control or load-store type of instruction. For example, in Table 1 (code 1.2), we will show how to perform addition of two operands using SBN instruction.

In this section we have given a brief idea about elliptic curves and OISC. In the next section, we will go into more details of OISC based on SBN instruction and will analyze it from the point of view of elliptic curve applications.

3   SBN-OISC and Elliptic Curve Scalar Multiplication

In the previous subsection, we have given a brief idea about the ECC and OISC, based on SBN instruction (from hereafter we will refer this as SBN-OISC).

In this section we will focus more on SBN-OISC in the context of ECC implementation. We will identify the critical challenges that the designer will face while implementing ECC using SBN-OISC and will provide the solutions to tackle those challenges. We will first describe a stand-alone SBN-OISC processor in the next subsection

3.1   Stand-Alone SBN-OISC Processor

A stand-alone SBN-OISC processor is shown in Fig. 2. The main components of a SBN processor are characterized below:

- **Instruction Memory:** Instruction memory stores the instructions to be executed and can be implemented on FPGA using *block RAMs*, configured as single port ROM. In the Fig. 2, the instruction memory can store up to 2^{11} number of instructions and each instruction is 21 bits wide. The format of the instruction is similar to Fig. 1, where address of both the operands are 5 bits wide and the length of the jump address is 11 bits.
- **Data Memory:** Data memory stores the final result of any computation, along with the input and all the temporary results, required during the computation. This has been implemented using *block RAM*, configured as true dual port RAM. The data memory has space of 32 entries, each of which are 260 bits wide. While implementing scalar multiplication in NIST P-256, the partially modular reduced output can be of size 259 bits which can be represented by 260 bits signed representation. Hence we have chosen the data path to be 260 bits wide.
- **ALU:** Arithmetical logical unit (ALU) of SBN-OISC contains a subtracter, which computes difference between the two inputs. If the result is negative, program counter gets updated by the jump address, specified in the instruction. Otherwise, the program counter gets updated by the immediate next instruction.

The above described architecture is simple and extremely lightweight, requiring 66 logic slices on a Virtex-5 platform. But, as we will show in the next subsection, further optimization of ECC operation can be achieved by introducing different variants of SBN instruction. In the next subsection, we will mainly concentrate on different variants of SBN instructions and will discuss how these different versions of SBN can accelerate ECC implementation.

3.2   Instruction Level Optimizations

Generally, though an OISC processor executes only a single instruction, it is possible to realize different versions of that single instruction to accelerate the desired operation. This approach helps us to reduce the size of instruction memory and consequently, results in faster execution of the aimed design. This is extremely helpful for computationally intensive ECC applications, as illustrated in the following discussion.

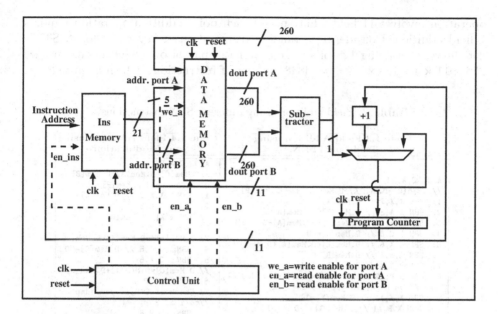

Fig. 2. Architecture of SBN-OISC

Switching Off Memory Write-Back. When we consider traditional SBN instruction $(SBN\ A, B, C)$, the memory location A always get updated by the result $D.Mem[A] - D.Mem[B]$ (D.Mem is the data memory). But we can reduce the required number of instruction count considerably, if we can switch off this memory write-back operation in some cases.

Let us consider a prime field addition operation. We assume that we need to add two operands stored at memory location A and B and the modulus of the field is stored at memory location P. In Table 2 (code 1.3) shows the realization of this operation using SBN. In this case, we can see that to implement prime field addition we will require 11 SBN instructions. Now, if each SBN instruction execution takes n clock cycles, total clock cycles requirement for field operation will be $11n$ clock cycles.

Now, let us consider the scenario shown in Table 2 (code 1.4), where we consider two variations of SBN instruction: SBN$_{nw}$ and SBN$_w$. $SBN_w\ A, B, C$ instruction is similar to normal SBN instruction, where memory location A get updated by the value $D.Mem[A] = D.Mem[A] - D.Mem[B]$. But in case of $SBN_{nw}\ A, B, C$, memory location A does not get updated and continue to store the previous value. If we use a combination of SBN_w and SBN_{nw} to implement prime field addition, we will need only 7 instruction as shown in Table 2 (code 1.4). Thus, we have a saving of 4 instructions if we use the strategy depicted in Table 2 (code 1.4). Similar saving can be obtained for field subtraction operation also. Now, in the case of ECC scalar multiplication, where for each key bit we need to do point doubling and if the key bit is 1, we need to do point addition, this saving translates into significant speed up. Each point doubling operation involves 11 field addition and each point addition operation requires 7 field addition. Considering a random distribution of key value for NIST P-256 curve, containing 128 bits of zero and 128 bits of one, we can save around $256 \times 11 \times 4 + 128 \times 7 \times 4 = 14848$ number of instructions, which is quite large.

Table 2. Field addition using different SBN instructions

Code 1.3 Field Addition using Traditional SBN	Code 1.4 Field Addition using our Modification
| ```
ADD_p C,A,B
//D.Mem[C]=D.Mem[A]+D.Mem[B] mod D.Mem[P]
1.  SBN X,X,2 // D.Mem[X]=0
2.  SBN X,A,3 // D.Mem[X]=-D.Mem[A]
3.  SBN X,B,4 // D.Mem[X]=-D.Mem[A]-D.Mem[B]
4.  SBN C,C,5 // D.Mem[C]= 0
5.  SBN C,X,6 // D.Mem[C]=D.Mem[A]+D.Mem[B]
6.  SBN R,R,7 // D.Mem[R]=0
7.  SBN R,X,8 // D.Mem[R]=D.Mem[A]+D.Mem[B]
8.  SBN R,P,12 // D.Mem[R]=D.Mem[R]-D.Mem[P],
        // on negative jump to ins. 12
9.  SBN X,X,10 // D.Mem[X]=0
10. SBN X,R,11 // D.Mem[X]= -D.Mem[R]
11. SBN C,X,12 // D.Mem[C]= D.Mem[R]
12. SBN ...   // Next Operation Code
``` | ```
ADD_p C,A,B
//D.Mem[C]=D.Mem[A]+D.Mem[B]
        mod D.Mem[P]
1.  SBN_w X,X,2 // D.Mem[X]=0
2.  SBN_w X,A,3
// D.Mem[X]=-D.Mem[A]
3.  SBN_w X,B,4
// D.Mem[X]=-D.Mem[A]-D.Mem[B]
4.  SBN_w C,C,5 // D.Mem[C]= 0
5.  SBN_w C,X,6
// D.Mem[C]=D.Mem[A]+D.Mem[B]
6.  SBN_nw C,P,8
// Check if C<P,if yes jump
//to ins. 8
7.  SBN_w C,P,8
// D.Mem[C]=D.Mem[C]-D.Mem[P]
8.  SBN ...
// Next Operation Code
``` |

Right Shift on SBN Processor. Right shift is an important operation for elliptic curve scalar multiplication execution as it is required during the field inversion operation. Right shift operation can be executed through SBN instruction by repeated subtraction of the operand. For example, if we wish to right shift an operand by 1 bit position, we need to subtract the operand by 2 until the subtraction result become less than 2. Now as we are concentrating on NIST P-256 curve, the operands are typically 256 bits long, making the sequence of repeated subtraction operation extremely time consuming. On the other hand, shifter design on the FPGA has zero LUT overhead if the number of bits to be shifted are fixed. Hence, it is better if we implement right shift operation using a dedicated right shifter module instead of using SBN.

To facilitate this in our architecture, we have introduced another flag (SBN_{rs} and $SBN_{\overline{rs}}$) in our instruction format. When this flag is set, the dedicated right shifter module reads the operand and shift it right by one bit position.

Shifting Key Register. As we have stated in Algorithm 1, the elliptic curve scalar multiplication operation involves point addition and point doubling operation. Point doubling happens for every key bit, but point addition happens only when the key bit value is one. Hence we need to scan the key value bit by bit to execute scalar multiplication operation. On a standard processor this can be implemented using shift and logical AND operation. However, executing logical operations using only SBN instruction is again time consuming and hence practically infeasible.

To solve this challenge, we have used a dedicated key register, separate from the data memory shown in Fig. 2. Also we have introduced another flag in our instruction format (SBN_{ks} and $SBN_{\overline{ks}}$), which when enabled will left shift the key register by one bit. The shifted out bit from the key register will decide whether point doubling or point addition will occur.

Multiplication Using SBN. Field multiplication using SBN is carried out by repeated addition. For example to multiply operand A with Operand B we need to add operand A, B times. Now we have already shown how to implement field addition using SBN in Table 2. To complete the multiplication operation, we need to run that code, B times using a loop. Now, in the worst case scenario, the operands value in NIST P-256 curve are in the range of 2^{256}, which makes repeated addition implementation impractical as the loop need to run 2^{256} times. Hence, we can not implement field multiplication using only SBN for ECC scalar multiplication.

To solve this problem, we have designed a lightweight multiplier using DSP blocks, which acts as an external multiplier core and execute the field multiplication operation. However, to reset this multiplier core and to provide operand data to the multiplier we need another variant of SBN instruction, which we refer as SBN_{mul} and $SBN_{\overline{mul}}$. The SBN_{mul} instruction resets the multiplier, whereas $SBN_{\overline{mul}}$ initiates the multiplication operation. The detailed description of this external multiplier core along with its interfacing with the SBN-OISC processor is provided in the next section.

In this section, we have discussed about different variations of SBN instruction, that is required for optimized ECC implementation. The list of these variants can be found in Table 3, where we have combined the discussed SBN instruction variations. It should be noted that when we reset the multiplier we don't need any memory write-back, as ALU output does not matter in that situation. Similarly when we are shifting the key register or doing the right shift operation, no memory write-back is needed.

Table 3. Different variant of SBN instruction

Instruction	Memory write-back	Multiplier reset	Key-shift	Right-shift
$SBN_{w\overline{mulksrs}}$	✓	x	x	x
$SBN_{nw\overline{mulksrs}}$	x	x	x	x
$SBN_{nw\overline{mulks}rs}$	x	x	✓	x
$SBN_{w\overline{mulks}rs}$	✓	x	x	✓
$SBN_{nwmul\overline{ksrs}}$	x	✓	x	x

To adopt these variations of SBN instructions in our architecture we also need to modify the instruction format. The modified instruction format is shown in Fig. 3. In the next section, we will discuss the modified SBN architecture which can support these instruction variants, along with field multiplier architecture. We would like to stress that though we are introducing different variants of SBN instruction, we are still using same ALU for each of this variant. Hence these variants are part of the same SBN instruction, with different flag values as shown in Table 3.

24	23	22	21	20		15		10	
w/$\overline{\text{nw}}$	mul/$\overline{\text{mul}}$	ks/$\overline{\text{ks}}$	rs/$\overline{\text{rs}}$	1st Operand Address		2nd Operand Address		Jump Address	
1	1	1	1	5		5		11	

Fig. 3. Modified instruction format of SBN-OISC

4   Lightweight Field Multiplier for SBN-OISC

As we have stated in the previous sections, we need to provide a dedicated light weight multiplier core to the SBN-OISC processor for efficient execution of the ECC operations. In this section we will focus on the architecture of this dedicated field multiplier and will describe the design strategies behind the proposed filed multiplier methodology.

The architecture of the field multiplier is shown in Fig. 4. As we can see, the architecture requires two DSP blocks, one for integer multiplication and another one for modular reduction operation. DSP blocks of Virtex-5 FPGA can support 25×18 signed multiplication. It can also provide 48 bit adder/accumulator support. For our implementation, we have used DSP block as 16×16 unsigned multiplier, configured in multiply and accumulate mode. Moreover, during addition operation, DSP block is configured as 32 bit adder.

We will first focus on the integer multiplier and will follow it with a discussion on the modular reduction operation.

4.1   Integer Multiplication

The integer multiplier receives two 256 bits long operands as input. The operands are divided into 16 bit words and are passed to the first DSP block through two multiplexers. The DSP block is configured in *multiply and accumulate mode* and support two different operations. In the first operation, DSP block computes $A * B + P$ where A and B are two multiplexer output and P is the accumulator output. This operation computes the summation of the partial products which are aligned with each other. Let us illustrate this with a small example in Eq. 2.

Let us consider a 32 bit multiplication of two operands $R(= r_1 2^{16} + r_0)$ and $S(= s_1 2^{16} + s_0)$, divided into 16 bit words. In this scenario the addition of

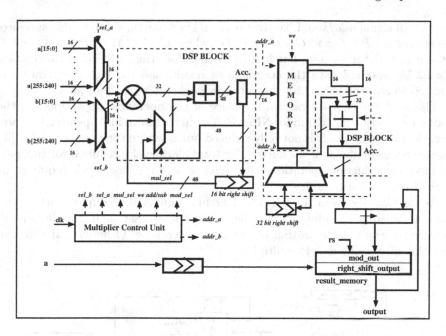

Fig. 4. Architecture of Lightweight Field Multiplier

partial products $r_1 s_0$ and $r_0 s_1$ are carried out by the operation $A * B + P$ as these partial products are aligned to each other. But for the partial products which are shifted, DSP blocks operate using the second instruction $A * B + C$, where $C = P >> 16$. The result is stored in memory of dimension 16×32 which is implemented using a block RAM configured as true dual port RAM.

$$R \times S = \sum_{j=0}^{1} r_j 2^{j*16} \times \sum_{i=0}^{1} s_i 2^{i*16} = r_0 s_0 + (r_1 s_0 + r_0 s_1) 2^{16} + r_1 s_1 2^{32} \quad (2)$$

The integer multiplication requires 256 iteration of the DSP block, along with three clock cycles for updating the data memory. Hence the total clock cycle count for integer multiplication is 259.

4.2   Modular Reduction

Once the memory is loaded with the integer multiplication result, modular reduction operation is initiated. For NIST curves, modular reduction operation requires a combination of addition and subtraction operation as shown in Algorithm 2 in Appendix A. Now in Algorithm 2, the modular reduction operation needs to add operands $T, S_1, S_2, S_3.S_4$ and subtract $D_1.D_2, D_3, D_4$ from them. We have separated the operands in 32 bit words and have used a DSP adder to execute the addition/subtraction operations. The memory produces 32 bits of output in a single clock cycle, which are added or subtracted depending on the control signal add/sub. Like the previous DSP blocks, this one also supports two operation: $P \pm C$ and $C + CONCAT$, where $CONCAT = P >> 32$ and P is the accumulator output. The first operation does the addition or subtraction of a 32 bit operand with the accumulator result, and the second operation is required to add the carry bits generated from the previous additions.

The addition and subtraction sequence of the operands are decided by the modular reduction algorithm for NIST P-256 curve, shown in Appendix A. Moreover, the produced result is not fully reduced but is within the range $[-4p, 5p]$ [3], where p is the modulus of the curve. The total clock cycles required for this partial modular reduction operation is 68, making the total clock cycle requirement for field multiplication 327.

As we have shown in the Fig. 4, our architecture is also coupled with a dedicated right shifter module. Now, when the rs flag is set high, the design will produce the right shifter output of the input operand a. Otherwise, it will produce the output of the field multiplier.

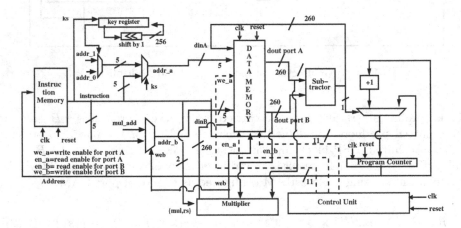

Fig. 5. ECC SBN-OISC processor architecture

5   Complete ECC SBN-OISC Processor

In this section, we will present the detailed description of our proposed ECC SBN-OISC processor. The complete architecture of the processor is shown in Fig. 5. The architecture and the working of the proposed processor is nearly similar to the stand alone SBN processor shown in Fig. 2 with some few modifications which are described below.

The ECC SBN-OISC processor is coupled with the multiplier core described in the previous section. Multiplier core is initiated by the mul flag of the instruction. As long as the mul flag is set to one, the multiplier stays in its initial stage. Once it is set low, the multiplier starts its operation and produces the partially reduced output along with signal *web* which indicates the completion of multiplication operation. In the stand alone SBN (Fig. 2), the data memory is updated only through port A. But in our case, we are also using the unused port B for writing the multiplier output into the memory. It must be noted that when the *rs* flag is set high, the multiplier module produces right shifted output of input, available through port A.

As we have mentioned earlier, we introduced a flag *ks* in our instruction format for shifting the key register. Key is stored in a different register which goes though a single bit left shift when *ks* flag is set high. If the MSB of the key bit is one, we select the address of the memory location containing value 1 (addr_1) and pass it to the data memory. Otherwise if the MSB bit is zero, we select the memory location containing value 0 (addr_0). Once this is done we can easily switch between point doubling and addition operation depending upon the memory location passed to the data memory.

The ALU of the proposed SBN-OISC processor is a subtracter, implemented through cascaded DSP blocks. The subtraction operation requires 6 clock cycles to be completed. Instruction fetch, memory read and memory write-back require single clock cycle for each operation. Hence total clock cycle required for a single SBN instruction requires 9 clock cycles.

6   Result and Comparison

In this section we will analyze the performance of the proposed ECC SBN-OISC processor in terms of timing and area. Table 4 shows the timing and area performance of the proposed processor. As we can see, the slice count required by the design for both Virtex-5 and Spartan-6 is very small. This is achieved by in-depth usage of block-RAMs and DSP blocks. The stand alone SBN processor is itself very lightweight, and the dedicated multiplier core is designed by judicious use of DSPs and block RAMs making the slice count extremely small. The block RAMs are used to implement both data and instruction memory of the SBN-OISC processor. Moreover all the temporary storages along with control units are also implemented through block RAMs which increases the block RAM consumption, but reduces the slice count considerably. A designer can choose a budget of slices and block RAMs and then can design the ECC crypto-processor according to that budget. In this paper, we wanted to explore the limit up-to which we can reduce the slice count by increasing the block RAM usage. The result in Table 4 shows that saving is significant in terms of slice usage and hence the objective of the paper is achieved.

Table 5 shows the comparison with the previous results. Among the previous work, the design proposed in [3] targets high speed architecture and is not intended for lightweight applications. Apart from that, the proposed ECC SBN-OISC processor shows comparable performance in terms of area and time product. But it is unfair to directly compare the proposed design and the previous designs [10, 11] as they were implemented on Virtex-II pro which is extremely inefficient in comparison with Virtex-5 device family. However as FPGA devices

Table 4. Area and timing performance of the proposed ECC SBN-OISC processor

Platform	Freq. (MHz)	Slices	LUTs	Flip-Flops	DSP for ALU	DSP for Multiplier	Block-RAM	Time (ms)
Virtex-5	171.5	81	212	35	6	2	22	11.1
Spartan-6	156.25	72	193	35	6	2	24	12.2

Table 5. Comparison of ECC SBN-OISC processor with existing designs

Reference	Slices	MULTs	BRAMs	Freq (MHz)	Latency (ms)	FPGA
Micro-ECC P-256 16 bit [10]	773	1	3	210	10.02	Virtex-II Pro
Micro-ECC P-256 32 bit [10]	1158	4	3	210	4.52	Virtex-II Pro
[11] 16 bit any prime curve	1832	2	9	108.20	29.83	Virtex-II Pro
[11] 32 bit any prime curve	2085	7	9	68.17	15.76	Virtex-II Pro
[3] P-256	1715	32 (DSP)	11	490	.62	Virtex-4
[15] P-256	221	1	3	Not shown	Not shown	Spartan-6
Present work, P-256	81	8(DSP)	22	171.5	11.1	Virtex-5
Present work, P-256	72	8(DSP)	24	156.25	12.2	Spartan-6

has evolved significantly in the last decade, there is a need to update design strategies which will be efficient on these modern FPGAs. Additionally, these old FPGA families are no longer recommended for new designs by Xilinx. Motivated by these reasons, we have chosen Virtex-5 and Spartan-6 as our implementation platform, as these two FPGA family though not much new, are equipped with most of the modern hard-IPs, present in the FPGAs. The proposed processor is also much faster when compared with lightweight software libraries for ECC like *TinyECC* [25]. The developed architecture is the first implementation which has reduced the slice requirement of an ECC processor to be less than 100 on Virtex-5 and Spartan-6 device family. The results shown here are obtained after post place and route analysis on Xilinx ISE.

7   Conclusion

In this paper we have merged two design strategies to create an extremely light-weight ECC crypto-processor for scalar multiplication in NIST P-256 curve. The first strategy was to use a single instruction processor (ECC SBN-OISC processor) to create lightweight framework for ECC scalar multiplication. Then we have equipped this processor with dedicated field multiplier along with some simple modification of the processor architecture and instruction format to make the scalar multiplication operation practical time feasible. The second strategy is to use the dedicated hard-IPs of the FPGA to reduce the slice consumption further. We have shown that by thorough usage of DSP blocks and block RAMs, the slice requirement decreases significantly. For Virtex-5 and Spartan-6, we have been able to achieve less than 100 slice consumption. To the best of our knowledge, this is the first implementation which has been able to achieve this feat.

A   Appendix 1

Here we will show two algorithm. The first algorithm is for ECC scalar multiplication using *double and add* methodology, shown in Algorithm 1.

Algorithm 1. Double-and-Add Algorithm

Data: Point P and scalar $k = k_{m-1}, k_{m-2}, k_{m-3}...k_2, k_1, k_0$, where $k_{m-1} = 1$
Result: $Q = kP$
1 $Q = P$
2 **for** $i = m - 2$ *to* 0 **do**
3 | $\quad Q = 2Q$ (Point Doubling)
4 | $\quad$ **if** $k_i = 1$ **then**
5 | $\quad\quad$ $Q = Q + P$ (Point Addition)

Next, we will present NIST specified fast algorithm for modular reduction in NIST P-256 curve, shown in Algorithm 2.

Algorithm 2. Fast Modular Reduction Algorithm for NIST P-256 Curve

---

Data: 512 bit product C represented as $C = C_{15}||C_{14}||...||C_0$, where each C_i is
 $\quad\quad$ a 32 bit integer, $i \in \{0, 15\}$
Result: $P = C$ mod P-256
1 $T = (C_7||C_6||C_5||C_4||C_3||C_2||C_1||C_0)$
2 $S_1 = (C_{15}||C_{14}||C_{13}||C_{12}||C_{11}||0||0||0)$
3 $S_2 = (0||C_{15}||C_{14}||C_{13}||C_{12}||0||0||0)$
4 $S_3 = (C_{15}||C_{14}||0||0||0||C_{10}||C_9||C_8)$
5 $S_4 = (C_8||C_{13}||C_{15}||C_{14}||C_{13}||C_{11}||C_{10}||C_9)$
6 $D_1 = (C_{10}||C_8||0||0||0||C_{13}||C_{12}||C_{11})$
7 $D_2 = (C_{11}||C_9||0||0||C_{15}||C_{14}||C_{13}||C_{12})$
8 $D_3 = (C_{12}||0||C_{10}||C_9||C_8||C_{15}||C_{14}||C_{13})$
9 $D_4 = (C_{13}||0||C_{11}||C_{10}||C_9||0||C_{15}||C_{14})$
10 $P = T + 2S_1 + 2S_2 + S_3 + S_4 - D_1 - D_2 - D_3 - D_4$ mod P-256

References

1. Daly, A., Marnane, W., Kerins, T., Popovici, E.: An FPGA implementation of a GF(p) ALU for encryption processors. Microprocess. Microsyst. **28**(56), 253–260 (2004). Special Issue on FPGAs: Applications and Designs
2. Batina, L., Mentens, N., Sakiyama, K., Preneel, B., Verbauwhede, I.: Low-cost elliptic curve cryptography for wireless sensor networks. In: Buttyán, L., Gligor, V.D., Westhoff, D. (eds.) ESAS 2006. LNCS, vol. 4357, pp. 6–17. Springer, Heidelberg (2006)

3. Güneysu, T., Paar, C.: Ultra high performance ECC over NIST primes on commercial FPGAs. In: Oswald, E., Rohatgi, P. (eds.) CHES 2008. LNCS, vol. 5154, pp. 62–78. Springer, Heidelberg (2008)

4. Satoh, A., Takano, K.: A scalable dual-field elliptic curve cryptographic processor. IEEE Trans. Comput. **52**, 449–460 (2003)

5. Orlando, G., Paar, C.: A scalable $GF(p)$ elliptic curve processor architecture for programmable hardware. In: Koç, Ç.K., Naccache, D., Paar, C. (eds.) CHES 2001. LNCS, vol. 2162, pp. 356–371. Springer, Heidelberg (2001)

6. Alrimeih, H., Rakhmatov, D.: Pipelined modular multiplier supporting multiple standard prime fields. In: 2014 IEEE 25th International Conference on Application-Specific Systems, Architectures and Processors (ASAP), pp. 48–56, June 2014

7. Roy, D.B., Mukhopadhyay, D., Izumi, M., Takahashi, J., Multiplication, T.B.: An efficient strategy to optimize DSP multiplier for accelerating prime field ECC for NIST curves. In: The 51st Annual Design Automation Conference, DAC 2014, San Francisco, CA, USA, 1–5 June 2014, pp. 177:1–177:6 (2014)

8. Kim, C.-J., Yun, S.-Y., Park, S.-C.: A lightweight ECC algorithm for mobile RFID service. In: Proceedings of the 5th International Conference on Ubiquitous Information Technologies and Applications (CUTE 2010), pp. 1–6, December 2010

9. He, D., Kumar, N., Chilamkurti, N., Lee, J.-H.: Lightweight ECC based RFID authentication integrated with an ID verifier transfer protocol. J. Med. Syst. **38**(10), 116 (2014)

10. Varchola, M., Güneysu, T., Mischke, O.: MicroECC: a lightweight reconfigurable elliptic curve crypto-processor. In: International Conference on Reconfigurable Computing and FPGAs, ReConFig 2011, Cancun, Mexico, November 30–December 2, 2011, pp. 204–210 (2011)

11. Vliegen, J., Mentens, N,. Genoe, J., Braeken, A., Kubera, S., Touhafi, A., Verbauwhede, I:. A compact FPGA-based architecture for elliptic curve cryptography over prime fields. In: 21st IEEE International Conference on Application-Specific Systems Architectures and Processors, ASAP 2010, Rennes, France, 7–9 July 2010, pp. 313–316 (2010)

12. Tawalbeh, L.A., Mohammad, A., Gutub, A.A.-A.: Efficient FPGA implementation of a programmable architecture for GF(p) elliptic curve crypto computations. Signal Process. Syst. **59**(3), 233–244 (2010)

13. Ghosh, S., Alam, M., Chowdhury, D.R., Gupta, I.S.: Parallel crypto-devices for GF(P) elliptic curve multiplication resistant against side channel attacks. Comput. Electr. Eng. **35**(2), 329–338 (2009)

14. Xilinx Inc.: Virtex-II and Virtex-II Pro X FPGA User Guide, 14 February 2011

15. Driessen, B., Güneysu, T., Kavun, E.B., Mischke, O., Paar, C., Pöppelmann, T.: IPSecco: a lightweight and reconfigurable IPSec core. In: International Conference on Reconfigurable Computing and FPGAs, ReConFig 2012, Cancun, Mexico, 5–7 December 2012, pp. 1–7 (2012)

16. Pöpper, C., Mischke, O., Güneysu, T.: MicroACP - a fast and secure reconfigurable asymmetric crypto-processor. In: Goehringer, D., Santambrogio, M.D., Cardoso, J.M.P., Bertels, K. (eds.) ARC 2014. LNCS, vol. 8405, pp. 240–247. Springer, Heidelberg (2014)

17. Himmighofen, A., Jungk, B., Reith, S.: On a FPGA-based method for authentication using edwards curves. In: 8th International Workshop on Reconfigurable and Communication-Centric Systems-on-Chip (ReCoSoC), Darmstadt, Germany, 10–12 July 2013, pp. 1–7 (2013)

18. Fan, J., Batina, L., Verbauwhede, I.: Light-weight Implementation options for curve-based cryptography: HECC is also ready for RFID. In: ICITST, pp. 1–6. IEEE (2009)
19. Kavun, E.B., Yalcin, T.: RAM-based ultra-lightweight FPGA implementation of PRESENT. In: International Conference on Reconfigurable Computing and FPGAs (ReConFig 2011), pp. 280–285, November 2011
20. Hankerson, D., Menezes, A.J., Vanstone, S.: Guide to Elliptic Curve Cryptography. Springer, New York (2003)
21. Mavaddat, F., Parhamt, B.: URISC: the ultimate reduced instruction set computer. Int. J. Electr. Eng. Educ. **25**, 327–334 (1988)
22. Gilreath, W.F., Laplante, P.A.: Computer Architecture : A Minimalist Perspective. The Springer International Series in Engineering and Computer Science. Springer, New York (2003)
23. Naccache, D.: Is theoretical cryptography any good in practice? In: CHES (2010)
24. Tsoutsos, N.G., Maniatakos, M.: Investigating the application of one instruction set computing for encrypted data computation. In: Gierlichs, B., Guilley, S., Mukhopadhyay, D. (eds.) SPACE 2013. LNCS, vol. 8204, pp. 21–37. Springer, Heidelberg (2013)
25. Liu, A., Ning, P., Tinyecc,: A configurable library for elliptic curve cryptography in wireless sensor networks. In: IPSN, pp. 245–256. IEEE Computer Society (2008)

Exploring Energy Efficiency of Lightweight Block Ciphers

Subhadeep Banik[1]([⊠]), Andrey Bogdanov[1], and Francesco Regazzoni[2]

[1] DTU Compute, Technical University of Denmark, Kongens Lyngby, Denmark
{subb,anbog}@dtu.dk
[2] ALARI, University of Lugano, Lugano, Switzerland
regazzoni@alari.ch

Abstract. In the last few years, the field of lightweight cryptography
has seen an influx in the number of block ciphers and hash functions
being proposed. One of the metrics that define a good lightweight design
is the energy consumed per unit operation of the algorithm. For block
ciphers, this operation is the encryption of one plaintext. By studying the
energy consumption model of a CMOS gate, we arrive at the conclusion
that the energy consumed per cycle during the encryption operation of an
r-round unrolled architecture of any block cipher is a quadratic function
in r. We then apply our model to 9 well known lightweight block ciphers,
and thereby try to predict the optimal value of r at which an r-round
unrolled architecture for a cipher is likely to be most energy efficient.
We also try to relate our results to some physical design parameters
like the signal delay across a round and algorithmic parameters like the
number of rounds taken to achieve full diffusion of a difference in the
plaintext/key.

Keywords: AES · Lightweight block cipher · Low power/energy circuits

1 Introduction

In the last few years, we have assisted to the pervasive diffusion of embedded
and smart devices, touching every aspect of our lives. These devices, are often
used for sensitive applications, such as the ones related to access control, bank-
ing and health, and they are often connected to create what is called internet
of things. The security needs of these applications lead to the creation of the
research area of Lightweight Cryptography, which aims at designing and implementing security primitives fitting the needs of extremely constrained devices.
Two main approaches can be followed to achieve this goal: designing new algorithms to be implemented into constrained devices, or trying to implement standards and known algorithms in a lightweight fashion, eventually relaxing the
performance constraints. Examples of the first approach are the large number of
algorithms proposed in recent years, such as HIGHT [16], KATAN [8], Klein [13],
LED [14], Noekeon [10], Present [6], Piccolo [21], Prince [7], Simon/Speck [3]

© Springer International Publishing Switzerland 2016
O. Dunkelman and L. Keliher (Eds.): SAC 2015, LNCS 9566, pp. 178–194, 2016.
DOI: 10.1007/978-3-319-31301-6_10

and TWINE [22]. Possible examples of the second approach are implementations of the Advanced Encryption Standard algorithm (AES) [11], SHA-256 [1], or Keccak [4].

Focusing on block ciphers in particular, it is important to notice that AES still remains the preferred choice for providing security also in constrained devices, even if some lightweight algorithms are now standardized. For this reason, several implementations of AES and its basic transformations (such as S-boxes) targeting low area and low power were proposed in the past, for example, the implementation of Feldhofer *et al.* [12] and the one of Moradi *et al.* [19]. The first design is based on a 8-bit datapath, and occupies approximately 3400 Gate Equivalents (GE). The second design features a mixed data path and requires approximately 2400 GE. The work of Hocquet *et al.* [15] discusses the silicon implementation of low power AES. The authors showed that by exploiting technological advances and algorithmic optimization the AES core, can consume as little as 740 pJ per encryption.

Despite a large number of previous works targeting area and power, only limited efforts were devoted to the optimization of the energy parameter. Energy and power are, for obvious reasons, correlated parameters. Power is the amount of energy consumed per unit time or simply the rate of energy consumption. More specifically, energy consumption is a measure of the total electrical effort expended during the execution of an operation, and the total energy consumed is essentially the time integral of power. However, being directly linked with the battery life or the amount of electrical work to be harvested, energy, rather than power, would become a more relevant parameter for evaluating the suitability of a design. In fact, energy is a much stricter constraint for future cyber-physical systems as well as for the next generation of implantable devices.

Designing for low energy can be significantly different than designing for low power. Furthermore, there is no guarantee that low power architectures would lead to low energy consumption. For instance, block ciphers implemented using smaller datapath and aggressively exploiting serialization to reuse components, result generally in smaller power consumption compared to round based designs having datapath as large as the blocksize of the cipher. However, serial implementations have high latency, which can be significantly larger compared to round based designs. As a result, the energy consumed per encryption for serial designs could be much higher than the corresponding figure for round based designs.

Starting with the AES algorithm, in this work, we carry out a complete exploration of the implementation choice of block ciphers concentrating on their energy consumption, discussing and evaluating the design choice of each round transformation, and the best trade-off between datapath and serialization. From the detailed analysis of this exploration, we extract a model for the energy consumption of a circuit, using as reference, a number of lightweight algorithms recently proposed.

The most significant previous works on this area are the one of Kerckhof *et al.* [17] and the one of Batina *et al.* [2]. The first work, addresses the problem of efficiency for lightweight designs. The authors present a comprehensive study comparing a number of algorithms using different metrics such as area, throughput, power, and energy, and applying state of the art techniques for reducing power consumption such as voltage scaling. However, the evaluation reported in the paper is at very high level and concentrates only on a specific implementation, without considering the effects on energy consumption of different design choices, such as size of the datapath, amount of serialization, or effects of architectural optimization applied at each stage of the algorithm. The second work explores area, power, and energy consumption of several recently-developed lightweight block ciphers and compares it with the AES algorithm, considering also possible optimization for the non linear transformation. However, no possible optimization was considered for the other transformations, and effects of other design choices, such as serialization were not considered in the work. In another work [18], a comparison of the energy consumptions of fully and partially unrolled circuits was done with respect to the latency in the circuit.

1.1 Contribution and Organization of the Paper

In this paper we complete the analysis started with these works, looking at all the parameters which might affect the energy consumption of a design. We start with the case of AES and investigate how the variation in (a) the architectural design of the individual components (S-box, MixColumn), (b) frequency of the clock signal and (c) serializing or unrolling the design can affect the energy consumption. Furthermore, starting with the detailed analysis of our exploration, we build an energy model for any r-round unrolled architecture of block ciphers. We prove that if all other factors are constant, then the total energy consumed per encryption in an r-round unrolled circuit is quadratic in r. We validate our model by estimating the energy consumed by several lightweight algorithms and comparing it with the figures obtained by simulating their implementations.

The remainder of the paper is organized as follows. Section 2 presents, as a motivating example a detailed study of the AES algorithm from the energy point of view. Section 3 presents our model for estimating the energy consumption of a block cipher, discussing how the contribution of each component is modeled and included into the overall energy consumption equation. Section 4 reports how our model is validated using a number of lightweight algorithms. Section 5, tabulates the final energy figures for all the block ciphers that we have considered, and relates these results to physical parameters like critical path and algorithmic parameters like the minimum number of rounds required to achieve full diffusion of a difference introduced in the plaintext or key. Section 6 concludes the paper.

2 A Case Study of Energy Consumption of AES 128

In this section, we investigate how the choice of architecture can affect the energy performance of implementations of AES 128. In our experiments, we considered three factors that would likely affect the energy metric of the encryption algorithm.

(a) **Architecture of S-Box/MixColumn:** It is known that the Canright architecture [9] is the one of the most compact representations of the AES S-box in terms of gate area. However it is unlikely to be the most efficient energy-wise. We then experimented with a Lookup table based S-box. However, we found that the Decoder-Switch-Encoder (DSE) architecture [5] is the most energy-efficient. We also considered two different variants of the MixColumn architecture. Considering AES MixColumn to be a linear map from $\{0,1\}^{32} \rightarrow \{0,1\}^{32}$, it can be composed with 152 xor gates by following the mathematical definition. However, as shown by [20], the outputs of several xor gates can be reused and it is possible to get a compact design in 108 gates. The 108 gate variant is likely to be more energy efficient as it provides a balanced datapath and also uses less gates. In Table 1, we present the area and energy per encryption figure for the round based designs for a combination of all the above choices of S-boxes/MixColumns at an operating frequency of 10 MHz (using the standard cell library of the STM 90 nm low leakage process). Clearly the DSE S-box and the 108 gate MixColumn is optimal in terms of energy efficiency.

(b) **Clock Frequency:** As already pointed out in [17], the energy consumption required to compute an encryption operation should be independent of frequency of operation, as energy is a metric which is a measure of the total switching activity of a circuit during the process. This is true for sufficiently high frequencies, where the total leakage energy consumed by the system is low over the total number of cycles required for encryption. However for the STM 90 nm low leakage process, at frequencies lower than 1 MHz, leakage energy naturally starts to play a significant role, thereby increasing the energy consumption. Furthermore, gates selected by synthesis tools for meeting a high clock frequency can be significantly different from the ones selected for achieving a low clock frequency. The selection of different gates, which is an indirect consequence of the clock frequency, would also affect the energy consumption. In Fig. 1, we present the variation in the energy consumption, for the round based AES architecture (using the DSE S-box and the 108 gate MixColumn) for frequencies ranging from 100 KHz to 100 MHz. We can see that for frequencies higher than 1 MHz, the energy consumption is more or less invariant with respect to frequency.

(c) **Width of the Data path/Unrolled Design:** We performed our experiments with numerous serialized implementations of AES in which the datapath width varies from 8 to 32 to 64 bits. For our experiments, we used the 8-bit serial architecture described in [19]. For the 32-bit serialized datapath we used three architectures described as follows

  1. A_1: In this architecture every round is completed in 9 cycles: 4 for the Substitution operation, 4 for the MixColumn operation and 1 for Shift row. This architecture takes 94 cycles to complete one encryption.
  2. A_2: In this architecture every round is completed in 5 cycles: 4 cycles are used for the combined Substitution operation and MixColumn operation and 1 is used for Shift row. This architecture takes 54 cycles to complete one encryption.

Table 1. Area, energy figures for round based AES 128 using different component architectures

#	S-box	MixColumn	Area (in GE)	Energy (in pJ)	Energy/bit (in pJ)
1	LUT	152 gates	13836.2	797.2	6.23
2	LUT	108 gates	13647.9	755.3	5.90
3	Canright	152 gates	8127.9	753.6	5.89
4	Canright	108 gates	7872.5	708.5	5.53
5	DSE	152 gates	12601.7	377.5	2.95
6	DSE	108 gates	12459.0	**350.7**	2.74

3. A_3: In this architecture every round is completed in 4 cycles. Extra multiplexers are used to ensure that each clock cycle performs the Shift Row, Substitution and MixColumn operation on a given chunk of 32 bit data. This architecture takes 44 cycles to complete one encryption.

Similarly, we used three architectures B_1, B_2 and B_3 for the 64-bit serial design that takes 52, 32 and 22 cycles respectively. Thereafter we continue to explore lower latency designs like the round based architecture and the 2, 3, 4, 5, 10 round unrolled architectures.

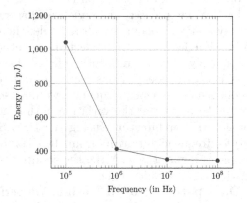

Fig. 1. Energy consumption for round based AES 128 over a range of clock frequencies

We present the area and energy per encryption figure for all the architectures using the DSE S-box, and the 108 gate MixColumn for designs synthesized with the standard cell library based on the STM 90 nm logic process, at a clock frequency of 10 MHz in Table 2. We found that the round based implementation of AES 128 is the most energy efficient. Since the serialized architectures take longer time to complete an encryption operation it was expected that they would consume more energy, but the fact that the round based design was better in terms of energy than its unrolled counterparts was certainly an interesting result. To understand the reason for this we first need to understand which components of the architecture are consuming the most energy. A breakdown of this energy consumption, by percentage of the total energy, for the various components is shown in Fig. 2.

Table 2. Area and energy figures for different AES 128 architectures

#	Design	Area (in GE)	#Cycles	Energy (pJ)	Energy/bit (pJ)
1	8-bit	2722.0	226	1913.1	14.94
2	32-bit (A_1)	4069.7	94	1123.3	8.77
	32-bit (A_2)	4061.8	54	819.2	6.40
	32-bit (A_3)	5528.4	44	801.7	6.26
3	64-bit (B_1)	6380.9	52	1018.7	7.96
	64-bit (B_2)	6362.6	32	869.8	6.79
	64-bit (B_3)	7747.5	22	616.2	4.81
4	Round based	12459.0	11	**350.7**	2.74
5	2-round	22842.3	6	593.6	4.64
6	3-round	32731.9	5	1043.0	8.15
7	4-round	43641.1	4	1416.5	11.07
8	5-round	53998.7	3	1634.4	12.77
9	10-round	101216.7	1	2129.5	16.64

We can see that the Substitution layer consumes the most part of the energy budget (24 % and 36.3 %) in the round based design and the 2-round unrolled designs respectively. However we also find that in the 2-round unrolled design, the second round functions (Substitution Layer, MixColumn, Add round key and round key logic) consume more energy than the first. To understand the reason for this trend, we need to study the energy consumption model in CMOS gates, and start to analyze the situation from there.

3   CMOS Energy Consumption Model

Currently, static CMOS is the dominant technology used for producing electronic devices. Two main reasons were behind the widespread diffusion of static CMOS: its robustness against noise and its limited static power consumption. With the shrinking of technologies, static power consumption of CMOS is increasing. Nevertheless, static CMOS is likely to continue to be the preferred technology for electronic fabrications in the foreseeable future.

Energy consumption of a static CMOS gate, is defined by the following equation:

$$E_{gate} = E_{load} + E_{sc} + E_{leakage}$$

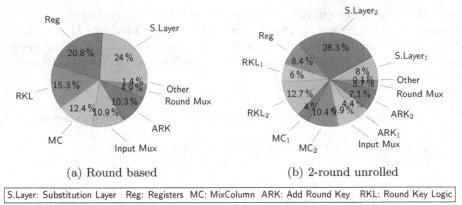

(a) Round based   (b) 2-round unrolled

S.Layer: Substitution Layer   Reg: Registers MC: MixColumn ARK: Add Round Key   RKL: Round Key Logic

Fig. 2. Energy shares for the round based and 2-round unrolled AES 128

where E_{sc} is the energy due to the short-circuit current. $E_{leakage}$ is the energy consumed due to the sub-threshold leakage current when the transistor is OFF. This component is usually small, but is gaining importance as the technology scaling makes the sub-threshold leakage more significant. E_{load} is the energy dissipated for charging and discharging the capacitive load C_L of a gate when output transitions occur.

Hardware implementations of any cryptographic primitive consist of a number of registers and logic blocks connected together as required by the specifications of the algorithm itself. Block ciphers based on SPN or Feistel designs, consist in particular of a round function and round key generation logic which transform a plaintext and key into a ciphertext by iterating the round function for a specific amount of rounds. Consider an ideal block cipher E operating on a plaintext space $\{0,1\}^{L_p}$ and a key space $\{0,1\}^{L_k}$. Its hardware implementation, illustrated in Fig. 3, would include:

A. A state register (SReg) and a key register (KReg) of L_p and L_k bits respectively, to store the intermediate states produced by the round function and the computed round keys.
B. Two input multiplexers placed before the state register and the key register respectively used to control the updating of the state or the loading of the plaintext or the initial key.
C. Depending on the choices of the designer, one or more instances of the round function (RF_i) and the round key generation logic (RK_i)
D. Additional logic needed to generate control signals, round constants etc.

A designer, depending on the specific requirements of the target application, can implement the algorithm following different strategies. One of the most important decisions in this respect is the number of instances of the round function to be replicated in hardware. The smallest amount of replication happens when a single instance of the round function and the round key are instantiated: implementations following this style are called round based architectures.

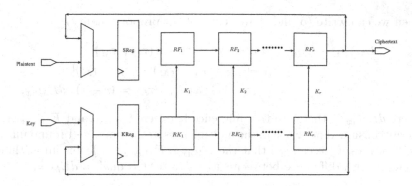

Fig. 3. Block cipher architecture

The round based architecture of a block cipher which has to be executed for R rounds, can be compute the result of an encryption in $R + 1$ clock cycles (1 cycle for the loading of the plaintext/key and the remaining R cycles for executing the R rounds). A designer can instantiate more than one round function, opting for an unrolled architecture. An r-round unrolled architecture consists of r instances of the round function and round key logic. Encryption on such a circuit would take $1 + \lceil \frac{R}{r} \rceil$ clock cycles.

The selection of the number of round functions instantiated depends on the specific optimization parameters. For instance, an r-round unrolled circuit for high values of r would require a smaller number of clock cycles to compute the encrypted ciphertext compared to a round based design. However, its power consumption is usually higher. It is thus interesting to investigate how the value of r affects the energy consumption for a given block cipher. To tackle this problem from a purely analytical point of view, one can make the following observations:

1. Assume that the designer has fixed the logic process and the frequency of operation of the circuit. Consider the input signal seen by the multiplexers i.e., outputs of RF_r and RK_r, respectively in an r-round design. If τ_F, τ_K represent the delays in each of the RF_i, RK_i blocks, then each multiplexer will see a signal that will be switching for around $r\tau_F, r\tau_K$ respectively in every clock cycle before stabilizing. If each of the muxes itself introduce a delay of τ_M, then their outputs will be switching for $r\tau_F + \tau_M, r\tau_K + \tau_M$ in every round. Since in a low leakage environment, the energy consumed is essentially the measure of the total number of logic switches, we can assume that the energy consumed in the each of the multiplexers is proportional to $r\tau_F + \tau_M, r\tau_K + \tau_M$ respectively. Let $E_{Mux,r}$ be used to denote the total energy drawn per cycle by the multiplexers in an r-round unrolled design.

Then we can write (α and β are constants of proportionality)

$$
\begin{aligned}
E_{Mux,r} - E_{Mux,1} &= \alpha \cdot [(r\tau_F + \tau_M) - (\tau_F + \tau_M)] \\
&\quad + \beta \cdot [(r\tau_K + \tau_M) - (\tau_K + \tau_M)] \\
&= (r-1) \cdot (\alpha\tau_F + \beta\tau_K) = (r-1) \cdot dE_{Mux},
\end{aligned}
$$

where dE_{Mux} is therefore the difference between $E_{Mux,r}$ and $E_{Mux,r-1}$, i.e. energy consumed per cycle in the multiplexers in the r and $r-1$ round unrolled architectures. One can see that the $E_{Mux,1}, E_{Mux,2}, \ldots$ forms an arithmetic sequence with difference between successive terms equal to dE_{Mux}.

2. Similarly, we can derive the energy drawn by the registers. However, registers switch only at the positive/negative edge of the clock (assuming a synchronous design). If $E_{Reg,r}$ is the total energy per cycle drawn by the registers in the r-round architecture, we have

$$
E_{Reg,r} = E_{Reg,1} + (r-1) \cdot dE_{Reg}.
$$

However the value of incremental energy dE_{Reg} when compared to $E_{Reg,1}$ is generally much smaller.

3. By similar arguments, the energy consumed in each successive logic block RF_i and similarly RK_i is likely to constitute two arithmetic sequences. The total energy drawn per cycle by the r round function blocks, given by E_{RF} is

$$
\begin{aligned}
E_{RF} = \sum_{i=1}^{r} E_{RF,i} &= \sum_{i=1}^{r} E_{RF,1} + (i-1) \cdot dE_{RF} \\
&= rE_{RF,1} + \frac{r(r-1)}{2} \cdot dE_{RF},
\end{aligned}
$$

and similarly, the total energy drawn per cycle by the r round key logic blocks is

$$
E_{RK} = rE_{RK,1} + \frac{r(r-1)}{2} \cdot dE_{RK}
$$

where $E_{RF,i}$ and $E_{RK,i}$ are the energy drawn per cycle by RF_i and RK_i respectively. dE_{RF}, dE_{RK} denote the incremental energy consumption per cycle between successive round function and round key logic blocks respectively. In deriving the above equations we have made the implicit assumption that the capacitive loads driven by the final blocks RF_r, RK_r are the same as the ones driven by the previous blocks RF_i, RK_i (for $i < r$). This, however, is not always true. For example, RF_r drives the multiplexer in front of the state register, and all of the previous RF_i blocks drive the subsequent RF_{i+1}. This may result in small deviation in the actual and the estimated energy consumed in the final block. However the deviation is usually negligible.

4. The energy drawn by the rest of the logic (E_{rem}) may or may not form a sequence with any special property for increasing values of r. This would

naturally depend on the specific algorithm of the block cipher. The value of this figure is usually a small fraction of the total energy drawn by the circuit: in the set of ciphers we have considered in this work, E_{rem} was never exceeding 5 % of the total energy budget.

Summing all the contributions, we can write the total energy E_r consumed per cycle in an r-round unrolled circuit as:

$$
\begin{aligned}
E_r &= E_{Mux,r} + E_{Reg,r} + E_{RK} + E_{RF} + E_{rem} \\
&= E_{Mux,1} + (r-1) \cdot dE_{Mux} + E_{Reg,1} + (r-1) \cdot dE_{Reg} \\
&\quad + r \cdot E_{RF,1} + \frac{r(r-1)}{2} \cdot dE_{RF} + r \cdot E_{RK,1} + \frac{r(r-1)}{2} \cdot dE_{RK} + E_{rem}
\end{aligned}
$$

E_r is a quadratic function in r in the form $Ar^2 + Br + C$, where

$$
A = \frac{dE_{RF} + dE_{RK}}{2}, \quad B = E_{RF,1} + E_{RK,1} + dE_{Reg} + dE_{Mux} - \frac{dE_{RF} + dE_{RK}}{2}
$$
$$
C = E_{Reg,1} + E_{Mux,1} + E_{rem} - dE_{Reg} - dE_{Mux}.
$$

To compute the total energy $\mathbf{E}_r$ which a particular implementation consumes to perform an encryption, the energy required for one round needs to be multiplied for total time required for the computation i.e. $\left(1 + \left\lceil \frac{R}{r} \right\rceil\right)$:

$$
\mathbf{E}_r = E_r \cdot \left(1 + \left\lceil \frac{R}{r} \right\rceil\right) = (Ar^2 + Br + C) \cdot \left(1 + \left\lceil \frac{R}{r} \right\rceil\right) \tag{1}
$$

As before, $\mathbf{E}_r$ is a function in r of the form $\alpha r^2 + \beta r + \gamma + \frac{\delta}{r}$. The analysis for a fully unrolled circuit is slightly different such circuits do not need registers used to store intermediate values. As a result, the total energy consumed by a fully unrolled design does not contain the $E_{Reg,r}$ component. Also, a fully unrolled circuit takes only a single clock cycle to complete an encryption.

4   Application of the Model

In this section we apply our model to determine the most energy efficient configuration for 9 lightweight block ciphers. For each algorithm, we measure **(a)** the parameters $E_{Reg}, E_{Mux}, E_{RF,1}, E_{RK_1}, E_{rem}$ and **(b)** the energy differentials $dE_{RF}, dE_{RK}, \ldots$ by simulating the energy consumption of the round based and 2-round unrolled design. Using this data, we predict the energy consumption required for one encryption by changing the number of unrolled rounds. Thereafter, we determine the value of r which achieves the highest energy efficiency. Finally, we compare our predictions with the actual energy consumption estimated using a well recognized gate level power simulator.

Estimation of power consumption (and, as a consequence, energy consumption) can be carried out at different levels. A designer has to trade the desired precision in the estimation with the time (and the level of circuit details) required for the simulation. A first approximation of power consumption can be achieved by simply counting the amount of switches which each node of the circuit makes during a given time period. This approach is extremely fast. However, the accuracy is very limited, as all the gates are assumed to consume the same amount of power. A better estimation can be obtained by collecting the switching activity of the circuit under test, obtained by simulating a significant and sufficiently large test bench, and annotating it back to the power estimation tool. In this way, the amount of switches is combined with the power fingerprint of each gate indicated in the technological library and produces a more precise estimation. The back-annotation of the switching activity can be carried out in different ways. The first and simpler, consists of annotating only the switching activity at the primary inputs. In this case, the tool estimates the switching of the internal gates. A more precise back annotation involves annotating the exact amount of switching of each gate, as produced by the simulation of a test bench. It is worth mentioning that most precise estimation of power consumption is obtained by simulating a circuit at SPICE level. This simulation, however, requires the availability of technological models (which are often not provided by the foundry) and needs significant amount of time to be carried out. For this reason, SPICE level simulation is not the preferred way to estimate power consumption. Power consumption estimation is also affected by the point in the design flow where it is carried out. A post-synthesized netlist contains all the information for estimating the power consumed by the gates, however it does not have information about the interconnecting wires. To obtain them, it is necessary complete the placement and the routing of the circuit, which is out of the scope of this work.

In this work, we are mainly concerned by the energy consumed by the gates. Hence, we carried out the energy estimation using the following design flow: The design was implemented at RTL level. A functional verification of the VHDL code was done using *Mentorgraphics ModelSim. Synopsys Design Compiler* was used to synthesize the RTL design. The switching activity of each gate of the circuit was collected by running post-synthesis simulation. The average power was obtained using *Synopsys Power Compiler*, using the back annotated switching activity. The energy was then computed as the product of the average power and the total time taken for one encryption.

For all the circuits, we set the operating Frequency at 10 MHz and the target library was the standard cell library of the STM 90 nm low leakage process. The operating frequency was fixed at 10 MHz since we have already established that at sufficiently high frequencies, the energy consumption of a circuit is invariant with frequency. We selected a set of 9 lightweight block ciphers of different design flavors. We classified them into two categories:

(a) **Iterated ciphers** are those all of whose round functions are similar. In this category we have AES 128, Noekeon, Present, Piccolo, TWINE and Simon 64/96. Such ciphers readily fit the model of energy consumption given by Eq. (1).

(b) **Non-iterated ciphers** are those whose round functions are not all similar. For example, in the cipher LED 128, the most significant bits and the least significant bits of the 128-bit key are alternately added to the state after every 4 rounds. So, in a round based design, to account for the addition of the round key once every four cycles, one needs to place a multiplexer/and gate to filter the key every fourth round. However, in a 2-round unrolled design, this filtering is not needed in the second round function. In a 3-round unrolled design, filtering would be needed in all the rounds, whereas in a 4-round unrolled design, filtering is not needed in any of the rounds. So, this is a cipher in which the structure of the round function varies widely from one architecture to another, we will call this a non-iterative cipher. Another example is Prince, in which 3 different type of round functions are used in the design itself. Finally we have the KATAN64 block cipher which is based on a bitwise Shift register. Since the cipher is based on a Shift register, its functioning is very different from the existing SPN/Feistel designs. Each round consists of the execution of a few simple Boolean Functions over only a limited number of bits of the current state. This is why we do not see a compounding of switching activity across rounds. Such ciphers do not readily fit the model for energy consumption as defined in Eq. (1). But the core logic remains the same, rounds further away from the register would consume more energy that the ones closer to it.

For all the iterated ciphers in our set we measured the values of $E_{Reg,1}, dE_{Reg}$, $E_{Mux,1}, dE_{Mux}, E_{RF,1}, dE_{RF}, E_{RK,1}, dE_{RK}\ E_{rem}$ and formulate the expression for $\mathbf{E}_r$ as given in Eq. (1). The results are shown in Table 3.

Table 3. Measured parameters for the iterated ciphers (all figures in pJ)

Cipher	Blocksize/Keysize	$E_{Reg,1}$	dE_{Reg}	$E_{Mux,1}$	dE_{Mux}	$E_{RF,1}$	dE_{RF}	$E_{RK,1}$	dE_{RK}	E_{rem}	$\mathbf{E}_r$
AES 128	128/128	6.75	1.49	3.65	3.20	32.70	8.26	16.10	8.26	0.42	$(6.13 + 6.03r + 20.48r^2) \cdot \left(1 + \lceil \frac{10}{r} \rceil\right)$
Noekeon	128/128	3.26	0.86	1.67	1.81	15.53	26.98	0.00	0.00	0.88	$(3.14 + 4.71r + 13.49r^2) \cdot \left(1 + \lceil \frac{17}{r} \rceil\right)$
Present	64/80	2.99	0.27	0.58	0.36	1.47	1.59	0.10	0.00	0.20	$(3.15 + 1.40r + 0.795r^2) \cdot \left(1 + \lceil \frac{32}{r} \rceil\right)$
Piccolo	64/80	1.56	0.39	0.61	1.00	3.93	7.87	0.74	0.00	0.70	$(1.48 + 2.13r + 3.93r^2) \cdot \left(1 + \lceil \frac{25}{r} \rceil\right)$
TWINE	64/80	3.08	0.37	0.76	0.67	1.56	2.16	0.48	0.25	0.42	$(3.23 + 1.82r + 1.25r^2) \cdot \left(1 + \lceil \frac{36}{r} \rceil\right)$
Simon 64/96	64/96	3.34	0.30	0.60	0.48	1.19	0.99	0.75	0.42	0.52	$(3.68 + 2.01r + 0.71r^2) \cdot \left(1 + \lceil \frac{42}{r} \rceil\right)$

Noekeon, when operated in direct mode, does not use a key schedule operation and hence $E_{RK,1}, dE_{RK}$ parameters are both zero for this cipher. Similarly, in Piccolo, the key schedule consists of selecting different portions of the key depending on the current round number and adding a round constant to it. This functionality can be achieved by a set of multiplexers and xor gates for any r-round unrolled architecture and so $dE_{RK} = 0$ for this cipher. The key schedule of Present is such that extremely slow diffusion occurs in the key path. So, the switching activity of the RK_1 block does not necessarily compound the switching activity in RK_2 and $E_{RK,1} = 0.1, dE_{RK} = 0$ is a reasonable approximation for analyzing less than 5-round unrolled designs.

By analyzing the expressions for $\mathbf{E}_r$ in Table 3, one can conclude that $r = 2$, is the optimal energy configuration for Present, TWINE and Simon 64/96. For AES 128, Piccolo and Noekeon, $r = 1$ is likely to be optimal in terms of energy. In Fig. 4, we compare our estimates for the energy consumption for upto the 4-round unrolled implementation calculated as per the Equation for $\mathbf{E}_r$ in Table 3, with the actual figures. It can be seen that for Present, TWINE and Simon 64/96, our prediction that $r = 2$ is the optimal energy configuration holds good. Similarly our prediction that the round based architecture is the most energy-efficient for AES 128, Noekeon and Piccolo also holds good.

For the non-iterated ciphers, although it is not possible to model the energy consumption of round unrolled designs, the concept holds that successive round functions consume more energy than the previous. The simulation results for all ciphers are given in Table 4. It can be seen that the round based configurations of LED 128 is most energy-efficient. Prince uses three types of round functions: Forward, Middle and Inverse. We implemented 3 architectures for Prince: the round based, Fully unrolled and a Half unrolled design in which Forward/Middle and the Inverse rounds are executed in one cycle each. Again, the round based design was found to be most energy efficient. Finally, we experimented with the round based and 2, 4, 8, 16 and 32 round unrolled versions of KATAN64. As can be seen in Table 4, the 16-round version was found to consume the least energy.

5   Discussion

Under a low leakage environment, the energy consumption in an circuit over a period of time is essentially a measure of the electrical work done by the voltage source in order to charge and discharge its gates. We have already seen that in any unrolled architecture, the gates in the later rounds of the design consume more energy, because the switching activity is compounded from one round to the next. Even then, intuitively it makes sense to investigate which degree of unrolling optimizes energy consumption, since an r-round unrolled design will inevitably reduce the total energy required to write updated states onto the state/key registers by a factor of almost r.

We know that the difference in the energy consumptions in any two successive rounds in any unrolled design will depend on the average number of gates that switch in the first round. A physical parameter that is closely related to the average number of gate switchings is the total signal delay in one round. The figures in Table 4 confirm that the ciphers which have low differential energies across successive unrolled architectures are also those in which a signal experiences low delay across a round. These are also the ciphers in which the 2-round unrolled design is more energy efficient. For example in Present, the critical path is composed of 1 S-box and 1 xor gate. In Simon 64/96, the critical path includes 3 xor gates and a single and gate. In Twine, the critical path is made up of 2 xor gates and an S-box. In all other ciphers, the critical path is comprised of atleast one S-box, multiple xor gates and MixColumn layers.

A design parameter that has considerable correlation with the differential energies, is the number of rounds required for full diffusion to take place. This

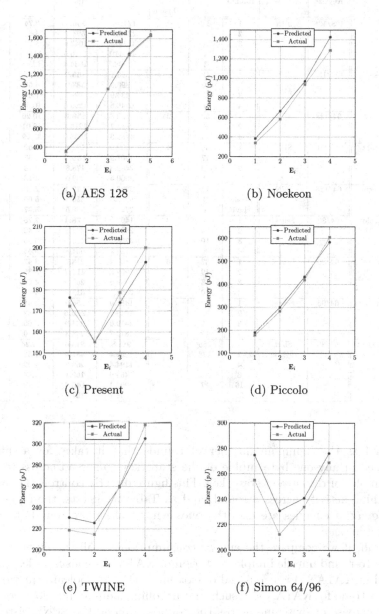

(a) AES 128

(b) Noekeon

(c) Present

(d) Piccolo

(e) TWINE

(f) Simon 64/96

Fig. 4. Actual and predicted energy consumptions

Table 4. Area, energy and related figures for all the ciphers

#	Cipher	Blocksize/ Keysize	Round Type	Unrolled Rounds	#Cycles	#Rounds for full diffusion	Area(in GE)	Energy (pJ)	Energy/bit (pJ)	Delay per round (ns)
1	AES 128	128/128	SPN	1	11	2	12459.0	**350.7**	2.74	3.32
				2	6		22842.3	593.4	4.64	
				3	5		32731.9	1043.0	8.15	
				4	4		43641.1	1416.5	11.07	
				5	3		53998.7	1634.4	12.77	
2	Noekeon	128/128	SPN	1	18	2	2348.1	**339.2**	2.65	3.41
				2	10		3890.3	583.0	4.55	
				3	7		5434.9	936.7	7.32	
				4	6		6946.6	1290.6	10.08	
3	LED 128	64/128	SPN	1	50	2	1830.8	**656.5**	10.26	5.25
				2	26		2864.7	1216.8	19.01	
				4	14		4780.3	1638.0	25.59	
4	Present	64/80	SPN	1	33	4	1439.9	172.3	2.69	2.09
				2	17		1967.9	**155.2**	2.43	
				3	12		2499.3	178.8	2.79	
				4	9		3000.4	200.0	3.13	
5	Prince	64/128	SPN	1	13	2	2286.5	**149.1**	2.33	4.06
				Half	3		8245.9	358.4	5.60	
				Full	1		7728.6	369.5	5.77	
6	Piccolo	64/80	Feistel	1	26	3	1492.0	**178.1**	2.78	3.28
				2	14		2385.5	282.8	4.42	
				3	10		3268.1	419.0	6.55	
				4	8		4124.7	604.8	9.45	
7	TWINE	64/80	Feistel	1	37	8	1408.2	218.7	3.42	3.10
				2	19		1902.8	**214.7**	3.35	
				3	13		2399.5	260.0	4.06	
				4	10		2850.8	318.0	4.97	
8	Simon 64/96	64/96	Feistel	1	43	4	1480.0	255.0	3.98	2.18
				2	22		1948.7	**212.5**	3.32	
				3	15		2419.0	234.0	3.65	
				4	12		2875.7	268.8	4.20	
9	KATAN64	64/80	Shift register	1	255		983.8	913.6	14.28	2.04
				2	128		1055.4	481.9	7.53	
				4	65		1194.4	269.8	4.22	
				8	33		1459.6	169.1	2.64	
				16	17		1992.4	**140.1**	2.19	
				32	9		3058.1	167.2	2.61	

is defined as the minimum number of rounds that it takes for a difference introduced in any one byte/nibble of the state/key to spread across to all the bytes/nibbles of the current state/key. This figure directly controls the quantum of switching activity across a round, and as Table 4 suggests, the ciphers with low differential energies are also the ones which take more rounds to achieve complete diffusion.

Overall, if we compare the energy consumptions of all the ciphers we find that the 16-round unrolled implementation of KATAN64 consumes least energy. A round in KATAN64 is composed of extremely simple Boolean equations, and hence the trend for KATAN64 is such that unrolling more rounds does not always lead to increase of switching across the rounds. Among the SPN/Feistel architectures, the round based implementation of Prince consumes the least energy, as it takes only 13 cycles to complete an encryption, and the fact that it does not employ any key schedule operation. Close second, is the 2-round unrolled implementation of Present, followed by the round based implementations of Present and Piccolo. Piccolo benefits from the fact that it does not have a key schedule operation, and hence does not expend any energy on writing values to a key register. Coming in next are the 3 and 4 round unrolled Present and the 2-round unrolled designs of Simon 64/96 and TWINE. It is also interesting to note that if we use a DSE based S-box, then the energy/bit figure of round based AES 128 is quite comparable to Prince and Present.

6 Conclusion

In this paper, we looked at the energy consumption figures of several lightweight ciphers with different degrees of unrolling. By constructing a model of energy consumption, we proved that the total energy consumed in a circuit during an encryption operation has roughly a quadratic relation with the degree of unrolling. In this respect we looked to apply our model to a number of lightweight ciphers and predict the most energy efficient architecture for the design. In the end, we tried to relate the energy consumption in an arbitrarily unrolled architecture of a circuit to physical parameters like critical path in a single round function and algorithmic parameters like number of rounds required to achieve full diffusion.

References

1. Descriptions of SHA-256, SHA-384, and SHA-512. http://csrc.nist.gov/groups/STM/cavp/documents/shs/sha256-384-512.pdf
2. Batina, L., Das, A., Ege, B., Kavun, E.B., Mentens, N., Paar, C., Verbauwhede, I., Yalçin, T.: Dietary recommendations for lightweight block ciphers: power, energy and area analysis of recently developed architectures. In: Hutter, M., Schmidt, J.-M. (eds.) RFIDsec 2013. LNCS, vol. 8262, pp. 101–110. Springer, Heidelberg (2013)
3. Beaulieu, R., Shors, D., Smith, J., Treatman-Clark, S., Weeks, B., Wingers, L.: The Simon and Speck Families of Lightweight Block Ciphers. IACR eprint archive. https://eprint.iacr.org/2013/404.pdf
4. Bertoni, G., Daemen, J., Peeters, M., Assche, G.V.: The Keccak Reference. http://keccak.noekeon.org/Keccak-reference-3.0.pdf
5. Bertoni, G., Macchetti, M., Negri, L., Fragneto, P.: Power-efficient ASIC synthesis of cryptographic S-boxes. In: Proceedings of the 14th ACM Great Lakes Symposium on VLSI. ACM, pp. 277–281(2004)
6. Bogdanov, A.A., Knudsen, L.R., Leander, G., Paar, C., Poschmann, A., Robshaw, M., Seurin, Y., Vikkelsoe, C.: PRESENT: an ultra-lightweight block cipher. In: Paillier, P., Verbauwhede, I. (eds.) CHES 2007. LNCS, vol. 4727, pp. 450–466. Springer, Heidelberg (2007)
7. Borghoff, J., et al.: PRINCE – a low-latency block cipher for pervasive computing applications. In: Wang, X., Sako, K. (eds.) ASIACRYPT 2012. LNCS, vol. 7658, pp. 208–225. Springer, Heidelberg (2012)
8. De Cannière, C., Dunkelman, O., Knežević, M.: KATAN and KTANTAN — a family of small and efficient hardware-oriented block ciphers. In: Clavier, C., Gaj, K. (eds.) CHES 2009. LNCS, vol. 5747, pp. 272–288. Springer, Heidelberg (2009)

9. Canright, D.: A very compact S-Box for AES. In: Rao, J.R., Sunar, B. (eds.) CHES 2005. LNCS, vol. 3659, pp. 441–455. Springer, Heidelberg (2005)
10. Daemen, J., Peeters, M., Assche, G.V., Rijmen, V.: Nessie Proposal: NOEKEON. http://gro.noekeon.org/Noekeon-spec.pdf
11. Daemen, J., Rijmen, V.: The Design of Rijndael: AES - The Advanced Encryption Standard. Springer, Heidelberg (2002)
12. Feldhofer, M., Wolkerstorfer, J., Rijmen, V.: AES implementation on a grain of sand. IEEE Proc. Inf. Secur. **152**(1), 13–20 (2005)
13. Gong, Z., Nikova, S., Law, Y.W.: KLEIN: a new family of lightweight block ciphers. In: Juels, A., Paar, C. (eds.) RFIDSec 2011. LNCS, vol. 7055, pp. 1–18. Springer, Heidelberg (2012)
14. Guo, J., Peyrin, T., Poschmann, A., Robshaw, M.: The LED block cipher. In: Preneel, B., Takagi, T. (eds.) CHES 2011. LNCS, vol. 6917, pp. 326–341. Springer, Heidelberg (2011)
15. Hocquet, C., Kamel, D., Regazzoni, F., Legat, J.-D., Flandre, D., Bol, D., Standaert, F.-X.: Harvesting the potential of nano-CMOS for lightweight cryptography: an ultra-low-voltage 65 nm AES coprocessor for passive RFID tags. J. Cryptograph. Eng. **1**(1), 79–86 (2011)
16. Hong, D., et al.: HIGHT: a new block cipher suitable for low-resource device. In: Goubin, L., Matsui, M. (eds.) CHES 2006. LNCS, vol. 4249, pp. 46–59. Springer, Heidelberg (2006)
17. Kerckhof, S., Durvaux, F., Hocquet, C., Bol, D., Standaert, F.-X.: Towards green cryptography: a comparison of lightweight ciphers from the energy viewpoint. In: Prouff, E., Schaumont, P. (eds.) CHES 2012. LNCS, vol. 7428, pp. 390–407. Springer, Heidelberg (2012)
18. Knežević, M., Nikov, V., Rombouts, P.: Low-latency encryption – is "Lightweight = Light + Wait"? In: Prouff, E., Schaumont, P. (eds.) CHES 2012. LNCS, vol. 7428, pp. 426–446. Springer, Heidelberg (2012)
19. Moradi, A., Poschmann, A., Ling, S., Paar, C., Wang, H.: Pushing the limits: a very compact and a threshold implementation of AES. In: Paterson, K.G. (ed.) EUROCRYPT 2011. LNCS, vol. 6632, pp. 69–88. Springer, Heidelberg (2011)
20. Satoh, A., Morioka, S., Takano, K., Munetoh, S.: A compact Rijndael hardware architecture with S-Box optimization. In: Boyd, C. (ed.) ASIACRYPT 2001. LNCS, vol. 2248, pp. 239–254. Springer, Heidelberg (2001)
21. Shibutani, K., Isobe, T., Hiwatari, H., Mitsuda, A., Akishita, T., Shirai, T.: *Piccolo*: an ultra-lightweight blockcipher. In: Preneel, B., Takagi, T. (eds.) CHES 2011. LNCS, vol. 6917, pp. 342–357. Springer, Heidelberg (2011)
22. Suzaki, T., Minematsu, K., Morioka, S., Kobayashi, E.: Twine: a lightweight block cipher for multiple platforms. In: Knudsen, L.R., Wu, H. (eds.) SAC 2012. LNCS, vol. 7707, pp. 339–354. Springer, Heidelberg (2013)

Short Papers

Forgery and Subkey Recovery on CAESAR Candidate iFeed

Willem Schroé[1,2], Bart Mennink[1,2(✉)], Elena Andreeva[1,2], and Bart Preneel[1,2]

[1] Department of Electrical Engineering, ESAT/COSIC,
KU Leuven, Leuven, Belgium
{bart.mennink,elena.andreeva,
bart.preneel}@esat.kuleuven.be
[2] iMinds, Ghent, Belgium

Abstract. iFeed is a blockcipher-based authenticated encryption design by Zhang, Wu, Sui, and Wang and a first round candidate to the CAESAR competition. iFeed is claimed to achieve confidentiality and authenticity in the nonce-respecting setting, and confidentiality in the nonce-reuse setting. Recently, Chakraborti et al. published forgeries on iFeed in the RUP and nonce-reuse settings. The latter attacks, however, do not invalidate the iFeed designers' security claims. In this work, we consider the security of iFeed in the nonce-respecting setting, and show that a valid forgery can be constructed after only one encryption query. Even more, the forgery leaks both subkeys $E_K(0^{128})$ and $E_K(PMN\|1)$, where K is the secret key and PMN the nonce used for the authenticated encryption. Furthermore, we show how at the price of just one additional forgery one can learn $E_K(P^*)$ for any freely chosen plaintext P^*. These design weaknesses allow one to decrypt earlier iFeed encryptions under the respective nonces, breaking the forward secrecy of iFeed, and leading to a total security compromise of the iFeed design.

Keywords: CAESAR · iFeed · Forgery · Subkey recovery · Breaking forward secrecy

1 Introduction

The CAESAR [3] competition was launched in 2014 to select a portfolio of recommended robust and efficient authenticated encryption algorithms by 2018. CAESAR is much in the spirit of the Advanced Encryption Standard [1] (AES) and SHA-3 hash [5] algorithm competitions, which were organized by the American institute of standards and technology (NIST). The main goal of this type of competitions is that they allow for public discussions, analysis, and therefore more comprehensive, secure and collectively agreed upon choices of cryptographic algorithms. In March, 2014, CAESAR received 57 submissions. In July 2015, 30 of them have advanced to the second round.

In this work we analyze the security of the first round CAESAR submission iFeed[AES] v1 – iFeed for short – by Zhang et al. [6]. iFeed is an AES

© Springer International Publishing Switzerland 2016
O. Dunkelman and L. Keliher (Eds.): SAC 2015, LNCS 9566, pp. 197–204, 2016.
DOI: 10.1007/978-3-319-31301-6_11

blockcipher-based design which combines PMAC-style authentication with dedicated encryption. Below we will treat iFeed generically, and consider it to be based on an arbitrary blockcipher E. iFeed processes the data in an on-line manner and it is inverse-free, meaning that both encryption and decryption only use the block cipher in forward direction. The design is inherently nonce-based: the authors claim *and* prove confidentiality and authenticity of iFeed in a setting where the adversary is not allowed to make repeated queries under the same nonce. We refer to this setting and type of adversary as *nonce-respecting*. The design is moreover claimed to achieve confidentiality under some conditions also in the *nonce-reuse* setting.

iFeed uses the secret key K to derive two subkeys: a nonce-independent $Z_0 = E_K(0^{128})$ and its multiples $Z_i = 2^i \cdot Z_0$, and a nonce-dependent $U = E_K(PMN\|10^*)$ where PMN is a variable public message number (nonce) and 10^* is the padding to a full 128-bit block with one 1 bit and appropriate number of 0 bits. The processing of the associated data and of the message are done independently of each other, both resulting in a subtag. The XOR of the two subtags produces the final tag. The general encryption and decryption procedures of iFeed are given in Figs. 1 and 2, respectively. The processing of the associated data is distinctively independent of the nonce: the inputs to the blockcipher E are masked only using Z_i. The data is encrypted using both the Z_i and U. One design choice of iFeed is that the computation of the subtags is performed using the same subkeys: Z_1 or Z_2 for the associated data, and $Z_1 \oplus U$ and $Z_2 \oplus U$ for the plaintext.[1]

Chakraborti et al. [4] recently presented forgeries on iFeed both in the nonce-reuse setting, and in a setting where the adversary is granted access to ciphertext decryptions irrespective of the verification result, also known as the *release of unverified plaintext* (RUP) setting from [2]. The iFeed designers, however, do not claim any security against these properties, and the attacks from Chakraborti et al. do not invalidate the security of iFeed.

Our Contribution. In this paper, we present a simple forgery attack on iFeed in the nonce-respecting model. Our attack exploits the unfortunate repetition of the Z_1 (respectively Z_2) subkeys at the finalization of the associated data and plaintext. Our attack on iFeed leads to a total security break and also invalidates the security claims and proofs posited by the designers of iFeed. The attack uses only one encryption query with no associated data and an arbitrary n-bit (or single block) plaintext. Then, we show how to construct a valid forgery with n-bit associated data and $2n$-bit ciphertext.

As a consequence of our attack the values of the subkeys Z_0 and U are also leaked. These subkeys are valuable information as they can be used to recover plaintext from old encryptions, even though they were performed under a different nonce. Namely, we show how at the price of just a single additional forgery, one learns $E_K(P^*)$ for any plaintext P^*. More concretely, the extra forgery allows an attacker to learn the $U' = E_K(PMN'\|10^*)$ values for some

---

[1] Z_1 is used if the last associated data block is fractional; Z_2 is used otherwise. Similarly for $Z_1 \oplus U$ and $Z_2 \oplus U$ with respect to the last plaintext block.

earlier used nonce PMN'. Once U' is obtained, the ciphertext with PMN' for any P^* can be retrieved: the inputs to E_K in the decryption algorithm can be computed offline (using Z_i, U', and the ciphertext), and forgeries can be performed to find out the plaintexts one by one.

Our forgeries not only invalidate the security claim and proof of iFeed [6], but they also break its forward secrecy. To gain a better understanding of the presented iFeed security weaknesses, we conclude by revisiting and pointing out the errata in the security proof of iFeed.

Outline. We introduce iFeed in Sect. 2. Our forgery and subkey recovery attack on iFeed is given in Sect. 3. Next, we show how the subkeys can be used to learn $E_K(P^*)$ for any P^* in Sect. 4. We elaborate on the key properties that caused this attack to work and look back at the proof of iFeed in Sect. 5. The paper is concluded in Sect. 6.

2   iFeed

iFeed [6] is a blockcipher-based authenticated encryption scheme. It is originally specified using the AES-128 blockcipher. Our attacks are generic and do not exploit any weakness of AES-128. Therefore, from now we simply describe iFeed based on any blockcipher $E : \{0,1\}^n \times \{0,1\}^n \to \{0,1\}^n$, where $n = 128$. iFeed is on-line, it only uses E in forward direction (inverse-free), and it allows for parallelized encryption.

The scheme is keyed via a key $K \in \{0,1\}^n$. It operates on public message numbers PMN of size between 1 and 127 bits, associated data A in $\{0,1\}^{\leq |A|_{\max}}$, and plaintexts/ciphertexts P/C from $\{0,1\}^{\leq |P|_{\max}}$, where $|A|_{\max}$ and $|P|_{\max}$ are some large values which sum to at most $2^{71} - 512$. The tag T is of size $32 \leq \tau \leq n = 128$. In the nonce-respecting setting, the public message number is required to be unique for every query to the iFeed encryption function $\mathcal{E}$. The iFeed encryption $\mathcal{E}$ and decryption $\mathcal{D}$ functions are depicted in Figs. 1 and 2, respectively. Here, $\mathrm{Pad}(X)$ equals X if $|X| = n$ and $X\|10^{n-1-|X|}$ if $|X| < n$, and the usage of Z_1 versus Z_2 (resp. Z_1' versus Z_2') depends on the last block of A (resp. P): Z_2/Z_2' is used if the last data block is of size n bits, Z_1/Z_1' is used if the last block is fractional. We remark that our attacks use integral blocks only, for which $\mathrm{Pad}(X) = X$ and Z_2 is used instead of Z_1.

For the presentation of the attacks, however, it suffices to only discuss a simplified version of iFeed. In more detail, in our attacks we will only query $\mathcal{E}$ on input of n-bit plaintext and no associated data. We also make the forgery queries to $\mathcal{D}$ for either n or $2n$-bit associated data and $2n$-bit ciphertext. All queries consider 127-bit PMN and tag size $\tau = n = 128$. In Algorithms 1 and 2, we give a formal description of iFeed's $\mathcal{E}$ and $\mathcal{D}$, respectively, for the input sizes relevant for our attacks. We refer to [6] for the general description of iFeed, and stress that our attacks easily translate to the general case.

The iFeed mode makes use of secret subkey $Z_0 = E_K(0^{128})$ which is used to derive additional subkeys $Z_i = 2 \cdot Z_{i-1}$. Here, the multiplication is performed

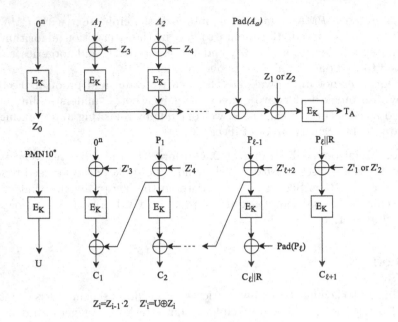

Fig. 1. iFeed encryption $\mathcal{E}$. All wires represent n-bit values. The output is the ciphertext $C_1 \cdots C_\ell$ and the tag $T = \text{left}_\tau(T_A \oplus C_{\ell+1})$

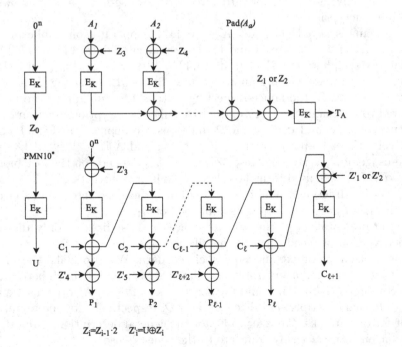

Fig. 2. iFeed decryption $\mathcal{D}$. All wires represent n-bit values. The output is the plaintext $P_1 \cdots P_\ell$ when T is $\text{left}_\tau(T_A \oplus C_{\ell+1})$

Algorithm 1. iFeed $\mathcal{E}$ for $|PMN| = 127$, $|A| = 0$, and $|P_1| = n$

1: $Z_0 = E_K(0^{128})$
2: **for** $i = 1, \ldots, 3$ **do**
3:     $Z_i = 2 \cdot Z_{i-1}$
4: $U = E_K(PMN\|1)$
5: $T_A = 0^{128}$
6: $C_1 = E_K(Z_3 \oplus U) \oplus P_1$
7: $C_2 = E_K(P_1 \oplus Z_2 \oplus U)$
8: **return** $(C, T) = (C_1, T_A \oplus C_2)$

Algorithm 2. iFeed $\mathcal{D}$ for $|PMN| = 127$, $|A'| = n$ or $2n$, and $|C'| = 2n$

1: $Z_0 = E_K(0^{128})$
2: **for** $i = 1, \ldots, 4$ **do**
3:     $Z_i = 2 \cdot Z_{i-1}$
4: $U = E_K(PMN\|1)$
5: **if** $|A'| = n$ **then**
6:     **define** $A'_1 = A'$
7:     $T'_A = E_K(A'_1 \oplus Z_2)$
8: **else**
9:     **parse** $A'_1 A'_2 = A'$
10:     $T'_A = E_K(E_K(A'_1 \oplus Z_3) \oplus A'_2 \oplus Z_2)$
11: **parse** $C'_1 C'_2 = C'$
12: $P'_1 = E_K(Z_3 \oplus U) \oplus C'_1 \oplus Z_4 \oplus U$
13: $P'_2 = E_K(P'_1 \oplus Z_4 \oplus U) \oplus C'_2$
14: $C'_3 = E_K(P'_2 \oplus Z_2 \oplus U)$
15: **if** $T' = T'_A \oplus C'_3$ **then**
16:     **return** $P' = P'_1 P'_2$
17: **else**
18:     **return** $\perp$

---

in the binary Galois Field $GF(2^{128})$ defined by the primitive polynomial $x^{128} + x^7 + x^2 + x + 1$.

3   Forgery and Subkey Recovery Attack on iFeed

Let $K \xleftarrow{\$} \{0,1\}^n$ be the secret key and consider $\tau = n$. We present our forgery attack on iFeed. It consists of one encryption query and the forgery itself. Upon successful verification, the forgery will disclose the subkeys $Z_0 = E_K(0^{128})$ and $U = E_K(PMN\|1)$.

- Fix arbitrary $PMN \in \{0,1\}^{127}$ and arbitrary $P_1 \in \{0,1\}^{128}$, and make **encryption query** with no associated data $A = \varepsilon$:

$$(C_1, T) = \mathcal{E}_K(PMN, \varepsilon, P_1);$$

- Write $C_1' = C_1 \oplus P_1 \oplus PMN\|1$, and fix an arbitrary $C_2' \in \{0,1\}^{128}$. Set $A = C_2'$, $T' = 0^{128}$, and output **forgery**:

$$\mathcal{D}_K(PMN, A, C_1'C_2', T').$$

We next demonstrate that the forgery attempt is successful. First, regarding the encryption query, note that

$$C_1 = E_K(Z_3 \oplus U) \oplus P_1. \tag{1}$$

Now, verification of the forgery (cf. Algorithm 2) succeeds if $T' = T_A' \oplus C_3' = 0^{128}$. Note that we have

$$
\begin{aligned}
P_1' &= E_K(Z_3 \oplus U) \oplus C_1' \oplus Z_4 \oplus U \\
     &= PMN\|1 \oplus Z_4 \oplus U, \\
P_2' &= E_K(P_1' \oplus Z_4 \oplus U) \oplus C_2' \\
     &= E_K(PMN\|1) \oplus C_2' = U \oplus C_2', \\
C_3' &= E_K(P_2' \oplus Z_2 \oplus U) \\
     &= E_K(C_2' \oplus Z_2).
\end{aligned}
$$

As we defined $A = C_2'$, this yields successful verification:

$$T_A' = E_K(A \oplus Z_2) = E_K(C_2' \oplus Z_2) = C_3'.$$

As verification is successful, $\mathcal{D}$ returns $P_1'P_2'$. The values $U = P_2' \oplus C_2'$ and $Z_0 = 2^{-4}(P_1' \oplus PMN\|1 \oplus U)$ are directly obtained.

4   Finding $E_K(P^*)$ for any Plaintext P^*

Below, we construct an additional forgery, which is based on the information gained from the main forgery attack. Namely, the goal is to use the knowledge of the subkeys U, Z_0, and the value $E_K(Z_3 \oplus U) = C_1 \oplus P_1$ from (1), to produce another forgery, which would allow one to learn $E_K(P^*)$ for any plaintext data $P^* \in \{0,1\}^n$.

- Let PMN be as before, define $(A_1'', A_2'') = (P^* \oplus Z_3, U)$, $(C_1'', C_2'') = (E_K(Z_3 \oplus U) \oplus P^*, 0^{128})$, and $T = 0^{128}$, and output **forgery**:

$$\mathcal{D}_K(PMN, A_1''A_2'', C_1''C_2'', T'').$$

The verification (cf. Algorithm 2) is successful if $T'' = T_A'' \oplus C_3'' = 0^{128}$. Note that we have

$$
\begin{aligned}
P_1'' &= E_K(Z_3 \oplus U) \oplus C_1'' \oplus Z_4 \oplus U \\
      &= P^* \oplus Z_4 \oplus U, \\
P_2'' &= E_K(P_1'' \oplus Z_4 \oplus U) \oplus C_2'' \\
      &= E_K(P^*), \\
C_3'' &= E_K(P_2'' \oplus Z_2 \oplus U) \\
      &= E_K(E_K(P^*) \oplus Z_2 \oplus U).
\end{aligned}
$$

On the other hand, T_A'' is computed from the two-block $(A_1'' A_2'')$ as follows

$$T_A'' = E_K(E_K(A_1'' \oplus Z_3) \oplus A_2'' \oplus Z_2)$$
$$= E_K(E_K(P^*) \oplus Z_2 \oplus U),$$

thus $C_3'' = T_A''$, and the verification is successful. The resulting plaintext satisfies $P_2'' = E_K(P^*)$. This attack works for any n-bit data block P^*, it can for instance be used to recover the subkeys of older iFeed encryptions (by putting $P^* = PMN'\|10^* \neq PMN\|1$), and indirectly to decrypt earlier encryptions without having possession of the key K.

5   Why the Attack Works

The attack of Sect. 3 is possible due to two main iFeed properties:

1. iFeed uses Z_2 as masking in both the T_A' and C_{l+1}';
2. The second last ciphertext block (C_1 in our encryption example) is not masked with $Z_4 \oplus U$ as other ciphertext blocks.

Using these properties, the forgery is constructed in such a way upon decryption, $PMN\|1$ is directly fed into E_K and the term U in the final mask $Z_2 \oplus U$ is canceled out.

The original submission document of iFeed [6] comes with an authenticity proof of security in the nonce-respecting setting. Our attack of Sect. 3, however, shows that this security claims is invalid. At a high level, the flaw is caused by an oversight that the subkeys for the associated data and the plaintext are dependent. Indeed, the associated data is masked with $Z_1, \ldots, Z_{a+2}$ and the plaintext with $Z_1 \oplus U, \ldots, Z_{\ell+2} \oplus U$ (where a and ℓ denote the number of associated data and plaintext blocks). For the decryption query, the proof claims that both T_A and $C_{\ell+1}$ is randomly generated except with a small probability.[2] These two cases independently of each other rely on the randomness of Z_0. In our attacks, T_A and $C_{\ell+1}$ are, indeed, both newly and randomly generated. However, their drawing is not independent, in fact, they satisfy $T_A = C_{\ell+1}$. Unfortunately, this is what in our analysis indicates a security problem which as exemplified leads to serious security problems.

6   Conclusion

Our attacks completely disprove and break the authenticity of the iFeed authenticated encryption scheme. In more detail, if an adversary has access to $\mathcal{D}$, it can forge *and* learn the secret subkeys. We furthermore show that with the knowledge of these data, an attacker can use $\mathcal{D}$ to decrypt earlier ciphertexts, even if they

---

[2] In fact, it is claimed that P_{w+1} is random, where w is the first block in the forgery that is different from the older encryption query with the same nonce, and that all subsequent values $P_{w+2}, \ldots, P_\ell, C_{\ell+1}$ are random.

were encrypted under a different nonce (because of Sect. 4). On the other hand, it appears that in the absence of a decryption mechanism, the confidentiality (CPA security) of iFeed still stands.

As a possible remedy to the design of iFeed and future work we suggest the exploration of the possibilities for applying different masks. One option may be to use $Z_i \oplus U$ as is for plaintext encryption but $3 \cdot Z_i$ for the associated data. We, however, do not recommend the replacement of nonce encryptions with encryptions of nonces masked with Z values since the same type of weaknesses surfaces.

Acknowledgments. The authors would like to thank Liting Zhang for verifying the attack. This work was supported in part by European Union's Horizon 2020 research and innovation programme under grant agreement No. 644052 HECTOR and grant agreement No. H2020-MSCA-ITN-2014-643161 ECRYPT-NET, and in part by the Research Council KU Leuven: GOA TENSE (GOA/11/007). Elena Andreeva and Bart Mennink are Postdoctoral Fellows of the Research Foundation – Flanders (FWO).

References

1. Announcing the Advanced Encryption Standard (AES), Federal Information Processing Standards Publication 197. United States National Institute of Standards and Technology (NIST), October 2001
2. Andreeva, E., Bogdanov, A., Luykx, A., Mennink, B., Mouha, N., Yasuda, K.: How to securely release unverified plaintext in authenticated encryption. In: Sarkar, P., Iwata, T. (eds.) ASIACRYPT 2014, Part I. LNCS, vol. 8873, pp. 105–125. Springer, Heidelberg (2014)
3. CAESAR: Competition for Authenticated Encryption: Security, Applicability, and Robustness, May 2014. http://competitions.cr.yp.to/caesar.html
4. Chakraborti, A., Datta, N., Minematsu, K., Gupta, S.S.: Forgery on iFeed[AES] in RUP and Nonce-Misuse Settings, CAESAR mailing list (2015)
5. SHA-3 Competition (2007–2012), United States National Institute of Standards and Technology (NIST). http://csrc.nist.gov/groups/ST/hash/sha-3/
6. Zhang, L., Wu, W., Sui, H., Wang, P.: iFeed[AES] v1, submission to CAESAR competition (2014)

Key-Recovery Attacks Against the MAC Algorithm Chaskey

Chrysanthi Mavromati[1,2]($\boxtimes$)

[1] R&D Lab, Capgemini-Sogeti, Paris, France
chrysanthi.mavromati@sogeti.com
[2] Laboratoire PRISM, Université de Versailles Saint-Quentin-en-Yvelines,
Versailles, France

Abstract. Chaskey is a Message Authentication Code (MAC) for 32-bit microcontrollers proposed by Mouha *et al.* at SAC 2014. Its underlying blockcipher uses an Even-Mansour construction with a permutation based on the ARX methodology. In this paper, we present key-recovery attacks against Chaskey in the single and multi-user setting. These attacks are based on recent work by Fouque, Joux and Mavromati presented at Asiacrypt 2014 on Even-Mansour based constructions. We first show a simple attack on the classical single-user setting which confirms the security properties of Chaskey. Then, we describe an attack in the multi-user setting and we recover all keys of 2^{43} users by doing 2^{43} queries per user. Finally, we show a variant of this attack where we are able to recover keys of two users in a smaller group of 2^{32} users.

Keywords: Message Authentication Code · Collision-based cryptanalysis · ARX · Even-Mansour · Chaskey · Multi-user setting

1 Introduction

A Message Authentication Code (MAC) algorithm is a basic component in many cryptographic systems and its goal is to provide integrity and data authentication. For this, it takes as input a message M and a n-bit secret key K, and outputs a tag τ which is usually appended to M. In general, MAC algorithms are built from universal hash functions, hash functions or block ciphers. The first security property of a MAC is that it should be impossible to recover the key K faster than 2^n operations (key recovery attack). It should also be impossible for an attacker to forge a tag for any unseen message without knowing the key (MAC forgery). Here, the attacker selects a message and simply guesses the correct MAC value without any knowledge of the key. The probability that the guess will be correct is 2^{-t} where t is the size of the tag τ. However, such attacks can be avoided by making the tag sufficiently large. MACs based on universal hash functions compute the universal hash of the message to authenticate and then encrypt this value. MAC algorithms based on block ciphers are usually given as a mode of operation for the block cipher. Typically, CBC-MAC is very similar to

© Springer International Publishing Switzerland 2016
O. Dunkelman and L. Keliher (Eds.): SAC 2015, LNCS 9566, pp. 205–216, 2016.
DOI: 10.1007/978-3-319-31301-6_12

the CBC encryption mode of operation and is one of the most commonly used MACs.

The implementation of a MAC algorithm on a typical microcontroller is a challenging issue as many problems may appear. More precisely, the use of a MAC based on a hash function or a block cipher might have a negative impact on the speed of the microcontroller due to the computational cost of the operations required by the underlying functions.

Recently, at SAC 2014, a new MAC algorithm named Chaskey has been introduced by Mouha *et al.* [8]. It is a permutation-based MAC algorithm and its underlying permutation relies on the ARX design. The designers claim that Chaskey is a lightweight algorithm which overcomes the implementation issues of a MAC on a microcontroller. The construction of Chaskey is based on some simple variants of the CBC-MAC proposed by Black *et al.* [3]. As CBC-MAC is not secure when used with messages of variable length, these variants were designed to authenticate arbitrary length messages. Alternatively, Chaskey can also be seen as an iterated Even-Mansour construction where the same subkey is xored in the input and output of the last permutation round.

In Asiacrypt 2014, Fouque *et al.* [6] presented new techniques to attack the Even-Mansour scheme in the multi-user setting. Based on this work, we present here some key-recovery attacks against the MAC algorithm Chaskey in the single and multi-user setting.

The paper is organized as follows: in Sect. 2 we recall the necessary background, present the multi-user setting, the Even-Mansour scheme and the general principle of collision-based attacks using the distinguished points method, in Sect. 3 we present the main specifications of the Chaskey MAC algorithm and finally, in Sect. 4, we describe new attacks against Chaskey.

2   Preliminaries

2.1   The Multi-user Setting

The security of most schemes in cryptography is usually studied when we have a single recipient of encrypted data: the single-user model. However, this setting ignores an important dimension of the real world where there are many users, who are all using the same algorithms, but each one has its own key. In this setting, the attacker tries to recover all or fraction of keys more efficiently than the complexity of the attack in the single-user model times the number of users. This can be done by amortizing the cost of the attack among the users.

In [7], Menezes studies the security of MAC algorithms in the multi-user setting. He shows that the security degrades when we pass from one to many users. The scenario attack against MAC algorithms in the multi-user setting can be seen as follows. Let $H_K : \{0,1\}^* \rightarrow \{0,1\}^t$ be a family of MACs where $K \in \{0,1\}^k$. In the single-user model, it should be hard for an attacker who has access to an oracle H_K to generate a valid message-tag pair (m, τ) without knowing the key K. In the multi-user setting, we suppose that we have L users with keys $K^{(i)} \in \{0,1\}^k$ where $0 \leq i \leq (L-1)$. Then for an attacker who has access to oracles for $H_{K^{(i)}}$, it should be difficult to produce a triplet (i, m, τ).

2.2 Collision-Based Attacks Using the Distinguished Point Technique Against the Even-Mansour Scheme

The Distinguished Point Technique. The distinguished point technique allows to find collisions in a very efficient way. For this, we first have to define a function f on a set S of size N and then define a distinguished subset S_0 of S with $\mathcal{D}$ distinguished points. The distinguished points should be easy to recognize and generate. For example, we can choose our distinguished points to be n-bits values with d zeros at the end. So, in a set of cardinality $N = 2^n$ our distinguished subset contains $\mathcal{D} = 2^{n-d}$ elements. Starting from a random point $x_0 \in S$ we build chains by evaluating the function f:

$$x_{i+1} = f(x_i).$$

When a distinguished point is detected, *i.e.* $x_\ell \in S_0$, the construction of the chain stops. To be able to recover the chain we need to store the starting point x_0, the distinguished point x_ℓ and the length of the chain ℓ which corresponds to the number of iterations of the function f. Two chains that pass through the same point necessarily end at the same distinguished point. To detect a collision we use the inverse result: two chains that end at the same distinguished point necessarily merge at some point unless one chain is a subchain of the other. Once a collision in the distinguished points set is detected, the real collision can easily be recovered. We assume that we have two colliding chains of length ℓ and ℓ' and that $\ell \geq \ell'$. Then, starting from the longer chain, we rebuild the chain by taking exactly $\ell - \ell'$ steps. From that point, it now suffices to build both chains in parallel until a collision is reached.

Collision-Based Attacks Against Even-Mansour. The Even-Mansour scheme is a minimalistic very efficient design of a block cipher proposed at Asiacrypt 91 [5]. The main idea is the construction of a keyed permutation family Π_{K_1, K_2} by a public permutation π that operates on n-bit values ($N = 2^n$):

$$\Pi_{K_1, K_2} = \pi(m \oplus K_1) \oplus K_2,$$

where m is the plaintext and K_1 and K_2 are the two whitening keys. There is also a simpler version presented by Dunkelman *et al.* [4], the *Single-key Even-Mansour*, where $K_1 = K_2$. Even and Mansour showed that this simple block-cipher is secure up to $\mathcal{O}(2^{n/2})$ queries of the adversary to the keyed permutation Π and to the public permutation π. It has also been proved [4] that the Single-key Even-Mansour has the same security bound as the original version.

   In [6], Fouque *et al.* describe a new technique for collision-based attacks using the distinguished point technique. They use this method to form attacks against the Even-Mansour scheme in the multi-user setting. The main idea is to find a collision between two chains, one constructed from the public permutation π and one from the keyed permutation Π. For this, as we cannot use permutations to detect collisions, we have to construct two functions F and f, one based on Π and one based on π. In previous attacks on Even-Mansour [2,4], these functions have been constructed by using the Davies-Meyer construction: $F(m) = \Pi(m) \oplus \Pi(m \oplus \delta)$ and $f(m) = \pi(m) \oplus \pi(m \oplus \delta)$. However, these functions cannot be used with the distinguished point technique as chains built by F and f defined as previously can eventually cross but they cannot merge as they consist of evaluations of two different functions. As a consequence, they use a different definition of F and f to solve this problem:

$$F(m) = m \oplus \Pi(m) \oplus \Pi(m \oplus \delta) \quad \text{and} \quad f(m) = m \oplus \pi(m) \oplus \pi(m \oplus \delta).$$

For two messages m and m' such as $m' = m \oplus K_1$, they remark that $F(m') = f(m) \oplus K_1$ and so they get two *parallel* chains, *i.e.* two chains that have a constant difference between them.

They also remark that, for two different users i and j, two chains constructed by using $F_{\Pi}^{(i)}$ and $F_{\Pi}^{(j)}$, where $F_{\Pi}^{(i)}(m) = m \oplus \Pi^{(i)}(m) \oplus \Pi^{(i)}(m \oplus \delta)$ and $F_{\Pi}^{(j)}(m) = m \oplus \Pi^{(j)}(m) \oplus \Pi^{(j)}(m \oplus \delta)$, can also become parallel and their constant difference would be equal to the XOR of the users keys, *i.e.* $K_1^{(i)} \oplus K_1^{(j)}$.

To attack Even-Mansour using the distinguished point technique in the multi-user setting, they build a set of chains for the public user using the function f and a small number of chains for every user by using the keyed permutation F. Whenever a collision $F^{(i)}(x) = F^{(j)}(y)$ for two points x and y, where $x = y \oplus K_1$, is detected between two users U_i and U_j, it yields $K_1^{(i)} \oplus K_1^{(j)}$. From these collisions it is possible to construct a graph whose vertices are the users and the edges represent the xor of the first keys of the users with two colliding chains, *i.e.* $K_1^{(i)} \oplus K_1^{(j)}$. When enough edges are present a giant component appears in the graph. Then, it suffices to find a single collision $F^{(i)}(x) = f(y)$ between a user and the public user to reveal all keys of the users in the giant component.

One of the main ideas of their technique is to detect parallel chains by simply testing if $\pi(y) \oplus \pi(y \oplus \delta) = \Pi(x) \oplus \Pi(x \oplus \delta)$ for two distinguished point x and y where $x = y \oplus K_1$. This does not add any extra cost, as values of $\pi(y) \oplus \pi(y \oplus \delta)$ and $\Pi(x) \oplus \Pi(x \oplus \delta)$ are needed to calculate the next element of the chain. Also, it does not require to go back and recompute the chains to detect the merging points.

3   The MAC Algorithm Chaskey

Chaskey was proposed at SAC 2014 by Mouha *et al.* It is a permutation-based MAC algorithm and its underlying permutation is based on the ARX design. Its design is similar to the permutation of the MAC algorithm SipHash [1]. However, in Chaskey, a state of 128-bits (instead of 256-bits in SipHash) is used which is decomposed in 4 words of 32-bits (instead of 64-bits in SipHash). Also, different rotation constants are used.

Chaskey takes as input a message M of arbitrary size and a 128-bit key K. The message M is split into ℓ blocks $m_1, m_2, \ldots, m_\ell$ of 128 bits each. If the last block is incomplete, a padding is applied. It outputs the t-bit tag τ (where $t \leq n$) that authenticates the message M. The underlying function is a permutation constructed using the ARX design.

Two subkeys K_1 and K_2 are generated from K as follows: K_1 is equal to the result of the multiplication by 2 (binary notation) of K and K_2 is equal to the multiplication by 2 of K_1. In general, we define $K_1 = \alpha K$ and $K_2 = \alpha^2 K$. To define multiplication in $GF(2^n)$, we need to specify the irreducible polynomial $f(x)$ of degree n that defines the representation of the field. The designers of Chaskey choose their irreductible polynomial to be $f(x) = x^{128} + x^7 + x^2 + x + 1$.

Chaskey operates in $\ell + 3$ steps. On the first step, we xor the first block m_1 with the key K. On the next step, we process the previous result through the permutation π and we xor the output with the next block. This procedure is repeated $\ell - 1$ times until all blocks are being processed. If the last block m_ℓ, which has been xored at the end of the $\ell - 1$ step, is a complete block, then, on the ℓ-th step, we simply xor the key K_1. If it is an incomplete block, it should be padded with $10^{n-|m_\ell|-1}$ before being used. Then, on the ℓ-th step, the key K_2 will be used instead of K_1. Then, the state will pass through the permutation π and will be xored with K_1 if m_ℓ is complete or with K_2 if incomplete. Finally, the t least significant bits are selected to be used as the tag τ. The whole procedure can be seen on Fig. 1.

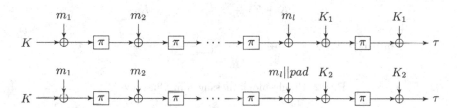

Fig. 1. The Chaskey MAC algorithm. First line when $|m_\ell| = n$ and second line when $0 \leq |m_\ell| \leq n$ where $pad = 10^{n-|m_\ell|-1}$.

There is also a variant of Chaskey, called Chaskey-B, where Chaskey can be seen as a MAC algorithm based on the Even-Mansour [5] block cipher. The authors define the block cipher E as $E_{X||Y}(m) = \pi(m \oplus X) \oplus Y$ and so it suffices to evaluate recursively the function $h_{i+1} = E_{K||K}(h_i \oplus m_i)$ for $i = 1, \ldots, l-1$ and $h_1 = 0^n$. If the last block is complete, we calculate $h_\ell = E_{K \oplus K_1||K_1}(h_\ell \oplus m_\ell)$. If not, we calculate $h_\ell = E_{K \oplus K_2||K_2}(h_\ell \oplus m_\ell||10^{n-|m_\ell|-1})$. Finally, the tag τ is equal to the t least significant bits. However, in this case, we can easily see that the key K, which is XORed after the application of the permutation π, vanishes at each iteration of the block cipher E. As a result, the key K intervenes only on the first block and the keys K_1 or K_2 on the last block:

$$\tau = \pi(\pi(\pi(\pi(m_1 \oplus K) \oplus m_2) \oplus \ldots) \oplus m_l \oplus K_1) \oplus K_1.$$

The permutation π consists of 8 rounds of a function that follows the ARX design: addition modulo 2^{32}, bit rotations and XOR. For constructing this round function, the authors use the same structure as SipHash [1] but instead of 64-bit words they are using words of 32 bits and different rotation constants. They consider that 8 applications of the round function is secure but they suggest that the 16 rounds version (Chaskey-LTS: long term security) should also be implemented in case of security issues.

The authors prove that Chaskey is secure up to $D = 2^{n/2}$ chosen plaintexts and $T = 2^n/D$ queries to π or π^{-1}.

4   Collision-Based Attacks Against Chaskey

In this section, we show that the collision-based attack described in [6] can be applied on Chaskey. Furthermore, we show that variants of this attack can be applied in the case of Chaskey. All attacks can be performed when we use single-block messages. Chaskey then becomes an Even-Mansour cipher:

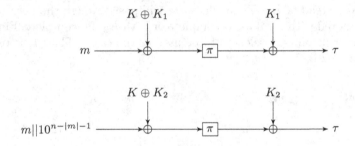

Fig. 2. Single-block messages in Chaskey

The main idea of all attacks is to build chains by using a function based on Chaskey and then search for collisions between them. For this, a function that can be used to build chains should be defined. However, in the case of Chaskey, for every user we can build two different chains depending on the subkey that we are using (Fig. 2).

To build the chains needed, we define the functions:

$$f_s(M) = K_s \oplus \pi(M \oplus (K_s \oplus K))$$

$$F_{f_s}(M) = f_s(M) \oplus f_s(M \oplus \delta) \oplus M$$

where K is the users key, K_s with $s \in \{1, 2\}$ represents the two subkeys generated as mentioned in Sect. 3 and δ is an arbitrary but fixed non zero constant.

In the multi-user setting we assume that L different users are all using the Chaskey MAC algorithm based on the same public permutation π. Each user U_i, with $0 \leq i \leq L$, chooses its own key $K^{(i)}$ at random and independently from all the other users and generates K_s with $s \in \{1, 2\}$. We define the functions $f_s(M)$ and $F_{f_s}(M)$ as above and we also define the function $F_\pi(M)$ for the public user as follows:

$$F_\pi(M) = M \oplus \pi(M) \oplus \pi(M \oplus \delta).$$

We remark that for two plaintexts M and M' where $M' = M \oplus K_s \oplus K$, we have $F_{f_s}(M \oplus K_s \oplus K) = F_\pi(M) \oplus (K_s \oplus K)$. So, two chains based on functions F_{f_s} and F_π may become parallel.

4.1 Key-Recovery Attack in the Single-User Setting

In this section, we show that an attack with complexity 2^{64} can be applied in the classical single-user scenario. This attack does not contradict the security bound showed in the original paper of Chaskey and has similar complexity with possible slide attacks based on [2]. However, with this attack we show a simple application on Chaskey of the parallel chains detection technique.

The attack is described below:

1. Create two chains constructed by using both:

$$F_{f_1}(M) = f_1(M) \oplus f_1(M \oplus \delta) \oplus M \text{ and } F_{f_2}(M') = f_2(M') \oplus f_2(M' \oplus \delta) \oplus M'$$

   until $f_1(M) \oplus f_1(M \oplus \delta)$ and $f_2(M') \oplus f_2(M' \oplus \delta)$ reach a distinguished point.
2. Store all endpoints of the constructed chains and search for collisions between the two different types of chains, *i.e.* for two plaintexts M and M' search for $(F_{f_1}(M))^a = (F_{f_2}(M'))^b$ where $a, b \geq 1$ is the number of iterations of the respective functions. (For the rest of this paper, to facilitate the reading, the use of a and b will be omitted from equations that represent collisions between chains.)
3. If a collision is found, recover the XOR of the two inputs $M \oplus M'$ which is expected to be equal to:

$$K_1 \oplus K \oplus K_2 \oplus K = (\alpha + \alpha^2)K$$

and thus recover the key K.

Analysis of the Attack. To find a collision between the two sets of endpoints, one constructed by using the function F_{f_1} and one constructed by using F_{f_2}, we need to construct two chains each of length 2^{64}. So, the total cost of the attack is 2^{64}. As said previously, this attack does not contradict the security claim of Chaskey. It is just an example which shows how to use the distinguished point technique to attack Chaskey in the single-user setting.

4.2 Key-Recovery Attack in the Multi-user Setting

To apply the attack of Fouque *et al.* we use the iteration functions f_s, F_{f_s} and F_π defined previously. The attack is described below:

1. In a set of L users, for every user U_i where $0 \leq i \leq (L-1)$, build a constant number of chains, starting from an arbitrary plaintext, using the function F_{f_s}. For every user, create some chains using the function $F_{f_1}(M) = f_1(M) \oplus f_1(M \oplus \delta) \oplus M$ until $f_1(M) \oplus f_1(M \oplus \delta)$ reaches a distinguished point and some chains using the function $F_{f_2}(M) = f_2(M) \oplus f_2(M \oplus \delta) \oplus M$ until $f_2(M) \oplus f_2(M \oplus \delta)$ reaches a distinguished poit.

2. Construct some chains for the unkeyed user starting from an arbitrary plaintext, by iterating the function F_π.

3. Store the endpoints and search for collisions between the users. In the case of Chaskey, we can have three types of collisions between the keyed users:
   - A collision between the two chains $F_{f_1}(M)$ and $F_{f_2}(M')$ of the same user i and, in this case, we recover $K_1^{(i)} \oplus K^{(i)} \oplus K_2^{(i)} \oplus K^{(i)}$. As $K_1^{(i)} = \alpha K^{(i)}$ and $K_2^{(i)} = \alpha^2 K^{(i)}$, we have that $M \oplus M' = (\alpha + \alpha^2)K^{(i)}$.
   - A collision between two similar chains of two different users i and j: $F_{f_1}^{(i)}(M) = F_{f_1}^{(j)}(M')$ or $F_{f_2}^{(i)}(M) = F_{f_2}^{(j)}(M')$. Then, we recover $K_1^{(i)} \oplus K^{(i)} \oplus K_1^{(j)} \oplus K^{(j)} = (1+\alpha)(K^{(i)} \oplus K^{(j)})$ or $K_2^{(i)} \oplus K^{(i)} \oplus K_2^{(j)} \oplus K^{(j)} = (1 + \alpha^2)(K^{(i)} \oplus K^{(j)})$.
   - A collision between two different type of chains between two different users i and j (cross collision): $F_{f_1}^{(i)}(M) = F_{f_2}^{(j)}(M')$. Then, we learn $K_1^{(i)} \oplus K^{(i)} \oplus K_2^{(j)} \oplus K^{(j)} = (1 + \alpha)K^{(i)} \oplus (1 + \alpha^2)K^{(j)}$.

   Also search for collisions between a chain of a keyed user i and the unkeyed user for whom we build chains by using the function F_π. A collision of this type may occur by using both F_{f_1} or F_{f_2}. From the first function, we learn $K_1^{(i)} \oplus K^{(i)}$ and from the second function $K_2^{(i)} \oplus K^{(i)}$. As a consequence, in both cases, we are able to recover $K^{(i)}$.

4. Build a graph where vertices represent the keyed and unkeyed users. More precisely, each user is represented by two vertices as for each user we use two different functions to build chains. Whenever a collision is obtained between two chains, add an edge between the corresponding vertices. This edge is labelled by the relation between the keys as explained in the previous step. An example of the graph can be seen in Fig. 3.

5. From the key relations found before, resolve the system and recover the users keys. With only a single collision for the unkeyed user, we learn almost all keys of the giant component. If we also find a collision for the unkeyed user, then we are able to learn all keys of our giant component.

Analysis of the Attack. From [6] we know that in a group of $N^{1/3}$ users we expect to recover almost all keys by doing $c \cdot N^{1/3}$ queries per user (where c is a small arbitrary constant) and $N^{1/3}$ unkeyed queries. So, for Chaskey, we are able to recover almost all keys of a group of 2^{43} users by doing 2^{43} queries to the unkeyed user and 2^{43} queries per user.

Improvement. We show here that in the case of Chaskey, the use of δ can be eliminated. This technique can be used when we search for a collision between two similar chains of two different users.

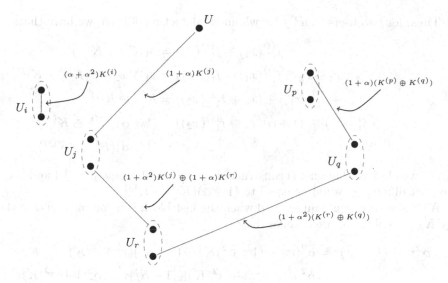

Fig. 3. Example of a giant component

If we use a full-block message x and its corresponding ciphertext y we observe that:

$$\alpha x \oplus (1+\alpha)y = \alpha(x' \oplus (1+\alpha)K) \oplus (1+\alpha)(y' \oplus \alpha K)$$
$$= \alpha x' \oplus \alpha(1+\alpha)K \oplus (1+\alpha)y' \oplus \alpha(1+\alpha)K$$
$$= \alpha x' \oplus (1+\alpha)y'$$

where x, x', y and y' are represented in Fig. 4.

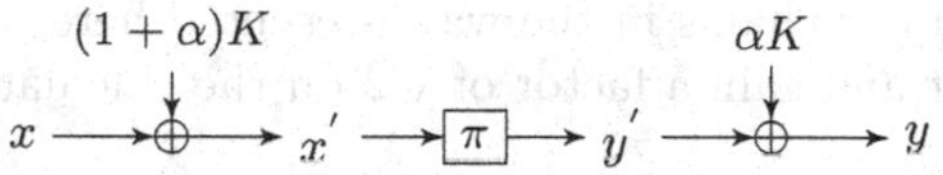

$$(1+\alpha)K \qquad\qquad \alpha K$$
$$x \longrightarrow \oplus \longrightarrow x' \longrightarrow \boxed{\pi} \longrightarrow y' \longrightarrow \oplus \longrightarrow y$$

Fig. 4. The MAC algorithm Chaskey when single-block full messages are used (the subkey $K_1 = \alpha K$ is used).

As previously, our goal is to define a function and build chains by using this function. Here, we define the function $F_{f_1}(x)$ as follows:

$$F_{f_1}(x) = x \oplus \alpha x \oplus (1+\alpha)f_1(x) = (1+\alpha)(x \oplus f_1(x))$$

where $f_1(x) = K_1 \oplus \pi(x \oplus (K_1 \oplus K))$. We also assume that two plaintexts x_1 and x_2 satisfy $x_1 \oplus x_2 = (1+\alpha)(K^{(i)} \oplus K^{(j)})$.

Thus, for two users i and j for whom we detect a collision, we have that:

$$f_1^{(i)}(x_1) \oplus f_1^{(j)}(x_2) = \alpha(K^{(i)} \oplus K^{(j)})$$

$$x_1 \oplus x_2 \oplus f_1^{(i)}(x_1) \oplus f_1^{(j)}(x_2) = x_1 \oplus x_2 \oplus \alpha(K^{(i)} \oplus K^{(j)})$$

$$(x_1 \oplus f_1^{(i)}(x_1)) \oplus (x_2 \oplus f_1^{(j)}(x_2)) = K^{(i)} \oplus K^{(j)}$$

$$(1+\alpha)(x_1 \oplus f_1^{(i)}(x_1)) \oplus (1+\alpha)(x_2 \oplus f_1^{(j)}(x_2)) = (1+\alpha)(K^{(i)} \oplus K^{(j)})$$

$$F_{f_1}^{(i)} \oplus F_{f_1}^{(j)} = (1+\alpha)(K^{(i)} \oplus K^{(j)})$$

and so we observe that chains constructed by F_{f_1} can become parallel and their constant difference would be equal to $(1+\alpha)(K^{(i)} \oplus K^{(j)})$.

A similar technique can be used when the last block is incomplete and so the key K_2 is used. Here, we observe that:

$$\alpha^2 m \oplus (1+\alpha^2)\tau = \alpha^2(m' \oplus (1+\alpha^2)K) \oplus (1+\alpha^2)(\tau' \oplus \alpha^2 K)$$

$$= \alpha^2 m' \oplus \underline{\alpha^2(1+\alpha^2)K} \oplus (1+\alpha^2)\tau' \oplus \underline{\alpha^2(1+\alpha^2)K}$$

$$= \alpha^2 m' \oplus (1+\alpha^2)\tau'.$$

Thus, in this case, we define the function F_{f_2} as follows:

$$F_{f_2}(x) = x \oplus \alpha^2 x \oplus (1+\alpha^2)f_2(x) = (1+\alpha^2)(x \oplus f_2(x))$$

where $f_2(x) = K_2 \oplus \pi(x \oplus (K_2 \oplus K))$. We also assume that two plaintexts x_1 and x_2 satisfy $x_1 \oplus x_2 = (1+\alpha^2)(K^{(i)} \oplus K^{(j)})$.
Equivalently, for two users i and j, we observe that:

$$F_{f_2}^{(i)} \oplus F_{f_2}^{(i)} = (1+\alpha^2)(K^{(i)} \oplus K^{(j)}).$$

So, chains constructed by F_{f_2} can also become parallel and their constant difference would be equal to $(1+\alpha^2)(K^{(i)} \oplus K^{(j)})$.

Thus, if we build our chains in the way presented here, we are able to have a small improvement and gain a factor of $\sqrt{2}$ on the calculation of our chains.

4.3   Variant of the Previous Attack with Cross Collisions

We show in this section that a variant of the previous attack is also possible. Indeed, we are able to apply a similar technique and we show that we can learn keys of two users when we detect one cross collision between them.
The attack works as follows:

1. For two users U_i and U_j in a set of L users, build a constant number of chains, starting from an arbitrary plaintext, using the function F_{f_s}, for $s \in \{1, 2\}$. For each user, create some chains using the function $F_{f_1}(M) = f_1(M) \oplus f_1(M \oplus \delta) \oplus M$ and some chains using the function $F_{f_2}(M) = f_2(M) \oplus f_2(M \oplus \delta) \oplus M$. The construction of the chains stops when $f_1(M) \oplus f_1(M \oplus \delta)$ and $f_2(M) \oplus f_2(M \oplus \delta)$ reach a distinguished point.

2. For each chain, store the endpoints and search for a cross collision. A cross collision for users U_i and U_j and for two plaintexts M and M' is detected when $F_{f_1}^{(i)}(M) = F_{f_2}^{(j)}(M')$. This indicates that the XOR of the two inputs $M \oplus M'$ is expected to be equal to:

$$K_1^{(i)} \oplus K^{(i)} \oplus K_2^{(j)} \oplus K^{(j)} = (1+\alpha)K^{(i)} \oplus (1+\alpha^2)K^{(j)}.$$

3. If a cross collision is detected, recover also the XOR of the corresponding outputs, which is equal to:

$$K_1^{(i)} \oplus K_2^{(j)} = \alpha K^{(i)} \oplus \alpha^2 K^{(j)}.$$

4. Solve the system:

$$(1+\alpha)K^{(i)} \oplus (1+\alpha^2)K^{(j)} = \Delta_1$$
$$\alpha K^{(i)} \oplus \alpha^2 K^{(j)} = \Delta_2$$

and thus recover $K^{(i)}$ and $K^{(j)}$.

Analysis of the Attack. This attack uses similar techniques to the attack described before. However, the innovation here consists of the fact that we are able to recover keys of two users, in a smaller group than before, by finding only a cross collision between them. More specifically, to find a cross collision between two users, it suffices to have a group of $\sqrt[4]{N}$ users and perform $\sqrt[4]{N}$ queries per user. So, in the case of Chaskey, we are able to recover two keys in a group of 2^{32} by doing 2^{32} queries per user. However, if we want to recover the keys of all users, then we need a group of 2^{43} users as previously.

5 Conclusion

In this paper, we presented key-recovery attacks against the MAC algorithm Chaskey. All attacks are using algorithmic ideas for collision based attacks of Fouque *et al.* presented in [6]. They all work when using single-block messages in the single or multi-user setting.

First, we show how to use these techniques to form an attack in the classical single-user setting. By applying this attack, we are able to recover the key of the user by doing 2^{64} operations. However, this attack does not contradict the security claim of Chaskey. Next, we presented two attacks in the multi-user setting. The first one is able to recover almost all keys of a group of 2^{43} users by doing 2^{43} queries per user. We also show that we can improve this attack and gain a factor of $\sqrt{2}$. Finally, the second attack in the multi-user setting, is able to recover the keys of 2 users in a smaller group of 2^{32} by doing 2^{32} queries per user. We are able to achieve this new result by exploiting the use of two different keys for the last block of Chaskey.

References

1. Aumasson, J.-P., Bernstein, D.J.: SipHash: a fast short-input PRF. In: Galbraith, S., Nandi, M. (eds.) INDOCRYPT 2012. LNCS, vol. 7668, pp. 489–508. Springer, Heidelberg (2012)
2. Biryukov, A., Wagner, D.: Advanced slide attacks. In: Preneel, B. (ed.) EURO-CRYPT 2000. LNCS, vol. 1807, pp. 589–606. Springer, Heidelberg (2000)
3. Black, J., Rogaway, P.: CBC MACs for arbitrary-length messages: the three-key constructions. J. Cryptology **18**(2), 111–131 (2005)
4. Dunkelman, O., Keller, N., Shamir, A.: Minimalism in cryptography: the even-mansour scheme revisited. In: Pointcheval, D., Johansson, T. (eds.) EUROCRYPT 2012. LNCS, vol. 7237, pp. 336–354. Springer, Heidelberg (2012)
5. Even, S., Mansour, Y.: A construction of a cipher from a single pseudorandom permutation. In: Matsumoto, T., Imai, H., Rivest, R.L. (eds.) ASIACRYPT 1991. LNCS, vol. 739, pp. 210–224. Springer, Heidelberg (1993)
6. Fouque, P.-A., Joux, A., Mavromati, C.: Multi-user collisions: applications to discrete logarithm, even-mansour and PRINCE. In: Sarkar, P., Iwata, T. (eds.) ASI-ACRYPT 2014. LNCS, vol. 8873, pp. 420–438. Springer, Heidelberg (2014)
7. Menezes, A.: Another look at provable security. In: Pointcheval, D., Johansson, T. (eds.) EUROCRYPT 2012. LNCS, vol. 7237, p. 8. Springer, Heidelberg (2012)
8. Mouha, N., Mennink, B., Van Herrewege, A., Watanabe, D., Preneel, B., Ver-bauwhede, I.: Chaskey: an efficient MAC algorithm for 32-bit microcontrollers. In: Joux, A., Youssef, A. (eds.) SAC 2014. LNCS, vol. 8781, pp. 306–323. Springer, Heidelberg (2014)

Differential Forgery Attack Against LAC

Gaëtan Leurent[(✉)]

Inria, project-team SECRET, Rocquencourt, France
Gaetan.Leurent@inria.fr

Abstract. LAC is one of the candidates to the CAESAR competition. In this paper we present a differential forgery attack on LAC. We study the collection of characteristics following a fixed truncated characteristic, in order to obtain a lower bound on the probability of a differential. We show that some differentials have a probability higher than 2^{-64}, which allows a forgery attack on the full LAC.

This work illustrates the difference between the probability of differentials and characteristics, and we describe tools to evaluate the probability of some characteristics.

Keywords: Differential cryptanalysis · Differentials · Characteristics · Forgery attack · Truncated differential · LBlock · LAC

1 Introduction

The CAESAR competition is an ongoing effort to identify new authenticated encryption primitives [3]. Authenticated encryption schemes provide both confidentiality and authenticity in a single primitive, instead of using an encryption scheme together with a MAC. The competition received 57 submissions in March 2014, and an important effort is now devoted to analyzing those candidates.

LAC is a CAESAR candidate designed by Zhang, Wu, Wang, Wu, and Zhang [10]. LAC uses the same structure as ALE [2]: it is based on a modified block cipher (the G function in LAC is based on LBlock [9]) that leaks part of its state. The main step of the algorithm is to encrypt the current state, and the leaked data is used as a keystream to produce the ciphertext. Meanwhile, a key schedule produces new keys for each encryption, and plaintext blocks are xored inside the state, so that the final state can be used to produce the tag T. This is depicted in Fig. 1.

In LAC, the main state is 64-bit wide, the key register is 80-bit wide, and the plaintext is divided in blocks of 48 bits. The G function is a modified version of LBlock. It uses 16 rounds of Feistel network, where the round function F applies a key addition, 8 parallel S-Boxes on the nibbles of the state, (the 8 S-Boxes are identical), and a nibble permutation. In addition, the inactive branch of the Feistel network is rotated by 2 nibbles; this is shown in in Fig. 2. The S-Box has a maximum differential probability of 2^{-2}, which is optimal for a 4-bit S-Box; it is defined as:

© Springer International Publishing Switzerland 2016
O. Dunkelman and L. Keliher (Eds.): SAC 2015, LNCS 9566, pp. 217–224, 2016.
DOI: 10.1007/978-3-319-31301-6_13

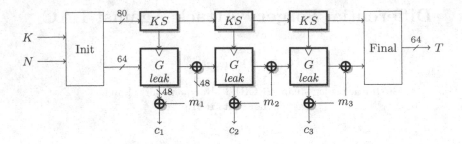

Fig. 1. LAC main structure

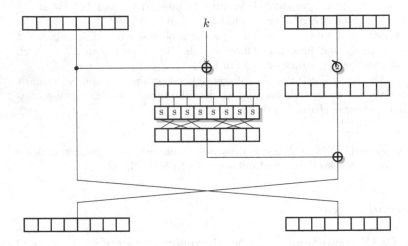

Fig. 2. A Feistel round of LAC (LBlock-s)

$$S = \{\mathsf{E}, 9, \mathsf{F}, 0, \mathsf{D}, 4, \mathsf{A}, \mathsf{B}, 1, 2, 8, 3, 7, 6, \mathsf{C}, 5\}.$$

We omit the description of the leak function and of the key schedule, because they don't affect our attack.

The security goals of LAC against forgery attacks is stated as:

Claim 2 (Integrity for the Plaintext). The security claim of integrity for the plaintext is that any forgery attack with an unused tuple $(PMN^*, \alpha^*, c^*, \tau^*)$ has a success probability at most 2^{-64}.

1.1 Description of the Attack

Our attack is a differential forgery attack: given the authenticated encryption (C, T) of a message M, we build a cipher-text $(C', T') = (C \oplus \Delta, T)$ that is valid with a probability higher than 2^{-64}.

More precisely, we use a two-block difference $\Delta = (\alpha, \beta)$ so that a difference α is first injected in the state, and we predict the difference β after one evaluation of G in order to cancel it. This will be successful if we can find a differential $\alpha \rightsquigarrow \beta$ in the function G with a probability p higher than 2^{-64}.

In Sect. 2.2, we give a differential with probability $p \approx 2^{-61.52}$. This yields a forgery attack using a single known ciphertext (of at least two blocks), with a success probability of $2^{-61.52}$.

In addition, the truncated characteristic we use does not affect the leaked output, so that, if the tag is valid, the plaintext corresponding to $(C \oplus \Delta, T)$ is $M \oplus \Delta$.

1.2   Characteristics and Differentials

We now introduce important notions for differential cryptanalysis.

A *differential* is given by an input difference α and an output difference β. The probability of the differential is the probability than a pair of plaintext with difference α gives a pair of ciphertext with difference β:

$$\Pr[\alpha \rightsquigarrow \beta] = \Pr_{K,x}[E(x \oplus \alpha) = E(x) \oplus \beta].$$

The probability of differentials is important to evaluate the security of a cipher against differential cryptanalysis, but it is quite challenging to compute this probability. Therefore, we introduce the notion of characteristics.

A *characteristic* is given by an input difference α, the difference α_i after each round, and the output difference β. Since all the intermediate difference are fixed, it is quite easy to evaluate the probability of a characteristic using the Markov cipher model (*i.e.* assuming that the rounds are independent). The probability of the differential $\alpha \rightsquigarrow \beta$ is the sum of the probability of all characteristics with input difference α and output difference β.

The designers of LAC studied its resistance against differential cryptanalysis using truncated characteristics. They show that any characteristic must have at least 35 active S-Boxes. Since the best transitions for the S-Box have a probability of 2^{-2}, any characteristic has a probability at most 2^{-70}. However, this does not imply a lower bound for the probability of *differentials*: if many good characteristics contribute to the same differential, the probability can increase significantly.

Proving an upper bound on the probability of differential is much harder than proving an upper bound for characteristics, and very few results are known in this setting. A notable example is the AES, for which an upper bound of 2^{-150} for any 4-round characteristic can easily be shown [4], and an upper bound of 1.881×2^{-114} for any 4-round differential was proved using significantly more advanced techniques [5].

In this work we give a more accurate estimation of the probability of differentials in the G function of LAC by considering more than one characteristic. Our results actually lead to a lower bound on the probability of some differential.

2   Characteristics Following the Same Truncated Trail

Truncated differential cryptanalysis was introduced by Knudsen in 1994 [6]. A truncated characteristic D does not specify the exact value of the differences at each step but uses partial information. The state is divided in words of a fixed size (usually bytes, but we use nibbles for LAC), and D only specifies whether the difference in each word in zero (inactive word) or non-zero (active word).

For a given truncated characteristic D, there exist many ways to instantiate the input/output differences and the intermediate differences. For a given input/output difference (α, β), we consider all the possible intermediate differences following D; this defines a collection of characteristics that all contribute to the same differential. If we can efficiently compute the sum of the probabilities of all those characteristics, this will give a more accurate lower bound of $\Pr[\alpha \rightsquigarrow \beta]$ than by considering a single characteristic.

2.1   Related Work

Recently, a technique was proposed to find differential characteristics using Mixed Integer Linear Programming (MILP) [7,8]. When a good characteristic $\alpha \rightsquigarrow \beta$ is found, this technique can also be used to find a collection of characteristics following the same differential $\alpha \rightsquigarrow \beta$, by adding this constraint to the MILP problem. This has been applied quite successfully, but it inherently requires to enumerate all the considered characteristics.

An analysis of TWINE by Biryukov, Derbez and Perrin is also based on clustering differential characteristics [1], using the same technique as presented here. TWINE is very similar to LBlock (and LAC), but the S-Box has a more uniform differential probability which limits this clustering effect compared to our results on LAC.

2.2   Computation of the Probability of a Truncated Characteristic

In this work we use a technique to compute the probability of a collection of characteristics without having to explicitly list all the characteristics. This allows to take into account a large number of characteristics. The collection of characteristics is defined by a truncated characteristic D, which specifies whether each word is active (*i.e.* with a non-zero difference), or inactive.

We denote by $\Pr[D : \alpha \rightsquigarrow \beta]$ the probability that a pair with input difference α gives an output difference β, in a way that all the intermediate differences follow the truncated characteristic D. We also denote the reduced version of D with only i rounds as D_i. We will compute exactly $\Pr[D : \alpha \rightsquigarrow \beta]$, *i.e.* we consider the collection of *all* characteristics corresponding to the truncated characteristics, with all possible choices of non-zero values.

In order to compute $\Pr[D : \alpha \rightsquigarrow \beta]$ for a given (α, β), we will first compute $\Pr[D_1 : \alpha \rightsquigarrow \alpha_j^{(1)}]$ for all the differences $\alpha_j^{(1)}$ following D_1. Then we iteratively build $\Pr\left[D_i : \alpha \rightsquigarrow \alpha_k^{(i)}\right]$ for all $\alpha_k^{(i)}$ following D_i using the results for D_{i-1}:

$$\Pr\left[D_i : \alpha \rightsquigarrow \alpha_k^{(i)}\right] = \sum_j \Pr\left[D_{i-1} : \alpha \rightsquigarrow \alpha_j^{(i-1)}\right] \times \Pr\left[\alpha_j^{(i-1)} \rightsquigarrow \alpha_k^{(i)}\right]$$

This corresponds to a matrix multiplication, where the matrix $M_i = \left[\Pr[\alpha_j^{(i-1)} \rightsquigarrow \alpha_k^{(i)}]\right]$ contains the probability of all the transitions $\alpha_j^{(i-1)} \rightsquigarrow \alpha_k^{(i)}$ corresponding to round i of the truncated characteristic. However, we don't explicitly perform a vector-matrix product, because the matrix is very sparse; instead we deduce the possible $\alpha_j^{(i-1)}$ with a non-zero $\Pr\left[\alpha_j^{(i-1)} \rightsquigarrow \alpha_k^{(i)}\right]$ from each $\alpha_k^{(i)}$.

2.3   Application to LAC

In order to apply this analysis to LAC, we first have to identify a good truncated characteristic. We use an automatic search for truncated characteristics, where we represent the state with a 16-bit vector, with zero and one to represent active and inactive nibbles. After computing all the possible transitions (there are at most $2^8 \times 2^{16}$ allowed transitions because of the Feistel structure), the problem of finding an optimal r-round truncated differential is reduced to the search of a shortest path in graph with $(r + 1) \times 2^{16}$ nodes, and at most $(r+1) \times 2^{24}$ edges. Moreover, the graph is structured with edges only from node of round i to nodes of round $i + 1$. This allows a very efficient search, round by round, with complexity $(r + 1) \times 2^{24}$.

We found several truncated characteristics with 35 active S-Boxes, and we use the one given in Fig. 3. Gray square denote active nibbles. When two active nibbles are xor-ed, the truncated characteristic specifies whether the sum should be zero (slashed square) or non-zero (black square). We note that this characteristic has at most 6 active nibbles at a given round; therefore there are at most 2^{24} possible differences $\alpha_j^{(i)}$ at round i, and the vector $\left[\Pr\left[D_i : \alpha \rightsquigarrow \alpha_j^{(i)}\right]\right]$ has at most 2^{24} entries. Moreover, each step has at most 3 active S-Boxes, therefore we have at most 2^9 possible transitions to consider for any fixed $\alpha_j^{(i)}$. Using this truncated characteristic, the algorithm can compute $\Pr[D : \alpha \rightsquigarrow \beta]$ for a fixed α and for all differences β following D with at most $16 \times 2^9 \times 2^{24} = 2^{37}$ simple operations.

After running this computation with all input differences α allowed by the truncated characteristic, we identified 17512 differentials with probability higher than 2^{-64}; the best differential identified by this algorithm has a probability $\Pr[D : \alpha \rightsquigarrow \beta] \approx 2^{-61.52}$. More precisely, the best differentials found are:

$$\Pr\left[0000000000004607 \overset{16}{\rightsquigarrow} 0000040000004400\right] \geq 2^{-61.52}$$

$$\Pr\left[0000000000004607 \overset{16}{\rightsquigarrow} 0000060000004400\right] \geq 2^{-61.52}$$

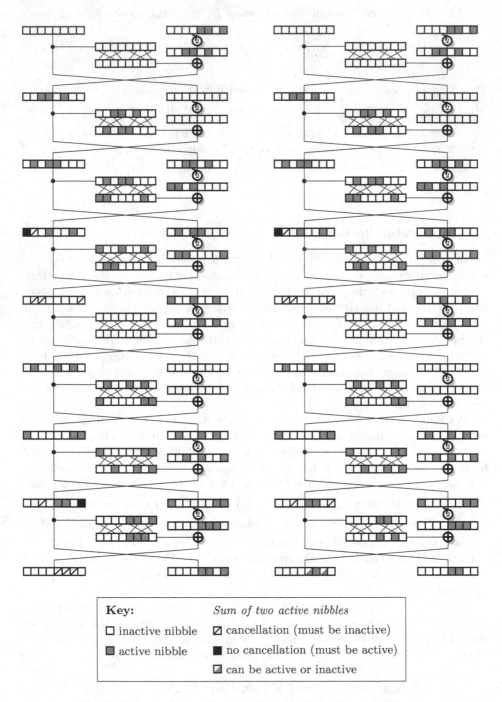

Fig. 3. Truncated characteristic for LAC with 35 active S-boxes.

These probabilities correspond to a collection of 302116704 truncated characteristics. The use of multiple characteristics allows to improve the estimation of the probability of the differential from 2^{-70} to $2^{-61.52}$.

2.4 Experimental Verification

In order to check that the algorithm is correct, we ran it with a reduced version of LAC with 8 rounds. We used the second half of the truncated differentials of Fig. 3, with 17 active S-boxes. We found that this leads to differentials with probability at least $2^{-29.76}$:

$$\Pr\left[0000000000006404 \xrightarrow{8} 0000040000004400\right] \geq 2^{-29.76}$$

$$\Pr\left[0000000000006404 \xrightarrow{8} 0000060000004400\right] \geq 2^{-29.76}$$

In this case, the use a multiple characteristics allows to improve the estimation of the probability of the differential from 2^{-34} (17 active S-Boxes) to $2^{-29.76}$.

For this reduced version, we ran experiments with 2^{40} random plaintext pairs following the first differential, and random round keys. We detected 1204 pairs with the expected output difference, which match very closely our prediction $(2^{40} \cdot 2^{-29.76} \approx 1209)$. This indicates that our computation is correct, and the lower bound is quite tight in this case.

3 Conclusion

Our analysis shows that there exists differentials for the full G function of LAC with probability higher than 2^{-64}. This allows a simple forgery attack with probability higher than 2^{-64} on the full version of LAC, contradicting the security claims. This shows that the security margin of LAC is insufficient.

Our analysis is based on aggregating a collection of characteristics following the same truncated characteristic. While each characteristic has a probability at most 2^{-70}, a collection of characteristics can have a probability as high as $2^{-61.52}$, giving a lower bound on the probability of the corresponding differential.

Since this technique is relatively simple, we recommend all designers to check whether it can be applied to their designs.

References

1. Biryukov, A., Derbez, P., Perrin, L.: Differential analysis and meet-in-the-middle attack against round-reduced TWINE. In: Leander, G. (ed.) FSE 2015. LNCS, vol. 9054, pp. 3–27. Springer, Heidelberg (2015)
2. Bogdanov, A., Mendel, F., Regazzoni, F., Rijmen, V., Tischhauser, E.: ALE: AES-based lightweight authenticated encryption. In: Moriai, S. (ed.) FSE 2013. LNCS, vol. 8424, pp. 447–466. Springer, Heidelberg (2014)
3. CAESAR: Competition for Authenticated Encryption: Security, Applicability, and Robustness. http://competitions.cr.yp.to/caesar.html

4. Daemen, J., Rijmen, V.: The Design of Rijndael: AES - The Advanced Encryption Standard. Information Security and Cryptography. Springer, Heidelberg (2002)
5. Keliher, L., Sui, J.: Exact maximum expected differential and linear probability for two-round Advanced Encryption Standard. IET Inf. Secur. 1(2), 53–57 (2007)
6. Knudsen, L.R.: Truncated and higher order differentials. In: Preneel, B. (ed.) FSE 1994. LNCS, vol. 1008, pp. 196–211. Springer, Heidelberg (1995)
7. Sun, S., Hu, L., Wang, M., Wang, P., Qiao, K., Ma, X., Shi, D., Song, L., Fu, K.: Constructing mixed-integer programming models whose feasible region is exactly the set of all valid differential characteristics of SIMON. IACR Cryptology ePrint Arch. **2015**, 122 (2015). http://eprint.iacr.org/2015/122
8. Sun, S., Hu, L., Wang, P., Qiao, K., Ma, X., Song, L.: Automatic security evaluation and (related-key) differential characteristic search: application to SIMON, PRESENT, LBlock, DES(L) and other bit-oriented block ciphers. In: Sarkar, P., Iwata, T. (eds.) ASIACRYPT 2014. LNCS, vol. 8873, pp. 158–178. Springer, Heidelberg (2014)
9. Wu, W., Zhang, L.: LBlock: a lightweight block cipher. In: Lopez, J., Tsudik, G. (eds.) ACNS 2011. LNCS, vol. 6715, pp. 327–344. Springer, Heidelberg (2011)
10. Zhang, L., Wu, W., Wang, Y., Wu, S., Zhang, J.: LAC: a lightweight authenticated encryption cipher. Submission to CAESAR, March 2014. http://competitions.cr.yp.to/round1/lacv1.pdf (v1)

Privacy Preserving Data Processing

Private Information Retrieval with Preprocessing Based on the Approximate GCD Problem

Thomas Vannet[✉] and Noboru Kunihiro

The University of Tokyo, Tokyo, Japan
tvannet@gmail.com

Abstract. PIR protocols allow clients to privately recover data on a server. While many protocols exist, none of them are practical due to their high computation requirement. We explain how preprocessing is a necessity to solve this issue, then we present two independent but related results. First, we show how Goldberg's robust multi-server PIR protocol is compatible with preprocessing techniques. We detail the theoretical computation/memory tradeoff and present practical implementation results. Then, we introduce a new single-server PIR protocol that is reminiscent of Goldberg's protocol in its structure but relies on the unrelated Approximate GCD assumption. We describe its performance and security, along with implementation results.

Keywords: Privacy · Distributed databases · Information-Theoretic Protocols · Sublinear computation · Sublinear communication · Approximate GCD

1 Introduction

Private Information Retrieval was introduced in 1995 [4] as a way to protect a client's privacy when querying public databases. Since there exists a trivial solution consisting in sending the entire database regardless of the query, all efforts were focused on building algorithms sending less data [3]. Eventually algorithms with average communication cost independent of the database [6] were proposed and the problem seemed solved. However a technical report published in 2007 by Sion [14] showed that in practice the algorithms suggested were all performing slower than the trivial one. Essentially, as network speeds increase just as fast as computing power does, the sending of the whole database to a client will never be much more than a constant times slower than the reading of said database by the server. As a consequence of this observation, subsequent PIR research mainly focused instead on improving the running time of the algorithms. Overall, it is fair to measure a PIR algorithm's efficiency as the sum of its computation time (which depends on the CPU) and communication time (which depends on the type of link between client and server). Asymptotically, the overall complexity will be the worse of the two.

© Springer International Publishing Switzerland 2016
O. Dunkelman and L. Keliher (Eds.): SAC 2015, LNCS 9566, pp. 227–240, 2016.
DOI: 10.1007/978-3-319-31301-6_14

It is well known that PIR algorithms must read at least as many bits as the entire database to provide information theoretical client privacy. As such, and regardless of the communication efficiency of the protocol, all those algorithms will asymptotically perform no better than the sending of the entire database with its overall linear complexity.

Clearly, a different approach was needed to achieve practicality. In 2004, Beimel et al. introduced [1] the notion of PIR with preprocessing in multi-server settings, allowing for faster running times. Other approaches include the batching of queries [9–11], where several queries are performed in a single read of the database, or the use of trusted hardware on the server [16, 17, 19]. All of those schemes come however with clear limitations and may only be used in specific settings. Namely, Beimel et al.'s scheme only works when several non colluding servers host the database, query batching only improves complexity by a constant factor and trusted hardware still requires some degree of trust on the client's side.

We argue that an alternative exists in single-server PIR with preprocessing based on computational assumptions. This type of scheme would have its limitation be the computational assumption, but so do most widely used cryptographic schemes. To the best of our knowledge, there is so far no scheme that manages to answer queries while reading less bits than the entire database size in a completely generic setting.

We point out strong structural similarities between a sublinear protocol by Beimel et al. [1] and a well performing and high-security multi-server protocol by Goldberg [7]. This allows us to combine the best of each protocol and turn Goldberg's protocol into a potentially sublinear one with a noticeably lower computational cost.

However this protocol, like any other information theoretically secure protocol, does not maintain client privacy when every server cooperates. Alternatively, the protocol cannot be used when there is only one server (or a single entity owning all the servers) hosting the database. While Goldberg also suggested a computationally secure protocol that solves this issue, it is particularly inefficient and our preprocessing technique cannot be used on it.

We thus also introduce a brand new single-server PIR protocol. Its computational security is derived from the Approximate GCD assumption [15], well-studied in the field of Fully Homomorphic Encryption. Its low communication rate is obtained using the common construction in Goldberg's and Beimel et al.'s protocols we pointed out earlier. Finally, its low computation complexity is possible using the precomputation techniques we introduce in the first sections of this paper.

The paper is organized as follows. First, we show how Goldberg's protocol is a perfect fit to allow for some preprocessing. Then we show how to do the actual preprocessing as efficiently as possible. In the second part, we present the Approximate GCD assumption and how it can be combined with Goldberg's approach to result in a new single-server PIR protocol. Then we show how precomputations can be achieved on this new protocol. We suggest and justify parameters to keep the scheme safe and present implementation performance.

1.1 Notations

In Sects. 2 and 3, we work in the usual multi-server PIR setting where several servers are hosting the same copy of a public database of size n. This database is $(b_1, \cdots, b_n)$, $b_i \in \{0, 1\}$. We call $t \geq 2$ the minimum number of servers the client has to query for the protocol to work. In Sect. 4, we work in a single-server setting $(t = 1)$.

We use the standard notation $\delta_{x,y} = 1$ if $x = y$, 0 otherwise.

For a variable x and a distribution $\mathcal{D}$, we use the notation $x \xleftarrow{\$} \mathcal{D}$ to indicate that x is picked at random from the distribution.

2 The 2-Dimension Database Construction

2.1 Motivation

In this section, we present the construction which will serve as the central building block for the next sections. It was independently used by Goldberg [7] and to some extent by Beimel et al. [1]. It is also found in Gasarch and Yerukhimovich's proposal [5]. Essentially, it is a very simple structure which can be declined in many different ways. It usually provides a communication complexity of roughly $O(\sqrt{n})$, along with a computational complexity of $O(\sqrt{n})$ on the client side. While this seems very high compared to other schemes achieving $O(n^\epsilon)$ for any $\epsilon > 0$ or even constant rate on average, we argue that this is more than enough. This is due to the fact that a query recovers an entire block of $O(\sqrt{n})$ bits. As it turns out, this value seems to correspond to what real-world systems would need. Indeed, for a medium-sized database of 2^{30} bits which would likely contain text data, a query recovering a couple kilobytes of data makes perfect sense. Similarly, in a large database of 2^{50} bits likely containing media files, recovering a megabyte or more of data seems standard.

Besides, its structure is the perfect candidate for precomputations as we will detail in Sect. 3. Finally, this structure can be used in recursive schemes as in Gasarch and Yerukhimovich's protocol [5].

2.2 Goldberg's Robust Protocol

Here we present Goldberg's version, on which we will be able to add a layer of precomputations in Sect. 3 and replace the security assumption in Sect. 4.

Recall that the database is $(b_1, \cdots, b_n)$, $b_i \in \{0, 1\}$. We assume the database can be split into **nb** blocks of **wpb** words of **bpw** bits each, such that **nb wpb bpw** $= n$. Note that **nb** stands for *number of blocks*, **wpb** for *words per block* and **bpw** for *bits per word*. If n cannot be decomposed in such a way, we can easily pad the database with a few extra bits to make it so. We set up the database as a 2-dimensional array of words where every word is characterized by two coordinates. In other words, we rewrite $\{b_i\}_{1 \leq i \leq n}$ as $\{w_{i,j} | 1 \leq i \leq \text{nb}, 1 \leq j \leq \text{wpb}\}$. We also call block X the set of words $\{w_{X,j} | 1 \leq j \leq \text{wpb}\}$. All $t \geq 2$ servers are using the same conventions for this 2-dimensional database.

Suppose the client wants to recover block X while keeping X secret. We set $\mathbb{S}$ a field with at least t elements. The client selects nb polynomials $P_1, \cdots, P_{nb}$ of degree $t-1$ with random coefficients in $\mathbb{S}$, except the constant term always set such that $P_X(0) = 1$ and $P_i(0) = 0$ when $i \neq X$. He also selects distinct values $\alpha_1, \cdots, \alpha_t \in \mathbb{S}$. These values can be anything and do not have to be kept secret. For instance $\alpha_k = k$ will work. The protocol has only one round, the client sends a request, the server responds and finally the client deduces the value of every bit in block X from the response.

During the first step, the client sends a request to every server involved in the protocol. Specifically, for $1 \leq k \leq t$, the client sends $(P_1(\alpha_k), \cdots, P_{nb}(\alpha_k))$ to server k.

We consider $w_{i,j}$ as an element of S. Now we define, for $1 \leq j \leq$ wpb, the degree $t-1$ polynomial Q_j with coefficients in S.

$$Q_j = \sum_{i=1}^{nb} P_i w_{i,j}$$

We will keep this notation through the paper. Note that it exclusively depends on the values the client sent and the database contents.

During the second step, every server answers. Specifically, server k computes the following values and sends them to the client.

$$\left\{ Q_j(\alpha_k) = \sum_{i=1}^{nb} P_i(\alpha_k) w_{i,j} \right\}_{1 \leq j \leq \text{wpb}}$$

After receiving the answer of all t servers, the client knows $Q_j(\alpha_1), \cdots,$ $Q_j(\alpha_t)$ for every $1 \leq j \leq$ wpb. Since Q_j is of degree $t-1$, the client can interpolate all of its coefficients and compute

$$Q_j(0) = \sum_{i=1}^{nb} P_i(0) w_{i,j} = w_{X,j}$$

Thus the client knows the value of every word in block X.

2.3   Security

Informally, as a variant on Shamir's secret sharing scheme [13], this scheme is information theoretically as secure as possible. It is a well known fact that a protocol cannot be information theoretically secure if every single server cooperates, thus the best we can hope for is information theoretical protection when at least one server does not collaborate, which is what is achieved here. More formally, the scheme is information theoretically secure against a coalition of up to $t-1$ servers. Clearly if it is safe against a coalition of $t-1$ servers, it is safe against any coalition of less than $t-1$ servers.

2.4   Communication Complexity

The client sends **nb** elements of $\mathbb{S}$ to each server. Similarly, each server returns **wpb** elements of $\mathbb{S}$. Total communication is $t(\mathbf{nb} + \mathbf{wpb}) \log |\mathbb{S}|$, which becomes $2t\sqrt{n} \log |\mathbb{S}| = O(t \log t \sqrt{n})$ when $\mathbf{nb} = \mathbf{wpb} = \sqrt{n}$, $\mathbf{bpw} = 1$ and $\mathbb{S}$ has exactly t elements. Note that the computational complexity is still linear in the size of the database here since every b_i has to be read by the servers. The seemingly high communication compared to protocols with polylogarithmic [3] or constant [6] communication rate is actually a non-issue, since asymptotically (and also in practice [14]), the server's computation time will be much higher than the data transmission time.

3   Adding Precomputations to Goldberg's Robust Protocol

In this section, we will show how we can improve the protocol from Sect. 2 by adding an offline step only executed once. First we use the exact same approach as Beimel et al., which is not practical in our setting, and then we show how to modify it to make it pratical and efficient.

3.1   Using Precomputations

We now show we can improve Goldberg's protocol's running time. To reduce the computational complexity on the server side we add a precomputation step before the algorithm is run. A similar idea was introduced in [1], which used essentially the same 2-dimensional database structure describe in Sect. 2. This step is only run once, has polynomial running time and can be done offline prior to any interaction with clients. Its complexity is thus irrelevant to the actual protocol running time.

Here is how we proceed. First, for every $1 \leq j \leq \mathbf{wpb}$ (every word in a block), we partition the $m = \mathbf{nb} \ \mathbf{bpw}$ bits $b_{1,j}, \cdots, b_{m,j}$ into m/r disjoint sets of r bits each (without loss of generality, we assume r divides m). There are m/r such sets for every j, or n/r sets in total. We call these sets $C_{i,j}$, $1 \leq i \leq m/r$ and $1 \leq j \leq \mathbf{wpb}$. For every $C_{i,j}$, the server computes all $|\mathbb{S}|^r$ possible linear combinations with cocfficients in $\mathbb{S}$. That is, every $Pre(C, \Delta) = \sum_{d=1}^{r} \delta_d C[d]$ for any $\Delta = (\delta_1 \cdots \delta_r) \in \mathbb{S}^r$. This requires a total of $\frac{n |\mathbb{S}|^r \log |\mathbb{S}|}{r}$ bits.

If we set $|\mathbb{S}| = t$ (which is the minimal number of distinct elements in $\mathbb{S}$, attained when **bpw** is 1) and $r = \epsilon \log n$ for some ϵ, this means a total of $\frac{n^{1+\epsilon \log t} \log t}{\epsilon \log n}$ bits have to be precomputed and stored on the servers. This value is potentially very large if t is too large. We will solve this issue in Sect. 3.2. Computing one $Pre(C, \Delta)$ requires r operations and overall the preprocessing of the whole database requires $O(n^{1+\epsilon \log t})$ operations.

During the online phase, each server only reads m/r elements of $\mathbb{S}$ to compute $Q_j(\alpha_k)$. Over all t servers, only $\frac{tn \log |\mathbb{S}|}{r}$ bits are read. Note that the precomputed database should be ordered in such a way that, for a fixed i, all $Pre(C_{i,j}, \Delta)$ are stored consecutively. This way Δ, which is sent by the client, only has to be read by the server once. The nature of the algorithm (reusing the same Δ for every $1 \leq i \leq$ nb) is what allows us to achieve sublinearity where different protocols would not benefit from the computations.

Once again, if we set $|\mathbb{S}| = t$, bpw $= 1$ and $r = \epsilon \log n$ for some ϵ, a total of $\frac{nt \log t}{\epsilon \log n}$ bits are read during the online phase. If $t \log t < \epsilon \log n$, that means the server's computation is sublinear in the size of the database.

Communication complexity and security are exactly the same as detailed in Sect. 2. Indeed, a coallition of servers receives exactly the same information from a client as before. This is interesting however if one notes in particular that the sublinear complexity would allow, from an information theoretical point of view, any individual server to distinguish some bits from the original database which are definitely not requested by the client. This could have potentially lead to an attack from a coallition of servers if it was not for the formal proof of Sect. 2. This shows yet again how fitting Goldberg's scheme is for this type of computational improvements through precomputation.

3.2   Decomposition over Base 2

In this section, we present another approach to make precomputations even more efficient. This potentially improves Goldberg's complexity by several orders of magnitude.

Clearly the most limiting factor in the construction from the previous section (and in Beimel et al.'s approach [1]) is the $|\mathbb{S}|^r$ factor. In some scenarios, we want t (and thus $|\mathbb{S}|$) to be large for redundancy purposes. Using words with more than a single bit can also be interesting. Besides, it is possible to create schemes based on a computational assumption which require very large groups $\mathbb{S}$, as we will show in Sect. 4. Alternatively, some implementations (including Goldberg's percy++ [8]) set a fixed large $\mathbb{S}$ regardless of how many servers the client decides to use. In all of those situations, the space requirements on the server become prohibitive. We assume here that $\mathbb{S} = \mathbb{Z}_\ell$ for a possibly large ℓ.

As before, in the first step the client sends to server k the following values.

$$(P_1(\alpha_k), \cdots, P_m(\alpha_k))$$

We can however decompose those values as follows, since elements of $\mathbb{S}$ have a standard representation as integers.

$$P_i(\alpha_k) = \sum_{c=0}^{\log \ell} P_{i,c} 2^c \text{ where } P_{i,c} \in \{0,1\}$$

Then the server's computation can be modified using the following equality.

$$\sum_{i=1}^{nb} P_i(\alpha_k) w_{i,j} = \sum_{c=0}^{\log \ell} 2^c \sum_{i=1}^{nb} P_{i,c} w_{i,j}$$

Now once again we partition $\{1, \cdots, m\}$ as defined earlier into m/r disjoint sets $C_{i,j}$ of r elements each, where $1 \leq i \leq m/r$ and $1 \leq j \leq \mathsf{wpb}$. For every such set C and $\delta_1, \cdots, \delta_r \in \{0,1\}$, the server precomputes $\sum_{d=1}^{r} \delta_d C[d]$. Each precomputed value requires only $\log r$ bits for storage.

If $r = \epsilon \log n$, the following holds regarding the precomputed database size.

$$\frac{m}{r}\mathsf{wpb}2^r \log r = n^{1+\epsilon}\frac{\log(\epsilon \log n)}{\epsilon \log n} < n^{1+\epsilon}$$

Thus less than $n^{1+\epsilon}$ bits are precomputed and stored in total. This value is independent of the size of $\mathbb{S}$ and thus indirectly independent of t. The one-time preprocessing requires $O(n^{1+\epsilon})$ elementary operations.

Now during the online phase, Q_j can be computed in $\frac{m}{r}\log l$ operations. Overall, the t servers will return their results in $n\frac{t\log \ell}{r} = n\frac{t\log \ell}{\epsilon \log n}$ operations, which is sublinear if $t\log \ell < \epsilon \log n$. Note that for real world values, $\log r = \log(\epsilon \log n)$ is small and summing m/r values should give a number that fits in 64 bits. This means that the online phase of the protocol can be efficiently implemented without using any large integer libraries for its most expensive step. Then only $\log \ell$ operations have to be performed using large integers. Because of this feature, the algorithm remains highly efficient even when $t\log l > \epsilon \log n$ as is the case with very large l, t or small ϵ (to save space).

3.3   A Note on Goldberg's Computationally Secure Scheme

If all t servers collaborate, it is very easy for them to recover the secret X. Furthermore, we know it is impossible to make an information theoretically secure protocol with sublinear communication when every server cooperates. However it is still possible to make a computationally secure protocol in this situation as shown in [7]. Note that because all of the servers can collaborate, we might as well consider that all t servers are the same, and that gives us a single server PIR protocol (although in some scenarios having several servers can also be convenient).

The basic idea behind the protocol is to encrypt the shares sent to the servers with an additively homomorphic scheme and to have the servers perform the computation on the ciphertexts. In practice, the Paillier cryptosystem is used. It has the interesting property that the product of ciphertexts is a ciphertext associated to the sum of the original plaintexts. It is not possible to add a pre-computing step in this situation because the parameters (the modulus) must be chosen privately by the client. As such, all hypothetical precomputations would be done in $\mathbb{Z}$ instead of some $\mathbb{Z}/\ell\mathbb{Z}$ and reading a hypothetical precomputed product would require reading as many bits as reading the individual components of the product.

Instead, we can choose to use another additive homomorphic scheme like one based on the Approximate GCD problem. The scheme has however to be changed in several ways which we describe in the next section.

4   Single Server PIR Using the Approximate GCD Assumption

4.1   Computational Assumptions

In this section we reuse the 2-dimensional construction of the database, but instead of securing it through Shamir's secret sharing scheme, we rely on the Approximate GCD assumption. The assumption, which has been the subject of widespread study thanks to its importance in the field of Fully Homomorphic Encryption, states that it is computationally hard to solve the Approximate GCD problem for sufficiently large parameters. This problem can be formulated as in Definition 1.

For any bit length λ_q and odd number p, we call $\mathcal{D}(\lambda_q, p)$ the random distribution of values $pq + \epsilon$ where q has λ_q bits and $\epsilon \ll p$. In addition, for a bit $b \in \{0, 1\}$, we call $\mathcal{D}(\lambda_q, p, b)$ the random distribution of values $pq + 2\epsilon + b$ where q has λ_q bits and $\epsilon \ll p$.

Definition 1 *(Approximate GCD)*. Let $\{z_i\}_i \xleftarrow{\$} \mathcal{D}(\lambda_q, p)$ be a polynomially large collection of integers. Given this collection, output p.

It was introduced in 2010 by Van Dijk et al. [15]. In the same paper, they also showed that the following problem can be reduced to the Approximate GCD problem.

Definition 2 *(Somewhat Homomorphic Encryption)*. Let $\{z_i\}_i \xleftarrow{\$} \mathcal{D}(\lambda_q, p)$ be a polynomially large collection of integers, $b \xleftarrow{\$} \{0, 1\}$ a secret random bit and $z \xleftarrow{\$} \mathcal{D}(\lambda_q, p, b)$. Given $\{z_i\}_i$ and z, output b.

Now here is how we construct the scheme.

As before we assume the database is $w_{i,j}$, $1 \leq i \leq \mathtt{nb}$, $1 \leq j \leq \mathtt{wpb}$, each $w_{i,j}$ containing $\mathtt{bpw}$ bits. Note that we have $\mathtt{nb\ wpb\ bpw} = n$, the total bit size of the database. First we consider the convenient case where $\mathtt{bpw} = 1$ (every word is a single bit).

Suppose the client wants to recover block X consisting of $\{w_{X,j}\}_{1 \leq j \leq \mathtt{wpb}}$. The client picks a large random odd number p which will be its secret key. He selects $\mathtt{nb}$ random large numbers q_i and ϵ_i and computes $P_i = pq_i + 2\epsilon_i + \delta_{i,X}$.

Then the server computes for every j from 1 to $\mathtt{wpb}$, $R_j = \sum_{i=1}^{\mathtt{nb}} b_{i,j} P_i$. He then sends $\{R_j\}_{1 \leq j \leq \mathtt{wpb}}$ to the client.

For every R_j received, the client can compute $(R_j \bmod p) \bmod 2 =$
$$\left(\sum_{i=1}^{\mathtt{nb}} b_{i,j}(pq_i + 2\epsilon_i + \delta_{i,X}) \bmod p\right) \bmod 2 = \sum_{i=1}^{\mathtt{nb}} b_{i,j}(2\epsilon_i + \delta_{i,X}) \bmod 2 = b_{X,j}.$$
As such, he retrieves the entire block X.

It is important to note that this construction is somewhat homomorphic and has been used to create fully homomorphic encryption [15], but here we only care about the additive property of the scheme. This is very important because without any multiplications, the noise will progress very slowly and we realistically do not have to worry about it becoming too large. Also, our construction is unrelated to other PIR protocols based off generic somewhat or fully homomorphic encryption systems like Yi et al.'s [18] or Boneh et al.'s [2].

4.2   Complexity

First let us look at the communication complexity. The client sends nb values P_i and the server returns wpb values R_j. We call λ_p the bit size of the secret p and λ_q the bit size of the q_is. We also write $\lambda = \lambda_p + \lambda_q$. The actual values of these security parameters will be discussed in Sect. 4.4. Each P_i is thus λ bits long. Since R_j is essentially a sum of nb P_is, its bit size is $\lambda + \log_2 nb$.

The overall communication of the protocol is $nb\lambda + wpb(\lambda + \log_2 nb)$ to retrieve a block of wpb bits. When $nb = wpb = \sqrt{n}$, and for fixed security parameters, the communication is $O(\sqrt{n} \log n)$ or $O(\log n)$ per bit recovered.

Now let us detail the computational complexity. For clarity's sake, we call operation any addition of two $\lambda + \log_2 nb$ or less bits integers, or a multiplication of a λ_p bit-long integer by a λ_q bit-long one. The client performs $O(nb)$ operations to send the query, the server in turn performs $O(nb\ wpb)$ operations to execute it. Finally the client performs $O(wpb)$ operations to recover the block values from the server's reply.

The overall computational complexity of the protocol is $O(nb\ wpb(\lambda + \log nb))$. When $nb = wpb = \sqrt{n}$, and for fixed security parameters, this becomes $O(n \log n)$.

4.3   Precomputations

As the scheme uses the 2-dimensional structure of the database, we can use precomputations in the exact same way we described in Sect. 3.2. We write $P_i = \sum_{k=0}^{\lambda-1} P_{i,k} 2^k$, where $P_{i,k} \in \{0,1\}$. The server selects a precomputing parameter r and computes every possible sum of r consecutive words in the database. The client sends the $P_{i,k}$s by groups of r bits for a fixed k. Now the server can compute the R_js r times faster than previously by working with small integers and only performing λ large integer operations per R_j. See Sect. 3.2 for details on this technique.

Communications are unchanged (the client sends the same amount of bits, simply changing their order) at $O(\log n)$ bits transmitted for every bit recovered.

Computations are unchanged on the client side. On the server side, each R_j requires $\lambda\ nb/r$ small integer operations and λ large integer operations. Overall computational complexity is $O(\lambda n/r + \lambda^2)$, which is asymptotically a r times improvement over the standard version.

4.4 Security

The security of a PIR scheme is defined by the following problem. Given two client queries for blocks X_1 and X_2 respectively, it should be computationally hard to distinguish them. In this scheme, this means that the server must distinguish two queries d_{X_1} and d_{X_2} where $d_X = \{P_i | P_i \overset{\$}{\leftarrow} \mathcal{D}(\lambda, p, 0) \text{ when } i \neq X, P_i \overset{\$}{\leftarrow} \mathcal{D}(\lambda, p, 1) \text{ otherwise}\}$. Let us call D_X the distribution of all possible queries d_X.

Now we can show that if an attacker can indeed distinguish these two queries when given access to as many samples from $\mathcal{D}(\lambda, p, 0)$ as it needs, it can break the Somewhat Homomorphic Encryption (Definition 2) and thus solve the Approximate GCD Problem (Definition 1).

Theorem 1. *Given access to values of $\mathcal{D}(\lambda, p, 0)$, if it is possible to distinguish two queries for distinct bits of a database in polynomial-time, then it is also possible to recover the value p in polynomial-time.*

Proof. Let us assume there exists a distinguishing algorithm $\mathcal{A}(d)$ that returns 1 with probability σ if $d \in D_{X_1}$ and returns 1 with probability $\sigma + \alpha$ if $d \in D_{X_2}$. Now given an encryption $e = pq + 2\epsilon + b$ of a secret bit b, we build an algorithm $\mathcal{B}(e)$ that recovers b with probability at least $1/2 + \alpha/4$.

First, with probability $\alpha/2$, we return 0 and stop immediately. Otherwise, we draw $r \overset{\$}{\leftarrow} \mathcal{D}(\lambda, p, 0)$ and $\mathbf{nb} - 2$ random elements from $\mathcal{D}(\lambda, p, 0)$. We generate a query d with the random elements in every position but X_1 and X_2. We randomly pick X_1 or X_2 and place e in this position, r in the other position. We then return $\mathcal{A}(d)$ if the picked position was X_1, $1 - \mathcal{A}(d)$ otherwise.

If b was equal to 0, then d contains only random encryptions of 0 and does not belong to some D_X. In this case, let us say $\mathcal{A}(d)$ returns 1 with probability σ'. Then $\mathcal{B}(e)$ returns 1 with probability $(1 - \alpha/2)(1/2 \cdot \sigma' + 1/2 \cdot (1 - \sigma')) = 1/2 - \alpha/4$. Now if b was equal to 1, then $\mathcal{B}(e)$ returns 1 with probability $(1 - \alpha/2)(1/2 \cdot (\sigma + \alpha) + 1/2(1 - \sigma)) = 1/2 + \alpha/2(3/4 - \alpha/4) \geq 1/2 + \alpha/4$ since $0 < \alpha \leq 1$.

Then we use the result from Van Dijk et al. [15] to show the reduction from recovering b to recovering the secret p and breaking the AGCD assumption. $\square$

Note that here we had to assume that the attacker has access to random encryptions of 0. In real life scenarios, this is a fair assumption as it is common for a server to know some metadata about the encrypted information which includes always-0 bits.

4.5 Multiple Bits Recovery

The process can easily be modified to recover words of more than 1 bit at no additional cost. Essentially, instead of picking the P_is as $pq_i + 2\epsilon + \delta_{i,X}$, we can pick $P_i = pq_i + 2^{\mathbf{bpw}}\epsilon + \delta_{i,X}$ where $\mathbf{bpw}$ is the number of bits per word. Since λ_p has to be large for security reasons and the noise only progresses linearly when processing the database, we can afford to start with a fairly large noise.

For instance, if λ_p is 1024 bits and we process databases with less than 2^{80} blocks, we could pick an ϵ with 512 bits and that would allow bpw to be as high as 432.

The server's response in this case is unchanged, $R_j = \sum_{i=1}^{nb} P_i w_{i,j}$ where each w_i is a word of bpw bits. The client then recovers block X consisting of all the $w_{X,j}$ by computing $w_{X,j} = (R_j \mod p) \mod 2^{bpw}$.

The overall communication of the protocol is still nb λ + wpb(λ + $\log_2$ nb) but we now retrieve a block of wpb bpw bits. Besides, we have bpw wpb nb $= n$, which means that when nb = wpb = $\sqrt{n/bpw}$ the communication is $O(\sqrt{\frac{n}{bpw}}(\lambda + \log\frac{n}{bpw}))$ or $O(\frac{1}{bpw}(\lambda + \log\frac{n}{bpw}))$ per bit recovered (note that λ has to be larger than bpw).

The overall computational complexity of the protocol is also unchanged at $O(nb\ wpb(\lambda + \log nb))$. When nb = wpb = $\sqrt{n/bpw}$ this becomes $O(\frac{n}{bpw}(\lambda + \log\frac{n}{bpw}))$. To sum up, both the communication and computational costs are improved by a factor of bpw, which can be several hundreds high. Note that the same improvement was possible on Goldberg's original scheme, and it comes at the cost of recovering blocks of bits bpw times larger.

The question this naturally brings up is whether or not this affects the security of the scheme. First, the scheme can be reduced to a version of the Somewhat Homomorphic Encryption scheme (Definition 2) on bpw bits as detailed in Sect. 4.4. Furthermore, we can show that this version of the Somewhat Homomorphic Encryption scheme can be reduced to the AGCD problem for numbers $pq + 2^k\epsilon + b$, $b \in \{0,1\}$. We do not detail the proof of this step as it follows exactly Van Dijk et al.'s [15].

This shows that this scheme is at least as secure as a version of the AGCD problem with bpw-1 known noise bits. Alternatively, any generic AGCD problem instance can be turned into one of those instances by bruteforcing the value of said bits. As such, for small enough word sizes (say, up to 32 bits), the security of the scheme is mostly unaffected.

5   Implementation

Goldberg provided an implementation called percy++ [8] that included the robust protocol from [7] along with that of other PIR protocols. Running tests showed that in multi-server settings, Chor et al.'s protocol [4] performed the fastest, followed by Goldberg's [7]. Note that Goldberg's, and by extension our version described in Sect. 3.2, contains much stronger security features. In single-server settings, Aguilar-Melchor et al.'s scheme [12] was deemed the best performing. The security of this scheme is based off the so called *Differential Hidden Lattice Problem* which the authors introduced and studied. In comparison, our single-server scheme defined in Sect. 4 relies on the computational Approximate GCD assumption on which the main candidate for fully homomorphic encryption is based and has thus received a lot of attention.

We made a straightforward implementation of our protocols without aiming for high levels of optimization. For multi-server protocols, we always picked $t = 2$ servers, the minimum to keep the protocol safe. In real world situations, a much higher number would be desirable, multiplying the overall time complexity by a non negligible constant. For our single server scheme, $\lambda = 1024$ was chosen with a noise ϵ always around $p/\mathrm{nb} > 2^{896}$. This is more than enough to stop all known lattice-based and other attacks against the assumption. In fact, much lower values may provide enough security if no other attacks are developped.

We only care about the computational complexity of the server since the communication and client computational complexities are asymptotically negligible. In situations where the preprocessing would generate files too large for our environment, we simulated reading from random files instead, which makes no difference from the server's point of view since the data received is indistinguishable from random data.

There are a lot of parameters that can be modified when running actual tests. For Goldberg's scheme, the preprocessing parameter r described in Sect. 3.1, the number of bits per word and the size of the database. Our implementation shows that, as expected, in identical settings the version with preprocessing performs r times faster than the original one. This is true for databases small enough that the preprocessed version would still fit in RAM and for databases large enough that it does not fit in RAM even without preprocessing. For middle-sized databases, the added preprocessing forces the algorithm to read data directly from the harddrive in a semi-random manner, which can slow it down compared to the original version depending on reading speeds.

Goldberg already showed that his multi-server robust scheme was several orders of magnitude faster than Aguilar-Melchor et al.'s heavily optimized single server scheme. In our implementation of our single-server from Sect. 4.5, we obtain speeds comparable to Golbeger's protocol. Adding precomputations, our scheme can be several times faster.

6   Conclusion

A completely practical PIR scheme has yet to be found. As long as the entire database has to be read for every query, such schemes will always perform too slowly. Based on the observation that preprocessing is a necessity to reach that goal, we first showed how such precomputations can be added to some already existing protocols and designed a new protocol compatible with this technique.

We presented the first PIR protocol allowing the recovery of a block of bits, protection against strong server collusion and Byzantine servers while performing in sublinear time. Its performance is theoretically better than other such algorithms and the design allows efficient implementations by mostly avoiding the need for large integers libraries. Actual implementation corroborates the theoretical findings. Our scheme is both a generalization of Goldberg's protocol and Beimel et al.'s protocol. Compared to Goldberg's, ours can perform many times faster but requires a polynomial expansion of the database. Even a quadratic expansion is sufficient to provide a much faster protocol. Compared to Beimel et al.'s, the scheme provides much stronger security against several servers cooperating, a major weakpoint of multi-server PIR protocols.

We also presented an efficient single-server scheme with a simple structure. It allows for precomputations and relies on a computational assumption which is fairly new but is receiving and will likely continue to receive a lot of attention from the research community. Furthermore, the way we use it allows for relatively small parameters compared to FHE schemes based on it.

Acknowledgment. This research was supported by CREST, JST.

References

1. Beimel, A., Ishai, Y., Malkin, T.: Reducing the servers' computation in private information retrieval: PIR with preprocessing. J. Cryptol. **17**(2), 125–151 (2004)
2. Boneh, D., Gentry, C., Halevi, S., Wang, F., Wu, D.J.: Private database queries using somewhat homomorphic encryption. Cryptology ePrint Archive, Report 2013/422 (2013). http://eprint.iacr.org/
3. Cachin, C., Micali, S., Stadler, M.: Computationally private information retrieval with polylogarithmic communication. In: Stern, J. (ed.) EUROCRYPT 1999. LNCS, vol. 1592, pp. 402–414. Springer, Heidelberg (1999)
4. Chor, B., Kushilevitz, E., Goldreich, O., Sudan, M.: Private information retrieval. J. ACM **45**(6), 965–981 (1998)
5. Gasarch, W., Yerukhimovich, A.: Computational inexpensive cPIR (2006). http://www.cs.umd.edu/arkady/pir/pirComp.pdf
6. Gentry, C., Ramzan, Z.: Single-database private information retrieval with constant communication rate. In: Caires, L., Italiano, G.F., Monteiro, L., Palamidessi, C., Yung, M. (eds.) ICALP 2005. LNCS, vol. 3580, pp. 803–815. Springer, Heidelberg (2005)
7. Goldberg, I.: Improving the robustness of private information retrieval. In: Proceedings of the IEEE Symposium on Security and Privacy (2007)
8. Goldberg, I., Devet, C., Hendry, P., Henry, R.: Percy++ project on sourceforge (2014). http://percy.sourceforge.net. (version 1.0. Accessed January 2015)
9. Henry, R., Huang, Y., Goldberg, I.: One (block) size fits all: PIR and SPIR with variable-length records via multi-block queries. In: 20th Annual Network and Distributed System Security Symposium, NDSS, San Diego, California, USA, 24–27 February 2013 (2013)
10. Ishai, Y., Kushilevitz, E., Ostrovsky, R., Sahai, A.: Batch codes and their applications (2004)
11. Lueks, W., Goldberg, I.: Sublinear scaling for multi-client private information retrieval (2015, to appear)
12. Melchor, C.A., Gaborit, P.: A lattice-based computationally-efficient private information retrieval protocol. Cryptology ePrint Archive, Report 2007/446 (2007). http://eprint.iacr.org/
13. Shamir, A.: How to share a secret. Commun. ACM **22**(11), 612–613 (1979)
14. Sion, R.: On the computational practicality of private information retrieval. In: Proceedings of the Network and Distributed Systems Security Symposium, Stony Brook Network Security and Applied Cryptography Lab Tech Report (2007)

15. van Dijk, M., Gentry, C., Halevi, S., Vaikuntanathan, V.: Fully homomorphic encryption over the integers. In: Gilbert, H. (ed.) EUROCRYPT 2010. LNCS, vol. 6110, pp. 24–43. Springer, Heidelberg (2010)
16. Wang, S., Ding, X., Deng, R., Bao, F.: Private information retrieval using trusted hardware. Cryptology ePrint Archive, Report 2006/208 (2006). http://eprint.iacr.org/
17. Yang, Y., Ding, X., Deng, R.H., Bao, F.: An efficient PIR construction using trusted hardware. In: Wu, T.-C., Lei, C.-L., Rijmen, V., Lee, D.-T. (eds.) ISC 2008. LNCS, vol. 5222, pp. 64–79. Springer, Heidelberg (2008)
18. Yi, X., Kaosar, M.G., Paulet, R., Bertino, E.: Single-database private information retrieval from fully homomorphic encryption. IEEE Trans. Knowl. Data Eng. 25(5), 1125–1134 (2013)
19. Yu, X., Fletcher, C.W., Ren, L., Van Dijk, M., Devadas, S.: Efficient private information retrieval using secure hardware

Dynamic Searchable Symmetric Encryption with Minimal Leakage and Efficient Updates on Commodity Hardware

Attila A. Yavuz[1]([⊠]) and Jorge Guajardo[2]

[1] The School of Electrical Engineering and Computer Science,
Oregon State University, Corvallis, OR 97331, USA
attila.yavuz@oregonstate.edu
[2] Robert Bosch Research and Technology Center, Pittsburgh, PA 15203, USA
Jorge.GuajardoMerchan@us.bosch.com

Abstract. Dynamic Searchable Symmetric Encryption (DSSE) enables a client to perform keyword queries and update operations on the encrypted file collections. DSSE has several important applications such as privacy-preserving data outsourcing for computing clouds. In this paper, we developed a new DSSE scheme that achieves the highest privacy among all compared alternatives with low information leakage, efficient updates, compact client storage, low server storage for large file-keyword pairs with an easy design and implementation. Our scheme achieves these desirable properties with a very simple data structure (i.e., a bit matrix supported with two hash tables) that enables efficient yet secure search/update operations on it. We prove that our scheme is secure and showed that it is practical with large number of file-keyword pairs even with an implementation on simple hardware configurations.

Keywords: Dynamic Searchable Symmetric Encryption · Privacy enhancing technologies · Secure data outsourcing · Secure computing clouds

1 Introduction

Searchable Symmetric Encryption (SSE) [8] enables a client to encrypt data in such a way that she can later perform keyword searches on it via "search tokens" [19]. A prominent application of SSE is to enable privacy-preserving keyword searches on cloud-based systems (e.g., Amazon S3). A client can store a collection of encrypted files at the cloud and yet perform keyword searches without revealing the file or query contents [13]. Desirable properties of a SSE scheme are as follows:

- *Dynamism*: It should permit adding or removing new files/keywords from the encrypted file collection securely after the system set-up.
- *Efficiency and Parallelization*: It should offer fast search/updates, which are parallelizable across multiple processors.

© Springer International Publishing Switzerland 2016
O. Dunkelman and L. Keliher (Eds.): SAC 2015, LNCS 9566, pp. 241–259, 2016.
DOI: 10.1007/978-3-319-31301-6_15

- *Storage Efficiency*: The SSE storage overhead of the server depends on the encrypted data structure (i.e., encrypted index) that enables keyword searches. The number of bits required to represent a file-keyword pair in the encrypted index should be small. The size of encrypted index should not grow with the number of operations. The persistent storage at the client should be minimum.
- *Communication Efficiency*: Non-interactive search/update a with minimum data transmission should be possible to avoid the delays.
- *Security*: The information leakage must be precisely quantified based on formal SSE security notions (e.g., dynamic CKA2 [12]).

Our Contributions. The preliminary SSEs (e.g.,[8,18]) operate on only static data, which strictly limits their applicability. Later, Dynamic Searchable Symmetric Encryption (DSSE) schemes (e.g., [4,13]), which can handle dynamic file collections, have been proposed. To date, there is no single DSSE scheme that outperforms all other alternatives for *all* metrics: privacy (e.g., info leak), performance (e.g., search, update times) and functionality. Having this in mind, we develop a DSSE scheme that achieves the highest privacy among all compared alternatives with low information leakage, non-interactive and efficient updates (compared to [12]), compact client storage (compared to [19]), low server storage for large file-keyword pairs (compared to [4,12,19]) and conceptually simple and easy to implement (compared to [12,13,19]). Table 1 compares our scheme with existing DSSE schemes for various metrics. We outline the desirable properties of our scheme as follows:

- *High Security*: Our scheme achieves a high-level of update security (i.e., *Level-1*), forward-privacy, backward-privacy and size pattern privacy simultaneously (see Sect. 5 for the details). We quantify the information leakage via leakage functions and formally prove that our scheme is dynamic CKA2-secure in random oracle model [3].
- *Compact Client Storage*: Compared to some alternatives with secure updates (e.g., [19]), our scheme achieves smaller client storage (e.g., 10–15 times with similar parameters). This is an important advantage for resource constrained clients such as mobile devices.
- *Compact Server Storage with Secure Updates*: Our encrypted index size is smaller than some alternatives with secure updates (i.e., [12,19]). For instance, our scheme achieves $4 \cdot \kappa$ smaller storage overhead than that of the scheme in [12], which introduces a significant difference in practice. Asymptotically, the scheme in [19] is more server storage efficient for small/moderate number of file-keyword pairs. However, our scheme requires only two bits per file-keyword pair with the maximum number of files and keywords.
- *Constant Update Storage Overhead*: The server storage of our scheme does not grow with update operations, and therefore it does not require re-encrypting the whole encrypted index due to frequent updates. This is more efficient than some alternatives (e.g., [19]), whose server storage grows linearly with the number of file deletions.

- *Dynamic Keyword Universe*: Unlike some alternatives (e.g., [8,12,13]), our scheme does not assume a fixed keyword universe, which permits the addition of new keywords to the system after initialization. Hence, the file content is not restricted to a particular pre-defined keyword but can be any token afterwards (encodings)[1].
- *Efficient, Non-interactive and Oblivious Updates*: Our basic scheme achieves secure updates non-interactively. Even with large file-keyword pairs (e.g., $N = 10^{12}$), it incurs low communication overhead (e.g., 120 KB for $m = 10^6$ keywords and $n = 10^6$ files) by further avoiding network latencies (e.g., 25–100 ms) that affect other interactive schemes (e.g., as considered in [4,12,16,19]). One of the variants that we explore requires three rounds (as in other DSSE schemes), but it still requires low communication overhead (and less transmission than that of [12] and fewer rounds than [16]). Notice that the scheme in [16] can only add or remove a file but cannot update the keywords of a file without removing or adding it, while our scheme can achieve this functionality intrinsically with a (standard) update or delete operation. Finally, our updates take always the same amount of time, which does not leak timing information depending on the update.
- *Parallelization*: Our scheme is parallelizable for both update and search operations.
- *Efficient Forward Privacy*: Our scheme can achieve forward privacy by retrieving not the whole data structure (e.g., [19]) but only some part of it that has already been queried.

2 Related Work

SSE was introduced in [18] and it was followed by several SSE schemes (e.g., [6,8,15]). The scheme of Curtmola et al. in [8] achieves a sub-linear and optimal search time as $O(r)$, where r is the number of files that contain a keyword. It also introduced the security notion for SSE called as *adaptive security against chosen-keyword attacks (CKA2)*. However, the static nature of those schemes limited their applicability to applications with dynamic file collections. Kamara et al. developed a DSSE scheme in [13] that could handle dynamic file collections via encrypted updates. However, it leaked significant information for updates and was not parallelizable. Kamara et al. in [12] proposed a DSSE scheme, which leaked less information than that of [13] and was parallelizable. However, it incurs an impractical server storage. Recently, a series of new DSSE schemes (e.g., [4,16,17,19]) have been proposed by achieving better performance and security. While being asymptotically better, those schemes also have drawbacks. We give a comparison of these schemes (i.e., [4,16,17,19]) with our scheme in Sect. 6.

Blind Seer [17] is a private database management system, which offers private policy enforcement on semi-honest clients, while a recent version [10] can also

---

[1] We assume the maximum number of keywords to be used in the system is pre-defined.

Table 1. Performance Comparison of DSSE schemes.

Scheme/Property	[13] Kamara 12'	[12] Kamara 13'	[19] Stefanov	[4] $\left(\prod_{2\text{lev}}^{\text{dyn,ro}}\right)$	This work
Size privacy	No	No	No	No	Yes
Update privacy	$L5$	$L4$	$L3$	$L2$	$L1$
Forward privacy	No	No	Yes	Yes	Yes
Backward privacy	No	No	No	No	Yes
Dynamic keyword	No	No	Yes	Yes	Yes
Client storage	4κ	3κ	$\kappa \log(N')$	$\kappa \cdot \mathcal{O}(m')$	$\kappa \cdot \mathcal{O}(n+m)$
Index size (server)	$z \cdot \mathcal{O}(m+n)$	$\mathcal{O}((\kappa+m) \cdot n)$	$13\kappa \cdot \mathcal{O}(N')$	$c''/b \cdot \mathcal{O}(N')$	$2 \cdot \mathcal{O}(m \cdot n)$
Grow with updates	No	No	Yes	Yes	No
Search time	$\mathcal{O}((r/p) \cdot \log n)$	$\mathcal{O}((r/p) \cdot \log^3(N'))$	$\mathcal{O}((r + d_w)/p)$	$1/b \cdot \mathcal{O}(r/p)$	$O(\frac{m}{p \cdot b})$
# Rounds update	1	3	3	1	1
Update bandwidth	$z \cdot \mathcal{O}(m'')$	$(2z\kappa)\mathcal{O}(m \log n)$	$z \cdot \mathcal{O}(m'' \log N')$	$z \cdot \mathcal{O}(m \log n + m'')$	$b \cdot \mathcal{O}(m)$
Update time	$\mathcal{O}(m'')$	$\mathcal{O}((m/p) \cdot \log n) + t$	$\mathcal{O}((m''/p) \cdot \log^2(N')) + t$	$\mathcal{O}(m''/p) + t$	$b \cdot \mathcal{O}(m/p)$
Parallelizable	No	Yes	Yes	Yes	Yes

- All compared schemes are *dynamic CKA2 secure* in Random Oracle Model (ROM) [3], and leak search and access patterns. The analysis is given for the worst-case (asymptotic) complexity.
- m and n are the maximum # of keywords and files, respectively. m' and n' are the current # of keywords and files, respectively. We denote by $N' = m' \cdot n'$ the total number of keywords and file pairs currently stored in the database. m'' is the # unique keywords included in an updated file (add or delete). r is # of files that contain a specific keyword.
- Rounds refer to the number of messages exchanged between two communicating parties. A *non-interactive* search and an *interactive* update operation require two and three messages to be exchange, respectively. Our main scheme, the scheme in [13] and some variants in [4] also achieve *non-interactive* update with only single message (i.e., an update token and an encrypted file to be added for the file addition) to be send from the client to the server. The scheme in [19] requires a transient client storage as $\mathcal{O}(N'^\alpha)$.
- κ is the security parameter. p is the # of parallel processors. b is the block size of symmetric encryption scheme. z is the pointer size in bits. t is the network latency introduced due to the interactions. α is a parameter, $0 < \alpha < 1$.
- Update privacy levels $L1,...,L5$ are described in Sect. 5. In comparison with Cash et al. [4], we took variant $\prod_{\text{bas}}^{\text{dyn,ro}}$ as basis and estimated the most efficient variant $\prod_{2\text{lev}}^{\text{dyn,ro}}$, where d_w, a_w, and c'' denote the total number of deletion operations, addition operations, the constant bit size required to store a single file-keyword pair, respectively (in the client storage, the worst case of $a_w = m$). To simplify notation, we assume that both pointers and identifiers are of size c'' and that one can fit b such identifiers/pointers per block of size b (also a simplification). The hidden constants in the asymptotic complexity of the update operation is significant as the update of [4] requires at least six PRF operations per file-keyword pair versus this work, which requires one.
- *Our persistent client storage is $\kappa \cdot \mathcal{O}(m+n)$. This can become 4κ if we store this data structure on the server side, which would cost one additional round of interaction.

handle malicious clients. Blind Seer focuses on a different scenario and system model compared to traditional SSE schemes: "The SSE setting focuses on data outsourcing rather than data sharing. That is, in SSE the data owner is the client, and so no privacy against the client is required" [17]. Moreover, Blind Seer requires three parties (one of them acts as a semi-trusted party) instead of two. Our scheme focuses on only basic keyword queries but achieves the highest update privacy in the traditional SSE setting. The update functionality of Blind Seer is not oblivious (this is explicitly noted in [17] on page 8, footnote 2). The Blind Seer solves the leakage problem due to non-oblivious updates by periodically re-encrypting the entire index.

3 Preliminaries and Models

Operators $||$ and $|x|$ denote the concatenation and the bit length of variable x, respectively. $x \xleftarrow{\$} S$ means variable x is randomly and uniformly selected from set S. For any integer l, $(x_0, \ldots, x_l) \xleftarrow{\$} S$ means $(x_0 \xleftarrow{\$} S, \ldots, x_l \xleftarrow{\$} S)$. $|S|$ denotes the cardinality of set S. $\{x_i\}_{i=0}^l$ denotes $(x_0, \ldots, x_l)$. We denote by $\{0,1\}^*$ the set of binary strings of any finite length. $\lfloor x \rfloor$ denotes the floor of x and $\lceil x \rceil$ denotes the ceiling of x. The set of items q_i for $i = 1, \ldots, n$ is denoted by $\langle q_1, \ldots, q_n \rangle$. Integer κ denotes the security parameter. $\log x$ means $\log_2 x$. $I[*, j]$ and $I[i, *]$ mean accessing all elements in the j'th column and the i'th row of a matrix I, respectively. $I[i, *]^T$ is the transpose of the i'th row of I.

An IND-CPA secure private key encryption scheme is a triplet $\mathcal{E} = (\mathsf{Gen}, \mathsf{Enc}, \mathsf{Dec})$ of three algorithms as follows: $k_1 \leftarrow \mathcal{E}.\mathsf{Gen}(1^\kappa)$ is a Probabilistic Polynomial Time (PPT) algorithm that takes a security parameter κ and returns a secret key k_1; $c \leftarrow \mathcal{E}.\mathsf{Enc}_{k_1}(M)$ takes secret key k_1 and a message M, and returns a ciphertext c; $M \leftarrow \mathcal{E}.\mathsf{Dec}_{k_1}(c)$ is a deterministic algorithm that takes k_1 and c, and returns M if k_1 was the key under which c was produced. A Pseudo Random Function (PRF) is a polynomial-time computable function, which is indistinguishable from a true random function by any PPT adversary. The function $F : \{0,1\}^\kappa \times \{0,1\}^* \rightarrow \{0,1\}^\kappa$ is a keyed PRF, denoted by $\tau \leftarrow F_{k_2}(x)$, which takes as input a secret key $k_2 \xleftarrow{\$} \{0,1\}^\kappa$ and a string x, and returns a token τ. $G : \{0,1\}^\kappa \times \{0,1\}^* \rightarrow \{0,1\}^\kappa$ is a keyed PRF denoted as $r \leftarrow G_{k_3}(x)$, which takes as input $k_3 \leftarrow \{0,1\}^\kappa$ and a string x and returns a key r. We denote by $H : \{0,1\}^{|x|} \rightarrow \{0,1\}$ a Random Oracle (RO) [3], which takes an input x and returns a bit as output.

We follow the definitions of [12,13] with some modifications: f_{id} and w denote a file with unique identifier id and a unique keyword that exists in a file, respectively. A keyword w is of length polynomial in κ, and a file f_{id} may contain any such keyword (i.e., our keyword universe is not fixed). For practical purposes, n and m denote the maximum number of files and keywords to be processed by application, respectively. $\mathbf{f} = (f_{id_1}, \ldots, f_{id_n})$ and $\mathbf{c} = (c_{id_1}, \ldots, c_{id_n})$ denote a collection of files (with unique identifiers $id_1, \ldots, id_n$) and their corresponding ciphertext computed under k_1 via Enc, respectively. Data structures δ and γ denote the index and encrypted index, respectively.

Definition 1. *A* DSSE *scheme is comprised of nine polynomial-time algorithms, which are defined as below:*

1. $K \leftarrow$ Gen(1^κ): *It takes as input a security parameter κ and outputs a secret key K.*
2. $(\gamma, \boldsymbol{c}) \leftarrow$ Enc$_K(\delta, \boldsymbol{f})$: *It takes as input a secret key K, an index δ and files $\boldsymbol{f}$, from which δ was constructed. It outputs encrypted index γ and ciphertexts $\boldsymbol{c}$.*
3. $f_j \leftarrow$ Dec$_K(c_j)$: *It takes as input secret key K and ciphertext c_j and outputs a file f_j.*
4. $\tau_w \leftarrow$ SrchToken(K, w): *It takes as input a secret key K and a keyword w. It outputs a search token τ_w.*
5. $\mathbf{id_w} \leftarrow$ Search(τ_w, γ): *It takes as input a search token τ_w and an encrypted index γ. It outputs identifiers $\mathbf{id_w} \subseteq \boldsymbol{c}$.*
6. $(\tau_f, c) \leftarrow$ AddToken(K, f_{id}): *It takes as input a secret key K and a file f_{id} with identifier id to be added. It outputs an addition token τ_f and a ciphertext c of f_{id}.*
7. $(\gamma', \boldsymbol{c}') \leftarrow$ Add$(\gamma, \boldsymbol{c}, c, \tau_f)$: *It takes as input an encrypted index γ, current ciphertexts $\boldsymbol{c}$, ciphertext c to be added and an addition token τ_f. It outputs a new encrypted index γ' and new ciphertexts $\boldsymbol{c}'$.*
8. $\tau_f' \leftarrow$ DeleteToken(K, f_{id}): *It takes as input a secret key K and a file f_{id} with identifier id to be deleted. It outputs a deletion token τ_f'.*
9. $(\gamma', \boldsymbol{c}') \leftarrow$ Delete$(\gamma, \boldsymbol{c}, \tau_f')$: *It takes as input an encrypted index γ, ciphertexts $\boldsymbol{c}$, and a deletion token τ_f'. It outputs a new encrypted index γ' and new ciphertexts $\boldsymbol{c}'$.*

Definition 2. *A* DSSE *scheme is correct if for all κ, for all keys K generated by* Gen(1^κ), *for all $\boldsymbol{f}$, for all $(\gamma, \boldsymbol{c})$ output by* Enc$_K(\delta, \boldsymbol{f})$, *and for all sequences of add, delete or search operations on γ, search always returns the correct set of identifier $\mathbf{id_w}$.*

Most known efficient SSE schemes (e.g., [4,5,12,13,16,19]) reveal the *access and search patterns* that are defined below.

Definition 3. *Given search query* Query $= w$ *at time t, the search pattern $\mathcal{P}(\delta, $ Query$, t)$ is a binary vector of length t with a 1 at location i if the search time $i \leq t$ was for w, 0 otherwise. The search pattern indicates whether the same keyword has been searched in the past or not.*

Definition 4. *Given search query* Query $= w_i$ *at time t, the access pattern $\Delta(\delta, \boldsymbol{f}, w_i, t)$ is identifiers $\mathbf{id_w}$ of files $\boldsymbol{f}$, in which w_i appears.*

We consider the following leakage functions, in the line of [12] that captures dynamic file addition/deletion in its security model as we do, but we leak much less information compared to [12] (see Sect. 5).

Definition 5. *Leakage functions $(\mathcal{L}_1, \mathcal{L}_2)$ are defined as follows:*

1. $(m, n, \mathbf{id_w}, \langle |f_{id_1}|, \ldots, |f_{id_n}| \rangle) \leftarrow \mathcal{L}_1(\delta, \boldsymbol{f})$: Given the index δ and the set of files $\boldsymbol{f}$ (including their identifiers), $\mathcal{L}_1$ outputs the maximum number of keywords m, the maximum number of files n, the identifiers $\mathbf{id_w} = (id_1, \ldots, id_n)$ of $\boldsymbol{f}$ and the size of each file $|f_{id_j}|, 1 \leq j \leq n$ (which also implies the size of its corresponding ciphertext $|c_{id_j}|$).

2. $(\mathcal{P}(\delta, \mathsf{Query}, t), \Delta(\delta, \boldsymbol{f}, w_i, t)) \leftarrow \mathcal{L}_2(\delta, \boldsymbol{f}, w, t)$: Given the index δ, the set of files $\boldsymbol{f}$ and a keyword w for a search operation at time t, it outputs the search and access patterns.

Definition 6. *Let $\mathcal{A}$ be a stateful adversary and $\mathcal{S}$ be a stateful simulator. Consider the following probabilistic experiments:*

Real$_\mathcal{A}(\kappa)$: *The challenger executes $K \leftarrow \mathsf{Gen}(1^\kappa)$. $\mathcal{A}$ produces $(\delta, \boldsymbol{f})$ and receives $(\gamma, \boldsymbol{c}) \leftarrow \mathsf{Enc}_K(\delta, \boldsymbol{f})$ from the challenger. $\mathcal{A}$ makes a polynomial number of adaptive queries $\mathsf{Query} \in (w, f_{id}, f_{id'})$ to the challenger. If $\mathsf{Query} = w$ then $\mathcal{A}$ receives a search token $\tau_w \leftarrow \mathsf{SrchToken}(K, w)$ from the challenger. If $\mathsf{Query} = f_{id}$ is a file addition query then $\mathcal{A}$ receives an addition token $(\tau_f, c) \leftarrow \mathsf{AddToken}(K, f_{id})$ from the challenger. If $\mathsf{Query} = f_{id'}$ is a file deletion query then $\mathcal{A}$ receives a deletion token $\tau'_f \leftarrow \mathsf{DeleteToken}(K, f_{id'})$ from the challenger. Eventually, $\mathcal{A}$ returns a bit b that is output by the experiment.*

Ideal$_{\mathcal{A},\mathcal{S}}(\kappa)$: *$\mathcal{A}$ produces $(\delta, \boldsymbol{f})$. Given $\mathcal{L}_1(\delta, \boldsymbol{f})$, $\mathcal{S}$ generates and sends $(\gamma, \boldsymbol{c})$ to $\mathcal{A}$. $\mathcal{A}$ makes a polynomial number of adaptive queries $\mathsf{Query} \in (w, f_{id}, f_{id'})$ to $\mathcal{S}$. For each query, $\mathcal{S}$ is given $\mathcal{L}_2(\delta, \boldsymbol{f}, w, t)$. If $\mathsf{Query} = w$ then $\mathcal{S}$ returns a simulated search token τ_w. If $\mathsf{Query} = f_{id}$ or $\mathsf{Query} = f_{id'}$, $\mathcal{S}$ returns a simulated addition token τ_f or deletion token τ'_f, respectively. Eventually, $\mathcal{A}$ returns a bit b that is output by the experiment.*

A DSSE is said $(\mathcal{L}_1, \mathcal{L}_2)$-secure against adaptive chosen-keyword attacks (CKA2-security) if for all PPT adversaries $\mathcal{A}$, there exists a PPT simulator $\mathcal{S}$ such that

$$|\Pr[\mathbf{Real}_\mathcal{A}(\kappa) = 1] - \Pr[\mathbf{Ideal}_{\mathcal{A},\mathcal{S}}(\kappa) = 1]| \leq \mathsf{neg}(\kappa)$$

Remark 1. In Definition 6, we adapt the notion of *dynamic CKA2-security* from [12], which captures the file addition and deletion operations by simulating corresponding tokens τ_f and τ'_f, respectively (see Sect. 5).

4   Our Scheme

We first discuss the intuition and data structures of our scheme. We then outline how these data structures guarantee the correctness of our scheme. Finally, we present our main scheme in detail (an efficient variant of our main scheme is given in Sect. 6, several other variants of our main scheme are given in the full version of this paper in [20]).

Intuition and Data Structures of Our Scheme. The intuition behind our scheme is to rely on a very simple data structure that enables efficient yet secure search and update operations on it. Our data structure is a bit matrix I that is augmented by two static hash tables T_w and T_f. If $I[i,j] = 1$ then it means keyword w_i is present in file f_j, else w_i is not in f_j. The data structure contains both the traditional index and the inverted index representations. We use static hash tables T_w and T_f to uniquely associate a keyword w and a file f to a row index i and a column index j, respectively. Both matrix and hash tables also maintain certain status bits and counters to ensure secure and correct encryption/decryption of the data structure, which guarantees a high level of privacy (i.e., $L1$ as in Sect. 5) with dynamic CKA2-security [12]. Search and update operations are encryption/decryption operations on rows and columns of I, respectively.

As in other index-based schemes, our DSSE scheme has an index δ represented by a $m \times n$ matrix, where $\delta[i,j] \in \{0,1\}$ for $i = 1, \ldots, m$ and $j = 1, \ldots, n$. Initially, all elements of δ are set to 0. I is a $m \times n$ matrix, where $I[i,j] \in \{0,1\}^2$. $I[i,j].v$ stores $\delta[i,j]$ in encrypted form depending on state and counter information. $I[i,j].st$ stores a bit indicating the state of $I[i,j].v$. Initially, all elements of I are set to 0. $I[i,j].st$ is set to 1 whenever its corresponding f_j is updated, and it is set to 0 whenever its corresponding keyword w_i is searched. For the sake of brevity, we will often write $I[i,j]$ to denote $I[i,j].v$. We will always be explicit about the state bit $I[i,j].st$. The encrypted index γ corresponds to the encrypted matrix I and a hash table. We also have client state information[2] in the form of two static hash tables (defined below). We map each file f_{id} and keyword w pair to a unique set of indices (i,j) in matrices (δ, I). We use static hash tables to associate each file and keyword to its corresponding row and column index, respectively. Static hash tables also enable to access the index information in (average) $\mathcal{O}(1)$ time. T_f is a static hash table whose key-value pair is $\{s_{f_j}, \langle j, st_j \rangle\}$, where $s_{f_j} \leftarrow F_{k_2}(id_j)$ for file identifier id_j corresponding to file f_{id_j}, index $j \in \{1, \ldots, n\}$ and st is a counter value. We denote access operations by $j \leftarrow T_f(s_{f_j})$ and $st_j \leftarrow T_f[j].st$. T_w is a static hash table whose key-value pair is $\{s_{w_i}, \langle i, \overline{st}_i \rangle\}$, where token $s_{w_i} \leftarrow F_{k_2}(w_i)$, index $i \in \{1, \ldots, n\}$ and $\overline{st}$ is a counter value. We denote access operations by $i \leftarrow T_w(s_{w_i})$ and $\overline{st}_i \leftarrow T_w[i].st$. All counter values are initially set to 1.

We now outline how these data structures and variables work and ensure the correctness of our scheme.

Correctness of Our Scheme. The correctness and consistency of our scheme is achieved via state bits $I[i,j].st$, and counters $T_w[i].st$ of row i and counters $T_f[j].st$ of column j, each maintained with hash tables T_w and T_f, respectively.

The algorithms SrchToken and AddToken increase counters $T_w[i].st$ for keyword w and $T_f[j].st$ for file f_j, after each search and update operations, respectively. These counters allow the derivation of a new bit, which is used to encrypt the corresponding cell $I[i,j]$. This is done by the invocation of random oracle

---

[2] It is always possible to eliminate client state by encrypting and storing it on the server side. This comes at the cost of additional iteration, as the client would need to retrieve the encrypted hash tables from the server and decrypt them. Asymptotically, this does not change the complexity of the schemes proposed here.

as $H(r_i||j||st_j)$ with row key r_i, column position j and the counter of column j. The row key r_i used in $H(.)$ is re-derived based on the value of row counter $\overline{st}_i$ as $r_i \leftarrow G_{k_3}(i||\overline{st}_i)$, which is increased after each search operation. Hence, if a search is followed by an update, algorithm AddToken derives a fresh key $r_i \leftarrow G_{k_3}(i||\overline{st}_i)$, which was not released during the previous search as a token. This ensures that AddToken algorithm securely and correctly encrypts the new column of added/deleted file. Algorithm Add then replaces new column j with the old one, increments column counter and sets state bits $I[*, j]$ to 1 (indicating cells are updated) for the consistency.

The rest is to show that SrchToken and Search produce correct search results. If keyword w is searched for the first time, SrchToken derives only r_i, since there were no past search increasing the counter value. Otherwise, it derives r_i with the current counter value $\overline{st}_i$ and $\overline{r}_i$ with the previous counter value $\overline{st}_i - 1$, which will be used to decrypt recently updated and non-updated (after the last search) cells of $I[i, *]$, respectively. That is, given search token τ_w, the algorithm Search step 1 checks if τ_w includes only one key (i.e., the first search) or corresponding cell value $I[i, j]$ was updated (i.e., $I[i, j].st = 1$). If one of these conditions holds, the algorithm Search decrypts $I[i, j]$ with bit $H(r_i||j||st_j)$ that was used for encryption by algorithm Enc (i.e., the first search) or AddToken. Otherwise, it decrypts $I[i, j]$ with bit $H(\overline{r}_i||j||st_j)$. Hence, the algorithm Search produces the correct search result by properly decrypting row i. The algorithm Search also ensures the consistency by setting all state bits $I[i, *].st$ to zero (i.e., indicating cells are searched) and re-encrypting $I[i, *]$ by using the last row key r_i.

Detailed Description. We now describe our main scheme in detail.

$K \leftarrow \mathsf{Gen}(1^\kappa)$: The client generates $k_1 \leftarrow \mathcal{E}.\mathsf{Gen}(1^\kappa)$, $(k_2, k_3) \xleftarrow{\$} \{0,1\}^\kappa$ and $K \leftarrow (k_1, k_2, k_3)$.

$(\gamma, \mathbf{c}) \leftarrow \mathsf{Enc}_K(\delta, \mathbf{f})$: The client generates $(\gamma, \mathbf{c})$:
  1. Extract unique keywords $(w_1, \ldots, w_{m'})$ from files $\mathbf{f} = (f_{id_1}, \ldots, f_{id_{n'}})$, where $n' \leq n$ and $m' \leq m$. Initially, set all the elements of δ to 0.
  2. Construct δ for $j = 1, \ldots, n'$ and $i = 1, \ldots, m'$:
     (a) $s_{w_i} \leftarrow F_{k_2}(w_i)$, $x_i \leftarrow T_w(s_{w_i})$, $s_{f_j} \leftarrow F_{k_2}(id_j)$ and $y_j \leftarrow T_f(s_{f_j})$.
     (b) If w_i appears in f_j set $\delta[x_i, y_j] \leftarrow 1$.
  3. Encrypt δ for $j = 1, \ldots, n$ and $i = 1, \ldots, m$:
     (a) $T_w[i].st \leftarrow 1$, $T_f[j].st \leftarrow 1$ and $I[i, j].st \leftarrow 0$.
     (b) $r_i \leftarrow G_{k_3}(i||\overline{st}_i)$, where $\overline{st}_i \leftarrow T_w[i].st$.
     (c) $I[i, j] \leftarrow \delta[i, j] \oplus H(r_i||j||st_j)$, where $st_j \leftarrow T_f[j].st$.
  4. $c_j \leftarrow \mathcal{E}.\mathsf{Enc}_{k_1}(f_{id_j})$ for $j = 1, \ldots, n'$ and $\mathbf{c} \leftarrow \{\langle c_1, y_1\rangle, \ldots, \langle c_{n'}, y_{n'}\rangle\}$.
  5. Output $(\gamma, \mathbf{c})$, where $\gamma \leftarrow (I, T_f)$. The client gives $(\gamma, \mathbf{c})$ to the server, and keeps (K, T_w, T_f).

$f_j \leftarrow \mathsf{Dec}_K(c_j)$: The client obtains the file as $f_j \leftarrow \mathcal{E}.\mathsf{Dec}_{k_1}(c_j)$.

$\tau_w \leftarrow \mathsf{SrchToken}(K, w)$: The client generates a token τ_w for w:
  1. $s_{w_i} \leftarrow F_{k_2}(w)$, $i \leftarrow T_w(s_{w_i})$, $\overline{st}_i \leftarrow T_w[i].st$ and $r_i \leftarrow G_{k_3}(i||\overline{st}_i)$.
  2. If $\overline{st}_i = 1$ then $\tau_w \leftarrow (i, r_i)$. Else (if $\overline{st}_i > 1$), $\overline{r}_i \leftarrow G_{k_3}(i||\overline{st}_i - 1)$ and $\tau_w \leftarrow (i, r_i, \overline{r}_i)$.

3. $T_w[i].st \leftarrow \overline{st}_i + 1$. The client outputs τ_w and sends it to the server.

$\mathbf{id_w} \leftarrow \mathsf{Search}(\tau_w, \gamma)$: The server finds indexes of ciphertexts for τ_w:

1. If $((\tau_w = (i, r_i) \vee I[i,j].st) = 1)$ hold then $I'[i,j] \leftarrow I[i,j] \oplus H(r_i\|j\|st_j)$, else set $I'[i,j] \leftarrow I[i,j] \oplus H(\overline{r}_i\|j\|st_j)$, where $st_j \leftarrow T_f[j].st$ for $j = 1, \ldots, n$.

2. Set $I[i,*].st \leftarrow 0$, $l' \leftarrow 1$ and for each j satisfies $I'[i,j] = 1$, set $y_{l'} \leftarrow j$ and $l' \leftarrow l' + 1$.

3. Output $\mathbf{id_w} \leftarrow (\mathbf{y_1}, \ldots, \mathbf{y_l})$. The server returns $(c_{y_1}, \ldots, c_{y_l})$ to the client, where $l \leftarrow l' - 1$.

4. After the search is completed, the server re-encrypts row $I'[i,*]$ with r_i as $I[i,j] \leftarrow I'[i,j] \oplus H(r_i\|j\|st_j)$ for $j = 1, \ldots, n$, where $st_j \leftarrow T_f[j].st$ and sets $\gamma \leftarrow (I, T_w)$.

$(\tau_f, c) \leftarrow \mathsf{AddToken}(K, f_{id_j})$: The client generates τ_f for a file f_{id_j}:

1. $s_{f_j} \leftarrow F_{k_2}(id_j)$, $j \leftarrow T_f(s_{f_j})$, $T_f[j].st \leftarrow T_f[j].st + 1$, $st_j \leftarrow T_f[j].st$.

2. $r_i \leftarrow G_{k_3}(i\|\overline{st}_i)$, where $\overline{st}_i \leftarrow T_w[i].st$ for $i = 1, \ldots, m$.

3. Extract $(w_1, \ldots, w_t)$ from f_{id_j} and compute $s_{w_i} \leftarrow F_{k_2}(w_i)$ and $x_i \leftarrow T_w(s_{w_i})$ for $i = 1, \ldots, t$.

4. Set $\overline{I}[x_i] \leftarrow 1$ for $i = 1, \ldots, t$ and rest of the elements as $\{\overline{I}[i] \leftarrow 0\}_{i=1, i \notin \{x_1, \ldots, x_t\}}^m$. Also set $I'[i] \leftarrow \overline{I}[i] \oplus H(r_i\|j\|st_j)$ for $i = 1, \ldots, m$.

5. Set $c \leftarrow \mathcal{E}.\mathsf{Enc}_{k_1}(f_{id_j})$ and output $(\tau_f \leftarrow (I', j), c)$. The client sends (τ_f, c) to the server.

$(\gamma', \mathbf{c}') \leftarrow \mathsf{Add}(\gamma, \mathbf{c}, c, \tau_f)$: The server performs file addition:

1. $I[*,j] \leftarrow (I')^T$, $I[*,j].st \leftarrow 1$ and $T_f[j].st \leftarrow T_f[j].st + 1$.

2. Output $(\gamma', \mathbf{c}')$, where $\gamma' \leftarrow (I, T_f)$ and $\mathbf{c}'$ is (c, j) added to $\mathbf{c}$.

$\tau'_f \leftarrow \mathsf{DeleteToken}(K, f)$: The client generates τ'_f for f:

1. Execute steps (1–2) of AddToken algorithm, which produce (j, r_i, st_j).

2. $I'[i] \leftarrow H(r_i\|j\|st_j)$ for $i = 1, \ldots, m^3$.

3. Output $\tau'_f \leftarrow (I', j)$. The client sends τ'_f to the server.

$(\gamma', \mathbf{c}') \leftarrow \mathsf{Delete}(\gamma, \mathbf{c}, \tau'_f)$: The server performs file deletion:

1. $I[*,j] \leftarrow (I')^T$, $I[*,j].st \leftarrow 1$ and $T_f[j].st \leftarrow T_f[j].st + 1$.

2. Output $(\gamma', \mathbf{c}')$, where $\gamma' \leftarrow (I, T_f)$, $\mathbf{c}'$ is (c, j) removed from $\mathbf{c}$.

Keyword Update for Existing Files: Some existing schemes (e.g., [16]) only permit adding or deleting a file, but do not permit updating keywords in an existing file. Our scheme enables keyword update in an existing file. To update an existing file f by adding new keywords or removing existing keywords, the client prepares a new column $\overline{I}[i] \leftarrow b_i$, $i = 1, \ldots, m$, where $b_i = 1$ if w_i is added and $b_i = 0$ otherwise (as in AddToken, step 4). The rest of the algorithm is similar to AddToken.

---

[3] This step is only meant to keep data structure consistency during a search operation.

5   Security Analysis

We prove that our main scheme achieves *dynamic adaptive security against chosen-keyword attacks (CKA2)* as below. It is straightforward to extend the proof for our variant schemes. Note that our scheme is secure in the Random Oracle Model (ROM) [3]. That is, $\mathcal{A}$ is given access to a random oracle $RO(.)$ from which she can request the hash of any message of her choice. In our proof, cryptographic function H used in our scheme is modeled as a random oracle via function $RO(.)$.

Theorem 1. *If* Enc *is IND-CPA secure,* (F, G) *are PRFs and* H *is a RO then our DSSE scheme is* $(\mathcal{L}_1, \mathcal{L}_2)$-*secure in ROM according to Definition 6 (CKA-2 security with update operations).*

Proof. We construct a simulator $\mathcal{S}$ that interacts with an adversary $\mathcal{A}$ in an execution of an **Ideal**$_{\mathcal{A},\mathcal{S}}(\kappa)$ experiment as described in Definition 6.

In this experiment, $\mathcal{S}$ maintains lists $\mathcal{LR}$, $\mathcal{LK}$ and $\mathcal{LH}$ to keep track the query results, states and history information, initially all lists empty. $\mathcal{LR}$ is a list of key-value pairs and is used to keep track $RO(.)$ queries. We denote value $\leftarrow \mathcal{LR}(\text{key})$ and $\perp \leftarrow \mathcal{LR}(\text{key})$ if key does not exist in $\mathcal{LR}$. $\mathcal{LK}$ is used to keep track random values generated during the simulation and it follows the same notation that of $\mathcal{LR}$. $\mathcal{LH}$ is used to keep track search and update queries, $\mathcal{S}$'s replies to those queries and their leakage output from $(\mathcal{L}_1, \mathcal{L}_2)$.

$\mathcal{S}$ executes the simulation as follows:

I. Handle RO(.) Queries: $b \leftarrow RO(x)$ takes an input x and returns a bit b as output. Given x, if $\perp = \mathcal{LR}(x)$ set $b \stackrel{\$}{\leftarrow} \{0, 1\}$, insert (x, b) into $\mathcal{LR}$ and return b as the output. Else, return $b \leftarrow \mathcal{LR}(x)$ as the output.

II. Simulate $(\gamma, \mathbf{c})$: Given $(m, n, \langle id_1, \ldots, id_{n'} \rangle, \langle |c_{id_1}|, \ldots, |c_{id_{n'}}| \rangle) \leftarrow \mathcal{L}_1(\delta, \mathbf{f})$, $\mathcal{S}$ simulates $(\gamma, \mathbf{c})$ as follows:

1. $s_{f_j} \stackrel{\$}{\leftarrow} \{0, 1\}^\kappa$, $y_j \leftarrow T_f(s_{f_j})$, insert (id_j, s_{f_j}, y_j) into $\mathcal{LH}$ and encrypt $c_{y_j} \leftarrow \mathcal{E}.\text{Enc}_k(\{0\}^{|c_{id_j}|})$, where $k \stackrel{\$}{\leftarrow} \{0, 1\}^\kappa$ for $j = 1, \ldots, n'$.
2. For $j = 1, \ldots, n$ and $i = 1, \ldots, m$
   (a) $T_w[i].st \leftarrow 1$ and $T_f[j].st \leftarrow 1$.
   (b) $z_{i,j} \stackrel{\$}{\leftarrow} \{0, 1\}^{2\kappa}$, $I[i, j] \leftarrow RO(z_{i,j})$ and $I[i, j].st \leftarrow 0$.
3. Output $(\gamma, \mathbf{c})$, where $\gamma \leftarrow (I, T_f)$ and $\mathbf{c} \leftarrow \{\langle c_1, y_1 \rangle, \ldots, \langle c_{n'}, y_{n'} \rangle\}$.

Correctness and Indistinguishability of the Simulation: $\mathbf{c}$ has the correct size and distribution, since $\mathcal{L}_1$ leaks $\langle |c_{id_1}|, \ldots, |c_{id_{n'}}| \rangle$ and Enc is a IND-CPA secure scheme, respectively. I and T_f have the correct size since $\mathcal{L}_1$ leaks (m, n). Each $I[i, j]$ for $j = 1, \ldots, n$ and $i = 1, \ldots, m$ has random uniform distribution as required, since $RO(.)$ is invoked with a separate random number $z_{i,j}$. T_f has the correct distribution, since each s_{f_j} has random uniform distribution, for $j = 1, \ldots, n'$. Hence, $\mathcal{A}$ does not abort due to $\mathcal{A}$'s simulation of $(\gamma, \mathbf{c})$. The probability that $\mathcal{A}$ queries $RO(.)$ on any $z_{i,j}$ before $\mathcal{S}$ provides I to $\mathcal{A}$ is negligible (i.e., $\frac{1}{2^{2\kappa}}$). Hence, $\mathcal{S}$ also does not abort.

III. Simulate τ_w: Simulator $\mathcal{S}$ receives a search query w on time t. $\mathcal{S}$ is given $(\mathcal{P}(\delta, \mathsf{Query}, t), \Delta(\delta, \mathbf{f}, w_i, t)) \leftarrow \mathcal{L}_2(\delta, \mathbf{f}, w, t)$. $\mathcal{S}$ adds these to $\mathcal{LH}$. $\mathcal{S}$ then simulates τ_w and updates lists $(\mathcal{LR}, \mathcal{LK})$ as follows:

1. If w in list $\mathcal{LH}$ then fetch corresponding s_{w_i}. Else, $s_{w_i} \xleftarrow{\$} \{0,1\}^\kappa$, $i \leftarrow T_w(s_{w_i})$, $\overline{st}_i \leftarrow T_w[i].st$ and insert $(w, \mathcal{L}_1(\delta, \mathbf{f}), s_{w_i})$ into $\mathcal{LH}$.
2. If $\bot = \mathcal{LK}(i, \overline{st}_i)$ then $r_i \leftarrow \{0,1\}^\kappa$ and insert $(r_i, i, \overline{st}_i)$ into $\mathcal{LK}$. Else, $r_i \leftarrow \mathcal{LK}(i, \overline{st}_i)$.
3. If $\overline{st}_i > 1$ then $\overline{r}_i \leftarrow \mathcal{LK}(i \| \overline{st}_i - 1)$, $\tau_w \leftarrow (i, r_i, \overline{r}_i)$. Else, $\tau_w \leftarrow (i, r_i)$.
4. $T_w[i].st \leftarrow \overline{st}_i + 1$.
5. Given $\mathcal{L}_2(\delta, \mathbf{f}, w, t)$, $\mathcal{S}$ knows identifiers $\mathbf{id_w} = (y_1, \ldots, y_l)$. Set $I'[i, y_j] \leftarrow 1$, $j = 1, \ldots, l$, and rest of the elements as $\{I'[i, j] \leftarrow 0\}_{j=1, j \notin \{y_1, \ldots, y_l\}}$.
6. If $((\tau_w = (i, r_i) \vee I[i, j].st) = 1)$ then $V[i, j] \leftarrow I[i, j]' \oplus I[i, j]$ and insert tuple $(r_i \| j \| st_j, V[i, j])$ into $\mathcal{LR}$ for $j = 1, \ldots, n$, where $st_j \leftarrow T_f[j].st$.
7. $I[i, *].st \leftarrow 0$.
8. $I[i, j] \leftarrow I'[i, j] \oplus RO(r_i \| j \| st_j)$, where $st_j \leftarrow T_f[j].st$ for $j = 1, \ldots, n$.
9. Output τ_w and insert (w, τ_w) into $\mathcal{LH}$.

Correctness and Indistinguishability of the Simulation: Given any $\Delta(\delta, \mathbf{f}, w_i, t)$, $\mathcal{S}$ simulates the output of $RO(.)$ such that τ_w always produces the correct search result for $\mathbf{id_w} \leftarrow \mathsf{Search}(\tau_w, \gamma)$. $\mathcal{S}$ needs to simulate the output of $RO(.)$ for two conditions (as in *III-Step 6*): (i) The first search of w_i (i.e., $\tau_w = (i, r_i)$), since $\mathcal{S}$ did not know δ during the simulation of $(\gamma, \mathbf{c})$. (ii) If any file f_{id_j} containing w_i has been updated after the last search on w_i (i.e., $I[i, j].st = 1$), since $\mathcal{S}$ does not know the content of update. $\mathcal{S}$ sets the output of $RO(.)$ for those cases by inserting tuple $(r_i \| j \| st_j, V[i, j])$ into $\mathcal{LR}$ (as in *III-Step 6*). In other cases, $\mathcal{S}$ just invokes $RO(.)$ with $(r_i \| j \| st_j)$, which consistently returns previously inserted bit from $\mathcal{LR}$ (as in *III-Step 8*).

During the first search on w_i, each $RO(.)$ output $V[i, j] = RO(r_i \| j | st_j)$ has the correct distribution, since $I[i, *]$ of γ has random uniform distribution (see *II-Correctness and Indistinguishability* argument). Let $J = (j_1, \ldots, j_l)$ be the indexes of files containing w_i, which are updated after the last search on w_i. If w_i is searched then each $RO(.)$ output $V[i, j] = RO(r_i \| j | st_j)$ has the correct distribution, since $\tau_f \leftarrow (I', j)$ for indexes $j \in J$ has random uniform distribution (see *IV-Correctness and Indistinguishability* argument). Given that $\mathcal{S}$'s τ_w always produces correct $\mathbf{id_w}$ for given $\Delta(\delta, \mathbf{f}, w_i, t)$, and relevant values and $RO(.)$ outputs have the correct distribution as shown, $\mathcal{A}$ does not abort during the simulation due to $\mathcal{S}$'s search token. The probability that $\mathcal{A}$ queries $RO(.)$ on any $(r_i \| j | st_j)$ before him queries $\mathcal{S}$ on τ_w is negligible (i.e., $\frac{1}{2^\kappa}$), and therefore $\mathcal{S}$ does not abort due to $\mathcal{A}$'s search query.

IV. Simulate (τ_f, τ_f'): $\mathcal{S}$ receives an update request $\mathsf{Query} = ((\langle \mathsf{Add}, |c_{id_j}|\rangle,$ $\overline{\mathsf{Delete}})$ at time t. $\mathcal{S}$ simulates update tokens (τ_f, τ_f') as follows:

1. If id_j in $\mathcal{LH}$ then fetch its corresponding (s_{f_j}, j) from $\mathcal{LH}$, else set $s_{f_j} \xleftarrow{\$}$ $\{0,1\}^\kappa$, $j \leftarrow T_f(s_{f_j})$ and insert (s_{f_j}, j, f_{id_j}) into $\mathcal{LH}$.
2. $T_f[j].st \leftarrow T_f[j].st + 1$, $st_j \leftarrow T_f[j].st$.
3. If $\bot = \mathcal{LK}(i, \overline{st}_i)$ then $r_i \leftarrow \{0,1\}^\kappa$ and insert $(r_i, i, \overline{st}_i)$ into $\mathcal{LK}$, where $\overline{st}_i \leftarrow T_w[i].st$ for $i = 1, \ldots, m$.
4. $I'[i] \leftarrow RO(z_i)$, where $z_i \xleftarrow{\$} \{0,1\}^{2\kappa}$ for $i = 1, \ldots, m$.
5. $I[*, j] \leftarrow (I')^T$ and $I[*, j].st \leftarrow 1$.
6. If $\mathsf{Query} = \langle \mathsf{Add}, |c_{id_j}| \rangle$, simulate $c_j \leftarrow \mathcal{E}.\mathsf{Enc}_k(\{0\}^{|c_{id}|})$, add c_j into $\mathbf{c}$, set $\tau_f \leftarrow (I', j)$ output (τ_f, j). Else set $\tau'_f \leftarrow (I', j)$, remove c_j from $\mathbf{c}$ and output τ'_f.

Correctness and Indistinguishability of the Simulation: Given any access pattern (τ_f, τ'_f) for a file f_{id_j}, $\mathcal{A}$ checks the correctness of update by searching all keywords $W = (w_{i_1}, \ldots, w_{i_l})$ included f_{id_j}. Since $\mathcal{S}$ is given access pattern $\Delta(\delta, \mathbf{f}, w_i, t)$ for a search query (which captures the last update before the search), the search operation always produces a correct result after an update (see *III-Correctness and Indistinguishability* argument). Hence, $\mathcal{S}$'s update tokens are correct and consistent.

It remains to show that (τ_f, τ'_f) have the correct probability distribution. In real algorithm, st_j of file f_{id_j} is increased for each update as simulated in *IV-Step 2*. If f_{id_j} is updated after w_i is searched, a new r_i is generated for w_i as simulated in *IV-Step 3* (r_i remains the same for consecutive updates but st_j is increased). Hence, the real algorithm invokes $H(.)$ with a different input $(r_i \| j \| st_j)$ for $i = 1, \ldots, m$. $\mathcal{S}$ simulates this step by invoking $RO(.)$ with z_i and $I'[i] \leftarrow RO(z_i)$, for $i = 1, \ldots, m$. (τ_f, τ'_f) have random uniform distribution, since I' has random uniform distribution and update operations are correct and consistent as shown. c_j has the correct distribution, since Enc is an IND-CPA cipher. Hence, $\mathcal{A}$ does not abort during the simulation due to $\mathcal{S}$'s update tokens. The probability that $\mathcal{A}$ queries $RO(.)$ on any z_i before him queries $\mathcal{S}$ on (τ_f, τ'_f) is negligible (i.e., $\frac{1}{2^{2\kappa}}$), and therefore $\mathcal{S}$ also does not abort due to $\mathcal{A}$'s update query.

V. Final Indistinguishability Argument: (s_{w_i}, s_{f_j}, r_i) for $i = 1, \ldots, m$ and $j = 1, \ldots, n$ are indistinguishable from real tokens and keys, since they are generated by PRFs that are indistinguishable from random functions. Enc is a IND-CPA scheme, the answers returned by $\mathcal{S}$ to $\mathcal{A}$ for $RO(.)$ queries are consistent and appropriately distributed, and all query replies of $\mathcal{S}$ to $\mathcal{A}$ during the simulation are correct and indistinguishable as discussed in *I-IV Correctness and Indistinguishability* arguments. Hence, for all PPT adversaries, the outputs of $\mathbf{Real}_\mathcal{A}(\kappa)$ and that of an $\mathbf{Ideal}_{\mathcal{A},\mathcal{S}}(\kappa)$ experiment are negligibly close:

$$|\mathsf{Pr}[\mathbf{Real}_\mathcal{A}(\kappa) = 1] - \mathsf{Pr}[\mathbf{Ideal}_{\mathcal{A},\mathcal{S}}(\kappa) = 1]| \leq \mathsf{neg}(\kappa)$$

$\square$

Remark 2. Extending the proof to Variant-I presented in Sect. 6 is straight-forward[4]. In particular, (i) interaction is required because even if we need to update a single entry (column) corresponding to a single file, the client needs to re-encrypt the whole b-bit block in which the column resides to keep consistency. This, however, is achieved by retrieving the encrypted b-bit block from the server, decrypting on the *client* side and re-encrypting using AES-CTR mode. Given that we use ROs and a IND-CPA encryption scheme (AES in CTR mode) the security of the DSSE scheme is not affected in our model, and, in particular, there is no additional leakage. (ii) The price that is paid for this performance improvement is that we need interaction in the new variant. Since the messages (the columns/rows of our matrix) exchanged between client and server are encrypted with an IND-CPA encryption scheme there is no additional leakage either due to this operation.

Discussions on Privacy Levels. The leakage definition and formal security model described in Sect. 3 imply various levels of privacy for different DSSE schemes. We summarize some important privacy notions (based on the various leakage characteristics discussed in [4,12,16,19]) with different levels of privacy as follows:

- *Size pattern*: The number of file-keyword pairs in the system.
- *Forward privacy*: A search on a keyword w does not leak the identifiers of files matching this keyword for (pre-defined) future files.
- *Backward privacy*: A search on a keyword w does not leak the identifiers of files matching this keywords that were previously added but then deleted (leaked via additional info kept for deletion operations).
- *Update privacy*: Update operation may leak different levels of information depending on the construction:

  - Level-1 ($L1$) leaks only the time t of the update operation and an index number. $L1$ does not leak the type of update due to the type operations performed on encrypted index γ. Hence, it is possible to hide the type of update via batch/fake file addition/deletion[5]. However, if the update is addition and added file is sent to the server along with the update information on γ, then the type of update and the size of added file are leaked.
  - Level-2 ($L2$) leaks $L1$ plus the identifier of the file being updated and the number of keywords in the updated file (e.g., as in [19]).
  - Level-3 ($L3$) leaks $L2$ plus when/if that identifier has had the same keywords added or deleted before, and also when/if the same keyword have been searched before (e.g., as in [4][6]).

---

[4] This variant encrypts/decrypts b-bit blocks instead of single bits and it requires interaction for add/delete/update operations.

[5] In our scheme, the client may delete file f_{id_j} from γ but still may send a fake file f'_{id_j} to the server as a fake file addition operation.

[6] Remark that despite the scheme in [4] leaks more information than that of ours and [19] as discussed, it does not leak the (pseudonymous) index of file to be updated.

- Level-4 ($L4$) leaks $L3$ plus the information whether the same keyword added or deleted from two files (e.g., as in [12]).
- Level-5 ($L5$) leaks significant information such as the pattern of all intersections of everything is added or deleted, whether or not the keywords were search-ed for (e.g., as in [13]).

Note that our scheme achieves the highest level of $L1$ update privacy, forward-privacy, backward-privacy and size pattern privacy. Hence, it achieves the highest level of privacy among its counterparts.

6 Evaluation and Discussion

We have implemented our scheme in a stand-alone environment using C/C++. By stand-alone, we mean we run on a single machine, as we are only interested in the performance of the operations and not the effects of latency, which will be present (but are largely independent of the implementation[7].) For cryptographic primitives, we chose to use the libtomcrypt cryptographic toolkit version 1.17 [9] and as an API. We modified the low level routines to be able to call and take advantage of AES hardware acceleration instructions natively present in our hardware platform, using the corresponding freely available Intel reference implementations [11]. We performed all our experiments on an Intel dual core i5-3320M 64-bit CPU at 2.6 GHz running Ubuntu 3.11.0-14 generic build with 4GB of RAM. We use 128-bit CCM and AES-128 CMAC for file and data structure encryption, respectively. Key generation was implemented using the expand-then-extract key generation paradigm analyzed in [14]. However, instead of using a standard hash function, we used AES-128 CMAC for performance reasons. Notice that this key derivation function has been formally analyzed and is standardized. Our use of CMAC as the PRF for the key derivation function is also standardized [7]. Our random oracles were all implemented via 128-bit AES CMAC. For hash tables, we use Google's C++ sparse hash map implementation [2] but instead of using the standard hash function implementation, we called our CMAC-based random oracles truncated to 80 bits. Our implementation results are summarized in Table 2.

Experiments. We performed our experiments on the Enron dataset [1] as in [13]. Table 2 summarizes results for three types of experiments: (i) Large number of files and large number of keywords, (ii) large number of files but comparatively small number of keywords and (iii) large number of keywords but small number of files. In all cases, the combined number of keyword/file pairs is between 10^9 and 10^{10}, which surpass the experiments in [13] by about two orders of magnitude and are comparable to the experiments in [19]. One key observation is that in

---

[7] As it can be seen from Table 1, our scheme is optimal in terms of the number of rounds required to perform *any* operation. Thus, latency will not affect the performance of the implementation anymore than any other competing scheme. This replicates the methodology of Kamara et al. [13].

contrast to [19] (and especially to [4] with very high-end servers), we do *not* use server-level hardware but a rather standard commodity Intel platform with limited RAM memory. From our results, it is clear that for large databases the process of generating the encrypted representation is relatively expensive, however, this is a one-time only cost. The cost per keyword search depends linearly as $\mathcal{O}(n)/128$ on the number of files in the database and it is not cost-prohibiting (even for the large test case of 10^{10} keyword/file pairs, searching takes only a few msec). We observe that despite this linear cost, our search operation is extremely fast comparable to the work in [13]. The costs for adding and deleting files (updates) is similarly due to the obliviousness of these operations in our case. Except for the cost of creating the index data structure, all performance data extrapolates to any other type of data, as our data structure is not data dependant and it is conceptually very simple.

Table 2. Execution times of our DSSE scheme. w.: # of words, f.: # of files

Operation	Time (ms)					
	w. $2 \cdot 10^5$	f. $5 \cdot 10^4$	w. 2000	f. $2 \cdot 10^6$	w. $1 \cdot 10^6$	f. 5000
Building searchable representation (offline, one-time cost at initialization)						
Keyword-file mapping, extraction	6.03 s		52 min		352 ms	
Encrypt searchable representation	493 ms		461 ms		823 ms	
Search and Update Operations (online, after initialization)						
Search for single key word	0.3 ms		10 ms		0.02 ms	
Add file to database	472 ms		8.83 ms		2.77 s	
Delete file from database	329 ms		8.77 ms		2.36 s	

Comparison with Existing Alternatives. Compared to Kamara et al. in [13], which achieves optimal $O(r)$ search time but leaks significant information for updates, our scheme has linear search time (for # of files) but achieves completely oblivious updates. Moreover, the [13] can not be parallelized, whereas our scheme can. Kamara et al. [12] relies on red-black trees as the main data structure, achieves parallel search and oblivious updates. However, it incurs impractical server storage overhead due to its very large encrypted index size. The scheme of Stefanov et al. [19] requires high client storage (e.g., 210 MB for moderate size file-keyword pairs), where the client fetches non-negligible amount of data from the server and performs an oblivious sort on it. We only require *one* hash table and four symmetric secret keys storage. The scheme in [19] also requires significant amount of data storage (e.g., 1600 bits) for per keyword-file pair at the server side versus 2 bits per file-keyword pair in our scheme (and a hash table[8]).

---

[8] The size of the hash table depends on its occupancy factor, the number of entries and the size of each entry. Assuming 80-bits per entry and a 50 % occupancy factor, our scheme still requires about $2 \times 80 + 2 = 162$ bits per entry, which is about a factor 10 better than [19]. Observe that for fixed m-words, we need a hash table with approximately $2m$ entries, even if each entry was represented by 80-bits.

The scheme in [4] leaks more information compared to [19] also incurring in non-negligible server storage. The data structure in [4] grows linearly with the number of deletion operations, which requires re-encrypting the data structure eventually. Our scheme does not require re-encryption (but we assume an upper bound on the maximum number of files), and our storage is constant regardless of the number of updates. The scheme in [16] relies on a primitive called "Blind-Storage", where the server acts only as a storage entity. This scheme requires higher interaction than its counterparts, which may introduce response delays for distributed client-server architectures. This scheme leaks less information than that of [4], but only support single keyword queries. It can add/remove a file but cannot update the content of a file in contrast to our scheme.

We now present an efficient variant of our scheme:

Variant-I: Trade-off Between Computation and Interaction Overhead.
In the main scheme, H is invoked for each column of I once, which requires $O(n)$ invocations in total. We propose a variant scheme that offers significant computational improvement at the cost of a plausible communication overhead.

We use counter (CTR) mode with a block size b for $\mathcal{E}$. We interpret columns of I as $d = \lceil \frac{n}{b} \rceil$ blocks with size of b bits each, and encrypt each block B_l, $l = 0, \ldots, d-1$, separately with $\mathcal{E}$ by using a unique block counter st_l. Each block counter st_l is located at its corresponding index a_l (block offset of B_l) in T_f, where $a_l \leftarrow (l \cdot b) + 1$. The uniqueness of each block counter is achieved with a global counter gc, which is initialized to 1 and incremented by 1 for each update. A state bit $T_f[a_l].b$ is stored to keep track the update status of its corresponding block. The update status is maintained only for each block but not for each bit of $I[i, j]$. Hence, I is a binary matrix (unlike the main scheme, in which $I[i, j] \in \{0, 1\}^2$). AddToken and Add algorithms for the aforementioned variant are as follows (DeleteToken and Delete follow the similar principles):

$(\tau_f, c) \leftarrow$ AddToken(K, f_{id_j}): The client generates τ_f for f_{id_j} as follows:
1. $s_{f_j} \leftarrow F_{k_2}(f_{id_j})$, $j \leftarrow T_f(s_{f_j})$, $l \leftarrow \lfloor \frac{j}{b} \rfloor$, $a_l \leftarrow (l \cdot b) + 1$ and $st_l \leftarrow T_f[a_l].st$. Extract $(w_1, \ldots, w_t)$ from f_{id_j} and compute $s_{w_i} \leftarrow F_{k_2}(w_i)$ and $x_i \leftarrow T_w(s_{w_i})$ for $i = 1, \ldots, t$. For $i = 1, \ldots, m$:
   (a) $r_i \leftarrow G_{k_3}(i \| \overline{st}_i)$, where $\overline{st}_i \leftarrow T_w[i].st$ [9].
   (b) The client requests l'th block, which contains index j of f_{id} from the server. The server then returns the corresponding block $(I[i, a_l], \ldots, I[i, a_{l+1} - 1])$, where $a_{l+1} \leftarrow b(l+1) + 1$.
   (c) $(\overline{I}[i, a_l], \ldots, \overline{I}[i, a_{l+1} - 1]) \leftarrow \mathcal{E}.\mathsf{Dec}_{r_i}(I[i, a_l], \ldots, I[i, a_{l+1} - 1], st_l)$.
2. Set $\overline{I}[x_i, j] \leftarrow 1$ for $i = 1, \ldots, t$ and $\{\overline{I}[i, j] \leftarrow 0\}_{i=1, i \notin \{x_1, \ldots, x_t\}}^m$.
3. $gc \leftarrow gc + 1$, $T_f[a_l].st \leftarrow gc$, $st_l \leftarrow T_f[a_l].st$ and $T_f[a_l].b \leftarrow 1$.
4. $(I'[i, a_l], \ldots, I'[i, a_{l+1} - 1]) \leftarrow \mathcal{E}.\mathsf{Enc}_{r_i}(\overline{I}[i, a_l], \ldots, \overline{I}[i, a_{l+1} - 1], st_l)$ for $i = 1, \ldots, m$. Finally, $c \leftarrow \mathcal{E}.\mathsf{Enc}_{k_1}(f_{id_j})$.
5. Output $\tau_f \leftarrow (I', j)$. The client sends (τ_f, c) to the server.

$(\gamma', \mathbf{c}') \leftarrow$ Add$(\gamma, \mathbf{c}, c, \tau_f)$: The server performs file addition as follows:

---

[9] In this variant, G should generate a cryptographic key suitable for the underlying encryption function $\mathcal{E}$ (e.g., the output of KDF is $b = 128$ for AES with CTR mode).

1. Replace $(I[*, a_l], \ldots, I[*, a_{l+1} - 1])$ with I'.
2. $gc \leftarrow gc + 1$, $T_f[a_l].st \leftarrow gc$ and $T_f[a_l].b \leftarrow 1$.
3. Output $(\gamma', \mathbf{c}')$, where $\gamma' \leftarrow (I, T_f)$ and $\mathbf{c}'$ is (c, j) added to $\mathbf{c}$.

Gen and Dec algorithms of the variant scheme are identical to that of main scheme. The modifications of SrchToken and Search algorithms are straightforward (in the line of AddToken and Add) and therefore will not be repeated. In this variant, the search operation requires the decryption of b-bit blocks for $l = 0, \ldots, d - 1$. Hence, $\mathcal{E}$ is invoked only $O(n/b)$ times during the search operation (in contrast to $O(n)$ invocation of H as in our main scheme). That is, the search operation becomes b times faster compared to our main scheme. The block size b can be selected according to the application requirements (e.g., $b = 64$, $b = 128$ or $b = 256$ based on the preferred encryption function). For instance, $b = 128$ yields highly efficient schemes if the underlying cipher is AES by taking advantage of AES specialized instructions in current PC platforms. Moreover, CTR mode can be parallelizable and therefore the search time can be reduced to $O(n/(b \cdot p))$, where p is the number of processors in the system. This variant requires transmitting $2 \cdot b \cdot O(m)$ bits for each update compared to $O(m)$ non-interactive transmission in our main scheme. However, one may notice that this approach offers a trade-off, which is useful for some practical applications. That is, the search speed is increased by a factor of b (e.g., $b = 128$) with the cost of transmitting just $2 \cdot b \cdot m$ bits (e.g., less than 2MB for $b = 128, m = 10^5$). However, a network delay t is introduced due to interaction.

Remark 3. The b-bit block is re-encrypted via an IND-CPA encryption scheme on the *client* side at the cost of one round of interaction. Hence, encrypting multiple columns does not leak additional information during updates over our main scheme.

We discuss other variants of our main scheme in the full version of this paper in [20].

References

1. The enron email dataaset. http://www.cs.cmu.edu/enron/
2. Sparsehash: an extemely memory efficient hash_map implementation, February 2012. https://code.google.com/p/sparsehash/
3. Bellare, M., Rogaway, P.: Random oracles are practical: a paradigm for designing efficient protocols. In: Proceedings of the 1st ACM Conference on Computer and Communications Security (CCS 1993), pp. 62–73. ACM, New York (1993)
4. Cash, D., Jaeger, J., Jarecki, S., Jutla, C., Krawcyk, H., Rosu, M.-C., Steiner, M.: Dynamic searchable encryption in very-large databases: data structures and implementation. In: 21th Annual Network and Distributed System Security Symposium – NDSS. The Internet Society, 23–26 February 2014
5. Cash, D., Jarecki, S., Jutla, C., Krawczyk, H., Roşu, M.-C., Steiner, M.: Highly-scalable searchable symmetric encryption with support for boolean queries. In: Canetti, R., Garay, J.A. (eds.) CRYPTO 2013, Part I. LNCS, vol. 8042, pp. 353–373. Springer, Heidelberg (2013)

6. Chang, Y.-C., Mitzenmacher, M.: Privacy preserving keyword searches on remote encrypted data. In: Ioannidis, J., Keromytis, A.D., Yung, M. (eds.) ACNS 2005. LNCS, vol. 3531, pp. 442–455. Springer, Heidelberg (2005)
7. Chen, L.: NIST special publication 800–108: recomendation for key derivation using pseudorandom functions (revised). Technical report NIST-SP800-108, Computer Security Division, National Institute of Standards and Technology, October 2009. http://csrc.nist.gov/publications/nistpubs/800-108/sp800-108.pdf
8. Curtmola, R., Garay, J., Kamara, S., Ostrovsky, R.: Searchable symmetric encryption: improved definitions and efficient constructions. In: Proceedings of the 13th ACM Conference on Computer and Communications Security, CCS 2006, pp. 79–88. ACM, New York (2006)
9. St Denis, T.: LibTomCrypt library. http://libtom.org/?page=features&newsitems=5&whatfile=crypt. Accessed 12 May 2007
10. Fisch, B., Vo, B., Krell, F., Kumarasubramanian, A., Kolesnikov, V., Malkin, T., Bellovin, S.M.: Malicious client security in blind seer: a scalable private DBMS. In: IEEE Symposium on Security and Privacy, SP. IEEE Computer Society, 18–20 May 2015
11. Gueron, S.: White Paper: Intel Advanced Encryption Standard (AES) New Instructions Set, Document Revision 3.01, September 2012. https://software. intel.com/sites/default/files/article/165683/aes-wp-2012-09-22-v01.pdf. Software Library https://software.intel.com/sites/default/files/article/181731/intel-aesni-sample-library-v1.2.zip
12. Kamara, S., Papamanthou, C.: Parallel and dynamic searchable symmetric encryption. In: Sadeghi, A.-R. (ed.) FC 2013. LNCS, vol. 7859, pp. 258–274. Springer, Heidelberg (2013)
13. Kamara, S., Papamanthou, C., Roeder, T.: Dynamic searchable symmetric encryption. In: Proceedings of the ACM Conference on Computer and Communications Security, CCS 2012, pp. 965–976. ACM, New York (2012)
14. Krawczyk, H.: Cryptographic extraction and key derivation: the HKDF scheme. In: Rabin, T. (ed.) CRYPTO 2010. LNCS, vol. 6223, pp. 631–648. Springer, Heidelberg (2010)
15. Kurosawa, K., Ohtaki, Y.: UC-secure searchable symmetric encryption. In: Keromytis, A.D. (ed.) FC 2012. LNCS, vol. 7397, pp. 285–298. Springer, Heidelberg (2012)
16. Naveed, M., Prabhakaran, M., Gunter, C.A.: Dynamic searchable encryption via blind storage. In: 35th IEEE Symposium on Security and Privacy, pp. 48–62, May 2014
17. Pappas, V., Krell, F., Vo, B., Kolesnikov, V., Malkin, T., Choi, S.G., George, W., Keromytis, A.D., Bellovin, S.: Blind seer: a scalable private DBMS. In: IEEE Symposium on Security and Privacy, SP, pp. 359–374. IEEE Computer Society, 18–21 May 2014
18. Song, D.X., Wagner, D., Perrig, A.: Practical techniques for searches on encrypted data. In: Proceedings of the IEEE Symposium on Security and Privacy, SP 2000, pp. 44–55. IEEE Computer Society, Washington, D.C. (2000)
19. Stefanov, E., Papamanthou, C., Shi, E.: Practical dynamic searchable encryption with small leakage. In: 21st Annual Network and Distributed System Security Symposium – NDSS. The Internet Society, 23–26 February 2014
20. Yavuz A.A., Guajardo, J.: Dynamic searchable symmetric encryption with minimal leakage and efficient updates on commodity hardware. IACR Cryptology ePrint Archive 107, March 2015

Side Channel Attacks and Defenses

Affine Equivalence and Its Application to Tightening Threshold Implementations

Pascal Sasdrich[✉], Amir Moradi, and Tim Güneysu

Horst Görtz Institute for IT Security, Ruhr-Universität Bochum,
Bochum, Germany
{pascal.sasdrich,amir.moradi,tim.gueneysu}@rub.de

Abstract. Motivated by the development of Side-Channel Analysis (SCA) countermeasures which can provide security up to a certain order, defeating higher-order attacks has become amongst the most challenging issues. For instance, Threshold Implementation (TI) which nicely solves the problem of glitches in masked hardware designs is able to avoid first-order leakages. Hence, its extension to higher orders aims at counteracting SCA attacks at higher orders, that might be limited to univariate scenarios. Although with respect to the number of traces as well as sensitivity to noise the higher the order, the harder it is to mount the attack, a d-order TI design is vulnerable to an attack at order $d+1$.

In this work we look at the feasibility of higher-order attacks on first-order TI from another perspective. Instead of increasing the order of resistance by employing higher-order TIs, we go toward introducing structured randomness into the implementation. Our construction, which is a combination of masking and hiding, is dedicated to TI designs and deals with the concept of "affine equivalence" of Boolean functions. Such a combination hardens a design practically against higher-order attacks so that these attacks cannot be successfully mounted. We show that the area overhead of our construction is paid off by its ability to avoid higher-order leakages to be practically exploitable.

1 Introduction

Side-channel analysis (SCA) attacks exploit information leakage related to cryptographic device internals e.g., by analyzing the power consumption [11]. Hence, integration of dedicated countermeasures to SCA attacks into security-sensitive applications is essential particularly in case of pervasive applications (see [9,17,20]). Amongst the known countermeasures, *masking* as a form of secret sharing scheme has been extensively studied by the academic communities [8,12]. Based on Boolean masking and multi-party computation concept, Threshold Implementation (TI) has been developed particularly for hardware platforms [15]. Since the TI concept is initially bases on counteracting only first-order attacks, trivially higher-order attacks, which make use of higher-order statistical moments to exploit the leakages, can still recover the secrets. Hence, the TI has been extended to higher orders [3] which might be limited to univariate

© Springer International Publishing Switzerland 2016
O. Dunkelman and L. Keliher (Eds.): SAC 2015, LNCS 9566, pp. 263–276, 2016.
DOI: 10.1007/978-3-319-31301-6_16

settings [18]. In addition to its area and time overheads, which increase with the desired security order, the minimum number of shares also naturally increases, e.g., 3 shares for the first-order, 5 shares for the second-order, and at least 7 shares for the third-order security.

Contribution: In this work we look at the feasibility of higher-order attacks on first-order secure TI designs from another perspective. Instead of increasing the resistance against higher-order attacks by employing higher-order TIs, we intend to introduce structured randomness into a first-order secure TI. Our goal is to practically harden designs against higher-order attacks that are known to be sensitive to noise.

Concretely, we investigate the PRESENT [7] S-box under first-order secure TI settings that is decomposed into two quadratic functions thereby allowing the minimum number of three shares. By changing the decompositions during the operation of the device we can introduce (extra) randomness to the implementation. In particular we present different approaches to find and generate these decompositions on an FPGA platform and compare them in terms of area and time overheads. More importantly, we examine and compare the practical evaluation results of our constructions using a state-of-the-art leakage assessment methodology [10] at higher orders.

Our proposed approach which can be considered as a *hiding* technique is combined with first-order TI which provides provably secure first-order resistance. Therefore, although such a combination leads to higher area overhead, it brings its own advantage, i.e., practically avoiding the feasibility of higher-order attacks.

Outline: The remainder of this article is organized as follows: Sect. 2 recapitulates the concept of TI. We also briefly introduce the S-box decomposition for TI and affine equivalence in case of the PRESENT S-box. In Sect. 3 different approaches to find and exchange affine equivalent functions are presented and compared. Practical evaluation of our construction is given in Sect. 4. Finally, we conclude our research in Sect. 5.

2   Background

2.1   Threshold Implementation

We use lower-case letters for single-bit random variables, bold ones for vectors, raising indices for shares, and lowering indices for elements within a vector. We represent functions with sans serif fonts, and sets with calligraphic ones.

Let us denote an intermediate value of a cipher by $\boldsymbol{x}$ made of s single-bit signals $\langle x_1, \ldots, x_s \rangle$. The underlying concept of Threshold Implementation (TI) is to use Boolean masking to represent $\boldsymbol{x}$ in a shared form $(\boldsymbol{x}^1, \ldots, \boldsymbol{x}^n)$, where $\boldsymbol{x} = \bigoplus_{i=1}^{n} \boldsymbol{x}^i$ and each $\boldsymbol{x}^i$ similarly denotes a vector of s single-bit signals $\langle x_1^i, \ldots, x_s^i \rangle$. A linear function $\mathsf{L}(.)$ can be trivially applied over the shares of $\boldsymbol{x}$ as $\mathsf{L}(\boldsymbol{x}) = \bigoplus_{i=1}^{n} \mathsf{L}(\boldsymbol{x}^i)$. However, the realization of non-linear functions, e.g., an S-box, over

Boolean masked data is challenging. Following the concept of TI, if the algebraic degree of the underlying S-box is denoted by t, the minimum number of shares to realize the S-box under the first-order TI settings is $n = t + 1$. Further, such a TI S-box provides the output $\boldsymbol{y} = \mathsf{S}(\boldsymbol{x})$ in a shared form $(\boldsymbol{y}^1, \ldots, \boldsymbol{y}^m)$ with $m \geq n$ shares (usually $m = n$) in case of Bijective S-boxes. In case of a bijective S-box (e.g., of PRESENT) the bit length of $\boldsymbol{x}$ and $\boldsymbol{y}$ (respectively of their shared forms) are the same.

Each output share $\boldsymbol{y}^{j \in \{1, \ldots, m\}}$ is given by a component function $\mathsf{f}^j(.)$ over a subset of the input shares. To achieve the first-order security, each component functions $\mathsf{f}^{j \in \{1, \ldots, m\}}(.)$ must be independent of at least one input share.

Since the security of masking schemes is based on the uniform distribution of the masks, the output of a TI S-box must be also uniform as it is used as input in further parts of the implementation (e.g., the SLayer output of one PRESENT cipher round which is given to the next SLayer round after being processed by the linear PLayer and key addition). To express the *uniformity* under the TI concept suppose that for a certain input $\mathbf{x}$ all possible sharings $\mathcal{X} = \left\{ (\boldsymbol{x}^1, \ldots, \boldsymbol{x}^n) | \mathbf{x} = \bigoplus_{i=1}^{n} \boldsymbol{x}^i \right\}$ are given to a TI S-box. The set made by the output shares, i.e., $\left\{ (\mathsf{f}^1(.), \ldots, \mathsf{f}^m(.)) | (\boldsymbol{x}^1, \ldots, \boldsymbol{x}^n) \in \mathcal{X} \right\}$, should be drawn uniformly from the set $\mathcal{Y} = \left\{ (\boldsymbol{y}^1, \ldots, \boldsymbol{y}^m) | \mathbf{y} = \bigoplus_{i=1}^{m} \boldsymbol{y}^i \right\}$ as all possible sharings of $\mathbf{y} = \mathsf{S}(\mathbf{x})$.

This process so-called <u>uniformity check</u> should be individually performed for $\forall \, \mathbf{x} \in \{0, 1\}^s$. We should note that if an S-box is a bijection and $m = n$, each $(\boldsymbol{x}^1, \ldots, \boldsymbol{x}^n)$ should be mapped to a unique $(\boldsymbol{y}^1, \ldots, \boldsymbol{y}^n)$. In other words, in this case it is enough to check whether the TI S-box forms also a bijection with $s \cdot n$ input (and output) bit length. For more detailed information we refer the interested reader to the original article [15].

2.2   S-Box Decomposition

Since the nonlinear part of most block ciphers, i.e., the S-box, has algebraic degree of $t > 2$, the number of input and output shares $n, m > 3$, which directly affects the circuit complexity and its area overhead. Therefore, it is preferable to decompose the S-box $\mathsf{S}(.)$ into smaller functions, e.g., $\mathsf{g} \circ \mathsf{f}(.)$, each of them with maximum algebraic degree of 2. It is noteworthy that if $\mathsf{S}(.)$ is a bijection, each of the smaller functions (here in this case $\mathsf{g}(.)$ and $\mathsf{f}(.)$) must also be a bijection. Such a trick helps keeping the number of shares for input and output at minimum, i.e., $n = m = 3$. However, it comes with the disadvantage of the necessity to place a register between each two consecutive TI smaller functions to avoid the glitches being propagated. Although such a composition is feasible in case of small S-boxes (let say up to 6-bit permutations [5]), it is still challenging to find such decompositions for 8×8 S-boxes. As stated before, the target of this work is an implementation of PRESENT cipher, which involves a 4×4 invertible cubic S-box (i.e., with the algebraic degree of 3) with

Truth Table C56B90AD3EF84712. Therefore, all the representations below are coordinated based on 4-bit bijections.

In [16], where the first TI of PRESENT is presented, the authors gave a decomposition of the PRESENT S-box by two quadratic functions, i.e., each of which with the algebraic degree of 2. Later the authors of [4,5] presented a systematic approach which allows deriving the TI of all 4-bit bijections. In their seminal work they provided 302 classes of 4-bit bijections, with the application that every 4-bit bijection is affine equivalent to only one of such 302 classes. Based on their classification, the PRESENT S-box belongs to the cubic class $\mathcal{C}_{266}^4$ with Truth Table 0123468A5BCFED97. It other words, it is possible to write the PRESENT S-box as $S : A' \circ \mathcal{C}_{266}^4 \circ A$, where $A'(.)$ and $A(.)$ are 4-bit bijective affine functions. Therefore, given the uniform TI representation of $\mathcal{C}_{266}^4$ one can easily apply $A(.)$ on all input shares and $A'(.)$ on all output shares to obtain a uniform TI of the PRESENT S-box.

As stated in [5] $\mathcal{C}_{266}^4$ can be decomposed into two 4-bit quadratic bijections belonging to the following combinations of classes: $(\mathcal{Q}_{12} \circ \mathcal{Q}_{12})$, $(\mathcal{Q}_{293} \circ \mathcal{Q}_{300})$, $(\mathcal{Q}_{294} \circ \mathcal{Q}_{299})$, $(\mathcal{Q}_{299} \circ \mathcal{Q}_{294})$, $(\mathcal{Q}_{299} \circ \mathcal{Q}_{299})$, $(\mathcal{Q}_{300} \circ \mathcal{Q}_{293})$, and $(\mathcal{Q}_{300} \circ \mathcal{Q}_{300})$. However, the uniform TI of the quadratic class $\mathcal{Q}_{300}$ with 3 shares can only be achieved if it is again decomposed in two parts. Therefore, the above decompositions in which $\mathcal{Q}_{300}$ is involved need to be implemented in 3 stages if the minimum number of 3 shares is desired. Excluding such decompositions we have four options to decompose the PRESENT S-box in two stages with 3-share uniform TI since the PRESENT S-box is affine equivalent to $\mathcal{C}_{266}^4$.

For the sake of simplicity – as an example – we consider the first decomposition, i.e., $\mathcal{Q}_{12} \circ \mathcal{Q}_{12}$, which indicates that it is possible to write the PRESENT S-box as $S : A'' \circ \mathcal{Q}_{12} \circ A' \circ \mathcal{Q}_{12} \circ A$, where all three $A''(.)$, $A'(.)$, and $A(.)$ are 4-bit affine bijections. Thanks to the classifications given in [5] a uniform first-order TI of $\mathcal{Q}_{12}$ can be achieved by *direct sharing*. For $\mathcal{Q}_{12}$:0123456789CDEFAB we can write

$$e = a, \qquad f = b + bd + cd, \qquad g = c + bd, \qquad h = d, \quad (1)$$

with $\langle a, b, c, d \rangle$ the 4-bit input, $\langle e, f, g, h \rangle$ the 4-bit output, and a and e the least significant bits.

The component functions of the uniform first-order TI of $\mathcal{Q}_{12}$ can be derived by $f_{\mathcal{Q}_{12}}^{i,j}(\langle a^i, b^i, c^i, d^i \rangle, \langle a^j, b^j, c^j, d^j \rangle) = \langle e, f, g, h \rangle$ as

$$e = a^i, \qquad\qquad f = b^i + b^j d^j + c^j d^j + d^j b^i + d^j c^i + b^j d^i + c^j d^i,$$
$$g = c^i + b^j d^j + d^j b^i + b^j d^i, \, h = d^i.$$
$$(2)$$

The three 4-bit output shares provided by $f_{\mathcal{Q}_{12}}^{2,3}(.,.)$, $f_{\mathcal{Q}_{12}}^{3,1}(.,.)$ and $f_{\mathcal{Q}_{12}}^{1,2}(.,.)$ make a uniform first-order TI of $\mathcal{Q}_{12}$. Since the affine transformations (A, A', A'') do not change the uniformity, by applying them on each 4-bit share separately we can construct a 3-share uniform first-order TI of the PRESENT S-box. Figure 1 shows the graphical view of such a construction, and the detailed formulas of the component functions are given in Appendix A.

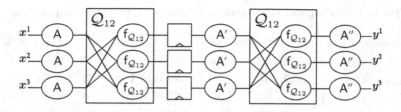

Fig. 1. A first-order TI of the PRESENT S-box

2.3 Affine Equivalence

In order to find such affine functions we give a pseudo code in Algorithm 1 which is mainly formed following [6]. The algorithm is based on precomputation of all 4×4 linear functions, i.e. $20\,160$ cases, each of which is represented by a 4×4 binary matrix with columns (c_0, c_1, c_2, c_3). Hence, each affine function $A(.)$ is considered as a matrix multiplication followed by a constant addition $A(x) = [c_0\ c_1\ c_2\ c_3] \cdot x \oplus c$.

Algorithm 1. Find affine equivalent triples

Input : $\mathcal{L}^4$: all 4×4 linear permutations,
       S: targeted S-box,
       F, G: targeted functions
Output: $\mathcal{A}$: all (A, A', A'') as $S : A'' \circ G \circ A' \circ F \circ A$

$\mathcal{A} \leftarrow \emptyset$
for $\forall L \in \mathcal{L}^4,\ \forall c \in \{0,1\}^4$ **do**
    form affine A by L and constant c
    for $\forall L' \in \mathcal{L}^4,\ \forall c' \in \{0,1\}^4$ **do**
        form affine A' by L' and constant c'
        $c'' \leftarrow G\left(A'\left(F\left(A\left(S^{-1}(0)\right)\right)\right)\right)$
        $c_1'' \leftarrow G\left(A'\left(F\left(A\left(S^{-1}(1)\right)\right)\right)\right) \oplus c''$
        $c_2'' \leftarrow G\left(A'\left(F\left(A\left(S^{-1}(2)\right)\right)\right)\right) \oplus c''$
        $c_3'' \leftarrow G\left(A'\left(F\left(A\left(S^{-1}(4)\right)\right)\right)\right) \oplus c''$
        $c_4'' \leftarrow G\left(A'\left(F\left(A\left(S^{-1}(8)\right)\right)\right)\right) \oplus c''$
        form affine A''^{-1} by columns $(c_1'', c_2'', c_3'', c_4'')$ and constant c''
        if $\forall y \in \{0,1\}^4 \setminus \{0,1,2,4,8\},\ G\left(A'\left(F\left(A\left(S^{-1}(y)\right)\right)\right)\right) \stackrel{?}{=} A''^{-1}(y)$ **then**
            derive affine A'' as the inverse of A''^{-1}
            $\mathcal{A} \leftarrow \mathcal{A} \cup \{(A, A', A'')\}$
        end
    end
end

Given the PRESENT S-box and $f = g = \mathcal{Q}_{12}$ the algorithm finds $147\,456$ such 3-tuple affine bijections (A, A', A''). Table 1 lists the number of found affine triples for each of the aforementioned decompositions.

Table 1. The number of existing affine triples for different compositions

Decomposition	No. of Triples	#(A)	#(A')	#(A'')	#(L)	#(L')	#(L'')
$\mathcal{Q}_{12} \circ \mathcal{Q}_{12}$	147 456	384	36 864	384	48	2 304	48
$\mathcal{Q}_{294} \circ \mathcal{Q}_{299}$	229 376	512	57 344	448	56	3 584	64
$\mathcal{Q}_{299} \circ \mathcal{Q}_{294}$	229 376	448	57 344	512	64	3 584	56
$\mathcal{Q}_{299} \circ \mathcal{Q}_{299}$	200 704	448	50 176	448	56	3 136	56

3 Design Considerations

This section briefly demonstrates the architecture the PRESENT TI which we have implemented. Afterwards, different approaches for generating and exchanging affine triples are presented and compared.

3.1 Threshold Implementation of PRESENT Cipher

PRESENT is a lightweight symmetric block cipher with a block size of 64 bits and either 80-bit or 128-bit security level (i.e., key size). The encryption of a plaintext is based on a Substitution-Permutation (S/P) network always taking 31 rounds and 32 sub-keys to compute the ciphertext (independently of the security level). The only difference between PRESENT-80 and PRESENT-128 is in the key schedule function to derive the sub-keys from the initial 80-bit or 128-bit key. Figure 2 gives an overview of our hardware architecture implemented on an Xilinx Spartan-6 FPGA. We opted to implement the PRESENT encryption scheme in a round-based manner along with the 128-bit key schedule variant. The sub-keys are derived on-the-fly. The substitution layer uses the first-order TI of the PRESENT S-box shown in Fig. 1 and implements 16 S-boxes in parallel before the permutation is applied bitwise to all 64-bit states. Due to the additional register stage within the TI S-box each round requires two clock cycles.

As stated in Sect. 2.3, given a certain decomposition there exist many triple affine functions to realize a uniform first-order TI of the PRESENT S-box. Our goal is to randomly change such affine functions on the fly, that it first does not affect the correct functionality of the S-box, and second randomizes the intermediate values – particularly the shared $\mathcal{Q}_{12}$ inputs – with the aim of hardening higher-order attacks. As shown in Fig. 2 all S-boxes share the same affine triple. In other words, at the start of each encryption an affine triple is randomly selected, and all S-boxes are configured accordingly. Although it is possible to change the affines more frequently, we kept the selected affines for an entire encryption process. To this end, we need an architecture to derive the affine triples randomly. Below we discuss about different ways to realize such a part of the design.

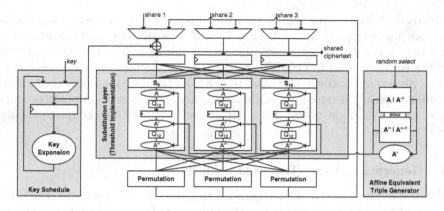

Fig. 2. Architecture of the PRESENT encryption design

3.2 Searching for the Affine Triples

At a first step, we decided to implement Algorithm 1 as a hardware circuit which searches for the affine triples in parallel to the encryption. The found affine triples are stored into a "First In, First Out" (FIFO) memory, and prior to each encryption one affine triple is taken from the FIFO with which the corresponding part of the TI S-boxes are configured. If the FIFO is empty, the previous affine triple is used again. Due to the fact that the search is not time-invariant, i.e., new affine triples are not found periodically, some affines are used multiple times in a row while others are only used once. Since the efficiency of SCA countermeasures depends on the uniformity of the used randomness, such an implementation may not achieve the desired goal (i.e., hardening the higher-order attacks) if certain affines are used more often that the others. One solution to find affine triples more often is to run the search circuit with a higher clock frequency compared to that of the encryption circuit. Although this measure is limited, it at least alleviates the problem of changing S-boxes not periodically. On the other hand, if affine triples are found too fast this may cause a FIFO overflow. In this case either some search results should be ignored or the search circuit should be stopped requiring some additional control logic.

3.3 Selecting Precomputed Affine Triples

As stated in Table 1, considering the decomposition $\mathcal{Q}_{12} \circ \mathcal{Q}_{12}$, there exist 147 456 triple affines (A, A', A''). Each single affine transformation is a 4-bit permutation, and it can be represented as a look-up table containing sixteen 4-bit entries which requires 64 bits of memory. This results in 27 Mbit memory in order to store all the affine triples. However, the employed Xilinx Spartan-6 FPGA (LX75) offers only 3 Mbit storage in terms of general purpose block memory (BRAM). Therefore, alternative approaches to generate the affine equivalent triples are necessary.

Instead of storing the affines in a look-up table, in the second option we represent an exemplary affine $\mathsf{A}(\boldsymbol{x}) = \boldsymbol{L} \cdot \boldsymbol{x} \oplus \boldsymbol{c}$, with $\boldsymbol{x}$ as a 4-bit vector, $\boldsymbol{L}$ a 4×4 binary matrix and $\boldsymbol{c}$ a 4-bit constant. In this case, only the binary matrix and the constant need to be stored which reduces the memory requirements to 20 bits per affine. However, still more than 8 Mbit memory are necessary to store all affine triples. Therefore, we could store only a fraction of all possible affine triples. As an example, 16 384 affine triples occupy 60 BRAMs of the Spartan-6 (LX75) FPGA.

3.4 Generating Affine Triples On-the-fly

A detailed analysis of the affine triples led to interesting observations. First, the number of affine triples depends on the components in the underlying decomposition. For instance, in case of $\mathcal{Q}_{299} \circ \mathcal{Q}_{299}$ 448×448 and in case of $\mathcal{Q}_{299} \circ \mathcal{Q}_{294}$ 448×512 affine triples exist (see Table 1). Second, the total number of affine triples is limited by the number of unique input affines A and the number of output affines A'' such that $|\mathsf{A}| \times |\mathsf{A}''|$ gives the number of corresponding affine triples. This means that all affine triples of a decomposition can be generated by combining all A with all A''. Furthermore, we have observed that all affines A (for each decomposition) consist of a few linear matrices combined with certain constants. In particular, in case of the decomposition $\mathcal{Q}_{12} \circ \mathcal{Q}_{12}$ the 384 input affines A are formed by 48 binary matrices $\boldsymbol{L}$ each of which combined with 8 different constants $\boldsymbol{c} \in \{0, \ldots, 7\}$ or $\boldsymbol{c} \in \{8, \ldots, 15\}$. Indeed the same holds for the 384 output affines A'' which are made of 48 binary matrices $\boldsymbol{L}''$ by constants $\boldsymbol{c} \in \{0, 1, 4, 5, 10, 11, 14, 15\}$ or $\boldsymbol{c} \in \{2, 3, 6, 7, 8, 9, 12, 13\}$. Therefore, it is sufficient to store only all relevant binary matrices $\boldsymbol{L}$ and $\boldsymbol{L}''$ in addition to a single bit indicating to which group their constants belong to. Hence, in total $48 \times 2 \times (16+1) = 1632$ bits of memory (fitting into a single BRAM) are required to store all necessary data. Even better, by arranging the binary matrices in the memory smartly the group of the corresponding constants can be derived from the address where the binary matrix is stored.

Given two input and output affines A and A'', we need to derive the middle affine A'. To this end, an approach similar to Algorithm 1 can be used. If we represent the middle affine as $\mathsf{A}'(\boldsymbol{x}) = \boldsymbol{L}' \cdot \boldsymbol{x} \oplus \boldsymbol{c}'$, the constant $\boldsymbol{c}$ and the columns (c_1', c_2', c_3', c_4') of the binary matrix $\boldsymbol{L}$ can be derived as

$$c' = \mathcal{Q}_{12}^{-1}\left(\mathsf{A}''^{-1}\left(\mathsf{S}\left(\mathsf{A}^{-1}\left(\mathcal{Q}_{12}^{-1}(0)\right)\right)\right)\right) \tag{3}$$

$$c_1' = \mathcal{Q}_{12}^{-1}\left(\mathsf{A}''^{-1}\left(\mathsf{S}\left(\mathsf{A}^{-1}\left(\mathcal{Q}_{12}^{-1}(1)\right)\right)\right)\right) \oplus c' \tag{4}$$

$$c_2' = \mathcal{Q}_{12}^{-1}\left(\mathsf{A}''^{-1}\left(\mathsf{S}\left(\mathsf{A}^{-1}\left(\mathcal{Q}_{12}^{-1}(2)\right)\right)\right)\right) \oplus c' \tag{5}$$

$$c_3' = \mathcal{Q}_{12}^{-1}\left(\mathsf{A}''^{-1}\left(\mathsf{S}\left(\mathsf{A}^{-1}\left(\mathcal{Q}_{12}^{-1}(4)\right)\right)\right)\right) \oplus c' \tag{6}$$

$$c_4' = \mathcal{Q}_{12}^{-1}\left(\mathsf{A}''^{-1}\left(\mathsf{S}\left(\mathsf{A}^{-1}\left(\mathcal{Q}_{12}^{-1}(8)\right)\right)\right)\right) \oplus c' \tag{7}$$

Obviously, this requires the inverse of both A and A''. Since it is not efficient to derive such inverse affines on the fly, we need to store all binary matrices $\boldsymbol{L}^{-1}$ and $\boldsymbol{L}''^{-1}$ in addition to all $\boldsymbol{L}$ and $\boldsymbol{L}''$. Fortunately, all such binary matrices (requiring 3 kbits) still fit into a single 16-kbit BRAM of Spartan-6 FPGA. It is noteworthy that the constant of each inverse affine can be computed by $L^{-1} \cdot c$.

In summary, at the start of each encryption two L and L'' (each of which from a set of 48 cases) are randomly selected, that needs $6 + 6$ bits of randomness[1]. In addition, $3 + 3$ random bits are also required to form constants c and c''. As exampled before, one bit of each constant should be additionally saved or derived from the address of the binary matrix. Therefore – excluding the masks required to represent the plaintext in a 3-share form for the TI design – in total 18 bits randomness is required for each encryption.

For ASIC platforms, where block memories are not easily available, an alternative is to derive the content of binary matrices L and L'' as Boolean functions over the given random bits. Hence, a fully combinatorial circuit can provide the input and output affines followed (as before) by a module which retrieves the middle affine.

3.5   Comparison

Table 2 gives an overview of the design of the three above-mentioned approaches to derive the affine triples. The table reports the area overhead, reconfiguration time, and coverage of the affines' space. Comparing the first naive approach (of searching the affine triples in parallel to the encryption) to the approach of precomputing affine triples, the logic requirements could be dramatically decreased at cost of additional memory. In addition, the amount of affine triples that are covered is limited potentially reducing the security gain. We should note that the 20 BRAMs used in the "Search" approach are due to the space required to store all 4×4 linear permutations $\mathcal{L}^4$ required to run Algorithm 1 (excluding those required for the FIFO). The last approach where the affine triples are generated on-the-fly seems to be the best choice. It not only leads to the least area overhead (both logic and memory requirements) but also covers the whole number of possible affine triples.

We should note that our design needs a single clock cycle to derive the middle affine A'. Indeed the 114 LUTs (reported in Table 2) are mainly due to realization of the Eqs. (3) and (7) in a fully combinatorial fashion.

Further, with respect to the design architecture of the encryption function (Fig. 2) the quadratic component functions of $\mathcal{Q}_{12}$ are implemented by look-up tables (LUTs), and the affine functions by fully combinatorial circuits realizing the binary matrix multiplication (AND operations) and XOR with the constant. Therefore, given $(16 + 4)$ bits as the content of the binary matrix and the constant, the circuit does not need any extra clock cycles for configuration. Table 2 also gives an overview of the area and speed overhead of our design compared to a similar designs. For the first reference, the TI S-box is implemented by the design of [16] (i.e., without any random affine). The second reference implements both a first-order and a second-order TI S-box for PRESENT in a similar fashion (using $\mathcal{Q}_{294}$ and $\mathcal{Q}_{299}$ instead of $\mathcal{Q}_{12}$) but with fixed affine transformations.

---

[1] For each selection $\in \{1, \ldots 48\}$ reject sampling with 6-bit random should be used.

Table 2. Area and time overhead of different design approaches

Section/Method/Module	Logic (LUT)	Memory (FF)	(BRAM16)	Reconfig. time (Cycles)	Affine coverage (Percent)	Max. freq. (MHz)	Order of TI
Section 3.2/Search	562	250	20	16	100.0	-	-
Section 3.3/Precompute	204	0	60	0	11.1	-	-
Section 3.4/Generate	114	20	1	1	100.0	-	-
Encryption [this work]	1720	722	0	-	-	112	1st
Encryption [16]	641	384	0	-	-	218	1st
Encryption [14]	808	384	0	-	-	207	1st
Encryption [14]	2245	1680	0	-	-	204	2nd

The numbers for the encryption function exclude the PRNG as well as the circuit which finds/derives the affines. Due to the extra logic to support arbitrary affines, our design is certainly larger and slower.

4   Evaluation

We employed a SAKURA-G platform [1] equipped with a Spartan-6 FPGA for practical side-channel evaluations using the power consumption of the device. The power consumption traces have been measured and recorded by means of a digital oscilloscope with a $1\,\Omega$ resistor in the V_{dd} path and capturing at the embedded amplifier of the SAKURA-G board. We sampled the voltage drop at a rate of $500\,\mathrm{MS/s}$ and a bandwidth limit of $20\,\mathrm{MHz}$ while the design was running at a low clock frequency of $3\,\mathrm{MHz}$ to reduce the noise caused by overlapping of the power traces.

4.1   Non-specific Statistical t-test

In order to evaluate the resistance or vulnerabilities of our designs against higher-order side-channel attacks we applied the well-known state-of-the-art leakage assessment metric called *Test Vector Leakage Assessment* (TVLA) methodology. This evaluation scheme is based on the Welch's (two-tailed) t-test and also known as *fix vs. random* or *non-specific* t-test. For further details, particularly how to apply this assessment tool for higher-order leakages as well as how to implement it efficiently in particular for large-scale investigations, we refer the reader to [19] giving detailed practical instructions. In short, we should note that such an assessment scheme examines the existence of leakage at a certain order without giving any reference to whether the detected leakage is exploitable by an attack. However, if the test reports no detectable leakage, it can be concluded that – with a high level of confidence – the device under test does not exhibit any exploitable leakage.

4.2   Results

In this section we present the result of the side-channel evaluations concerning the efficiency of our introduced approaches to avoid higher-order leakages. In

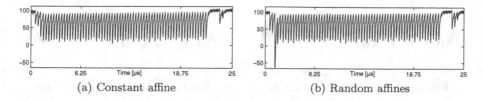

(a) Constant affine

(b) Random affines

Fig. 3. Sample traces of the PRESENT encryption function

order to solely evaluate the influence of randomly exchanging the affine triples we considered a single design in our evaluations. As a reference, the design is kept running with a constant affine triple[2], and its evaluation results are compared to the case where the affine triples are randomly changed prior to each encryption. Note that in both cases (constant affine and random affine) the PRNG which provides masks for the initial second-order masking (with three shares) is kept active. In other words, both designs – based on the TI concept – are expected to provide first-order resistance, and their difference should be in exhibiting higher-order leakages.

In Sect. 3 we introduced three different approaches to derive affine triples. Due to the issues and limitation of both first approaches, we have included the practical evaluation results of only the third option in Sect. 3.4, i.e., generating affine triples on-the-fly, which covers all possible affine triples.

Figure 3 shows two sample traces corresponding to the cases where the affine triple is constant or random. The main difference between these two traces can be seen by a large power peak at the beginning of the trace belonging to the random affines. Such a peak indicates the corresponding clock cycle where the random affine is selected and the middle affine is computed (as stated in Sect. 3.5, it is implemented by a fully combinatorial circuit). The first-order, second-order and third-order t-test results are shown in Figs. 4, 5 and 6 respectively for both constant and random affine. As expected, both designs do not exhibit any first-order leakage confirming the validity of our setup and designs. However, changing the affine triples randomly could avoid the second- and third-order leakage from being detectable. This can be seen in Figs. 5 and 6. We should highlight that the evaluations of the design with a constant affine have been performed by 50 million traces while we continued the measurements and evaluations of the design with random affines up to 200 million traces.

5   Discussions

The scheme, which we have introduced here to harden higher-order attacks, at the first glance seems to just add more randomness to the design. We should stress that our approach is not the same as the concept of *remasking* applied in [2,5,13]. Remasking (or mask refreshing) can be done e.g., by adding two

---

[2] This has been easily done by fixing the corresponding 18 random bits.

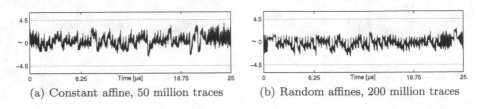

(a) Constant affine, 50 million traces   (b) Random affines, 200 million traces

Fig. 4. Non-specific t-test: first-order evaluation results

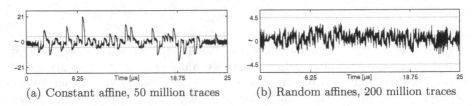

(a) Constant affine, 50 million traces   (b) Random affines, 200 million traces

Fig. 5. Non-specific t-test: second-order evaluation results

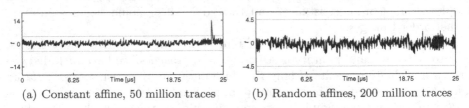

(a) Constant affine, 50 million traces   (b) Random affines, 200 million traces

Fig. 6. Non-specific t-test: 3rd-order evaluation results

new fresh random masks r^1 and r^2 to the input of the TI S-box in Fig. 1 as $(x^1 \oplus r^1, x^2 \oplus r^2, x^3 \oplus r^1 \oplus r^2)$. Since our construction of the PRESENT TI S-box fulfills the uniformity, such a remasking does not have any effect on the practical security of the design as both (x^1, x^2, x^3) and $(x^1 \oplus r^1, x^2 \oplus r^2, x^3 \oplus r^1 \oplus r^2)$ are 3-share representations of x. In contrast, in our approach e.g., the input affine A randomly changes. Hence the input of the first $\mathcal{Q}_{12}$ function is a 3-share representation of $A(x)$. Considering a certain x, random selection of the input affine leads to random $A(x)$ which is also represented by three Boolean shares. Therefore, the intermediate values of the S-box (at both stages) are not only randomized but also uniformly shared. As a result, hardening both second- and third-order attacks which make use of the leakage of the S-box can be justified. Note that since the S-box output stays valid as a Boolean shared representation of $S(x)$ and random affine triples do not affect the PLayer (of the PRESENT cipher), the key addition and the values stored in the state register, our approach is not expected to harden third-order attacks that target the leakage of these modules. However, our construction (which is a combination of masking and hiding) allows to achieve the presented efficiencies with low number of (extra) required randomness, i.e., 18 bits per encryption. Indeed, our approach might be seen as a form of shuffling which can be applied on the order of S-box executions in a serialized architecture. However, our construction is independent of the underlying architecture (serialized versus round-based) and allows hiding the exploitable higher-order leakages in a systematic way.

References

1. Side-channel AttacK User Reference Architecture. http://satoh.cs.uec.ac.jp/SAKURA/index.html
2. Bilgin, B., Gierlichs, B., Nikova, S., Nikov, V., Rijmen, V.: A more efficient AES threshold implementation. In: Pointcheval, D., Vergnaud, D. (eds.) AFRICACRYPT. LNCS, vol. 8469, pp. 267–284. Springer, Heidelberg (2014)
3. Bilgin, B., Gierlichs, B., Nikova, S., Nikov, V., Rijmen, V.: Higher-order threshold implementations. In: Sarkar, P., Iwata, T. (eds.) ASIACRYPT 2014, Part II. LNCS, vol. 8874, pp. 326–343. Springer, Heidelberg (2014)
4. Bilgin, B., Nikova, S., Nikov, V., Rijmen, V., Stütz, G.: Threshold implementations of all 3 × 3 and 4 × 4 S-boxes. In: Prouff, E., Schaumont, P. (eds.) CHES 2012. LNCS, vol. 7428, pp. 76–91. Springer, Heidelberg (2012)
5. Bilgin, B., Nikova, S., Nikov, V., Rijmen, V., Tokareva, N., Vitkup, V.: Threshold implementations of small S-boxes. Crypt. Commun. $7(1)$, 3–33 (2015)
6. Biryukov, A., Cannière, C.D., Braeken, A., Preneel, B.: A toolbox for cryptanalysis: linear and affine equivalence algorithms. In: Biham, E. (ed.) EUROCRYPT 2003. LNCS, vol. 2656, pp. 33–50. Springer, Heidelberg (2003)
7. Bogdanov, A., Knudsen, L.R., Leander, G., Paar, C., Poschmann, A., Robshaw, M.J.B., Seurin, Y., Vikkelsoe, C.: PRESENT: an ultra-lightweight block cipher. In: Paillier, P., Verbauwhede, I. (eds.) CHES 2007. LNCS, vol. 4727, pp. 450–466. Springer, Heidelberg (2007)
8. Chari, S., Jutla, C.S., Rao, J.R., Rohatgi, P.: Towards sound approaches to counteract power-analysis attacks. In: Wiener, M. (ed.) CRYPTO 1999. LNCS, vol. 1666, pp. 398–412. Springer, Heidelberg (1999)
9. Eisenbarth, T., Kasper, T., Moradi, A., Paar, C., Salmasizadeh, M., Shalmani, M.T.M.: On the power of power analysis in the real world: a complete break of the KEELOQ code hopping scheme. In: Wagner, D. (ed.) CRYPTO 2008. LNCS, vol. 5157, pp. 203–220. Springer, Heidelberg (2008)
10. Goodwill, G., Jun, B., Jaffe, J., Rohatgi, P.: A testing methodology for side channel resistance validation. In: NIST Non-invasive Attack Testing Workshop (2011). http://csrc.nist.gov/news_events/non-invasive-attack-testing-workshop/papers/08_Goodwill.pdf
11. Kocher, P.C., Jaffe, J., Jun, B.: Differential power analysis. In: Wiener, M. (ed.) CRYPTO 1999. LNCS, vol. 1666, p. 388. Springer, Heidelberg (1999)
12. Mangard, S., Oswald, E., Popp, T.: Power Analysis Attacks - Revealing the Secrets of Smart Cards. Springer, New York (2007)
13. Moradi, A., Poschmann, A., Ling, S., Paar, C., Wang, H.: Pushing the limits: a very compact and a threshold implementation of AES. In: Paterson, K.G. (ed.) EUROCRYPT 2011. LNCS, vol. 6632, pp. 69–88. Springer, Heidelberg (2011)
14. Moradi, A., Wild, A.: Assessment of hiding the higher-order leakages in hardware. In: Güneysu, T., Handschuh, H. (eds.) CHES 2015. LNCS, vol. 9293, pp. 453–474. Springer, Heidelberg (2015)
15. Nikova, S., Rijmen, V., Schläffer, M.: Secure hardware implementation of nonlinear functions in the presence of glitches. J. Cryptology $24(2)$, 292–321 (2011)

16. Poschmann, A., Moradi, A., Khoo, K., Lim, C., Wang, H., Ling, S.: Side-channel resistant crypto for less than 2,300 GE. J. Cryptology **24**(2), 322–345 (2011)
17. Rao, J.R., Rohatgi, P., Scherzer, H., Tinguely, S.: Partitioning attacks: or how to rapidly clone some GSM cards. In: IEEE Symposium on Security and Privacy, pp. 31–41. IEEE Computer Society (2002)
18. Reparaz, O.: A note on the security of higher-order threshold implementations. Cryptology ePrint Archive, Report 2015/001 (2015). http://eprint.iacr.org/
19. Schneider, T., Moradi, A.: Leakage assessment methodology - a clear roadmap for side-channel evaluations. Cryptology ePrint Archive, Report 2015/207 (2015). http://eprint.iacr.org/
20. Zhou, Y., Yu, Y., Standaert, F.-X., Quisquater, J.-J.: On the need of physical security for small embedded devices: a case study with COMP128-1 implementations in SIM cards. In: Sadeghi, A.-R. (ed.) FC 2013. LNCS, vol. 7859, pp. 230–238. Springer, Heidelberg (2013)

A Necessary Component Functions for a First-Order TI of PRESENT S-box

$$y^1 = f_{\mathcal{Q}_{12}}^{2,3}(\langle a^2, b^2, c^2, d^2 \rangle, \langle a^3, b^3, c^3, d^3 \rangle) = \langle e, f, g, h \rangle$$

$$e = a^2, \qquad\qquad f = b^2 + b^3 d^3 + c^3 d^3 + d^3 b^2 + d^3 c^2 + b^3 d^2 + c^3 d^2,$$

$$g = c^2 + b^3 d^3 + d^3 b^2 + b^3 d^2, \quad h = d^2. \tag{8}$$

$$y^2 = f_{\mathcal{Q}_{12}}^{3,1}(\langle a^3, b^3, c^3, d^3 \rangle, \langle a^1, b^1, c^1, d^1 \rangle) = \langle e, f, g, h \rangle$$

$$e = a^3, \qquad\qquad f = b^3 + b^1 d^1 + c^1 d^1 + d^1 b^3 + d^1 c^3 + b^1 d^3 + c^1 d^3,$$

$$g = c^3 + b^1 d^1 + d^1 b^3 + b^1 d^3, \quad h = d^3. \tag{9}$$

$$y^3 = f_{\mathcal{Q}_{12}}^{1,2}(\langle a^1, b^1, c^1, d^1 \rangle, \langle a^2, b^2, c^2, d^2 \rangle) = \langle e, f, g, h \rangle$$

$$e = a^1, \qquad\qquad f = b^1 + b^2 d^2 + c^2 d^2 + d^2 b^1 + d^2 c^1 + b^2 d^1 + c^2 d^1,$$

$$g = c^1 + b^2 d^2 + d^2 b^1 + b^2 d^1, \quad h = d^1. \tag{10}$$

Near Collision Side Channel Attacks

Barış Ege[1]([⊠]), Thomas Eisenbarth[2], and Lejla Batina[1]

[1] Radboud University, Nijmegen, The Netherlands
b.ege@cs.ru.nl
[2] Worcester Polytechnic Institute, Worcester, MA, USA

Abstract. Side channel collision attacks are a powerful method to exploit side channel leakage. Otherwise than a few exceptions, collision attacks usually combine leakage from distinct points in time, making them inherently bivariate. This work introduces the notion of near collisions to exploit the fact that values depending on the same sub-key can have similar while not identical leakage. We show how such knowledge can be exploited to mount a key recovery attack. The presented approach has several desirable features when compared to other state-of-the-art collision attacks: Near collision attacks are truly univariate. They have low requirements on the leakage functions, since they work well for leakages that are linear in the bits of the targeted intermediate state. They are applicable in the presence of masking countermeasures if there exist distinguishable leakages, as in the case of leakage squeezing. Results are backed up by a broad range of simulations for unprotected and masked implementations, as well as an analysis of the measurement set provided by DPA Contest v4.

Keywords: Side channel collision attack · Leakage squeezing · Differential power analysis

1 Introduction

Side channel analysis and countermeasures belong to the most active research areas of applied cryptography today. Many variants are known and all kinds of attacks and defenses are introduced since the seminal paper by Kocher et al. [13]. The assumptions for attacks, power and adversary models vary, but all together it can be said that the challenges remain to defend against this type of attacks as an adversary is assumed to always take the next step.

For example, side channel collision attacks exploit the fact that identical intermediate values consume the same power and hence similar patterns can be observed in power/EM measurements. More in detail, an internal collision attack exploits the fact that a collision in an algorithm often occurs for some intermediate values. This happens if, for at least two different inputs, a function within the algorithm returns the same output. In this case, the side channel traces are assumed to be very similar during the time span when the internal collision persists. Since their original proposal [21], a number of works have

© Springer International Publishing Switzerland 2016
O. Dunkelman and L. Keliher (Eds.): SAC 2015, LNCS 9566, pp. 277–292, 2016.
DOI: 10.1007/978-3-319-31301-6_17

improved on various aspects of collision attacks, such as collision finding [5] or effective key recovery [10].

There are also different approaches in collision detection. Batina et al. introduce Differential Cluster Analysis (DCA) as a new method to detect internal collisions and extract keys from side channel signals [2]. The new strategy includes key hypothesis testing and the partitioning step similar to those of DPA. Being inherently multivariate, DCA as a technique also inspired a simple extension of standard DPA to multivariate analysis. The approach by Moradi et al. [17] extends collision attacks by creating a first order (or higher order in [15]) leakage model and comparing it to the leakage of other key bytes through correlation. The approach is univariate only if leakages for different sub-keys occur at the same time instance, i.e. for parallel implementations, as often found in hardware. When software implementations are considered, these two sensitive values would leak in different times, therefore other papers pursued the possibility to pursue a similar attack for software implementations in a bivariate setting [8, 23]. Although finding the exact time samples which leak information about the intended intermediate variables increases the attack complexity, this type of attacks are especially favourable when the leakage function is unknown, or it is a non-linear function of the bits of the sensitive variable [10].

In general, it is desirable for attacks to apply to a wide range of leakage functions. Some strategies are leakage model agnostic, e.g. Mutual Information Analysis [11]. In contrast to this assumption-less leakage model approach, there is also an alternative in choosing a very generic model as in stochastic models approach [20]. We follow this direction in terms of restricting ourselves to leakages that are linear functions of the contributing bits. Nevertheless, in our scenario this is considered merely as a ballpark rather than a restriction.

When univariate attacks are considered such as the one that is proposed in this work, the best way to mitigate is to implement a masking scheme. However, one of the biggest drawbacks of masking schemes is the overhead introduced into implementations. Recently there has been a rising interest in reducing the entropy needed and thereby the implementation overhead by cleverly choosing masks from a reduced set. These approaches are commonly referred to as Low entropy masking schemes (LEMS) or leakage squeezing. In fact, LEMS are a low-cost solution proposed to at least keep or even enhance the security over classical masking [7, 18, 19]. Since the proposal, LEMS have been analyzed from different angles, including specific attacks [24], a detailed analysis of the applicability of made assumptions [12] and problems that may occur during its implementation [16]. Attention to LEMS has been stipulated to a specific version of LEMS, the Rotating S-box Masking (RSM) [18], since it has been used for both DPA contest v4 and v4.2 [3].

Our Contributions. The contribution of this work can be summarised as follows:

– We introduce a new way of analysing side channel measurements which is void of strong assumptions on the power consumption of a device.

- The attack that we propose is a non-profiled univariate attack which only assumes that the leakage function of the target device is linear.
- We further extend this idea to analyse a low entropy masking scheme by improving on [24], and we show that our technique is more efficient to recover the key than generic univariate mutual information analysis.
- The proposed attack is applicable to any low entropy mask set that is a binary linear code [4].

Structure. The rest of the paper is structured as follows. Section 2 introduces the notation used throughout the work and also the ideas in the literature that leads to our new attack. Section 3 introduces the near collision attack and present simulated results in comparison to other similar attacks in the literature. Section 4 introduces the extension of our idea to a low entropy masking scheme together with a summary of the previous work that it is improved upon. This section also presents comparative results of the extended attack and other attacks similar to it in the literature, and a discussion on the attack complexity. Finally, Sect. 5 concludes the paper with some directions for further research.

2 Backgound and Notation

In this section we briefly summarize side channel attacks and also introduce the notation used throughout the paper.

Side channel analysis is a cryptanalysis method aiming to recover the secret key of a cryptographic algorithm by analyzing the unintended information leakages observed through various methods. In this work, we focus on the information leakage on the power consumption or electro-magnetic leakage of an implementation. Further, we use the Advanced Encryption Standard (AES) to explain our new attack and run experiments as it is a widely deployed crypto algorithm around the world. This ensures comparability with other works in the literature that use AES for presenting results, but does not hinder generalization to other block ciphers in a natural way.

Correlation based power or EM analysis (CPA) against AES implementations usually focuses on the output of the S-box operation which is the only non-linear element of the algorithm. This non-linearity ensures a good distinguishability between the correct and incorrect key guesses for CPA; the correlation between the observed and the predicted power or EM leakage will be (close to) zero if the key guess is incorrect, due to the highly nonlinear relation between the predicted state and the key. To run a CPA the analyst observes the power (or EM) leakages of the device for each input $x \in X$ and stores it as an observed value $o^x \in O^X$. The next step is to reveal the relation between o^x and x through estimating the power consumption of the target device. Assume that the analyst would like to estimate power consumption with the Hamming weight function ($\mathrm{HW}(x)$) which returns the number of ones in the bit representation of a given variable. In this case, the power estimation for the input value x becomes $P(x, k_g) = \mathrm{HW}(S(x \oplus k_g))$, where k_g is a key guess for the part of the key related to x. Proceeding this way, the analyst forms 256 sets $P_{k_g} = \{P(x, k_g) : x \in X\}$ from the known input values $x_i \in X$ for each key guess $k_g \in \mathbb{F}_2^8$. What remains is to compare the estimated power consumptions P_{k_g} with the set of observations O^X on the power consumption through a distinguisher, in this case through computing the Pearson correlation coefficient $\rho(P_{k_g}, O^X)$, $\forall k_g \in \mathbb{F}_2^8$. If the analyst has sufficient data and if the modelled leakage P is close enough to the actual leakage function L of the device (i.e. a linear representative of L), then the correct key k_c should result in a distinguishing value for the Pearson correlation when compared to the wrong key guesses k_w. In case P is not a linear representative of L however, then the correct key may not be distinguishable with this technique. Therefore, the choice of power model determines the strength of CPA.

Collision attacks aim to amend this problem by removing the requirement to estimate L in an accurate manner. Linear collision attacks against AES use the fact that if there are two S-box outputs equal to each other, then their power consumption should be the same [5]. If two S-box outputs for inputs x_i and x_j are equal to each other, then

$$S(x_i \oplus k_i) = S(x_j \oplus k_j) \tag{1}$$

$$\Rightarrow \quad x_j \oplus k_j = x_j \oplus k_j \tag{2}$$

$$\Rightarrow \quad x_i \oplus x_j = k_i \oplus k_j \tag{3}$$

when S is an injective function. Since the AES S-box is bijective, collisions as above reveal information about the relation of a pair of key bytes. Detecting collisions can be a challenging task, and therefore Moradi et al. [17] proposes to use Pearson correlation for collision detection by comparing pairs of vectors which have a fixed difference in their input bytes, which in turn represents the difference between the corresponding key bytes as explained above. After running a linear collision attack (referred to as the *'correlation enhanced collision attack'*), the analyst can reduce the key space radically and solve the remaining linear system of equations to recover the entire key. Hence the only challenge remains is finding the time instances where the targeted leakages occur, which can be a time consuming task depending on the amount of samples the analyst acquires for analysis.

Although the resulting work load for brute forcing the key is reduced significantly by a linear collision attack, to further reduce the work load, Ye et al. [23] proposes a new collision attack (namely the *'non-linear collision attack'*) to directly recover a key byte rather than a linear relation of two key bytes. Rather than looking for a collision in the same power measurement, non-linear collision attack looks for a linear relation between the input of the S-box for a plaintext value x and an S-box output value of another input x' which are related to the same key byte as:

$$x' \oplus k = S(x \oplus k) \tag{4}$$

$$x' = S(x \oplus k) \oplus k. \tag{5}$$

Therefore, the collision can be tested by building a hypothesis for k and whenever a collision is detected, the correct key for that byte is immediately revealed.

Even though the key byte can be recovered directly with this attack, the intrinsic problem here remains and that is the challenge to find the two leaking samples which refer to the leakage of such values. Next section presents the univariate solution to this problem which removes the requirement of strong leakage assumptions to be able to mount a side channel attack similar to side channel collision attacks.

3   Side Channel Near Collision Attack

In this chapter we introduce the univariate non-profiled attack, namely the side channel near collision attack (NCA) with an example to an AES implementation. NCA is very similar to other collision attacks in the sense that a priori knowledge of the leakage function is not required to mount it. However, unlike collision attacks proposed up until now, near collision attack exploits the existence of very similar but yet distinct values that are computed when the inputs are assumed to be selected uniformly at random from the entire set of inputs: $\mathbb{F}_2^8$. This brings up an implicit power model assumption that the power consumption should be linearly related to the bits of the sensitive value that is computed in the device. In comparison to the popular Hamming weight model, this implicit power model assumption is a much weaker one and therefore makes the attack more powerful against a wider range of platforms and devices with different leakage functions.

The main idea of a near collision attack (NCA) is to separate the measurements into two vectors and statistically compare how these two vectors are related to each other. Assuming that the S-box output leaks in the measurements, for any input byte x_0, another input value x_1 is computed for the same byte and for a key guess k_g as:

$$S(x_1 \oplus k_g) = S(x_0 \oplus k_g) \oplus \Delta(t) \tag{6}$$
$$x_1 = S^{-1}(S(x_0 \oplus k_g) \oplus \Delta(t)) \oplus k_g \tag{7}$$

where $\Delta(t)$ is an 8-bit value with a '1' at the tth bit position and '0' elsewhere. If the key guess is correct, then the S-box outputs have only one bit (XOR) difference. If the key guess is not correct, then the outputs will have a random (XOR) difference in between. Note that this property holds due to AES S-box's strength against differential cryptanalysis. Proceeding in this way, one can form a pair of vectors, X_0 and X_1 such that

$$X_0 = [x_0^i \in \mathbb{F}_2^8 : i \in \{1, \ldots, 128\}, \ S(x_0^i \oplus k_g) \wedge \Delta(t) = 0] \tag{8}$$

where $\wedge$ is the bit-wise AND operation, and X_1 is formed from each element of X_0 through the relation given in Eq. (7). This way the whole set of values in $\mathbb{F}_2^8$ are separated into two vectors and now they can be used to generate a statistic for k_g which in turn can be used to distinguish the correct key from others.

For $t = 8$, the observed values corresponding to the sets X_0 and X_1 can be visualized in Fig. 1 for an incorrect and a correct key guess under the assumption that the Hamming weight of a value leaks in observations. The difference between the observed values is also included in the plot for ease of comparison. As it is clearly visible from Fig. 1, when the key guess is correct, there is a clear linear relation between the vectors of observed values O^{X_0} and O^{X_1} corresponding to X_0 and X_1 respectively. Therefore Pearson correlation coefficient $(\rho(O^{X_0}, O^{X_1}))$ can be used as a statistical distinguisher in this case to recover the key.

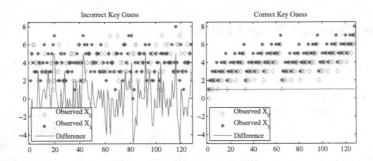

Fig. 1. Simulation values in sets X_0 and X_1 for incorrect (left) and correct (right) key guesses.

For real measurements, this attack can be implemented in a known plaintext setting by computing the mean of the observed values $\mu(O^{x_0^i})$ and $\mu(O^{x_1^i})$ for each input value x_0^i and x_1^i to reduce noise as in [17]. Furthermore, the attack can be run on larger than 128 value vectors to reduce multiple times for different values of t to compute the byte difference $\Delta(t)$, and the resulting correlation coefficients can be added together for each key guess k_g to constitute one final value to better distinguish the key.

3.1   Simulated Experiments on Unprotected AES Implementation

We have run simulated experiments to assess the capabilities of the near collision attack (NCA) and its efficiency in comparison to other similar attacks in the literature. To evaluate how our attack reacts to noise, we have fixed the number of traces and conducted experiments with various signal to noise ratio values (SNR $= \frac{var(signal)}{var(noise)}$, where signal and noise are computed as defined in [14]).

An important note here is that we have use scaled simulated values to mimic the measurements collected from an oscilloscope. Usually when simulated measurements are analysed, the fact that the simulations provide unnaturally optimistic measurements is neglected. Since this may lead to misleading simulation results which cannot be reproduced in real life, we have chosen to filter the simulated traces and scale them to the resolution of an 8-bit oscilloscope, therefore producing 256 unique values for traces. Note that, depending on the noise level the simulated traces can cover a large range of values. Therefore we have chosen to scale the values in a way such that the maximum and minimum values (128 and -127) are assigned to values ($\mu + 3 \times \sigma$) and ($\mu - 3 \times \sigma$) respectively, where

μ is the mean, and σ is the standard deviation of the simulated traces. The rest of the values are distributed equally over the sub-ranges which are of equal size.

As to measure the robustness of our technique against different linear leakage functions, we have used two ways to compute the simulated traces:

(a) The first method computes the Hamming weight of the S-box output (HW model).
(b) The second method is a weighted linear function of the bits of the S-box output, where the weight values are picked uniformly at random in the range $[-1, 1] \subset \mathbb{R}$ (Random linear model).

For comparison, we have selected the popular non-profiling univariate attacks, namely: correlation power analysis (CPA) [6], absolute sum DPA (AS-DPA) [1], non-profiled linear regression attack (NP-LRA) [9], and univariate mutual information analysis (UMIA) [11]. We have included CPA with Hamming weight model to have a basis for comparison as it is a popular choice for doing side channel analysis. The choice of AS-DPA and NP-LRA are to have a comparison with attacks which also have weak assumptions on the leakage model; AS-DPA assumes that each bit of the sensitive variable contribute significantly to the power consumption, where NP-LRA usually limits the algebraic order of the leakage function. For this work, we have restricted the basis functions of NP-LRA to linear relations (the case $d = 1$ in [9]), so that it would be a fair comparison to our work. Furthermore, we have included the leakage model dependent UMIA with Hamming weight model (UMIA-(HW)), and the leakage model agnostic variant UMIA which measures the mutual information between the least significant 7 bits of the sensitive variable and power measurements (UMIA-(7 LSB)).

We have run the experiments with 10 000 traces to put all methods on fair ground. Note that MIA requires a large number of traces as its distinguishing ability depends on the accuracy of the joint probability distribution estimations between the sensitive variable and power traces. We have computed the guessing entropy [22] over 100 independent experiments for each SNR value considered. Figure 2 presents the results of these experiments. As it is visible in Fig. 2(a) which shows results for Hamming weight leakage function, CPA has an obvious advantage over all other methods. When the second leakage function is considered however (Fig. 2(b)), the attacks using relaxed assumptions on the leakage function outperforms CPA. We also clearly see that MIA cannot deal with high levels of noise as efficiently as NCA, AS-DPA and NP-LRA.

Finally, if we only consider the attacks which have fewer assumptions on the leakage function, absolute sum DPA and the non-profiled linear regression attack seems to be able to deal with noise more efficiently when compared to NCA in an unprotected setting. Section 4 explains how the near collision approach of looking for small differences in the sensitive values can lead to a significant improvement over the state of the art attack against low entropy masking schemes.

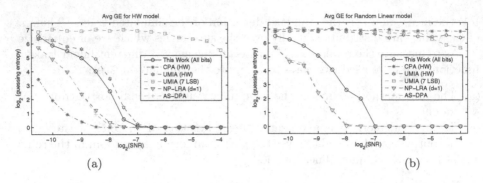

Fig. 2. SNR vs Guessing Entropy values computed over 100 independent experiments with 10 000 traces for perfect HW leakage (a), and random linear leakage (b).

3.2 Implementation Efficiency of NCA

Although near collision attack has the advantage of having reduced assumptions on the target device, this comes at a price, namely in computation time. For each key guess, the analyst should find the measurements which have a particular value in its corresponding plaintext. Although this can be a cumbersome operation, it does not scale up when the analyst has to run the analysis on multiple samples of collected measurements. To give a more accurate idea on the timing cost of NCA, Table 1 summarizes the average running time of each attack that is run in the previous section. The table presents average running time of each attack on 10 000 traces. All attacks are implemented as Matlab scripts executed in Matlab 2015a running on a PC with a Xeon E7 CPU. Note that the performance numbers assume the traces to be already loaded into memory in all cases.

Table 1. Average timing results from 100 independent experiments.

Technique	Time (s)
NCA	5.2727
CPA (HW)	1.0285
AS-DPA	1.1568
NP-LRA ($d = 1$)	2.7621
UMIA (HW)	1.4130
UMIA (7 LSB)	6.6153

Looking at the results presented in Table 1 and also Fig. 2, AS-DPA seems to be the best choice for the analyst in the tested cases in terms of running time and the ability to deal with Gaussian noise. However, even AS-DPA and NP-LRA are more efficient in the unprotected case, these techniques are not applicable in a univariate attack setting against low entropy masking schemes.

4 Near Collision Attack Against LEMS

A rather effective countermeasure against first order attacks such as introduced in the previous sections of this work is to use a masking scheme. However, one of the biggest drawbacks of masking schemes is the overhead introduced to the implementations. Low entropy masking schemes (LEMS) are a solution proposed to keep the security of classical masking [7,18,19] but reducing the implementation costs significantly. In this section, we argue how near collision attack idea can be extended to low entropy masking schemes. In particular, we focus on the mask set that is also used in the DPA Contest v4 traces:

$$M_{16} = \{00, 0F, 36, 39, 53, 5C, 65, 6A, 95, 9A, A3, AC, C6, C9, F0, FF\}.$$

4.1 Leaking Set Collision Attack

Leaking set collision attack is based on the observation that two sensitive variables which are masked with the mask set M_{16} lead to the same 16 leaking (masked) values if they are the bit-wise complement of each other [24]. Following this observation, the authors propose to compare the so-called leaking sets, the measurements corresponding to the 16 masked values, for each input x and it's pair x' computed as

$$x' = S^{-1}(S(x \oplus k) \oplus (\text{FF})_{16}). \tag{9}$$

Once the input pairs per key guess are computed, the analyst collects the observed values O^x and $O^{x'}$ corresponding to x and x'. If the key guess is correct, O^x and $O^{x'}$ should have the same distribution. If the key guess is not correct however, the resulting distributions will differ significantly, thanks to the AES S-box's good resistance against differential attacks. For comparing the distributions of these two sets, authors of [24] propose to use the 2-sample Kolmogorov-Smirnov (KS) test statistic. As KS test measures the distance between two distributions, the correct key guess should result in a lower KS test statistic than the incorrect key guesses do.

4.2 Leaking Set Near Collision Attack

We now define the 'leaking set near collision attack' (LS-NCA) as a combination of the LSCA idea proposed in [24] and the near collision attack (NCA) introduced in Sect. 3. Leaking set near collision attack can be summarized as an extension of the idea explained in Sect. 4.1 to the entire mask set of M_{16}. Similarly, we use the same observation that some input values lead to the same distribution in the S-box output as a direct result of the properties of the mask set that is used. As the authors of [24] point out in their work, whenever a sensitive value x is protected with the mask set M_{16}, the value $x \oplus (\text{FF})_{16}$ also results in the same values after applying the mask set. A further observation on the mask set M_{16} is that it is a closed set with respect to the XOR operation. In other words, XOR of any two elements in the set M_{16} results in another element of the mask set:

$$m_i \oplus m_j \in M_{16}, \forall m_i, m_j \in M_{16}. \tag{10}$$

This means that a sensitive variable x protected with the mask set M_{16}, and another sensitive variable $y(i) = x \oplus m_i$, $m_i \in M_{16}$ leads to the same masked values. It is easy to see that the property exploited in [24] is one particular case of the observation given in Eq. (10). Therefore, rather than directly comparing two similar distributions, if one collects all data from the input values which lead to the same distribution in the same set, this will lead to an equally reliable statistical analysis with less data. One should note that once all data that contribute to the same distribution are collected together, it is no longer possible to make a comparison between different sets and expect the same distribution. Therefore, we utilize a similar approach as we have done in Sect. 3 and look for sensitive values with 1-bit differences for comparison.

Leaking set near collision attack can be summarized as follows:

1. Generate the (disjoint) subsets of inputs (x) of which the S-box outputs contribute to the same distribution:

$$D^{x_i}_{M_{16}} = \{x : x = S^{-1}(S(x_i) \oplus m), \ \forall \ m \in M_{16}\}.$$

2. Make a key guess k_g.
3. For each input byte $x \oplus k_g$ which contribute to the same distribution (e.g. $x \oplus k_g \in D^{x_i}_{M_{16}}$), collect the corresponding measurement sample in a set $O^{x_i}(k_g)$.
4. Use 2-sample Kolmogorov-Smirnov (KS) test to check how similar the distributions of $O^{x_i}(k_g)$ and $O^{x_j}(k_g)$ are, where $S(x_i) \oplus S(x_j) = \Delta(t), \forall t \in \{1, ..., 8\}$.
5. Store sum of all 2-sample KS test statistics for each k_g.

Note that in Step 4, only the sets which have a 1-bit difference in between are used for 2-sample KS-test statistic calculation. In fact, sets with more than one bit difference in their S-box outputs might have the same Hamming weight, which in turn leads to similar (but not the same) distributions. Therefore, we expect the correct key to lead to a large distance between the two distributions. In case of an incorrect key guess however, each of the 16 elements in the set $D^{x_i}_{M_{16}}$ will lead to 16 distinct values after the S-box, therefore resulting in a distribution which spans the entire space $\mathbb{F}_2^8$. Sets with only one bit difference however will always result in different distributions. For instance, if the device leaks the Hamming weight of a value it computes, comparing sets with more than one bit difference would introduce noise in the cumulative KS-test statistic as values $(05)_{16}$ and $(03)_{16}$ have a 2-bit XOR difference in between, but have the same Hamming weight. Further note that doing the analysis on 1-bit different sensitive values limits the analysis to 64 calls to the 2-sample KS-test, therefore saves running time when the device leaks the Hamming weight of the sensitive variable. On the other hand, if the leakage function is an injection, all $\binom{16}{2} = 120$ combinations should be compared cumulatively. Here, using only the sets with

1-bit difference for comparison can be thought of a method similar to using mutual information analysis (MIA) by estimating power consumption with the Hamming weight model, since there is an implicit leakage function assumption that there is no inter-bit interaction in the leaking variable. In fact, if the leakage function is non-linear, the improvement gained through using only 1-bit different sets for comparing distributions would be less pronounced.

Unlike LSCA (outlined in Sect. 4.1), the cumulative test statistic now results in much smaller values for the incorrect key guesses. This is due to the fact that a wrong key guess (k_w) results in a random sampling of the set $\mathbb{F}_2^8$ and taking into account that 16 masks in M_{16} results in 16 distinct values for each sample in the set, the resulting $O^{x_i}(k_w)$ has a cardinality much closer to $|\mathbb{F}_2^8|$. However, this does not diminish the distinguishability of the correct key from other candidates. In the case where the key guess is correct (k_c), the set $O^{x_i}(k_c)$ will have around 16 unique values (the exact number can increase due to noise in the measurement setup). Now that we have a much smaller sampling of the set $\mathbb{F}_2^8$, a comparison of distinct sets $([O^{x_i}(k_c), O^{x_j}(k_c)], i \neq j)$ is more meaningful, and in fact the cumulative 2-sample KS-test statistic value results in a larger value than the one obtained for a wrong key guess as the distributions are definitely different.

Note that this mask set is an example of the mask sets that are generated as a linear code [4]. As binary linear codes have the intrinsic property of being closed sets with respect to the XOR operation, any mask set that is a binary linear code is vulnerable to the attack explained in this section.

4.3   Simulated Experiments on AES Implementation with LEMS

In this section we present results of our simulated experiments with different SNR values and also with two different leakage functions as it is done in Sect. 3.1. In our simulations, we compare our attacks to the previously proposed univariate, non-profiled attacks: univariate MIA (UMIA) and LSCA that is recalled in Sect. 4.1. To compare the efficiency of the attacks in terms of the expected remaining work to find the key, we use the guessing entropy metric [22]. For each experiment using a random linear model as the leakage function, we generate 8 values which are picked uniformly at random from $[1, 10] \subset \mathbb{R}$. Unlike the simulations presented in Sect. 3.1, we have chosen to use a linear leakage function which is slightly different, and favour the attacks assuming that each bit of the sensitive value contributes to the leakages. Although this may not always be the case in real life, we choose to use this leakage function as it is a favourable leakage model for MIA using the Hamming weight model. We present results that show the proposed technique in Sect. 4.2 is more efficient than MIA in terms of handling the noise in an unknown linear leakage model setting even when the leakage model favours MIA.

The experiments are carried out with various SNR values and the results are presented in Fig. 3 for both Hamming weight model and the random linear leakage model we computed for each experiment. Note that the attack which assumes a linear leakage model and computes only 64 comparisons is marked as 'Linear', and the attack which computes all possible 120 comparisons is marked

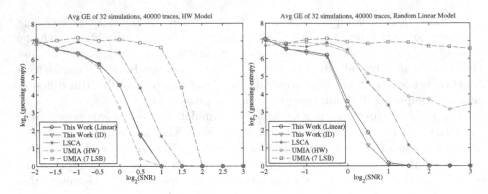

Fig. 3. Efficiency of the proposed attacks in comparison with previous works for various SNR values.

as 'ID' for identity model. A quick look at Fig. 3 shows that, similar to the case in near collision attack proposed in Sect. 3, the attack is indifferent to changes in the leakage model as long as it stays linear. Moreover, if the leakage model is a random linear function of the bits of the sensitive variable, univariate MIA fails to recover the key for leakage models with a high variance of its weight values. However when the leakage model follows a strict Hamming weight leakage, then univariate MIA seems to handle noise more efficiently than both of our proposed approaches to analyse the low entropy masking scheme at hand.

4.4   Experiments on DPA Contest V4 Traces

This section shows the efficiency of our attacks compared to the similar previously proposed methods in the context of a real world scenario. For reproducibility of our results, we used DPA Contest v4 traces which are EM measurements collected from a smart card having an ATMega-163 microcontroller which implements an LEMS against first and second order attacks [19]. The implementation uses the mask set M_{16} given in Sect. 4.

Figure 4 presents an analysis in time domain which reveals that even with 1000 traces, it is possible to recover the key with high confidence. To further test the reliability of our technique on the DPA Contest v4 traces, we have focused our analysis on the time samples where each of the 16 S-box outputs lead to the highest signal-to-noise ratio (SNR computed following the definition in [14]) for computing guessing entropy and the results of the analysis are presented in Fig. 5.

First thing to notice in the figure is that LSCA has some room for improvement even when compared to a generic univariate MIA (UMIA (7-bit) in Fig. 5). On the other hand, when MIA is applied with a more accurate power model (in this case the Hamming weight model), the gap is rather large. When the leaking set near collision attack proposed in this work is considered, it is easy to see that the one which does not assume any power model ('ID') performs twice as efficient

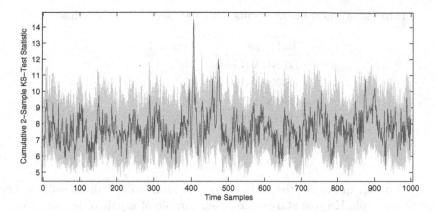

Fig. 4. Leaking set near collision attack results VS time samples.

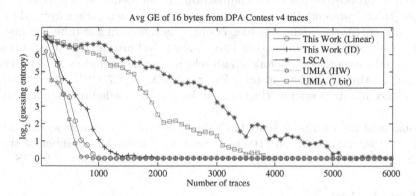

Fig. 5. Logarithm of average partial guessing entropy for various univariate attacks on DPA Contest v4 traces.

in terms of the number of traces required to recover the full key when compared to the generic univariate MIA ('7 LSB'). Moreover, when the leakage function is assumed to have a linear relation with respect to the bits of the leaking value, the results are almost identical to the ones from a univariate MIA which models the power consumption as the Hamming weight of the leaking value. One should note that Hamming weight model is rather accurate in this case as SNR values (computed with Hamming weight model) vary between 3 and 5 for the points taken into consideration for the analysis.

4.5 Implementation Efficiency of the Attack

Similar to the near collision attack, leaking set near collision attack also requires to find the traces in the measurement set which correspond to a set of input bytes. Although this operation is computationally heavy when applied to a large trace set, it does not get worse when multiple samples are needed to be analysed.

Table 2. Average timing results from 100 independent experiments.

Technique	Time (s)
LS-NCA (Linear)	13.7144
LS-NCA (ID)	15.4003
LSCA	8.6176
UMIA (HW)	1.5648
UMIA (7 LSB)	7.6951

The analyst can group all the traces corresponding to a leaking set and then compute 2-sample KS test statistic for each sample of a pair of leaking sets.

As in Sect. 3, we have run simulated experiments to assess the time required to run the proposed attacks in comparison to the other attacks run in this section. Table 2 presents the average running times of each attack applied to the chosen low entropy masking scheme. Timings presented in the table are average running times over 100 independent experiments that are run over 10 000 traces. Similar to the experiments before, all attacks are implemented as Matlab scripts executed in Matlab 2015a run on a PC with a Xeon E7 CPU. Note that the performance numbers assume the traces to be already loaded into memory in all cases.

Looking at the results presented in Table 2 and taking into consideration that the leaking set near collision attacks (LS-NCA) require less number of traces, they are the strongest attacks against software implementations of LEMS.

5   Conclusions

In this work, we introduced a new way of analysing side channel traces, namely the side channel near collision attack (NCA). Unlike the collision attacks proposed in the literature, NCA is intrinsically univariate and only assumes the leakage function to be linear. Simulations show that NCA is indifferent to changes in the linear leakage function.

Furthermore, we present a new attack, leaking set near collision attack, against the low entropy masking scheme used in DPA Contest v4 [19]. This attack improves the attack proposed in [24] by fully exploiting the properties of the used mask set, and combining it with the NCA approach. As the proposed attack is univariate, it is especially of interest for software implementations of low entropy masking schemes. Simulations show that in case the leakage function diverges from a perfect Hamming weight leakage but yet stays a linear function, our attack overpowers univariate MIA.

It should be noted that not only the mask set M_{16}, but all mask sets which have a linear relation in between (as proposed in [4]) are vulnerable to the attack presented in this paper.

Application of the proposed analysis methods to non-linear leakage functions remains a research direction to follow as a future work.

Acknowledgements. This work was supported in part by the Technology Foundation STW (project 12624 - SIDES), The Netherlands Organization for Scientific Research NWO (project ProFIL 628.001.007), the ICT COST actions IC1204 TRUDEVICE, and IC1403 CRYPTACUS. LB is supported by NWO VIDI and Aspasia grants. TE is supported by the National Science Foundation under grant CNS-1314770.

References

1. Agrawal, D., Rao, J.R., Rohatgi, P.: Multi-channel attacks. In: Walter, C.D., Koç, Ç.K., Paar, C. (eds.) CHES 2003. LNCS, vol. 2779, pp. 2–16. Springer, Heidelberg (2003). http://dx.doi.org/10.1007/978-3-540-45238-6_2

2. Batina, L., Gierlichs, B., Lemke-Rust, K.: Differential cluster analysis. In: Clavier, C., Gaj, K. (eds.) CHES 2009. LNCS, vol. 5747, pp. 112–127. Springer, Heidelberg (2009). http://dx.doi.org/10.1007/978-3-642-04138-9_9

3. Bhasin, S., Bruneau, N., Danger, J.-L., Guilley, S., Najm, Z.: Analysis and improvements of the DPA contest v4 implementation. In: Chakraborty, R.S., Matyas, V., Schaumont, P. (eds.) SPACE 2014. LNCS, vol. 8804, pp. 201–218. Springer, Heidelberg (2014). http://dx.doi.org/10.1007/978-3-319-12060-7_14

4. Bhasin, S., Carlet, C., Guilley, S.: Theory of masking with codewords in hardware: low-weight dth-order correlation-immune Boolean functions. Cryptology ePrint Archive, Report 2013/303 (2013). http://eprint.iacr.org/

5. Bogdanov, A.: Multiple-differential side-channel collision attacks on AES. In: Oswald, E., Rohatgi, P. (eds.) CHES 2008. LNCS, vol. 5154, pp. 30–44. Springer, Heidelberg (2008). http://dx.doi.org/10.1007/978-3-540-85053-3_3

6. Brier, E., Clavier, C., Olivier, F.: Correlation power analysis with a leakage model. In: Joye, M., Quisquater, J.-J. (eds.) CHES 2004. LNCS, vol. 3156, pp. 16–29. Springer, Heidelberg (2004). http://dx.doi.org/10.1007/978-3-540-28632-5_2

7. Carlet, C., Danger, J.-L., Guilley, S., Maghrebi, H.: Leakage squeezing of order two. In: Galbraith, S., Nandi, M. (eds.) INDOCRYPT 2012. LNCS, vol. 7668, pp. 120–139. Springer, Heidelberg (2012). http://dx.doi.org/10.1007/978-3-642-34931-7_8

8. Clavier, C., Feix, B., Gagncrot, G., Roussellet, M., Verneuil, V.: Improved collision-correlation power analysis on first order protected AES. In: Preneel, B., Takagi, T. (eds.) CHES 2011. LNCS, vol. 6917, pp. 49–62. Springer, Heidelberg (2011). http://dx.doi.org/10.1007/978-3-642-23951-9_4

9. Doget, J., Prouff, E., Rivain, M., Standaert, F.X.: Univariate side channel attacks and leakage modeling. J. Cryptograph. Eng. **1**(2), 123–144 (2011). http://dx.doi.org/10.1007/s13389-011-0010-2

10. Gérard, B., Standaert, F.-X.: Unified and optimized linear collision attacks and their application in a non-profiled setting. In: Prouff, E., Schaumont, P. (eds.) CHES 2012. LNCS, vol. 7428, pp. 175–192. Springer, Heidelberg (2012). http://dx.doi.org/10.1007/978-3-642-33027-8_11

11. Gierlichs, B., Batina, L., Tuyls, P., Preneel, B.: Mutual information analysis. In: Oswald, E., Rohatgi, P. (eds.) CHES 2008. LNCS, vol. 5154, pp. 426–442. Springer, Heidelberg (2008). http://dx.doi.org/10.1007/978-3-540-85053-3_27

12. Grosso, V., Standaert, F.X., Prouff, E.: Leakage Squeezing, Revisited (2013)

13. Kocher, P.C., Jaffe, J., Jun, B.: Differential power analysis. In: Wiener, M. (ed.) CRYPTO 1999. LNCS, vol. 1666, pp. 388–397. Springer, Heidelberg (1999)

14. Mangard, S., Oswald, E., Popp, T.: Power Analysis Attacks: Revealing the Secrets of Smart Cards (Advances in Information Security). Springer, New York, Secaucus (2007)

15. Moradi, A.: Statistical tools flavor side-channel collision attacks. In: Pointcheval, D., Johansson, T. (eds.) EUROCRYPT 2012. LNCS, vol. 7237, pp. 428–445. Springer, Heidelberg (2012). http://dx.doi.org/10.1007/978-3-642-29011-4_26

16. Moradi, A., Guilley, S., Heuser, A.: Detecting hidden leakages. In: Boureanu, I., Owesarski, P., Vaudenay, S. (eds.) ACNS 2014. LNCS, vol. 8479, pp. 324–342. Springer, Heidelberg (2014). http://dx.doi.org/10.1007/978-3-319-07536-5_20

17. Moradi, A., Mischke, O., Eisenbarth, T.: Correlation-enhanced power analysis collision attack. In: Mangard, S., Standaert, F.-X. (eds.) CHES 2010. LNCS, vol. 6225, pp. 125–139. Springer, Heidelberg (2010). http://dx.doi.org/10.1007/978-3-642-15031-9_9

18. Nassar, M., Guilley, S., Danger, J.-L.: Formal analysis of the entropy/security trade-off in first-order masking countermeasures against side-channel attacks. In: Bernstein, D.J., Chatterjee, S. (eds.) INDOCRYPT 2011. LNCS, vol. 7107, pp. 22–39. Springer, Heidelberg (2011). http://dx.doi.org/10.1007/978-3-642-25578-6_4

19. Nassar, M., Souissi, Y., Guilley, S., Danger, J.L.: RSM: a small and fast countermeasure for AES, secure against 1st and 2nd-order zero-offset SCAs. In: Proceedings of the Conference on Design, Automation and Test in Europe, DATE 2012, EDA Consortium, San Jose, CA, USA, pp. 1173–1178 (2012). http://dl.acm.org/citation.cfm?id=2492708.2492999

20. Schindler, W., Lemke, K., Paar, C.: A stochastic model for differential side channel cryptanalysis. In: Rao, J.R., Sunar, B. (eds.) CHES 2005. LNCS, vol. 3659, pp. 30–46. Springer, Heidelberg (2005). http://dx.doi.org/10.1007/11545262_3

21. Schramm, K., Wollinger, T., Paar, C.: A new class of collision attacks and its application to DES. In: Johansson, T. (ed.) FSE 2003. LNCS, vol. 2887, pp. 206–222. Springer, Heidelberg (2003). http://dx.doi.org/10.1007/978-3-540-39887-5_16

22. Standaert, F.-X., Malkin, T.G., Yung, M.: A unified framework for the analysis of side-channel key recovery attacks. In: Joux, A. (ed.) EUROCRYPT 2009. LNCS, vol. 5479, pp. 443–461. Springer, Heidelberg (2009). http://dx.doi.org/10.1007/978-3-642-01001-9_26

23. Ye, X., Chen, C., Eisenbarth, T.: Non-linear collision analysis. In: Sadeghi, A.-R., Saxena, N. (eds.) RFIDSec 2014. LNCS, vol. 8651, pp. 198–214. Springer, Heidelberg (2014). http://dx.doi.org/10.1007/978-3-319-13066-8_13

24. Ye, X., Eisenbarth, T.: On the vulnerability of low entropy masking schemes. In: Francillon, A., Rohatgi, P. (eds.) CARDIS 2013. LNCS, vol. 8419, pp. 44–60. Springer, Heidelberg (2014). http://dx.doi.org/10.1007/978-3-319-08302-5_4

Masking Large Keys in Hardware: A Masked Implementation of McEliece

Cong Chen[1](✉), Thomas Eisenbarth[1], Ingo von Maurich[2],
and Rainer Steinwandt[3]

[1] Worcester Polytechnic Institute, Worcester, MA, USA
{cchen3,teisenbarth}@wpi.edu
[2] Ruhr-Universität Bochum, Bochum, Germany
ingo.vonmaurich@rub.de
[3] Florida Atlantic University, Boca Raton, USA
rsteinwa@fau.edu

Abstract. Instantiations of the McEliece cryptosystem which are considered computationally secure even in a post-quantum era still require hardening against side channel attacks for practical applications. Recently, the first differential power analysis attack on a McEliece cryptosystem successfully recovered the full secret key of a state-of-the-art FPGA implementation of QC-MDPC McEliece. In this work we show how to apply masking countermeasures to the scheme and present the first masked FPGA implementation that includes these countermeasures. We validate the side channel resistance of our design by practical DPA attacks and statistical tests for leakage detection.

Keywords: Threshold implementation · McEliece cryptosystem · QC-MDPC codes · FPGA

1 Motivation

Prominent services provided by public-key cryptography include signatures and key encapsulation, and their security is vital for various applications. In addition to classical cryptanalysis, quantum computers pose a potential threat to currently deployed asymmetric solutions, as most of these have to assume the hardness of computational problems which are known to be feasible with large-scale quantum computers [18]. Given these threats, it is worthwhile to explore alternative public-key encryption schemes that rely on problems which are believed to be hard even for quantum computers, which might become reality sooner than the sensitivity of currently encrypted data expires [5]. The McEliece cryptosystem [12] is among the promising candidates, as it has withstood more than 35 years of cryptanalysis. To that end, efficient and secure implementations of McEliece should be available even nowadays. The QC-MDPC variant of the McEliece scheme proposed in [13] is a promising efficient alternative to prevailing schemes, while maintaining reasonable key sizes. The first implementations

© Springer International Publishing Switzerland 2016
O. Dunkelman and L. Keliher (Eds.): SAC 2015, LNCS 9566, pp. 293–309, 2016.
DOI: 10.1007/978-3-319-31301-6_18

of QC-MDPC McEliece were presented in [10], and an efficient and small hardware engine of the scheme was presented in [19]. However, embedded crypto
cores usually require protection against the threat of physical attacks when used
in practice. Otherwise, side channel attacks can recover the secret key quite
efficiently, as shown in [6].

Our Contribution. In this work we present a masked hardware implementation of QC-MDPC McEliece. Our masked design builds on the lightweight
design presented in [19]. We present several novel approaches of dealing with
side channel leakage. First, our design implements a hybrid masking approach,
to mask the key and critical states, such as the syndrome and other intermediate
states. The masking consists of Threshold Implementation (TI) based Boolean
masking for bit operations and arithmetic masking for needed counters. Next, we
present a solution for efficiently masking long bit vectors, as needed to protect
the McEliece keys. This optimization is achieved by generating a mask on-the-fly
using a LFSR-derived PRG. Through integration of PRG elements, the amount
of external randomness needed by the engine is considerably reduced when compared to other TI-based implementations. Our design is fully implemented and
analyzed for remaining side channel leakage. In particular, we validate that the
DPA attack of [6] is no longer feasible. We further show that there are also no
other remaining first-order leakages nor other horizontal leakages as exploited
in [6].

After introducing necessary background in Sect. 2, we present the masked
McEliece engine in Sects. 3 and 4. Performance results are presented in Sect. 5.
A thorough leakage analysis is presented in Sect. 6.

2   Background

In the following we introduce moderate-density parity-check (MDPC) codes and
their quasi-cyclic (QC) variant with a focus on decoding since we aim to protect
the secret key. Afterwards we summarize how McEliece is instantiated with QC-
MDPC codes as proposed in [13]. As our work extends an FPGA implementation
of QC-MDPC McEliece that is unprotected against side channel attacks [19], we
give a short overview of the existing implementation and summarize relevant
works on the masking technique of threshold implementations.

2.1   Moderate-Density Parity-Check Codes

MDPC codes belong to the family of binary linear $[n, k]$ error-correcting codes,
where n is the length, k the dimension, and $r = n - k$ the co-dimension of a
code C. Binary linear error-correcting codes are equivalently described either
by their generator G or by their parity-check matrix H. The rows of generator
matrix $G \in \mathbb{F}_2^{k \times n}$ form a basis of C while $H \in \mathbb{F}_2^{r \times n}$ describes the code as the
kernel $C = \{c \in \mathbb{F}_2^n \mid Hc^T = 0^\perp\}$ where $0^\perp$ represents an all-zero column vector.
The syndrome of any vector $x \in \mathbb{F}_2^n$ is defined as $s = Hx^T \in \mathbb{F}_2^r$. Hence, the code
C is comprised of all vectors $x \in \mathbb{F}_2^n$ whose syndrome is zero for a particular

parity-check matrix H. MDPC codes are defined by only allowing a *moderate* Hamming weight $w = O(\sqrt{n \log(n)})$ for each row of the parity-check matrix. By an (n, r, w)-MDPC code we refer to a binary linear $[n, k]$ code with such a constant row weight w.

A code C is called quasi-cyclic (QC) if for some positive integer $n_0 > 0$ the code is closed under cyclic shifts of its codewords by n_0 positions. Furthermore, it is possible to choose the generator and parity-check matrix to consist of $p \times p$ circulant blocks if $n = n_0 \cdot p$ for some positive integer p. This allows to completely describe the generator and parity-check matrices by their first row. If an (n, r, w)-MDPC code is quasi-cyclic with $n = n_0 \cdot r$, we refer to it as an (n, r, w)-QC-MDPC code.

Several t-error-correcting decoders have been proposed for (QC-)MDPC codes [1,8,10,11,13,21]. The implementation that we base our work on implements the optimized decoder presented in [10], which in turn is an extended version of the bit-flipping decoder of [8]. Decoding a ciphertext $x \in \mathbb{F}_2^n$, is achieved by:

1. Computing the syndrome $s = Hx^T$.
2. Computing the number of unsatisfied parity checks $\#_{upc}$ for every ciphertext bit.
3. If $\#_{upc}$ exceeds a precomputed threshold b, invert the corresponding ciphertext bit and add the corresponding column of the parity-check matrix to the syndrome.
4. In case $s = 0^\perp$, decoding was successful, otherwise repeat Steps 2/3.
5. Abort after a defined maximum of iterations with a decoding error.

2.2 McEliece Public Key Encryption with QC-MDPC Codes

The McEliece cryptosystem was introduced using binary Goppa codes [12]. Instantiating McEliece with t-error-correcting (QC-)MDPC codes was proposed in [13], mainly to significantly reduce the size of the keys while still maintaining reasonable security arguments. The proposed parameters for an 80-bit security level are $n_0 = 2, n = 9602, r = 4801, w = 90, t = 84$, which results in a much more practical public key size of 4801 bit and a secret key size of 9602 bit compared to binary Goppa codes which require around 64 kByte for public keys at the same security level.

The main idea of the McEliece cryptosystem is to encode a plaintext into a codeword using the generator matrix of a code selected by the receiver and to add a randomly generated error vector of weight t to the codeword which can only be removed by the intended receiver. We summarize QC-MDPC McEliece in the following by introducing key-generation, encryption and decryption.

Key-Generation. The parity-check matrix H is the secret key in QC-MDPC McEliece. As the code is quasi-cyclic, the parity-check matrix consists of n_0 concatenated $r \times r$ blocks $H = (H_0 | \ldots | H_{n_0-1})$. We denote the first row of each of these blocks by $h_0, \ldots, h_{n_0-1} \in \mathbb{F}_2^r$. The public key in QC-MDPC McEliece is the corresponding generator matrix G, which is computed from H in standard form as $G = [I_k \,|\, Q]$ by concatenation of the identity matrix $I_k \in \mathbb{F}_2^{k \times k}$ with

$$Q = \begin{pmatrix} (H_{n_0-1}^{-1} \cdot H_0)^T \\ (H_{n_0-1}^{-1} \cdot H_1)^T \\ \cdots \\ (H_{n_0-1}^{-1} \cdot H_{n_0-2})^T \end{pmatrix}.$$

The key generation starts by randomly selecting first row candidates $h_0, \ldots, h_{n_0-1} \in_R \mathbb{F}_2^r$ such that the overall row weight (wt) sums up to $w = \sum_{i=0}^{n_0-1} \mathrm{wt}(h_i)$. Since we intend to generate a code which is quasi-cyclic, the n_0 blocks of the parity-check matrix are generated from the first rows by cyclic shifts. The resulting parity-check matrix belongs to an (n, r, w)-QC-MDPC code with $n = n_0 \cdot r$. If the last block H_{n_0-1} is non-singular, i. e., if $H_{n_0-1}^{-1}$ exists, the public key is computed as $G = [I_k \,|\, Q]$. Otherwise new candidates for h_{n_0-1} are generated until a non-singular H_{n_0-1} is found.

Encryption. A plaintext $m \in \mathbb{F}_2^k$ is encrypted by encoding it into a codeword using the recipient's public key G and by adding a random error vector $e \in \mathbb{F}_2^n$ of weight $\mathrm{wt}(e) \leq t$ to it. Hence, the ciphertext is computed as $x = (m \cdot G \oplus e) \in \mathbb{F}_2^n$.

Decryption. Given a ciphertext $x \in \mathbb{F}_2^n$, the intended recipient removes the error vector e from x using the secret code description H and a QC-MDPC decoding algorithm Ψ_H yielding mG. Since $G = [I_k \,|\, Q]$, the first k positions of mG are equal to the k-bit plaintext.

2.3   FPGA Implementation of QC-MDPC McEliece

Our work extends on the lightweight implementation of McEliece based on QC-MDPC code for reconfigurable devices by [19]. Their resource requirements are 64 slices and 1 block RAM (BRAM) to implement encryption and 159 slices and 3 BRAMs to implement decryption on a Xilinx Spartan-6 XC6SLX4 FPGA. The goal of our work is to protect the secret key. Hence, only the decryption engine is discussed in this paper. From a high-level point of view, decryption works as follows: at first the syndrome of the ciphertext is computed. Then for each ciphertext bit the number of unsatisfied parity-checks $\#_{\mathrm{upc}}$ are counted and if they exceed a defined threshold, the ciphertext bit is inverted. When a ciphertext bit is inverted, the corresponding row of the parity-check matrix is added to the syndrome. The DPA presented in [6] shows that the described architecture is vulnerable to an efficient horizontal key recovery attack, since neither the key nor internal states are masked.

2.4   Threshold Implementation

Threshold implementation (TI) is a masking-based technique to prevent first order and higher order side channel leakage. Since its introduction in [15], many symmetric cryptosystems have been implemented in TI [2–4, 14, 17]. More importantly, most of these works have performed thorough leakage analysis and have shown that TI actually prevents the promised order leakage (if carefully implemented). Even higher-order leakage, while not prevented, usually comes at a highly increased cost of needed observations. TI performs Boolean secret sharing on all sensitive variables. Computations on the shares are then performed in a way that ensures *correctness*, maintains *uniformity* of the shares, and ensures *non-completeness* of the computation, that is, each sub-operation can only be performed on a strict sub-set of the inputs. A detailed description of TI is available in [15].

We choose TI for McEliece because TI is fairly straightforward to apply and to implement, yet it is effective. Furthermore, large parts of McEliece are linear, and hence cheap to mask using TI. The decoder part, while not linear, is also fairly efficient to mask using TI, as shown in Sect. 4. At the same time, our implementation avoids several of the disadvantages of TI: Unlike [16], we convert our addition to arithmetic masking once the values get larger, yielding a much more efficient addition engine than one solely relying on TI. By including the pseudorandom mask generation in the crypto core, we significantly cut both the required memory space usually unavoidably introduced by TI as well as the required overhead of random bits consumed by TI engines. Note that the TI-AES engines presented in [3,14] consume about 8000 bits of randomness per encryption, while our engine only consumes 160 bits per decryption.

3 Masking QC-MDPC McEliece

An effective way to counteract side channel analysis is to employ masking. Masking schemes aim to randomize sensitive intermediate states such that the leakage is independent of processed secrets. In QC-MDPC McEliece, the key bits and the syndrome are sensitive values that need to be protected and therefore they must be masked whenever they are manipulated. Similarly, since the decoding operation processes the sensitive syndrome, leakage of the decoder needs to be masked as well.

3.1 Masked Syndrome Computation

As described in Sect. 2.1, the decoding algorithm begins with the syndrome computation $s = Hx^T$. Both the parity-check matrix H and the syndrome s are sensitive values and can cause side channel leakage. However, since the syndrome computation is a linear operation, masking this operation is simple and efficient. Intuitively, H can be split into two shares, H_m and M such that $H = H_m \oplus M$, by Boolean masking. The mask matrix M is created in correspondence to H, by first generating uniformly distributed random masks for h_i, $m_0, \ldots, m_{n_0-1} \in \mathbb{F}_2^r$ of the n_0 blocks, which then comprise the first row of mask matrix M. Each bit in the m_i is uniformly set to 0 or 1. Next, the remaining rows of the mask matrix M are obtained by quasi-cyclic shifts of the first row, according to the construction of H. The masked syndrome s_m and the syndrome mask m_s can be computed independently as $s_m = H_m x^T$ and $m_s = M x^T$. The syndrome s is available as the combination of the two shares $s = s_m \oplus m_s$.

Algorithm 1. Masked Error Correction Decoder

Input: H_m, M_1, M_2, s_m, m_{s_1}, m_{s_2}, x, $B = b_0, ..., b_{max-1}$, max
Output: Error free codeword x or DecodingFailure
 1: **for** $i = 0$ to max-1 **do**
 2:     **for** every ciphertext bit x_j **do**
 3:         $\#_{upc} = \mathsf{SecHW}(\mathsf{SecAND}(s_m, m_{s_1}, m_{s_2}, H_{m,j}, M_{1,j}, M_{2,j}))$
 4:         $d = (\#_{upc} > b_i)$, $d \in \{0, 1\}$
 5:         $x = x \oplus (d \cdot 1_j)$ ▷ Flip the jth bit of x
 6:         $s_m = s_m \oplus (d \cdot H_{m,j} \oplus \bar{d} \cdot M_{2,j})$ ▷ Update syndrome
 7:         $m_{s_1} = m_{s_1} \oplus M_{1,j}$ ▷ Update masks
 8:         $m_{s_2} = m_{s_2} \oplus M_{2,j} \oplus (\bar{d} \cdot M_{1,j})$
 9:     **end for**
10:     **if** $\mathsf{SecHW}(s_m, m_{s_1}, m_{s_2}) == 0$ **then**   ▷ Check for remaining errors
11:         **return** x
12:     **end if**   ▷ For constant run time, this if-statement can be moved after the
         for-loop
13: **end for**
14: **return** DecodingFailure

3.2 Masked Decoder

After syndrome computation, the error correction decoder computes the number of unsatisfied parity check equations between the sensitive syndrome and one row of the sensitive parity check matrix. By comparing that number with a predefined threshold (usually denoted b), the decoder decides whether to flip the corresponding bit in the ciphertext. Masking the actual decoding steps is more complex, since both inputs, namely the syndrome and the parity check matrix, as well as the control flow of the decoder can leak sensitive information and thus need to be protected. Unlike the syndrome computation, the decoder performs a binary AND and a Hamming weight computation on sensitive data. Both operations are non-linear and thus need more elaborate protection than just a straightforward Boolean masking. In the following we explain how these operations can be implemented. Algorithm 1 describes the masked version of the decoder. Note that the algorithm has been formulated with a constant execution flow to better represent the intended hardware implementation. Further note that the algorithm and its FPGA implementation exhibit a constant timing behavior (except the number of decoder iterations) and that all key-related variables are masked. The number of decoder iterations can be set to maximum by simply moving the if-statement out of the loop. For the chosen 9602/4801 parameter set, max would be set to 5, increasing the average run time roughly by a factor 2 (cf. [21]).

In Algorithm 1, we make use of two special functions. Function SecAND computes the bitwise AND operation between syndrome s and secret key H in a secure way without leaking any sensitive information. The other function SecHW computes the Hamming Weight of a given vector. Both functions are explained in detail in the following. An all-zero vector with the jth bit equal to 1 is indicated by 1_j.

Secure AND Computation. One important step when decoding a QC-MDPC code is to compute the unsatisfied parity-check equations which starts with a non-linear bitwise AND operation between the syndrome and one row of the secret key matrix. Our function SecAND performs a bitwise AND operation between two bit vectors, namely $s \wedge h$. Since the AND is a non-linear operation, simple two-share Boolean masking is not applicable. Instead, we follow the concept of Threshold Implementation as described in Sect. 2.4. We adopt the bitwise AND operation from [15], which provides first-order security when applied to *three* Boolean shares. This means that the two-share representations of the two inputs, i. e., the syndrome and parity check matrix, need to be extended to a three-share representation.

To achieve a three-share representation of both syndrome and parity check matrix, the masking is expanded in the following way: After syndrome computation as explained in Sect. 3.1, the syndrome is represented as $s_m \oplus m_s$ and the secret key is represented as $H_{m,j} \oplus M_j$. Next, the syndrome representation is extended as $s_m \oplus m_{s_1} \oplus m_{s_2}$ and the key as $H_{m,j} \oplus M_{1,j} \oplus M_{2,j}$. Here, m_{s_2} and $M_{2,j}$ are two new uniformly distributed random mask vectors and m_{s_1} is derived as $m_{s_1} = m_s \oplus m_{s_2}$ and $M_{1,j} = M_j \oplus M_{2,j}$. The following equations show how to achieve a TI version of $s \wedge h$ that satisfies correctness and non-completeness, but not uniformity.

$$
\begin{aligned}
s \wedge h &= (s_m \oplus m_{s_1} \oplus m_{s_2}) \wedge (H_{m,j} \oplus M_{1,j} \oplus M_{2,j}) \\
&= (s_m \wedge H_{m,j}) \oplus (s_m \wedge M_{1,j}) \oplus (H_{m,j} \wedge m_{s_1}) \oplus \\
&\quad (m_{s_1} \wedge M_{1,j}) \oplus (m_{s_1} \wedge M_{2,j}) \oplus (M_{1,j} \wedge m_{s_2}) \oplus \\
&\quad (m_{s_2} \wedge M_{2,j}) \oplus (m_{s_2} \wedge H_{m,j}) \oplus (M_{2,j} \wedge s_m)
\end{aligned}
\tag{1}
$$

As pointed out in [15], in order to fulfill uniformity, one can introduce additional uniform random masks to mask each share. By introducing two more uniformly random vectors r_1 and r_2, the three output shares can be computed as follows. Let sh denote the result of the TI version of the AND operation. Using the equations above, sh can be split into three shares sh_i, which are now uniformly distributed thanks to the r_i and are given as:

$$
\begin{aligned}
sh_1 &= (s_m \wedge H_{m,j}) \oplus (s_m \wedge M_{1,j}) \oplus (H_{m,j} \wedge m_{s_1}) \oplus r_1 \\
sh_2 &= (m_{s_1} \wedge M_{1,j}) \oplus (m_{s_1} \wedge M_{2,j}) \oplus (M_{1,j} \wedge m_{s_2}) \oplus r_2 \\
sh_3 &= (m_{s_2} \wedge M_{2,j}) \oplus (m_{s_2} \wedge H_{m,j}) \oplus (M_{2,j} \wedge s_m) \oplus r_1 \oplus r_2
\end{aligned}
\tag{2}
$$

Secure Hamming Weight Computation. In the unprotected FPGA implementation of [19], the Hamming weight computation of sh is performed by looking up the weight of small chunks of sh from a precomputed table and then accumulating those weights to get the Hamming weight of sh. However, the weight of a chunk is always present in plain and the computation of it can result in side channel leakage that will lead to the recovery of the Hamming weight.

Even though the knowledge of the weight does not necessarily recover the chunk value, it still yields information about sh and thus the secret key h.

For a side-channel secure implementation, both the input and the output of a Hamming weight computation for each chunk must be masked. Since the weight of all chunks needs to be accumulated, it is preferable to use Arithmetic masking instead of Boolean masking. For example, the Hamming weight of sh can be calculated using the following equation:

$$\text{wt}(sh) = \sum_{i=1}^{|sh|} sh_{1,i} \oplus sh_{2,i} \oplus sh_{3,i} \tag{3}$$

where subscript i refers to the i-th bit of each share and $|sh|$ is the length of sh in bits. Using a secure conversion function from Boolean masking to Arithmetic masking [7], each Boolean mask tuple $(sh_{1,i}, sh_{2,i}, sh_{3,i})$ can be converted to an Arithmetic mask pair $(A_{1,i}, A_{2,i})$ such that $sh_{1,i} \oplus sh_{2,i} \oplus sh_{3,i} = A_{1,i} + A_{2,i}$. Then, the Hamming weight of sh can be computed as:

$$\text{wt}(sh) = \sum_{i=1}^{|sh|} A_{1,i} + A_{2,i} = \sum_{i=1}^{|sh|} A_{1,i} + \sum_{i=1}^{|sh|} A_{2,i} \tag{4}$$

According to Eq. (4), we only accumulate $A_1 = \sum_{i=1}^{|sh|} A_{1,i}$ and $A_2 = \sum_{i=1}^{|sh|} A_{2,i}$, respectively, and sum them up in the end to obtain the total Hamming weight $\text{wt}(sh) = A_1 + A_2$.

Secure Syndrome Checking. In order to test whether decoding of the input vector was successful, the syndrome has to be tested for zero. If the Hamming weight of the syndrome is zero, then all bits of the syndrome must be zero. Otherwise, there must be some bits set as 1 and the number of set bits equals the Hamming weight of the syndrome. Note that we perform SecHW operation over the three shares of syndrome s in order to prevent the leakage.

4   Implementing a Masked QC-MDPC McEliece

This section presents more details of the masked FPGA implementation of QC-MDPC McEliece decryption based on the unprotected one in [19]. We follow the structure of the original design, including the same security parameters, but replace vulnerable logic circuits with masked circuits.

4.1   Overview of the Masked Implementation

Each time before the decryption is started, both the ciphertext and the masked secret keys h_{0m}, h_{1m} are written into the BRAMs of the decryption engine. As shown in Fig. 1, one BRAM stores the $2 \cdot 4801$-bit ciphertext, the second BRAM stores the $2 \cdot 4801$-bit masked secret key and third BRAM stores the 4801-bit masked syndrome and the 4801-bit syndrome mask. Note that the secret keys are

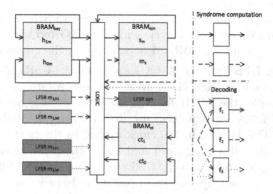

Fig. 1. Abstract block diagram of the masked QC-MDPC McEliece decryption implementation.

masked before being transferred to the crypto core. The seeds for the internal PRG are transferred with the masked key. Each BRAM is dual-ported, offers $18/36$ kBit, and allows to read/write two 32-bit values at different addresses in one clock cycle.

Computations are performed in the same order as in [19]: To compute the masked syndrome s_m, set bits in the ciphertext x select rows of the masked parity-check matrix blocks that are accumulated. In parallel, the syndrome mask m_s is computed in the same manner. Rotating the two parts of the secret key is implemented in parallel, as in the unprotected implementation. Efficient rotation is realized using the READ_FIRST mode of Xilinx's BRAMs which allows to read the content of a 32-bit memory cell and then to overwrite it with a new value, all within one clock cycle.

An abstraction of this implementation is depicted in Fig. 1. The three block RAMs are used to store the masked keys (h_{0m} and h_{1m}), the shared syndrome (s_m and m_s) and the ciphertext (ct_0 and ct_1). The LFSR blocks are used to generate the missing masks on-the-fly. The logic blocks for the two phases of the McEliece decryption are shown on the left side of Fig. 1.

4.2   Masking Syndrome Computation

The syndrome computation is a linear operation and requires only two shares for sensitive variables. Once the decryption starts, 32-bit blocks of the masked secret keys h_{0m}, h_{1m} are read from the secret key BRAM at each clock cycle and are XORed with the 32-bit block of s_m read from the syndrome BRAM depending on whether the corresponding ciphertext bits are 1. Then the result will be written back into the syndrome BRAM at the next clock cycle and at the same time the rotated 32-bit blocks of the masked keys will be written back into the secret key BRAM. Meanwhile, we need to keep track of the syndrome mask m_s. Since syndrome computation is a linear operation, we can similarly add up the secret key masks synchronously to generate the syndrome mask.

In our secure engine, we use two 32-bit leap forward LFSRs to generate random 32-bit secret key masks each clock cycle which are XORed with the 32-bit block of m_s read from the syndrome BRAM depending on the ciphertext.

Cyclic Rotating LFSRs. Our 32-bit leap forward LFSRs not only generate a 32-bit random mask at each clock cycle but also rotate synchronously with the key. For example, the LFSR for h_{0m} first needs to generate the 4801-bit mask m_{h_0} in the following sequence: $m_{h_0}[0:31], m_{h_0}[32:63], \ldots, m_{h_0}[4767:4799], m_{h_0}[4800]$. This is done in 150 clock cycles. In the next round, the secret key is rotated by one bit as $h_{0m} \ggg 1$ and hence the mask sequence should be: $m_{h_0}[4800:30], m_{h_0}[31:62], \ldots, m_{h_0}[4766:4798], m_{h_0}[4799]$. After 4801 rounds of rotation, the LFSR ends up with its initial state. In order to construct a cyclic rotating PRG with a period of 4801 bits, we combine a common 32-bit leap forward LFSR with additional memory and circuits, based on the observation that the next state of the LFSR either completely relies on the current state or actually sews two ends of the sequence together, e.g., $m_{h_0}[4800:30]$. As shown in Fig. 2, five 32-bit registers are employed instead of just one. The combinational logic circuit computes the next 32-bit random mask given the input stored in INTSTATEREG. The following steps describe the functionality of our LFSR:

1. Initially, the 32-bit seed *seed* $[0:31]$ of the sequence is stored in register IVREG and the first 32 bits of the sequence, e.g., $m_{h_0}[0:31]$ are stored in the other registers.
2. During the rotation, the combinational logic circuits output the new 32-bit result and feed it back. If the new result is part of the 4801-bit sequence, then it will go through the MUX, overwriting the current state registers INTSTATEREG and EXTSTATEREG at the next clock cycle.
3. If the new result contains bits that are not part of the sequence, then those bits will be replaced. For example, when $m_{h_0}[4767:4799]$ is in INTSTATEREG, the new result will be $m_{h_0}[4800:4831]$ in which only bit $m_{h_0}[4800]$ is in the mask sequence and $m_{h_0}[4801:4831]$ will be dropped. The MUX gate will only let $m_{h_0}[4800]$ go through together with $m_{h_0}[0:30]$ stored in EXTBIT$_{0\text{-}31}$ and the concatenation $m_{h_0}[4800:30]$ will overwrite register EXTSTATEREG.
4. $m_{h_0}[4800:30]$ will not be written into register INTSTATEREG because given $m_{h_0}[4800:30]$ as input, the combinational logic circuit will not output the next valid state $m_{h_0}[31:62]$. Therefore, we concatenate part of the seed in IVREG and part of the first 32-bits in INTBIT$_{0\text{-}31}$, e.g., $\{seed[31], m_{h_0}[0:30]\}$ and overwrite INTSTATEREG. Then, the new output will be $m_{h_0}[31:62]$. The concatenation is implemented as a cyclic bit rotation as shown in Fig. 2. After 32 rotations, the seed is rotated to INTBIT$_{0\text{-}31}$ and the first 32-bit $m_{h_0}[0:31]$ is rotated to IVREG. Hence, they will be swapped back in the next clock cycle.

To sum up, EXTSTATEREG always contains the valid 32-bit mask while INTSTATEREG always contains 32-bit input that results in the next valid state. The rotated secret key is generated in 150 clock cycles. After 4801×150 clock cycles, the LFSR returns to its initial state and idles.

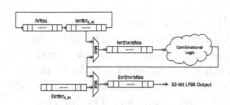

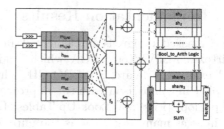

Fig. 2. The structure of the cyclic rotating LFSR that is used to generate the masks on-the-fly.

Fig. 3. Layout of our pipelined QC-MDPC McEliece decoder for the first part of the secret key, h_0.

4.3 Masking the Decoder

As mentioned in Sect. 3, the masked secret keys and the syndrome are extended to three shares. Hence, more LFSRs are instantiated to generate the additional shares as shown in Fig. 1. Two LFSRs generate the third shares of h_0 and h_1, another LFSR generates the third share of the syndrome.

We use h_0 as example to describe the decoder, since h_1 is processed in parallel using identical logic circuits. We split h_0 into three shares: h_{0m} stored in the BRAM and m_{1,h_0} and m_{2,h_0} generated by two LFSRs. The syndrome is split into s_m and m_{s_1} which are stored in BRAM and m_{s_2} which is generated by an LFSR. After decoding is started, each 32-bit share is read or generated at each clock cycle and then SecAND and SecHW are performed. This is implemented using a pipelined approach as shown in Fig. 3.

The left part of Fig. 3 illustrates the bitwise SecAND operation using Eq. (2). The 32-bit shares are fed into shared functions f_1, f_2, f_3, and the outputs are three 32-bit shares of the result. As mentioned before, two additional random vectors r_1, r_2 are required to mask the outputs in order to achieve uniformity. Our design uses only two fresh random bits b_1, b_2 together with the shifted input shares as the random vectors because the neighboring bits are independent of each other. That is $r_1 = \{b_1, m_{1,h_0}[0:30]\}$ and $r_2 = \{b_2, m_{2,h_0}[0:30]\}$. Both $m_{1,h_0}[31]$ and $m_{2,h_0}[31]$ are shifted out and are used as b_1 and b_2 in the next clock cycle. The right part shows the structure of SecHW. To compute the Hamming weight of the unmasked result $sh_1 \oplus sh_2 \oplus sh_3$ without leaking side channel information, a parallel counting algorithm is applied to accumulate the weight of each bit position of the word. We use 32×2 6-bit Arithmetic masked counters[1] and each bit in the word $sh_1 \oplus sh_2 \oplus sh_3$ will be added into the corresponding counter during each clock cycle. More specifically, the three shares of each bit of sh are converted and added into the two Arithmetic masked counters. After 150 clock cycles, we sum the overall Arithmetic masked Hamming weight. To convert and accumulate the masked weights, we employ the secure conversion method developed in [7].

---

[1] Note that the Hamming weight of $s \wedge H$ is bounded to the weight of h_i, i.e., wt$(s \wedge h_i) \leq w/2 = 45$, i.e., 6-bit registers are always sufficient.

5   Implementation Results

The masked design is implemented in VHDL and is synthesized for Xilinx Virtex-5 XC5VLX50 FPGA which holds the crypto engine in the side channel evaluation board SASEBO-GII. The implementation results are listed in Table 1 in comparison with the unprotected implementation of [19]. In terms of Flip-Flops (FFs) and Look-Up Tables (LUTs), the masked implementation uses 8 times as many resources as the unprotected implementation. The increase is mainly due to the masked Hamming weight computation which requires many registers to store the Hamming weights of small chunks. Moreover, the leap forward LFSR also utilizes many Flip-Flops and has to be instantiated five times in our design. The number of occupied BRAMs remains constant, only the occupied memory within the syndrome BRAM increases by a factor of 2 in the masked implementation because the syndrome masks are also stored in this BRAM. The performance of the masked design is compromised for security and the maximum clock frequency is reduced by a factor of 4.3. This is mainly because the addition of 32 6-bit weight registers in SecHW is done in one clock cycle resulting a long critical path and in turn a low clock frequency. Shortening the critical path can be an interesting goal in future work. Note that the number of clock cycles remains the same as for the unprotected implementation, unless the early termination of the decoder is disabled, in which case the average run time doubles compared to [19] (assuming that the maximum number of iterations is set to 5 similarly to [20], with early termination enabled the decoder requires 2.4 iterations on average as was shown in [10,21]). The resulting mean overhead of our implementation is 4, which is in line with other masked implementations[2]. The TI AES engine in [14] introduces an area overhead of a factor 4 as well, but that implementation does not include the pseudorandom generators needed to generate the 48 bits of randomness consumed per cycle, while ours does.

Table 1. Resources usage comparison between the unprotected and masked implementations on Xilinx Virtex-5 XC5VLX50 FPGAs.

Implementation	FFs	LUTs	Slices	BRAMs	Frequency
Unprotected [19]	412	568	148	3	318 MHz
Masked	3045	4672	1549	3	73 MHz
Overhead factor	7.4x	8.2x	10.5x	1x	4.3x

6   Leakage Analysis

Next we analyze the implementation for remaining leakage. We first apply the DPA presented in [6] on the protected implementation. Next we use the leakage detection methodology developed in [9] to detect any other potentially

---

[2] When computing the geometric mean of the overhead of the three hardware components (LUTs, FFs, and BRAMs), the resulting area overhead is actually 3.9.

exploitable leakages. The evaluated implementation is placed on the Xilinx Virtex-5 XC5VLX50 FPGA of the SASEBO-GII board. The power measurements are acquired using a Tektronix DSO 5104 oscilloscope. The board was clocked at 3 MHz and the sampling rate was set to 100 M samples per second. In order to quantify the resilience of our masked implementation to power analysis attacks, we collected 10,000 measurements using the same ciphertext but two different sets of secret keys. The first set is actually 5,000 repetitions of a fixed key while the second set contains 5,000 random keys. The two sets of keys are fed into the decryption engine alternatingly.

6.1 Differential Power Analysis

A Differential Power Analysis on the FPGA implementation of QC-MDPC McEliece of [19] was presented in [6]. The attack exploits the leakage caused by the key rotation in the syndrome computation phase. The 4801-bit keys h_0 and h_1 rotate in parallel for 4801 rounds, each round lasts for 150 clock cycles. Thus during one decryption, each key bit is rotated into the one bit carry register 150 times which results in a strong leakage. By averaging the 150 leakage samples for each key bit, one can generate the 4801-sample differential trace which contains features caused by the set key bits and then one can recover the value of the key bits by interpreting the features. For the unprotected implementation, the secret key can be completely recovered using the average differential trace of only 10 measurements. For more details about the key recovery we refer to [6]. In contrast to the unprotected implementation, no features are present in the differential trace of the fixed secret key (red line) even with 500 times more traces, as shown in Fig. 4. Hence, the key bit value cannot be recovered. The peaks in the trace are not the features caused by set key bits because in the differential trace of the random secret keys where the key bits are randomly set as 1 the same peaks appear. Thus, they cannot be used as features to recover secret key bits as done in [6]. The two differential traces almost overlap, showing that the leakage is indistinguishable between fixed key and random key when using a masked implementation.

6.2 Leakage Detection

We employ Welch's T-test suite to quantify the leakage indistinguishability between two sets of secret keys. Welch's T-test is a statistical hypothesis test used to decide whether the means of two distributions are the same. T-statistic t can be computed as:

$$t = \frac{\overline{X} - \overline{Y}}{\sqrt{\frac{s_X{}^2}{N_X} + \frac{s_Y{}^2}{N_Y}}} \tag{5}$$

where $\overline{X}, \overline{Y}$ are the sample means of random variables X, Y, s_X, s_Y are the sample variances and N_X, N_Y are the sample sizes. The pass fail criteria is

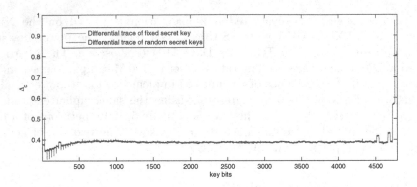

Fig. 4. Comparison between two differential traces of two sets of secret keys.

defined as $[-4.5, 4.5]$ as developed in [9]. In our case, we obtained two groups of leakage samples, one for the fixed key set and the other for the random key set. Each group has $5,000$ power traces as well as $5,000$ derived differential traces. We first performed the T-test using the original power traces and Fig. 5.1 shows the t-statistics along the whole decryption. The t-statistics are within the range of $[-4.5, 4.5]$ which implies a confidence of more than 99.999% for the null hypothesis showing that the two sample groups are indistinguishable.

To assess the vulnerability to first-order horizontal attacks, we also performed a T-test on the derived differential traces. The results are shown in Fig. 5.2. Similarly, the t-statistics are also within the predefined range and it validates the indistinguishability between the two sets of secret keys. Hence, it can be concluded that the design does not contain any remaining first-order leakage of the key.

6.3   Masking the Ciphertext?

The decoder corrects errors in the ciphertext x, eventually yielding the plaintext derived value $m \cdot G$ and thereby implicitly the error vector e. Similarly, the values d and $\#_{\mathrm{upc}}$ assigned in line 3 and 4 of Algorithm 1 are not masked and can potentially reveal the error locations, hence e. In either case, the equivalent leakage of information of e or $m \cdot G$ is possible. In our implementation, we chose not to mask x and its intermediate state, nor d and $\#_{\mathrm{upc}}$. This choice is justifiable for two reasons. First, both e or $m \cdot G$ are key-independent and will not reveal information about the secret key. Furthermore, e or $m \cdot G$ are ciphertext dependent, that is, any information that can be revealed will be only valid for the specific encrypted message. Hence, if such information is to be discovered, it must be recovered using SPA-like approaches. More explicitly, the only possible attack is a message recovery attack, and that requires SPA techniques, as, e.g., applied in [20]. Nevertheless, d and $\#_{\mathrm{upc}}$ are variables that have dependence on both the ciphertext and the key, just as the number of decoding iterations that might be revealed by a non-constant time implementation, i. e., if the decoding algorithm

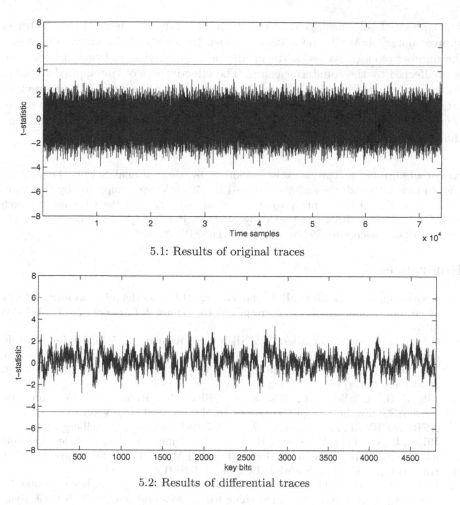

5.1: Results of original traces

5.2: Results of differential traces

Fig. 5. T-test between the two groups of *original* power traces (5.1) and *differential* power traces (5.2) corresponding to the two sets of secret keys. Both cases indicate the absence of leakage for the given number of traces.

tests the syndrome for zero after each decoding iteration and exits when this condition is reached. However, up to now there is no evidence suggesting that their information can be used to perform key recovery attacks. We leave this as an open question for future research.

7   Conclusion

This work presents the first masked implementation of a McEliece cryptosystem. While masking the syndrome computation is straightforward and comes at a low overhead, the decoding algorithm requires more involved masking techniques.

Through on-the-fly mask generation, the area overhead is limited to a factor of approximately 4. While the maximum clock frequency of the engine decreases, the number of clock cycles for the syndrome computation and each decoder run is unaffected by the countermeasures. The effectiveness of the applied masking has been analyzed by leakage detection methods and by showing that previous attacks do not succeed anymore. Exploring if any information about the secret key can be derived from the number of decoding iterations leaves an interesting challenge for future work.

Acknowledgments. This work is supported by the National Science Foundation under grant CNS-1261399 and grant CNS-1314770. IvM was supported by the European Union H2020 PQCrypto project (grant no. 645622) and the German Research Foundation (DFG). RS is supported by NATO's Public Diplomacy Division in the framework of "Science for Peace", Project MD.SFPP 984520.

References

1. Berlekamp, E.R., McEliece, R.J., van Tilborg, H.C.: On the inherent intractability of certain coding problems (corresp.). IEEE Trans. Inf. Theor. **24**(3), 384–386 (1978)
2. Bilgin, B., Daemen, J., Nikov, V., Nikova, S., Rijmen, V., Van Assche, G.: Efficient and first-order DPA resistant implementations of Keccak. In: Francillon, A., Rohatgi, P. (eds.) CARDIS 2013. LNCS, vol. 8419, pp. 187–199. Springer, Heidelberg (2014)
3. Bilgin, B., Gierlichs, B., Nikova, S., Nikov, V., Rijmen, V.: A more efficient AES threshold implementation. In: Pointcheval, D., Vergnaud, D. (eds.) AFRICACRYPT. LNCS, vol. 8469, pp. 267–284. Springer, Heidelberg (2014)
4. Bilgin, B., Gierlichs, B., Nikova, S., Nikov, V., Rijmen, V.: Higher-order threshold implementations. In: Sarkar, P., Iwata, T. (eds.) ASIACRYPT 2014, Part II. LNCS, vol. 8874, pp. 326–343. Springer, Heidelberg (2014)
5. Bos, J.W., Costello, C., Naehrig, M., Stebila, D.: Post-quantum key exchange for the TLS protocol from the ring learning with errors problem. In: 36th IEEE Symposium on Security and Privacy (2015)
6. Chen, C., Eisenbarth, T., von Maurich, I., Steinwandt, R.: Differential power analysis of a McEliece cryptosystem. In: Malkin, T., et al. (eds.) ACNS 2015. LNCS, vol. 9092, pp. 538–556. Springer, Heidelberg (2015). doi:10.1007/978-3-319-28166-7_26. Preprint http://eprint.iacr.org/2014/534
7. Coron, J.-S., Großschädl, J., Vadnala, P.K.: Secure conversion between boolean and arithmetic masking of any order. In: Batina, L., Robshaw, M. (eds.) CHES 2014. LNCS, vol. 8731, pp. 188–205. Springer, Heidelberg (2014)
8. Gallager, R.: Low-density Parity-check Codes. IRE Trans. Inf. Theor. **8**(1), 21–28 (1962)
9. Gilbert Goodwill, B.J., Jaffe, J., Rohatgi, P., et al.: A testing methodology for side-channel resistance validation. In: NIST Non-invasive Attack Testing Workshop (2011)
10. Heyse, S., von Maurich, I., Güneysu, T.: Smaller keys for code-based cryptography: QC-MDPC McEliece implementations on embedded devices. In: Bertoni, G., Coron, J.-S. (eds.) CHES 2013. LNCS, vol. 8086, pp. 273–292. Springer, Heidelberg (2013)

11. Huffman, W.C., Pless, V.: Fundamentals of Error-Correcting Codes. cambridge University Press, Cambridge (2010)
12. McEliece, R.J.: A public-key cryptosystem based on algebraic coding theory. Deep Space Netw. Prog. Rep. **44**, 114–116 (1978)
13. Misoczki, R., Tillich, J.-P., Sendrier, N., Barreto, P.: MDPC-McEliece: new McEliece variants from moderate density parity-check codes. In: Proceedings of the IEEE International Symposium on Information Theory (ISIT), pp. 2069–2073. IEEE (2013)
14. Moradi, A., Poschmann, A., Ling, S., Paar, C., Wang, H.: Pushing the limits: a very compact and a threshold implementation of AES. In: Paterson, K.G. (ed.) EUROCRYPT 2011. LNCS, vol. 6632, pp. 69–88. Springer, Heidelberg (2011)
15. Nikova, S., Rechberger, C., Rijmen, V.: Threshold implementations against side-channel attacks and glitches. In: Ning, P., Qing, S., Li, N. (eds.) ICICS 2006. LNCS, vol. 4307, pp. 529–545. Springer, Heidelberg (2006)
16. Schneider, T., Moradi, A., Güneysu, T.: Arithmetic addition over boolean masking. In: Malkin, T., et al. (eds.) ACNS 2015. LNCS, vol. 9092, pp. 559–578. Springer, Heidelberg (2015). doi:10.1007/978-3-319-28166-7_27
17. Shahverdi, A., Taha, M., Eisenbarth, T.: Silent SIMON: a threshold implementation under 100 slices. In: Proceedings of IEEE Symposium on Hardware Oriented Security and Trust (HOST) (2015)
18. Shor, P.W.: Polynomial-time algorithms for prime factorization and discrete logarithms on a quantum computer. SIAM J. Comput. **26**(5), 1484–1509 (1997)
19. von Maurich, I., Güneysu, T.: Lightweight code-based cryptography: QC-MDPC McEliece encryption on reconfigurable devices. In: Design, Automation and Test in Europe - DATE, pp. 1–6. IEEE (2014)
20. von Maurich, I., Güneysu, T.: Towards side-channel resistant implementations of QC-MDPC McEliece encryption on constrained devices. In: Mosca, M. (ed.) PQCrypto 2014. LNCS, vol. 8772, pp. 266–282. Springer, Heidelberg (2014)
21. von Maurich, I., Oder, T., Güneysu, T.: Implementing QC-MDPC McEliece encryption. ACM Trans. Embed. Comput. Syst. **14**(3), 44:1–44:27 (2015)

Fast and Memory-Efficient Key Recovery in Side-Channel Attacks

Andrey Bogdanov[1], Ilya Kizhvatov[2(✉)], Kamran Manzoor[1,2,3],
Elmar Tischhauser[1], and Marc Witteman[2]

[1] Department of Applied Mathematics and Computer Science,
Technical University of Denmark, Kongens Lyngby, Denmark
{anbog,ewti}@dtu.dk
[2] Riscure B.V., Delft, The Netherlands
{kizhvatov,witteman}@riscure.com
[3] Aalto University, Espoo, Finland
kamran.manzoor@aalto.fi

Abstract. Side-channel attacks are powerful techniques to attack implementations of cryptographic algorithms by observing its physical parameters such as power consumption and electromagnetic radiation that are modulated by the secret state. Most side-channel attacks are of divide-and-conquer nature, that is, they yield a ranked list of secret key chunks, e.g., the subkey bytes in AES. The problem of the key recovery is then to find the correct combined key.

An optimal key enumeration algorithm (OKEA) was proposed by Charvillon et al. at SAC'12. Given the ranked key chunks together with their probabilities, this algorithm outputs the full combined keys in the optimal order – from more likely to less likely ones. OKEA uses plenty of memory by its nature though, which limits its practical efficiency. Especially in the cases where the side-channel traces are noisy, the memory and running time requirements to find the right key can be prohibitively high.

To tackle this problem, we propose a score-based key enumeration algorithm (SKEA). Though it is suboptimal in terms of the output order of candidate combined keys, SKEA's memory and running time requirements are more practical than those of OKEA. We verify the advantage at the example of a DPA attack on an 8-bit embedded software implementation of AES-128. We vary the number of traces available to the adversary and report a significant increase in the success rate of the key recovery due to SKEA when compared to OKEA, within practical limitations on time and memory. We also compare SKEA to the probabilistic key enumeration algorithm (PKEA) by Meier and Staffelbach and show its practical superiority in this case.

SKEA is efficiently parallelizable. We propose a high-performance solution for the entire conquer stage of side-channel attacks that includes SKEA and the subsequent full key testing, using AES-NI on Haswell Intel CPUs.

© Springer International Publishing Switzerland 2016
O. Dunkelman and L. Keliher (Eds.): SAC 2015, LNCS 9566, pp. 310–327, 2016.
DOI: 10.1007/978-3-319-31301-6_19

1   Introduction

In a Nutshell. Side-channel attacks are powerful techniques to extract cryptographic keys from secure chips. With a few exceptions, they are divide-and-conquer attacks recovering individual small chunks of the key. For every chunk of the key, the attack's divide phase called a *distinguisher* will assign a ranking value to every possible candidate. Depending on the attack technique, this value can be a correlation coefficient or a probability value. The output of the distinguisher is a set of lists of all key chunk candidates and their ranking values.

If the attacker can perform a sufficient number of measurements to reduce noise, the *conquer* phase is trivial. The correct full key is obtained by taking the top-ranking candidate. However, this is rare in practical settings with low signal-to-noise ratio due to target complexity and countermeasures, and limited time to perform the attack. In such practical settings, the correct key chunks are not likely to be ranked on the top of the list, but somewhere close to the top. Hence, to recover the full key, the attacker needs to perform the search over different combinations of key chunk candidates. By taking into account the ranking values coming from the distinguisher, the search can be made more efficient. The problem is known as *key enumeration*. More generally, besides key enumeration, the conquer phase also includes key verification.

Related Work and Motivation. Though most research in side-channel attacks focuses on the divide phase, there are several papers tackling the issue of key enumeration. The most recent line of relevant research is the optimal key enumeration algorithm (OKEA) proposed Charvillon et al. in SAC'12 [14,15]. It is optimal with respect to the order of the full keys by their probabilities. OKEA is superior to the very similar algorithm of Pan et al. from SAC'10 [10] in terms of memory use, and to the probabilistic key enumeration algorithm (PKEA) suggested by Meier and Staffelbach in EUROCRYPT'91 [9] (and well applicable in the context of DPA and DFA) in terms of the amount of candidates tried. A dedicated solution for the case of an attack recovering individual bits was proposed by Dichtl in COSADE'10 [3]. The papers [8,16] include solutions to tackle the conquer phase as well.

The optimal key enumeration algorithm from [15] has some practical limitations though. First, for practical cases, it consumes a lot of memory and its performance is hampered by random memory access pattern. Second, it is difficult to efficiently parallelize on a multi-core CPU[1]. Thus, we are more interested in the practical optimality, i.e., an optimality in terms of how much time on average it takes to find the key on a modern CPU. Previous papers lack a systematic comparison with respect to this metric.

---

[1] We focus on a common non-specialized platform, namely, on a workstation with a multi-core CPU. We are fully aware that dedicated platforms, in particular GPUs and FPGAs, may bring faster smart brute force solutions; they are out of scope of this work.

In an independent work [7], a parallelizable score-based key enumeration algorithm is presented. It casts the key enumeration problem as a solution to a multi-dimensional integer knapsack, which is achieved by listing all valid paths in the respective graph representation. The algorithm is shown to be more efficient than OKEA. The approach also provides efficient key rank estimation by counting paths in the graph.

Quality Versus Speed. Two aspects are of great importance when searching for a key: *quality* and *speed*. With *quality* we mean the relation between candidate generation and the correctness probability of a key candidate. A brute force algorithm that tries every key candidate in the plain numerical order is likely to have a bad quality. This is because probable key candidates will not get precedence over unprobable candidates. With *speed* we mean how fast subsequent key candidates are generated. Typically the key verification can be done with optimized hardware or software implementations of the cryptographic algorithm. Ideally, the enumeration should be faster than the verification to benefit from the optimization of the cryptographic implementation.

The overall performance of the approach, i.e. *the time it takes to find the correct key*, depends on both quality and speed. In this paper we look at both factors and find that a high quality may reduce speed and suboptimality may minimize the search time.

Our Contributions. The contributions of this paper are as follows. First, we propose a new key enumeration algorithm, namely, the score-based key enumeration algorithm (SKEA), that can find the full key in practice faster than OKEA and can be efficiently parallelized. Second, we provide a comprehensive comparison of OKEA, PKEA and SKEA in terms of the practical runtime metric for DPA on an embedded 8-bit software target. Finally, we propose a full solution for the conquer stage (including key enumeration and testing) for AES that includes an efficient key verification implementation using AES-NI extensions on Haswell CPUs.

Thus, our work aims to bridge the gap between theory and practice in key enumeration and puts forward the importance of the practical metrics for estimating implementation attack complexity.

Part of this work has been first presented in [1,6].

2   Background on Key Enumeration Algorithms

Most side-channel attacks are based on a divide-and-conquer approach in which a side-channel distinguisher is used to obtain ranking information about parts (chunks) of the key ("divide") which is then used to determine candidates for the complete key ("conquer"). This complete key is the result of a function, e.g. concatenation, of all the key parts. A key part candidate is a possible value of a key part that is chosen because the attack suggests a good probability for that value to be correct.

For instance, in a side-channel attack on the S-box step of a block cipher, such a key part comprises all key bits corresponding to a single S-box. The complete key recovered by the attack would be the concatenation of all such key parts in one round.

The problem of finding (candidates for) the complete key based on the observed side-channel information for the key parts is called the *key enumeration problem*: For each of the m key parts, we are given a list of key part candidates, each of which has a certain associated side-channel information (e.g., leakage). We are then to enumerate candidates for the complete key, ideally in order of likelihood according to the likelihood of the individual key part candidates.

Several key enumeration algorithms have been proposed, among them the classic probabilistic algorithm (PKEA) due to Meier and Staffelbach [9], peak distribution analysis by Pan et al. [10], and the recent optimal key enumeration (OKEA) by Veyrat-Charvillon et al. from SAC 2012 [15].

Ranking vs. Likelihood. Both PKEA and OKEA require the discrete probability distribution of each subkey part to be known and computable by the attacker. In profiled side-channel attacks, the training phase allows the attacker to rank the key part candidates k for each of the m key parts according to their actual probabilities $\Pr(k \mid L_i)$, with L_i denoting the complete available side-channel information for the i-th key part. In non-profiled attacks, the ranking of key part candidates is performed according to certain statistics extracted from the side-channel information, which might not reflect their actual likelihood. In [15], a method called *Bayesian extension* has been proposed in order to convert the ranking statistics into an equivalent probability value based on a certain leakage model. It essentially does a Bayesian model comparison to obtain the conditional probabilities $\Pr(k \mid L_i)$ from the probabilities $\Pr(L_i \mid k)$ which are obtained according to the leakage model.

2.1 Probabilistic Key Enumeration (PKEA)

At EUROCRYPT 1991, Meier and Staffelbach [9] describe a probabilistic key enumeration algorithm which we denote PKEA. In their work, the attacker is assumed to have no knowledge about the key part distributions, but is able to sample from them. The algorithm then simply samples key parts according to their individual distributions and tests the resulting complete key candidates.

Besides its simplicity, one of its main advantages is that it only requires a small constant amount of memory. On the other hand, the most likely key candidates will occur many times due to the stateless nature of this algorithm, which leads to unnecessary work duplication.

2.2 Optimal Key Enumeration (OKEA)

At SAC 2012, Veyrat-Charvillon et al. proposed a deterministic key enumeration algorithm which guarantees to output the key candidates in nondecreasing order of their likelihood [15]. Due to this property, it is referred to as an optimal key enumeration algorithm, or OKEA.

This algorithm is based on the idea that if the key part candidates for the m key parts are given in decreasing order of their probability, the m-dimensional space of all complete key candidates can be enumerated in order of their likelihood by only keeping track of the next most likely candidate in each dimension (this set of key part candidates is called the *Frontier set*). By pruning already enumerated complete key candidates and following a recursive decomposition for higher dimensions than two, this algorithm generates key candidates in optimal order without having to keep all possible key part combinations in memory at any time.

While constituting a significant improvement over both PKEA and the naive enumeration approach, OKEA is usually memory-bound, which means that even though keys are enumerated in optimal order, the memory required to keep track of the (nested) Frontier sets can be prohibitive in practice before the correct key is found, especially for higher dimensions such as $m = 16$ for attacks on the AES. Furthermore, the way the recursive decomposition is performed and the intermediate results are recombined also makes effective parallelization of OKEA difficult.

3   Our Score-Based Algorithm (SKEA)

In this section, we introduce SKEA, our proposed score-based algorithm for the key enumeration problem. Like OKEA, it is a deterministic algorithm. Instead of guaranteeing an optimal enumeration order, it is intended to be optimized for practical applicability by only employing a fixed maximum amount of memory (as opposed to OKEA's dynamically increasing memory usage) and by being easily parallelizable.

3.1   Score Concepts

SKEA works with *scores* which are assigned to each key part candidate based on the likelihood that it will be part of the correct complete key according to the available side-channel information.

Scores are integer values, and can be derived for instance from a correlation value in a side-channel analysis attack, or the number of faults that match with a candidate key part value in differential fault attack. SKEA can deal with scores assigned from an arbitrary integer interval $[s_{\min}, s_{\max}]$, and the size of this interval will typically be chosen according to the required precision, since a larger interval means more distinctive possible scores.[2]

---

[2] For a standard power analysis attack on an 8-bit S-box in AES, computing the score as $\lfloor 50 |\rho| + 0.5 \rfloor$ for a correlation coefficient of ρ, resulting in an interval size of 50, was found to yield good results (see Sect. 5).

We furthermore define the *cumulated score* as the sum of all the key part scores comprising one complete key. SKEA generates candidate keys by descending cumulated scores. It starts by generating the key with the largest possible cumulated score S_{max}, and proceeds by finding all keys with a score equal to $S_{max} - 1$, and so forth. This property guarantees a certain quality, i.e. good key candidates will be enumerated earlier than worse ones. Note that candidates with the same score are generated in no particular order, so a lack of precision in scores will lead to some decrease of quality.

An Example with $m = 4$. For simplicity, we will explain the score concept for $m = 4$ key parts with 4 candidates each (corresponding to 4 key parts of 2 bits each). This will also serve as a running example to illustrate the SKEA algorithm. Figure 1 shows a simple example of a score table. The key consists of columns representing 4 key parts that can have only 4 values. The cells in the columns contain the sorted scores. Row 0 shows the best scores for all key parts, while row 3 shows the worst scores. A complete key candidate consists of the concatenation of selected cells for all columns, having a cumulated score equal to the sum of the individual scores for each key part. We directly see that the best cumulated score would be 19, resulting from the sequence $\{0, 0, 0, 0\}$, while the worst cumulated score would be 3, resulting from the sequence $\{3, 3, 3, 3\}$.

Key part $\longrightarrow$

Score $\downarrow$	0	1	2	3
0	5	5	4	5
1	2	3	2	4
2	1	1	0	3
3	0	1	0	2

Fig. 1. Example of a score table for $m = 4$ and a 2-bit S-box.

3.2 Enumerating Score Paths

SKEA now makes use of the score table in order to enumerate key candidates in order of decreasing cumulated scores. We refer to a combination of key part candidates with cumulated score equal to s to a possible *score path* for score s. Given a best possible cumulated score of S_{max}, we then must find all possible score paths leading to a cumulated score of S_{max}, followed by $S_{max} - 1$, and so on.

In order to find valid score paths conforming to a certain cumulated score, SKEA makes use of a simple backtracking strategy. This algorithm essentially performs a depth-first search, which is efficient because impossible paths can be pruned early. The pruning is controlled by the score that must be reached for the solution to be accepted.

19

5	5	4	5
2	3	2	4
1	1	0	3
0	1	0	2

18

5	5	4	5
2	3	2	4
1	1	0	3
0	1	0	2

17

5	5	4	5		5	5	4	5		5	5	4	5
2	3	2	4		2	3	2	4		2	3	2	4
1	1	0	3		1	1	0	3		1	1	0	3
0	1	0	2		0	1	0	2		0	1	0	2

16

5	5	4	5		5	5	4	5		5	5	4	5		5	5	4	5
2	3	2	4		2	3	2	4		2	3	2	4		2	3	2	4
1	1	0	3		1	1	0	3		1	1	0	3		1	1	0	3
0	1	0	2		0	1	0	2		0	1	0	2		0	1	0	2

Fig. 2. Possible score paths conforming to cumulated scores of $19, 18, 17$ and 16.

We continue our toy example with $m = 4$ and 2-bit key parts from Sect. 3.1 by listing all possible score paths s for cumulated scores of 19, 18, 17, and 16 in Fig. 2.

3.3 Efficiently Eliminating Incompatible Score Paths

To achieve a fast decision process during the backtracking, SKEA makes use of precomputed tables containing possible ranges of key part selection indices that may lead to score paths with the desired cumulated score. Suppose we are given a score table $S[i][k]$ containing the scores of candidate k for the i-th key part.

Cumulated Minimum and Maximum Scores. First we compute tables for minimal and maximal cumulated scores that can be reached by completing a path to the right:

$$\text{cMin}[i] = \min_k \sum_{j=i}^{m-1} S[j][k], \quad 0 \le i < m,$$

$$\text{cMax}[i] = \max_k \sum_{j=i}^{m-1} S[j][k], \quad 0 \le i < m.$$

When completing a score path from left to right, the cumulated minimum and maximum score table gives the minimum and maximum score that can be added by including cells from the current column, and to the right.

cMin

3	3	2	2

cMax

19	14	9	5

Fig. 3. Minimum and maximum cumulated scores

For our running example, these tables are given in Fig. 3. They show that by adding only the rightmost column, a minimum of 2, and a maximum of 5 can be added to the score. Likewise, by adding the 2nd, 3rd, and 4th columns a minimum of 3, and a maximum of 14 can be added to the cumulated score.

Key Part Index Limits. Using the cumulated minimum and maximum tables, we create two additional tables that hold the indices for completing the path successfully for a certain desired cumulated score, when proceeding from a score path up to a given key part index:

$$iMin[i][s] = \arg\min_k (S[i][k] + cMin[i+1]) \geq s,$$

$$iMax[i][s] = \arg\max_k (S[i][k] + cMax[i+1]) \leq s,$$

$$0 \leq i < m, cMin[i] \leq s \leq cMax[i], cMin[m] = cMax[m] = 0.$$

Cumulated score	Key part 0			Key part 1			Key part 2			Key part 3		
		Min	Max		Min	Max		Min	Max		Min	Max
	19	0	0	14	0	0	9	0	0	5	0	0
	18	0	0	13	0	0	8	0	0	4	1	1
	17	0	0	12	0	1	7	0	1	3	2	2
	16	0	1	11	0	1	6	0	1	2	3	3
	15	0	2	10	0	3						

Fig. 4. Minimum and maximum indices

In other words, these tables represent for a desired score which minimal and maximal cell indices in the given column can be used.

For our running example, the corresponding tables are given in Fig. 4. For instance, if a score of 12 is desired from key part 1 onward, only the candidate key parts in row 0 and 1 will match.

Altogether, the preparation of these tables enables a fast pruning of impossible paths, and speeds up the selection of proper candidates.

3.4   The SKEA Algorithm

The full SKEA algorithm can then be presented as follows. Given the number of key parts m and a score table $S[i][k]$ for all key part candidates k for key part number i, we first use Algorithm 3.1 to precompute the tables containing the cumulated minima, maxima and corresponding index selection limits. These tables are then used for efficient pruning in a depth-first search identifying all possible score paths conforming to a cumulated score equal to $S_{max}, S_{max} - 1, \ldots$ and so forth, see Algorithm 3.2, where line 2 can be efficiently parallelized.

Algorithm 3.1. Generating score bounds for SKEA

Input: Number m of key parts
Input: Score table $S[i][k]$ holding score of candidate k for key part i $(0 \leq i < m)$
Output: $cMin[i]$, the global cumulative minima per key part index: For $0 \leq i < m$,
$cMin[i] = \min_k \sum_{j=i}^{m-1} S[j][k]$, $cMin[m] = 0$.
Output: $cMax[i]$, the global cumulative maxima per key part index: For $0 \leq i < m$,
$cMax[i] = \max_k \sum_{j=i}^{m-1} S[j][k]$, $cMax[m] = 0$.
Output: $iMin[i][s]$, the minimum index for attaining a certain minimum score s start-
ing from key part i: $iMin[i][s] = \arg\min_k (S[i][k] + \text{cMin}[i+1]) \geq s$.
Output: $iMax[i][s]$, the maximum index for attaining a certain minimum score s start-
ing from key part i: $iMax[i][s] = \arg\max_k (S[i][k] + \text{cMax}[i+1]) \leq s$.

Algorithm 3.2. Scored-based key enumeration algorithm (SKEA)

Input: Number m of key parts
Input: Score table $S[i][k]$ holding score of candidate k for key part i $(0 \leq i < m)$
Output: List $\mathcal{L}$ of complete key candidates
1:   $cMin, cMax, iMin, iMax \leftarrow$ Bounds(S) (using Alg. 3.1)
2:   **for** $s = cMax[0]$ **downto** $cMin[0]$ **do**
3:     $i \leftarrow 0$
4:     $k[0] \leftarrow iMin[0][s]$
5:     $cs \leftarrow s$
6:     **while** $i \geq 0$ **do**
7:       **while** $i < m - 1$ **do**
8:         $cs \leftarrow cs - S[i][k[i]]$
9:         $i \leftarrow i + 1$
10:        $k[i] \leftarrow iMin[i][cs]$
11:       **end while**
12:       add complete key candidate k to $\mathcal{L}$
13:       **while** $i \geq 0$ **and** $k[i] \geq iMax[i][cs]$ **do**
14:         $i \leftarrow i - 1$
15:         **if** $i \geq 0$ **then**
16:           $cs \leftarrow cs + S[i][k[i]]$
17:         **end if**
18:       **end while**
19:       **if** $i \geq 0$ **then**
20:         $k[i] \leftarrow k[i] + 1$
21:       **end if**
22:     **end while**
23: **end for**

---

4   Comparison Methodology and Setting

The primary goal of the comparison is to understand which enumeration algorithm finds the full key faster on a typical workstation.

4.1   Methodology

We perform comparison between OKEA, PKEA, and SKEA in an experimental setting. We run the algorithm implementations on a sample of lists obtained from CPA of a SW AES implementation. We register the running time for finding the correct full key. From the collected running times we estimate success rate [12] of the full key recovery as a function of the running time and the number of traces used to obtain the lists. This is our main comparison metric.

To make the comparison feasible, we limit the maximum running time and maximum memory available to the algorithm implementations. The maximum running time is set to 10 days on a single CPU core. The maximum memory per core is set to 12 GB (the limitation imposed by the configuration of the computing cluster we have at hand).

We note that the imposed limits introduce right censoring to the data we measure. Therefore, computing obvious statistics such as average time to find the key is not straightforward. This is why we choose to use success rate for comparison. Additionally, we analyze and compare the histograms of the obtained data.

In addition to the execution time, we also register and analyze other parameters: the number of key candidates tried and the amount of memory required. This lets us better understand the behavior of the algorithms.

At this point we compare only the enumeration running time. Full key verification in the experiment is done by comparison to the known correct full key. Full key verification for the real-life unknown key setting is addressed in Sect. 6.

4.2   Experimental Setting

1. **Obtaining the lists.** We obtain the lists of correlation coefficients from CPA of a SW AES implementation RijndaelFurious [11] running on an 8-bit AVR microcontroller[3]. We acquire 2000 power consumption traces and perform 20 independent attacks with up to 100 traces. Each attack is performed in an incremental way: we obtain the lists for the number of traces varying from 10 to 100. Each attack for a given number of traces results in a set of 16 lists with 256 correlation values each, together with the corresponding key byte candidate values.
2. **Converting the lists.** We convert the correlation values according to the algorithm requirements. For OKEA and PKEA we use the Bayesian extension (see Sect. 2). For SKEA we use conversion to integer scores (see Sect. 3.1).

---

[3] The target is chosen only for the sake of efficient comparison; real enumeration on this target is hardly needed because very small amount of traces is required for successful CPA.

3. **Algorithm implementations.** For OKEA, we use the open-source implementation [13]. For PKEA and SKEA, we use our own implementations in C. All the three implementations are single-threaded. Through a unified wrapper, an implementation takes a set of 16 per-key lists and returns the running time, the maximum memory used, and the number of key candidates tried in case the correct key is found within the defined time and memory limits. Otherwise, an out-of-time or out-of-memory notification is returned. We build all the three algorithm implementations for x86_64 Linux with `gcc` using full optimization (`-O3`).

4. **Running the algorithms.** We deploy the algorithms to a computing cluster. The cluster consists of 42 IBM NeXtScale nx360 M4 nodes each having 2 Intel Xeon E5-2680 v2 CPUs and 128 GB RAM, and 64 HP ProLiant SL2x170z G6 nodes each having 2 Intel Xeon X5550 CPUs and 24 GB RAM. To ensure sufficient amount of memory available, we run one experiment instance per node. We run enumeration experiments for the number of traces starting from 60 down to 42 in steps of 2 or 3. With more than 60 traces, enumeration is not required as the full key is successfully recovered by combining the topmost key byte candidates from the lists. With less than 42 traces, the algorithms fail to recover the key in less than 10 core-days for all the 20 experimental samples. For OKEA, all the unsuccessful cases are due to out-of-memory error and not due to the 10-day limit.

5   Comparison Results

First, we visualize the success rate of the full key recovery versus number of traces for the attack in Fig. 5. The figure shows success rate for different time boundaries: 10 days (the maximum we could run), 1 day, and 1 h. It can be seen that SKEA provides distinctly higher success rate, especially when it is possible to run key enumeration for longer time.

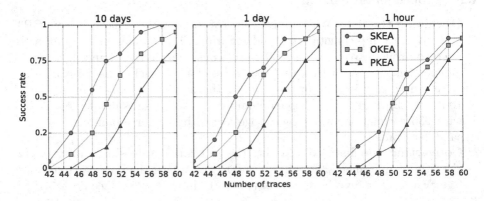

Fig. 5. Success rate of the full key recovery

Next, we compare memory use and number of key tries for SKEA and OKEA. Figure 6 shows the empirical dependency of memory use on the number of enumerated keys. As expected, OKEA memory use grows with the increase in the number of key tires, becoming prohibitive in our setting when the full key is in positions deeper than 2^{32}. SKEA memory use does not depend on enumeration depth. Additional figures are given in Appendix A.

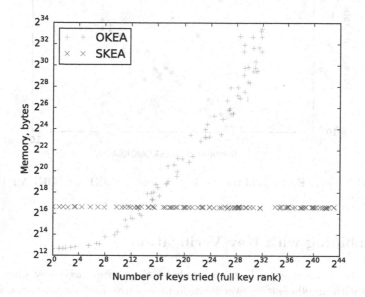

Fig. 6. Memory use versus rank for successful cases

Finally, to understand the relation between the full key rank and the running time, for the experiments where both SKEA and OKEA succeed we calculate SKEA-to-OKEA ratio of the full key rank and SKEA-to-OKEA ratio of the running time. We present the scatter plot of the ratios in Fig. 7. The gray lines are boundaries where the corresponding ratios are equal to 1, i.e. the algorithms take the same time resp. make the same number of tries to find the correct key. The quadrants of the plot relative to the $(1, 1)$ point have the following interpretation.

- Bottom right: SKEA takes more time and makes less trials.
- Top right: SKEA takes more time and makes more trials.
- Bottom left: SKEA takes less time and makes less trials.
- Top left: SKEA takes less time and makes more trials.

It can bee seen that even when SKEA makes more trials, in most of the cases it still finds the key faster than OKEA.

From our empirical comparison we conclude that in practice SKEA appears to be more efficient than OKEA. In particular, it requires less memory, especially when the number of trials (i.e. the full key rank) is high.

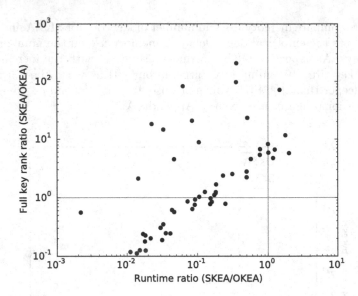

Fig. 7. SKEA-to-OKEA ratio of the full key rank versus SKEA-to-OKEA ratio of the running time

6   Combining with Key Verification

In this section, we discuss how to combine the score-based key enumeration algorithm with an efficient key verification procedure. For concreteness, we focus on AES-128 as the block cipher, and on general-purpose Intel CPUs equipped with the AES-NI instruction set [4,5].

The Setting. We consider an architecture in which both SKEA and the key verification process are implemented in software on general-purpose CPUs. The concrete CPU used in our experiments is an Intel Core i5-2400 (Sandy Bridge microarchitecture) with four cores running at 3.1 GHz. For consistency, Turbo Boost was disabled during our measurements.

The key candidates enumerated by SKEA are verified by trial encryption of a small number of known plaintexts and comparing the encryption results with the ciphertexts produced by the correct but unknown key. It is expected that one or two blocks of known plaintext will typically be enough to uniquely identify the key. We note that both the key enumeration (SKEA) and the key verification can be parallelized.

In order to make optimal use of the available resources, it is desirable to have a balanced throughput (measured in keys per second) between both the key enumeration and key verification phases. On our platform, an optimized implementation of SKEA achieves an enumeration throughput of about $2^{25.7}$ keys per second. On the other hand, a straightforward implementation of AES encryption using AES-NI instructions results in a key verification throughput of only about $2^{24.1}$ keys per second. This means that roughly three cores doing key verifications would be necessary to match the performance of the key enumeration on a single core.

The reason for this low AES throughput is that Intel's AES instructions are designed for encrypting large amounts of data with the same key. As a result, the AES round function instructions are heavily pipelined, while the key schedule helper instructions are not: On all microarchitectures from Sandy/Ivy Bridge to Haswell, the `aeskeygenassist` instruction has a latency of 10 cycles with an inverse throughput of 8. Since expanding an AES-128 master key requires 10 applications of `aeskeygenassist` plus some shifts and XORs, this amounts to a latency of around 100 cycles per key.

6.1 Optimized Parallel Key Verification

A solution to overcome this performance bottleneck is based on the observation that one can avoid using `aeskeygenassist` by using the `aesenclast` instruction plus `pshufb` for the S-box step of the key schedule, and then computing the LFSR manually using SSE instructions [2]. By doing this for four keys in parallel, all operations can be carried out on 128-bit XMM registers. In pseudocode, the AES key schedule for four user-supplied keys $k_1, \ldots, k_4$ is then computed as follows:

```
# register w0: first 32 bits of k1,...,k4
# ...
# register w3: last 32 bits of k1,...,k4
# rk1,...,rk4 contain first round keys corresponding to k1,...,k4
loop from r=1 to 11:
  pshufb     tmp, w3, ROL8_MASK
  aesenclast tmp, ZERO
  aesenc     block1, rk1
  aesenc     block2, rk2
  aesenc     block3, rk3
  aesenc     block4, rk4
  pshufb     tmp, SHIFTROWS_INV
  pxor       tmp, RC[r]
  pxor       w0, tmp
  pxor       w1, w0
  pxor       w2, w1
  pxor       w3, w2
# ... combine first 32 bits of w0,w1,w2,w3 in rk1
#     using punpck(l/h)dq and shifts on Sandy/Ivy bridge
#     or vpunpck(l/h)dq and vpblend on Haswell
# ... likewise for rk2, rk3, rk4
```

This way, one can keep the instruction pipeline at least partially filled. On Sandy/Ivy Bridge, `aesenc` and `aesenclast` have a latency of 4; on Haswell, the latency is 7. The reciprocal throughput is always 1. This means that the pipeline can be filled completely on Sandy/Ivy Bridge, and to 5/7 on Haswell. This optimized implementation of the AES key schedule results in an improved throughput of about $2^{25.5}$ keys per second on our test platform, yielding a factor of 2.6 speed-up for the key verification process.

With this optimized key verification, we have roughly equal throughput for both steps, which means that one core for key enumeration can be paired to one core for key verification.

6.2 Combined Performance Measurements

We provide an overview of the performance of key enumeration, verification and their combination on our test platform in Table 1. All measurements we averaged over 200 executions of the algorithm.

Table 1. Performance of combined key enumeration and verification using AES-NI on a Core i5-2400. All numbers are given in keys per second.

	1 core	2 cores	4 cores
(1) SKEA key enumeration	$2^{25.68}$	$2^{26.62}$	$2^{27.59}$
(2) Key verification (naive)	$2^{24.10}$	$2^{25.08}$	$2^{26.09}$
(3) Key verification (optimized)	$2^{25.46}$	$2^{26.44}$	$2^{27.41}$
(1) + (3) combined	$2^{23.87}$	$2^{25.39}$	$2^{26.41}$

Discussion. One can observe that the key enumeration throughput of SKEA scales basically linearly with the number of available cores. The same essentially holds for the key verification, both in the straightforward and the optimized AES-NI implementations. When combining enumeration with verification on a single core, the overall performance becomes limited by the extent to which AES instructions and general purpose instructions needed for SKEA (ALU, memory accesses) can be carried out concurrently while contending for the memory interface. When both enumeration and verification can be assigned to separate cores, however, the combined performance is essentially equal to the minimum of the single-core performances of the two individual tasks. Being able to run two copies of these in the 4-core setup again roughly doubles the achieved throughput. This indicates that off-the-shelf general purpose CPUs can offer an attractive platform for the combined key enumeration and verification.

Acknowledgements. We would like to thank Nicolas Veyrat-Charvillon and François-Xavier Standaert for providing detailed explanations about the Bayesian extension of non-profiled side-channel attacks introduced in [15].

References

1. Bogdanov, A., Kizhvatov, I., Manzoor, K., Witteman, M.: Efficient practical key recovery for side channel attacks. MCrypt Seminar in Cryptography, Les Deux Alpes (2014). http://mcrypt.org/pub/kizhvatov_mcrypt.pdf

2. Bogdanov, A., Mendel, F., Regazzoni, F., Rijmen, V., Tischhauser, E.: ALE: AES-based lightweight authenticated encryption. In: Moriai, S. (ed.) FSE 2013. LNCS, vol. 8424, pp. 447–466. Springer, Heidelberg (2014)
3. Dichtl, M.: A new method of black box power analysis and a fast algorithm for optimal key search. J. Cryptograph. Eng. 1(4), 255–264 (2011). Presented in COSADE'10
4. Gueron, S.: Intel's new AES instructions for enhanced performance and security. In: Dunkelman, O. (ed.) FSE 2009. LNCS, vol. 5665, pp. 51–66. Springer, Heidelberg (2009)
5. Gueron, S.: Intel Advanced Encryption Standard (AES) New Instructions Set. Intel Corporation (2010)
6. Manzoor, K.: Efficient practical key recovery for side-channel attacks. Master's thesis, Aalto University, June 2014. http://cse.aalto.fi/en/personnel/antti-yla-jaaski/msc-thesis/2014-msc-kamran-manzoor.pdf
7. Martin, D.P., O'Connell, J.F., Oswald, E., Stam, M.: Counting keys in parallel after a side channel attack. Cryptology ePrint Archive, Report 2015/689 (2015). http://eprint.iacr.org/2015/689
8. Mather, L., Oswald, E., Whitnall, C.: Multi-target DPA attacks: pushing DPA beyond the limits of a desktop computer. In: Sarkar, P., Iwata, T. (eds.) ASIACRYPT 2014. LNCS, vol. 8873, pp. 243–261. Springer, Heidelberg (2014)
9. Meier, W., Staffelbach, O.: Analysis of pseudo random sequences generated by cellular automata. In: Davies, D.W. (ed.) EUROCRYPT 1991. LNCS, vol. 547, pp. 186–199. Springer, Heidelberg (1991)
10. Pan, J., van Woudenberg, J.G.J., den Hartog, J.I., Witteman, M.F.: Improving DPA by peak distribution analysis. In: Biryukov, A., Gong, G., Stinson, D.R. (eds.) SAC 2010. LNCS, vol. 6544, pp. 241–261. Springer, Heidelberg (2011)
11. Poettering, B.: AVRAES: The AES block cipher on AVR controllers. http://point-at-infinity.org/avraes/
12. Standaert, F.-X., Malkin, T.G., Yung, M.: A unified framework for the analysis of side-channel key recovery attacks. In: Joux, A. (ed.) EUROCRYPT 2009. LNCS, vol. 5479, pp. 443–461. Springer, Heidelberg (2009)
13. Veyrat-Charvillon, N.: Key enumeration and key rank estimation software. http://people.irisa.fr/Nicolas.Veyrat-Charvillon/software.html
14. Veyrat-Charvillon, N., Gérard, B., Renauld, M., Standaert, F.: Efficient implementations of a key enumeration algorithm. In: COSADE 2012 (2012, in press)
15. Veyrat-Charvillon, N., Gérard, B., Renauld, M., Standaert, F.-X.: An optimal key enumeration algorithm and its application to side-channel attacks. In: Knudsen, L.R., Wu, H. (eds.) SAC 2012. LNCS, vol. 7707, pp. 390–406. Springer, Heidelberg (2013)
16. Veyrat-Charvillon, N., Gérard, B., Standaert, F.-X.: Soft analytical side-channel attacks. In: Sarkar, P., Iwata, T. (eds.) ASIACRYPT 2014. LNCS, vol. 8873, pp. 282–296. Springer, Heidelberg (2014)

A Additional Figures

Figure 8 shows empirical distributions (histograms) of the algorithm running time versus the number of traces for the experiment described in Sect. 4. Bins are spanned along the vertical axis showing the running time, the value of each bin is represented by the color intensity, there 200 bins of width 4320 s

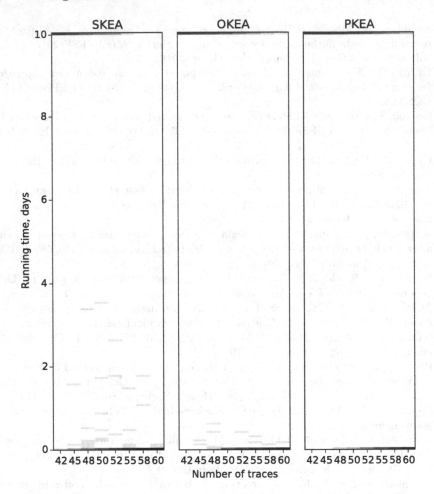

Fig. 8. Empirical distributions of the algorithm running time

(1 h 12 min). The last (topmost) bin accumulates cases that ran out of time or (for OKEA) out of memory.

Figures 9 and 10 show memory use and the number of full key candidates tried before the correct key is found.

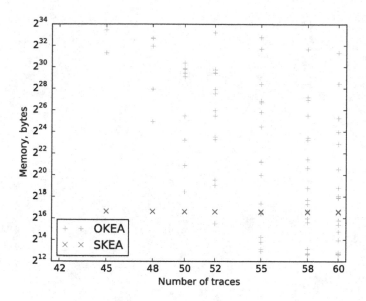

Fig. 9. Memory use for successful cases

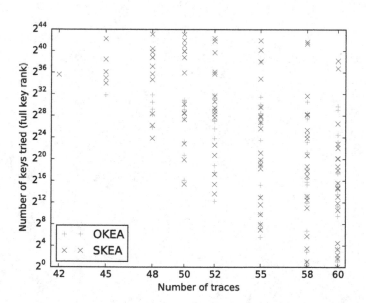

Fig. 10. Number of key tries for successful cases

New Cryptographic Constructions

An Efficient Post-Quantum One-Time Signature Scheme

Kassem Kalach[1]([✉]) and Reihaneh Safavi-Naini[2]

[1] Department of Combinatorics and Optimization and Institute
for Quantum Computing, University of Waterloo, Waterloo, ON, Canada
k2kalach@uwaterloo.ca

[2] Department of Computer Science, University of Calgary, Calgary, Canada

Abstract. One-time signature (OTS) schemes are important cryptographic primitives that can be constructed using one-way functions, and provide post-quantum security. They have found diverse applications including forward security and broadcast authentication. OTS schemes are time-efficient, but their space complexity is high: the sizes of signatures and keys are linear in the message size. Regular schemes (e.g., based on discrete logarithm or factoring problems, and their variants), however, have very low space complexity but are not time-efficient. Therefore, in particular, they are not suitable for resource-constraint devices. Many widely used signature schemes are not post-quantum. In this paper, we give a signature scheme that has the advantages of the previous two approaches. It provides constant-size short signatures, and is much more time-efficient than schemes in the second approach. We prove that our scheme is post-quantum secure as long as the SVP in ideal lattices is hard in the presence of a quantum computer. We use SWIFFT: a family of provable collision-resistant functions, which have efficient implementations, comparable with that of SHA-256, hence our proposed scheme could be used on resource-constrained devices.

Keywords: Post-quantum cryptography · Broadcast authentication · Digital signatures

1 Introduction

Digital signatures schemes may be grouped into two main categories or types depending on the two fundamentally distinct approaches of Lamport [25] and Diffie-Hellman [14], each with its own features and target applications. Here is a brief overview.

The first type, known as *one-time signature* (OTS) schemes, is typically founded on general one-way functions (e.g., SHA-256), and allows to sign one

---

K. Kalach–The bulk of this research was done when the author was a Post-doctoral Fellow at the University of Calgary. The author has been a member of the CryptoWorks21 program.

O. Dunkelman and L. Keliher (Eds.): SAC 2015, LNCS 9566, pp. 331–351, 2016.
DOI: 10.1007/978-3-319-31301-6_20

message per secret/public key pair. The first scheme of this type was described in 1976 by Diffie and Hellman [14], based on a suggestion of Lamport, and eventually published in 1979 [25]. More generally, fixed-time or t-time signature schemes allow to sign at most t messages using the same key pair; signing more messages makes the scheme insecure. The parameter t is a positive integer that must be chosen properly during the setup phase.

The second type involves one-way functions whose security is based on the computational hardness of some strongly structured mathematical problems (e.g., the discrete logarithm and factoring problems), and allows to sign an *undetermined* number (polynomially bounded) of messages using the same key pair. They are known as ordinary or regular schemes because the most deployed schemes, which are variants of ElGamal [18]–DSA, and ECDSA [24]– and RSA [41], belong to this category. Regular schemes have usually short keys and signatures. However, they involve time-consuming arithmetic operations (modular multiplications and exponentiations), particularly inconvenient for applications on devices with limited resources.

OTS schemes has attractive properties: (i) they can be constructed from general one-way functions, thus increasing the possible problems on which the security can be based [4]; (ii) they are usually very computationally efficient; and (iii) if the one-way function became insecure, then one could replace the function (e.g. SHA-1 with SHA-2). This is not possible in regular schemes; (iv) importantly, such one-way functions are quantum secure; one may only need to double the size of the domain to overcome attacks based on Grover's algorithm [21].

Applications of OTS schemes are many and diverse. They have been used to (i) construct on-line/off-line schemes that combine OTS and regular schemes [19]; (ii) transform regular unforgeable schemes into strong ones [23]; (iii) construct chosen-ciphertext composed encryption schemes [15]; (iv) construct fail-stop schemes [22]; and (v) provide forward-secure signatures [1] and one-time proxy signatures [28]. They are also used in broadcast authentication [38,40]. Importantly, they are used to construct post-quantum digital signatures [8,9].

A signature scheme *must* provide (0) some well defined security; and *should* afford at least some of the following features: (1) short signatures; (2) fast key-generation, signing and verification algorithms; (3) possibility to sign many messages per key pair; and (4) short keys. These properties, whose order depends on the target application, have driven the research on signature schemes.

Important typical shortcomings of OTS schemes are (i) the number of messages to be signed per key pair is predetermined; (ii) the signature is long; and (iii) the secret/public keys are also long. Overcoming the last two has motivated much research over the years [4,6,9,16,30,35,39,40,48]. Here we focus on schemes that are based on (general) post-quantum one-way functions, which are evolutions of Lamport scheme [25], and have a number of attractive properties.

Here is a review of some schemes that are relevant to our work. Throughout this work, let $f \colon \{0,1\}^\gamma \to \{0,1\}^\gamma$ a one-way function with a security parameter γ, and M an ℓ-bit message. Lamport scheme to sign a 1-bit message b works as follows. Choose randomly two secrets, x_0, x_1, and compute their corresponding images under f, $y_0 = f(x_0)$ and $y_1 = f(x_1)$. The signing key is $SK := (x_0, x_1)$ and the verification key is $PK := (y_0, y_1)$. The signature is then x_b. Any party can verify the signature by evaluating f on x_b and comparing the result with y_b in the public key. The scheme is provably secure, efficient, and simple. However, there are two main disadvantages: the signature size and keys size are linear in the message size. Subsequently, this construction has been improved or generalized over decades [4–6,16,30–32,39,40,48].

The first improvement (M-OTS) is due to Merkle [30,32] who reduced the keys and signature sizes to almost half on the price of only logarithmic additional time overhead. Here is briefly the idea. The key generation consists in choosing random secrets x_i and computing the public values $y_i = f(x_i)$ for $1 \leq i \leq t$ where $t = \ell + \lfloor \log \ell \rfloor + 1$. To sign a message M, form $M' = M||c = (b_1, \ldots, b_t)$ where c is the binary representation of the number of zeros in M, then find the positions $i_1 < \cdots < i_k$ such $b_{i_j} = 1$ where $1 \leq j \leq k$. The signature is then $(s_1, \ldots, s_k)$ with $s_j = x_{i_j}$. To verify a signature $\sigma = (s_1, \ldots, s_k)$ on M, again form M', determine the positions of ones, then accept σ if and only if $y_j = f(s_j)$ for all $1 \leq j \leq k$. More details can be found in [29,46].

Winternitz [30,32] proposed to Merkle the idea, called W-OTS, of signing several bits simultaneously at the expense of more evaluations of the one-way function, thus trading time for space. More details are given in Sect. 5.4.

Implicitly based on cover-free families (defined in Sect. 3.1), Bos and Chaum gave a scheme able to reduce the public key size to almost half, or to decrease the signature size and verification time at the expense of increasing the public key length [6]. Signing is by revealing k out of $e = 2k$ secret elements. Reyzin and Reyzin [40] presented a slight generalization of BC [6], considering a general k instead of $k = e/2$. They also proposed an r-time signature scheme HORS (for Hash to Obtain Random Subset) [40]. For time and space efficiency, they use random structures instead of explicit CFF constructions. However, the security holds using a strong additional assumption: the existence of subset-resilient functions. The signature scheme can be used r times, for small values of r. However, the security decreases as r increases. Pieprzyk et al. [39] proposed a t-time signature scheme (HORS++) that achieves security against t-adaptive chosen-message attacks, using explicitly a t-CFF. They gave t-CFF constructions based on polynomials, error-correction codes, etc. However, it "is only of theoretical interest" [39]. They also extended the scheme to increase the number of messages to sign, using one-way hash chains.

Based on the discrete log assumption [37], van Heyst and Pedersen gave a fail-stop scheme (vHP) with signature twice as long as the security parameter [22]. They also gave methods to transform OTS scheme into t-time one. Groth [20] employed an NIZK proof system to construct a OTS scheme similar to that van Heyst and Pedersen. Inspired by Pedersen commitment scheme [37], Zaverucha and Stinson gave a OTS scheme (ZS) with signature size about 273 bits for 128-bit security, thanks to the algebraic properties of the DLP. However, the latter has an efficient quantum algorithm [44], and the scheme verification algorithm is slower than most OTS schemes and ECDSA.

1.1   Contribution

We present a new OTS (and t-time signature) scheme that combines the advantages of both reviewed approaches. Indeed, it provides signatures independent of the message size, thus much shorter signatures, compared with all schemes based on general one-way functions [3,5,6,25,30,40] (including Winternitz one) while maintaining the same time complexity.

The scheme is much more time-efficient than regular schemes, which are based on discrete logarithm or factoring problems and their variants. In particular, it has very simple and fast signing, requiring only regular additions, and a very fast verifying, requiring only additions and *one* evaluation of a hash function. We use a family of hash functions (SWIFFT), which can have efficient implementations comparable with that of SHA-256, hence our scheme could be used even on resource-constrained devices.

The security of the scheme is strongly unforgeable under adaptive chosen-message attacks and reduced to the security of knapsack functions (SWIFFT), which are provably collision-resistant assuming the hardness of the shortest vector problem in cyclic lattices. We give a formal security proof based on cover-free families and hash functions. On top of that, the scheme is post-quantum secure, meaning resistant against quantum attacks.

We use a perceived disadvantage of SWIFFT, not pseudo-random due to linearity, as a feature and use its "conditional" linearity (when the input is not binary vector) to compress several secrets into one, resulting in a short signatures. However, there is a main challenge: SWIFFT security holds if the input is a vector of small coefficients. Adding component-wise many vectors may result in a vector of large coefficients, and render the function easy to invert or collide. We give solutions to overcome this issue using some new technical ideas.

1.2   The Scheme High-Level Description

The scheme is based on 1-cover-free families and one-way functions. A w-cover-free family is a collection $\mathcal{B}$ of subsets of a set E with the property that the union of any w subsets in $\mathcal{B}$ will not include ("cover") any other subset in $\mathcal{B}$. As observed by Merkle, the one-way function can be viewed as a commitment scheme. During key generation, the signer commits to randomly chosen secret elements associated with E by computing and publishing their corresponding images. This makes the public key. Each message corresponds to a unique subset in $\mathcal{B}$. To sign a message, the signer finds its corresponding subset in $\mathcal{B}$ and typically reveals the secret elements corresponding to this subset. Using the homomorphic property of SWIFFT, we can add these secrets and effectively compress them into a single value, making up a short signature. This also provides faster verification: to verify a signed message, again find the associated subset in $\mathcal{B}$ and add mod p the corresponding public values. Finally, compare the result with the provided signature.

1.3   Paper Organization

The remaining material is organized as follows. Related work are given in the next section while preliminaries are given in Sect. 3. Our new scheme is the subject of Sects. 4, and 5 contains our conclusions and future work.

2   Related Work

The literature on OTS schemes and their variants is immense, more details about this topic can be found in [16,29,46]. Here is a review of some additional related schemes.

One main disadvantage of OTS schemes is the authentication and management of many public keys. Using a complete binary tree and a collision-resistant hash function, Merkle [30,32] introduced a solution to this problem, which allows to transform any OTS into a scheme that allows to sign many messages using one public key.

Naor et al. [36] studied whether OTS became practical by applying recent improvements in hash tree traversal to M-OTS scheme. In order to provide a practical efficient broadcast authentication protocol, Perrig proposed an r-time signature scheme [38] having fast verification and relatively short signatures when compared with the earlier related schemes. Both efficiency and security are based on the *birthday problem*, thus the name "BiBa" (for Bins and Balls). BiBa's disadvantages are the signing time, which is longer than that of most previously similar schemes, and its security considered in the *random oracle model* and decreases as r increases.

Bleichenbacher and Maurer formalized the concept of one-time signatures in terms of acyclic graphs. In particular, they unified the schemes of Lamport, Merkle, Winternitz, Vaudenay [47], and Even et al. [19]. To solve the problem of packet source authentication for multicast, Rohatgi [42] presented a hybrid signature mainly based on a collision-resistant hash function and a commitment scheme. The scheme improves over the scheme of [19]. In terms of power consumption, Seys and Preneel [43] evaluated ECDSA, M-OTS, W-OTS, HORS with or without Merkle trees.

Lyubashevsky and Micciancio gave an "asymptotically efficient" one-time signature scheme [26], which we will try to compare with our scheme. Mohassel gave a general construction for transforming any chameleon hash function to strongly unforgeable OTS schemes, in addition to instantiations based on the hardness of factoring, discrete-log, and worst-case lattice-based assumptions [35].

3   Preliminaries

The symbol '$\otimes$' is used for vector convolution, and '$\cdot$' to emphasize scalar multiplication, and $[k] = \{0, \ldots, k\}$ for integer k. The logarithm base 2 is denoted by log.

3.1   Cover-Free Families

Definition 1 (Cover-free Families). A set system $(E, \mathcal{B})$ is called a k-uniform w-cover-free family if E is a finite set of e elements (or points), $\mathcal{B}$ is a collection of s subsets (or blocks) of size k, and for all $\Delta \subset E$ with $|\Delta| = w$ and all $i \notin \Delta$, it holds that

$$\left| B_i \setminus \bigcup_{j \in \Delta} B_j \right| \geq 1.$$

A w-cover-free family with e elements and s subsets is denoted by $w\text{-}CFF(e, s)$.

In this work, we consider k-uniform 1-CFF where each subset has size k, and any two distinct subsets in $\mathcal{B}$ differ on at least one element. Importantly, a 1-CFF has optimal construction, which consists of setting $k = \lfloor e/2 \rfloor$. Indeed, there is a simple encoding algorithm that requires k subtractions (or additions) and about e comparisons using some pre-computation; it is well explained in [3,48]. In practice, encoding is twice faster than MD5 hashing, assuming MD5 requires 500 arithmetic operations [3]. The algorithm and more details are given in Sect. 5.3

3.2   Compact Knapsack Functions (SWIFFT)

A family of SWIFFT functions [27] is described by three main parameters: power of 2 security parameter n, small integer $m > 0$, and modulus $p > 0$. Define $R = \mathbb{Z}_p[\alpha]/(\alpha^n + 1)$ to be the ring of polynomials in α having integer coefficients modulo p and $\alpha^n + 1$. Any element of R can be written as a polynomial of degree smaller than n with coefficients in $\mathbb{Z}_p = \{0, 1, \ldots, p-1\}$. An instance of the family is specified by m fixed elements $a_1, \ldots, a_m \in R$. These elements (or multipliers) must be chosen uniformly and independently. The function output then corresponds to the following algebraic expression:

$$f(x) = \sum_{i=1}^{m} (a_i \otimes x_i) \in \mathbb{Z}_p^n$$

where $x_1, \ldots, x_m$ are polynomials in R with *binary* coefficients, and correspond to the input $x \in \{0, 1\}^{n \times m}$. These functions are provably collision-resistant.

The security of the functions depends on the choice of the parameters, in particular the domain. In [26], $R = \mathbb{Z}_p[\alpha]/(\alpha^n + 1)$ where n is power of 2, the domain $D = \{y \in R : \|y\|_\infty \leq d\}$ for some d; $m > \frac{\log p}{\log 2d}$; and $p > 4dmn^{1.5} \log n$. Originally [34], the domain was $D = \{0, \ldots, \lfloor p^\delta \rfloor\}^n$ for positive δ. In practice, for most efficient FFT, p will be a prime such that $p - 1$ is multiple of $2n$ [27]. Here $\|y\|_\infty$ is the infinite norm of y. Unfortunately, the exact value of d is not known yet, it is a work in progress [33].

3.3   OTS Security Definition

The standard security notion for digital signatures is the (existential) *unforge-ability under adaptive chosen-message attacks*, which can be modelled using the following security game between a challenger and an adversary. Consider DSS $=$ (Gen, Sign, Ver) with some security parameter η, message space, and polynomial-time (in η) quantum adversary $\mathcal{A}$.

1. Run Gen(1^η) to obtain a signing secret key SK and corresponding verification public key PK.
2. Give $\mathcal{A}$ the public key PK and a signature for at most *one* message. Let M_1 be the query asked by $\mathcal{A}$ and σ_1 its corresponding signature.
3. $\mathcal{A}$ finally outputs (M, σ).

An adversary $\mathcal{A}$ (existentially) forges a signature, or wins the unforgeability game (uf-cma) for DSS, if $\text{Ver}(PK, M, \sigma) = 1$ and $M \neq M_1$. Let $\text{PROB}_{\text{DSS},\mathcal{A}}^{\text{uf-cma}}$ denote the probability that $\mathcal{A}$ forges a signature taken over the random bits of the challenger and adversary; symbolically

$$\text{PROB}_{\text{DSS},\mathcal{A}}^{\text{uf-cma}} = \Pr[(M \neq M_1) \wedge \text{Ver}(PK, M, \sigma) = 1].$$

An adversary $\mathcal{A}$ is said to (t, ε)-win the security game if (M, σ) is output in time t such that $\text{PROB}_{\text{DSS},\mathcal{A}}^{\text{uf-cma}} = \varepsilon$.

Definition 2 (uf-cma OTS post-quantum security). A signature scheme is (existentially) unforgeable under chosen-message quantum attacks, or secure, if every polynomial-time quantum adversary forges a signature, or wins the unforgeability experiment, only with probability negligible in the security parameter.

This definition is the same as the standard classical one except that it considers quantum attacks. These attacks are captured in the security game, and model the real life scenarios of (classical) OTS schemes. More precisely, $\mathcal{A}$ can make use of any quantum computation (or communication) such as evaluating the hash function in superpositions. On the other hand, $\mathcal{A}$ cannot do queries in superpositions to the signing oracle, since the signing algorithm is controlled by a key only known to the signer, and signs one message at a time. In this definition, we assume that the scheme is implemented on a classical computer, otherwise it may require a different security model.

In this framework, systems based on DLP or factoring are not post-quantum because of Shor's quantum polynomial time algorithm for these problems [44]. However, this is not the case for SVP on which the security of SWIFFT functions are based.

Dedicated discussions about the requirements for the quantum security of classical schemes, which are satisfied in this work, are established by Song [45].

4   The New Scheme

In this section we describe our new signature scheme, give its formal proof of security, analyze its time and space complexity, and compare it with selected work. We also provide concrete parameters for a practical security level.

4.1   Scheme Description

We consider without loss of generality signatures of ℓ-bit messages M; longer messages can be adapted to this length using some secure hash function. We first give our OTS signature scheme, which can easily be extended into t-time signature one using standard techniques.

Setup: Consider an optimal k-uniform 1-CFF$(E, \mathcal{B})$ with $E = \{1, 2, \ldots, e\}$, and choose parameters e and k such that $\binom{e}{k} \geq 2^\ell$. An optimal 1-CFF requires that $k = \lfloor e/2 \rfloor$. Let $S \colon \{0,1\}^\ell \to \mathcal{B}$ be a bijection that maps a message $0 \leq M < \binom{e}{k}$ into the Mth element of $\mathcal{B}$ denoted by B_M. A very efficient constructive algorithm of this bijective is discussed in Sect. 3.1.

Let $f \colon \{0,1\}^{nm} \to \mathbb{Z}_p^n$ be a SWIFFT function where m, n, and p are the main parameters. The function also maintains its security even in a domain for small entries (see Sect. 3.2), this is needed for the security proof.

Key Generation: Choose randomly secret elements $x_1, \ldots, x_e \in \{0,1\}^{nm}$ and compute their images $y_i = f(x_i)$ for $i = 1, \ldots, e$. Output the secret signing key $SK := (x_1, \ldots, x_e)$ and the public verification key $PK := (y_1, \ldots, y_e)$.

Signing: To sign a message M using SK, compute the k-subset $B_M \in \mathcal{B}$, then

$$\sigma := \left(\sum_{i \in B_M} x_i \right) \in [k]^{nm} .$$

Verification: Given (M, σ, PK) as input, to verify whether σ is a valid signature for M using PK, compute B_M, check that $\sigma \in [k]^{nm}$ and output 1 if and only if

$$f(\sigma) = \left(\sum_{i \in B_M} y_i \mod p \right) .$$

Output 0 otherwise.

Remark 1. Since $k < p$ is always the case in this work, the signature is actually the component-wise addition of k binary vectors of dimension nm, thus a vector in $[k]^{nm}$. Consequently, signing becomes much faster being reduced to component-wise counting, which can be done in parallel.

Remark 2. The functions are linear if addition is done in $\mathbb{Z}_p$ but not in $\mathbb{Z}_2$. This is straightforward to verify viewing the algebraic expression of f with nm multipliers $a_1, \ldots, a_m$ and input $x \in \{0,1\}^{nm}$ as the matrix-vector product Ax. The matrix A is obtained from the skew-circulant matrices of a_i for $1 \le i \le m$ (Sect. 3.2). Consequently, Ver will output 1 whenever presented with a valid signature. Mathematically, the verification process is easy to validate:

$$f(\sigma) = A\sigma = A\left(\sum_{i \in B_M} x_i\right) = \sum_{i \in B_M} Ax_i = \sum_{i \in B_M} f(x_i) = \sum_{i \in B_M} y_i \mod p.$$

4.2 Security Proof

Theorem 1. *If $\mathcal{A}$ is a quantum adversary that (t, ε)-wins the unforgeability security game for our one-time signature scheme, then $\mathcal{A}$ can be used to $(t+c, \varepsilon c')$ find function collisions where c, c' are constants.*

Proof. Let DSS denote our one-time signature scheme, and assume that $\mathcal{A}$ queries a signature σ_1 for one message M_1 and outputs (M, σ) with $M \ne M_1$, which can be verified using PK; meaning $\Pr[\text{Ver}(PK, M, \sigma) = 1] \ge \varepsilon$.

Informally, forging a signature σ for a new M after obtaining a valid signature for M_1 is reduced to find collisions in SWIFFT functions. Considering the post-quantum security of the function, this cannot be done in polynomial time even using quantum computers. Otherwise, we could use the quantum forger to find collisions in the functions. Note that a quantum forger can always have some speedup over a classical one using Grover's search [21], reducing the adversary complexity by a square root at most. However, this is not considered in theory.

We now proceed with the formal proof. We devise a (quantum) algorithm $\mathcal{A}'^{\mathcal{A}}$ that uses $\mathcal{A}$ as a subroutine to find function collisions in time in the order of t with probability "close" to ε. Given access to f and other public parameters, the collision-finding algorithm $\mathcal{A}'^{\mathcal{A}}$ works as follows.

1. Create an instance of DSS using f
   (a) Choose $x_i \in \{0,1\}^{nm}$ and set $y_i = f(x_i)$ for all $i \in [e]$.
   (b) Run $\mathcal{A}$ on $PK = (y_1, \ldots, y_e)$ and all other system parameters.
2. When $\mathcal{A}$ queries a signature for M_1 do
   (a) Compute B_{M_1};
   (b) Return $\sigma_1 := \sum_{i \in B_{M_1}} x_i$.
3. When $\mathcal{A}$ outputs (M, σ)
   (a) Return σ.

We now analyze the behaviour of $\mathcal{A}'^{\mathcal{A}}$. First of all, it runs in time in the order of t (running time of $\mathcal{A}$). Indeed, the steps 1, 2 and 3 take a constant time. In Step 1, PK is exactly distributed as in the real execution. In Step 2, $\mathcal{A}'^{\mathcal{A}}$ answers similarly to a real execution the adaptive query made by $\mathcal{A}$ since it knows all the secrets. Therefore, $\mathcal{A}'^{\mathcal{A}}$ can answer any signature query with probability one.

Next, we prove that the reduction succeeds in outputting a collision when $\mathcal{A}$ makes a forgery, and compute the probability of finding a collision. Let $\bar{\sigma} = \sum_{i \in B_{M_1}} \bar{x}_i$ be the legitimate signature, and $\sigma = \sum_{i \in B_M} x_i$. There are the following cases in which $f(\sigma) = f(\bar{\sigma}) = \sum_{i \in B_M} y_i \mod p$:

1. $\sigma = \bar{\sigma}$ with $\sigma \neq \sigma_1$;
2. $\sigma = \bar{\sigma}$ with $\sigma = \sigma_1$;
3. $\sigma \neq \bar{\sigma}$.

The objective now is to show that the probability of the first two cases is negligible. Case (1) happens with probability smaller than 2^{-nm}. Indeed, $\bar{\sigma}$ is not known to the adversary, and it requires at least one uniformly distributed secret value in $\{0,1\}^{nm}$ since $|B_M \setminus B_{M_1}| \geq 1$. The only thing $\mathcal{A}$ knows about $\bar{\sigma}$ is that it is the sum of at least one uniformly secret nm-dimensional binary vector, which was not considered beforehand, and other values in $[k]^{nm}$, which may be part of σ_1 (see Lemma of Sect. 5.1).

Case (2) essentially happens with the same probability as the first case since it reduces to the problem of finding a different k-subset of elements in the SK that sums to σ. Again, this requires at least one element in $\{0,1\}^{mn}$. The probability of this even is upper bounded by one over the number of possible distinct k-subset sums, meaning in the order of 2^{-nm}.

Case (3) happens with complementary probability exponentially close to 1. Accordingly, $\mathcal{A'}^{\mathcal{A}}$ can find a collision whenever $\mathcal{A}$ makes a forgery. The probability of this event is

$$\mathrm{PROB}_f(\mathcal{A'}^{\mathcal{A}}) \geq \mathrm{PROB}_{\mathrm{DSS},\mathcal{A}}^{\mathrm{uf\text{-}cma}} \cdot (1 - \mathrm{negl}(nm)).$$

Given that f is collision-resistant, $\mathrm{PROB}_f(\mathcal{A'}^{\mathcal{A}})$ is negl in nm when $t = poly(n)$. Therefore, $\mathrm{PROB}_{\mathrm{DSS},\mathcal{A}}^{\mathrm{uf\text{-}cma}}$ is negligible. In conclusion, the security of the signature scheme is reduced to the collision-resistance. This ends the proof.

Keep in mind that the security proof holds for quantum adversaries. The only difference may arise in the concrete sense where the success probability may be slightly larger because of some non-significant quantum search speed-up.

Remark 3. We now discuss a tricky point related to the function security, which depends on the domain (see Sect. 3.2). The coefficients of σ are in $[k/2]$ on the average. Now, if $k/2$ is so large that σ violates the domain constraint, then the signature may be a vector in $\mathbb{Z}_p^{nm}$ for which the function is easy to invert or have collisions. If this is the case, then an adversary may choose some M, compute B_M, add the corresponding values in the public key, giving $y = \sum_{i \in B_M} y_i \mod p$, then output the inverse of y as a forgery. To avoid this concern, it is sufficient to choose k appropriately. Fortunately, this is always possible by several means. First of all, k is smaller than p for any set of parameters, in particular $k < p/2$ since $k \leq e/2$. Second, we can always trade large p for time (or space); the domain is $D = \{0, \ldots, \lfloor p^{\delta} \rfloor\}^n$ for positive δ.

Strong Unforgeability. In the strong unforgeability game means that a new signature $\sigma \neq \sigma_1$ on a previously signed message M_1 is also a forgery. Our scheme is strongly unforgeable, and the proof is reduced to the collision case. Note that this is not the standard definition, but may be useful in some cryptographic applications.

4.3 Asymptotic Evaluation

We give in this section an asymptotic comparison with the most related work, summarized in Table 1, and a concrete evaluation is given in Sect. 4.4. We compare with (stand-alone) OTS schemes because they can be used as building blocks (without Merkle tree), even beyond digital signatures, and determine the overall efficiency when used with Merkle trees [3,7,9,16,30,36]. We classify the schemes into categories, compare the schemes in each category, then compare the categories together.

In order to provide a reasonable evaluation, the time complexity is measured in terms of the number of evaluations of a general one-way function (OWF), SWIFFT function (our case) or explicit arithmetic operations. To give the reader a more concrete feeling about the timing, it is useful to recall of the following. Although SWIFFT competes with SHA-3 (see Sect. 5.2), we assume that it is multiple times slower. On the other hand, an implementation with the crypto++ library indicates that an exponentiation in a 160-bit group costs about 3 300 hashes [2]. Therefore, it is fair to assume that working in a 224-bit or 256-bit group still requires few thousands hashes of (OWF or SWIFFT) function. Finally, message encoding is twice faster than MD5 hash [3], see Sect. 5.3 for more details.

OWF-based Schemes. Efficient schemes based on general one-way functions essentially have the same time and space complexity. Indeed, the schemes BC, BCC, RR, BTT, etc. [3,6,39,40] are more general and more efficient than L-OTS [25]. We refer to them as BC (category) since they have the same time and space complexity when using 1-CFF, and BC scheme was the first to use the optimal setting. They provide post-quantum security and very efficient time complexity.

M-OTS and BC also have the same space and time complexity essentially. Indeed, we found that $\ell < e$ because $2^\ell \leq \binom{e}{k}$ is necessary to sign ℓ-bit messages, and it is known that $\binom{e}{k} < 2^e$. Therefore, we don't loose much by assuming that $e \sim \tilde{\ell} = \ell + \lfloor \log \ell \rfloor + 1$. M-OTS signature size is $\gamma\tilde{\ell}/2$ bits (on the average) while BC signature is $\gamma e/2$.

It is also relevant to discuss W-OTS, whose signature size (or communication cost) drops linearly in its parameter w while the computational cost grows exponentially. Therefore, any performance gain, if any, is only possible for small values of w. Indeed, a theoretical result [16] claims that W-OTS is most efficient when $w = 2$, and practical one recommends $w = 4$ since it is fast and give relatively short signatures. Another result [43] says that the minimum power consumption cost occurs when $w = 2$. In Table 1, the key generation costs is

$C_{gen} = (2^w - 1)\tilde{\ell}/w$ while the signing and verification are $C_{gen}/2$ on the average. However, W-OTS is unique in that the public key is not needed to be a part of the signature using Merkle tree.

The main disadvantage of such schemes is the space complexity, which is linear in the security parameter. However, the secret key can be reduced to a single seed using a pseudo-random number generator (PRNG), and the public key can be reduced to a single value using some hash function. These are common techniques [4, 8, 16, 36]. Thus, the most challenging limitation for these schemes is the signature size, which we improve significantly in this work.

DLP-based Schemes. The scheme of van Heyst and Pedersen (vHP) (and Groth [20]) essentially provides the best balanced performance using the minimal assumption (DLP). Bellare and Shoup scheme [2] provides the shortest keys and best time complexity. However, they use a collision-resistant hash function and DLP, and did not improve the signature size. Mohassel scheme provides a nice theoretical constructions, but using a target collision-resistant function (TCRF) and without practical advanatge. Zaverucha and Stinson scheme provides the shortest signatures using only DLP, but on the expense of much slower key-generation phase and longer public-keys. None of them is post-quantum.

DLP-based Schemes vs BC. BC is post-quantum secure and provides very efficient time complexity; this is true even compared with any signature scheme. DLP-based ones provide much better space complexity, however, they are much slower and not post-quantum.

Our Scheme vs the Others. Our scheme is time and space efficient, providing the advantages of both approaches. First of all our signature size is independent of the message size, which is not the case for all schemes based on general one-way functions. Consequently, it is much shorter, which improves the main limitation of this category (BC). The key generation and signing algorithms are essentially as efficient as those of BC. The time of key generation is equivalent to e evaluations of an efficient hash function. The only difference is that we use SWIFFT functions, which are competitive with SHA-3 (see Sect. 5.2). Now, the signing is very efficient requiring only encoding, and regular additions of k binary vectors, which is very fast and can be done in parallel. Thus, it is dominated by the encoding algorithm, which is very efficient (see Sect. 5.3). Our verification algorithm requires k additions in $\mathbb{Z}_p^n$, one evaluation of SWIFFT and one comparison, in contrast with k evaluations and k comparisons in BC.

Comparing with DLP-based schemes, our scheme has (i) a much faster key generation, requiring e evaluations, in contrast with at least *two* modular exponentiations in any of the DLP-based schemes; in particular it is much faster ZS one; (ii) a faster signing, requiring regular additions, in contrast with group additions or multiplications; (iii) a much faster verification, requiring k modular additions and *one* function evaluation, in contrast with modular exponentiations. Further more, our scheme is post-quantum assuming the shortest vector

Table 1. Asymptotic comparison: γ and λ are security parameters for OWF and DL, respectively, and ℓ is the message length. Here $\tilde{x} \leq x + \log x + 2$, seed is 128 bits, and $C_{gen} = (2^w - 1)\ell/w$. Arithmetic: count (add binary vectors), add (modular addition), mult (modular multiplication), exp (modular exponentiation).

| | Security | | | Time (function evaluations) | | | Space (bits) | |
Scheme	Func	PQ	Gen	Sign	Ver	SK	PK	σ
L-OTS	OWF	Yes	2ℓ	–	ℓ	seed	$2\ell\gamma$	$\ell\gamma$
M-OTS	OWF	Yes	$\tilde{\ell}$	–	$\tilde{\ell}$	seed	$\tilde{\ell}\gamma$	$\tilde{\ell}\gamma/2$
W-OTS	OWF	Yes	C_{gen}	$\approx C_{gen}/2$	$\approx C_{gen}/2$	seed	$(\ell/w)\gamma$	$(\ell/w)\gamma$
BC ...	OWF	Yes	$e \approx \tilde{\ell}$	encode	k eval	seed	$e\gamma$	$k\gamma$
vHP	DLP	No	4 exp	2 mult	3 exp	4λ	2λ	2λ
			2 mult	2 add	2 mult			
BS	DLP	No	2 exp	1 mult	1 exp	2λ	λ	2λ
	CRHF							
Moh	DLP,	No	5+exp	2 mult	5+exp	5λ	4λ	2λ
	TCR			4 add				
ZS	DLP	No	2e exp	k add	2 exp	seed	$O(\lambda\ell)$	$\lambda + c$
			e mult	encode	k mult			
This work	SWIFFT	Yes	e eval	k count	1 eval	seed	$en \log p$	$mn \log \frac{k}{2}$
				encode	k add			

problem in idea lattices is hard for quantum computers. However, our signature is much longer because of SWIFFT input size.

Note that there is an encoding step during verification (Ver) whenever there is one during signing (Sign). However, we do not show it in Table 1 for convenience because it is so efficient that it is dominated by the other operations.

Lattice-Based OTS Schemes. For completeness, we compare with related OTS schemes based on knapsack functions. Lyubashevsky and Micciancio gave a direct construction of OTS scheme, which is asymptotically efficient [26] and whose idea is the following. The scheme is parametrized by integers m, n, k; a ring R; subsets of matrices $\mathcal{H} \subset R^{n \times m}$, $\mathcal{K} \subseteq R^{m \times k}$; and vectors $\mathcal{M} \subseteq R^k$, $\mathcal{S} \subseteq R^m$. The parameters should satisfy certain properties for the scheme to be secure. The underlying hardness assumption is the collision resistance of a linear hash function family mapping $\mathcal{S}$ into R^n. The secret key is a matrix $\mathbf{K} \in R^{m \times k}$ while the public key consists of a matrix $\mathbf{H} \in R^{n \times m}$ along with the matrix product $\mathbf{HK}$. To sigh a message $v \in R^k$, compute $\sigma = \mathbf{K}v$. To verify (σ, v), check that $v \in \mathcal{S}$ and $\mathbf{H}\sigma = \mathbf{HK}v$. Choosing $R = \mathbb{Z}[x]/(x^n + 1)$ and $R = \mathbb{Z}_p$ produces a scheme based on the Ring-SIS and SIS problem, respectively.

There is some connection with our scheme, however, it is not easy to provide a comparison since the paper does not contain concrete or asymptotic evaluation, or a comparison with any previous work. We can still observe the following. Our signing algorithm is simply the component-wise addition of binary vectors, instead of matrix multiplication, and our verification algorithm requires only one matrix multiplication and ring elements additions instead of two matrix multiplication. Our scheme use a different encoding technique, and able to sign a message without being encoded as a vector. It is not clear how these differences may affect the concrete efficiency. Thus, a future careful comparison is important.

SWIFFT was suggested in [10] to implement the general one-way function in W-OTS, arguing that it has provable security. However, SWIFFT requires input at least 4 times larger than SHA-3 for the same security level, thus making the signature at least 4 times more.

4.4 Concrete Parameters

For a concrete security level and more accurate comparison with other work, we select parameters to sign messages of size $\ell = 224$ bits, and provide *classical* security level of 112 bits; quantum ones may be part of a future work. Therefore, we can use SHA-224 or SHA-256, which are suitable collision-resistant hash functions for digital signatures. While 80-bit security level is disallowed after 2014, 112-bit level is acceptable until 2030 according to NIST recommendation in July 2012 [17]. We will consider two possible implementations of the OWF depending on the output lengths $\gamma = 224$ or 128.

As with elliptic curve cryptography in general, the bit size of the public key believed to be needed for ECDSA is about twice the size of the security level, thus the group order q should have 224 bits.

For SWIFFT with $n = 64$, $m = 16$ and $p = 257$, the known algorithms to invert a function have about 2^{128} time and space complexity, and those to find collisions takes time at least 2^{106} and requires almost as much space. Since the security of our scheme reduces to the collision resistance, we assume randomized hashing [12] is used so that known algorithms to find collisions require at least 2^{112} time complexity. We can always increase n, but it would be too much unless a security level much larger than 112 bits is desired. Finally, the output length is about 512 bits, which "can easily be represented using 528 bits" [27].

Considering 224-bit messages to sign, optimal 1-CFF is obtained by setting $e = 229$ and $k = 114$ so that $\binom{229}{114} > 2^{224}$. However, we consider the minimum k satisfying this inequality, which is $k = 107$. This relaxation mainly allows to have small coefficients and slightly shorter signatures.

Table 2 shows that our scheme provides signatures of 6 144 bits (0.75 KB) on the average, which is the shortest among all schemes based on general one-way functions, including Winternitz, and essentially keeps the same time efficiency. Our signing (and verification) is faster than W-OTS even for its typical value $w = 3$. In any case, our scheme provides multiple times shorter signatures (even using OWF with 128-bit output length), and the most efficient verification algorithm, which is dominated by one evaluation of SWIFFT. Our PK is 14.76 KB, which is the longest among all schemes. This fact is due to the function output length.

Table 2. Concrete evaluation with the most related efficient work. Security level 112 bits; W-OTS parameter $w = 2, 3$; $\ell = 224$ bits.

Scheme	Security OWF	PQ	Time (function evaluation) Gen	Sign	Ver	Space (bits) SK	PK	σ
W-OTS	SHA-224	Yes	351	176	176	seed	26,208	26,208
		Yes	553	277	277	seed	17,696	17,696
BC	SHA-224	Yes	229	encode	107	seed	51,296	23,968
vHP	DLP	No	4 exp	2 mult	3 exp	896	448	448
			2 mult	2 add	2 mult			
ZS	DLP	No	458 exp	encode	2 exp,	seed	51,296	241
			229 mult	107 add	107 mult			
Ours	SWIFFT	Yes	229	107 count	107 add,	seed	120,912	6,144
				encode	1 eval			

However, the public key size is not a main concern and can be reduced using a standard technique.

Comparing with DLP-based schemes, our scheme has much better time complexity, but longer signatures due to the SWIFFT input length. The ZS signature size is 241 bits. However, the key generation time takes 458 exponentiations.

Note that our verification algorithm is the fastest one among all schemes considered in this work.

Finally, observe that the signature coefficients are now in $[0, 54]$ on the average, which are considered to be small mod 257. If smaller values are needed, then we use the solutions of Remark 3. For example, we can use $p = 641$, which slightly increases the PK size.

Using AES128-Based OWF. The OWF may be implemented using primitives with smaller output length, using AES-128 for instance. Even in this case, our scheme improves on the most efficient scheme, W-OTS with $w = 3$, where the signature is $10\,112$ bits (1.65 times longer than our signature), and the key generation is 553 evaluations.

4.5   General CFF

There are essentially two methods to convert a OTS scheme into t-time signature scheme, Merkle hash tree and w-CFF with $w > 1$. The first one was mentioned earlier and is a standard technique, so we only comment on the general type of CFF.

A w-CFF with $w > 1$ allows to turn a OTS scheme into many/multiple-time scheme with relatively efficient time-complexity. However, the keys size becomes so large (hundreds of Mb to sign 1000 messages) that they become impractical. In theory too, it turns out that signing w messages using w instances of a 1-CFF instead of using a single w-CFF would reduce storage by a factor of $w/\log w$ [48]. The main advantage of 1-CFFs is their simple and efficient optimal construction, which also give easier reduction. A drawback is that the number of public keys to manage increases, but this can be solved using Merkle hash tree.

5   Conclusion and Future Work

We gave a one-time (and fixed-time) signature scheme that keeps the useful properties of those based on general one-way functions (post-quantum security and time-efficiency) while providing shorter signatures, thus improving significantly the main limitation of such schemes. Our verification algorithm is dominated by one evaluation of the hash function, which is a unique feature among all other related schemes. Accordingly, our scheme may be convenient for applications running on devices of limited resources.

Regular schemes (based on DLP and factoring) are insecure against quantum adversaries while our scheme is reduced to the security of SWIFFT functions, which are post-quantum collision-resistant as long as the SVP in ideal lattices is hard for quantum computers. Besides, our scheme is much more time-efficient. On the other hand, our signatures (and keys) are multiple times longer because of the function input and output length.

This work arises several possible extensions. A first one may be implementing the scheme, with or without Merkle tree, in order to provide more concrete evaluations. It is also important to provide a more accurate analysis of the concrete parameters involved in the security level. In particular, SWIFFT security level was estimated using the "best known" classical attack. Therefore, a future work should accomplish more detailed analysis of the parameters, and consider the best known quantum attacks.

An important work would be to improve further the signature (and preferably the keys) size. A straightforward method would be to find some family of functions having some homomorphic properties with small input (and output) size. The current function has input size that is 4 to 8 times larger than the general functions. Another important work would be to design an efficient t-time signature scheme, even for small t, without using Merkle tree.

Acknowledgments. The authors would like to thank Anne Broadbent for her comments on an earlier version of this paper, Andreas Hülsing for helping in the security proof, John Schanck for discussions on lattices, and Fang Song for discussions on the reduction of an earlier version. The authors would also like to thank the anonymous reviewers for their valuable comments.

This work is in part supported by Natural Sciences and Engineering Research Council (NSERC) of Canada, and CryptoWorks21.

Appendix: Other Useful Material

5.1 Technical Lemma

Lemma 1 ([48]). *Let χ_n be the probability distribution on $[n2^{\ell}]$ defined as $\chi_n = X_1 + \ldots X_n$ where X_i is the uniform distribution on $[2^{\ell}]$. The min-entropy of χ_n is then at least ℓ bits.*

5.2 SWIFFT Implementation

SWIFFT has efficient software implementations using number-theoretic or modular arithmetic FFT algorithm and its inherent parallelism for implementing multiplication. Importantly, the more the numbers are large the faster the FFT is when compared with the most efficient multiplications algorithms, which is in favour of FFT in the context of cryptography [13]. It was implemented using C and compiled using gcc version 4.1.2 on a PC running under Linux kernel 2.6.18 [27]. Tests on a 3.2 GHz Intel Pentium 4 show that the basic compression function can be evaluated in 1.5 μs on the above system, yielding a throughput close to 40 MB/s in a standard chaining mode of operation. For comparison, SHA-256 was tested on the same system using the highly optimized implementation in opcnssl version 0.9.8 (using the openssl speed benchmark), yielding a throughput of 47 MB/s when run on 8 KB blocks.

5.3 Encoding Algorithm

Cover-free family constructions and encoding algorithms have been studied in detail, and several approaches have been proposed [6,11,39,40]. In this work, we consider k-uniform 1-CFF where each subset has size k, and any two distinct subsets in $\mathcal{B}$ differ on at least one element. Importantly, 1-CFFs have an optimal construction, which consists of setting $k = \lfloor e/2 \rfloor$. Indeed, there is a simple and very efficient "ranking" algorithm, which is described in [3,48] and due to Cover [11]. The encoding algorithm requires k subtractions (or additions) and about e comparisons using some pre-computation. Here is the pseudo-code by Bicakci, Tung, and Tsudik [3] after their quotation; "To put things in perspective, consider that a single MD5 hash computation requires approximately 500 arithmetic operations. Thus, our mapping (in both directions) costs less than one MD5 hash.".

```
Input : Message m, set E=[1,e], subset size k;
Output: Unique k-subset of E (k-dimensional vector a);
q:=1;
for i=1 to k do
    while m > Binomial(e-q,k-i) do
        m:=m - Binomial(e-q,k-i);
        q:=q+1;
    end while;
```

```
   a[i]:=q;
   q:=q+1;
end for;
```

For example, to encode 3-bit messages, we need a set of size 5 and subsets of size 2 so that the total number of subsets is at least $2^3 = 8$.

```
S( 1 )=  [ 1, 2 ]
S( 2 )=  [ 1, 3 ]
S( 3 )=  [ 1, 4 ]
S( 4 )=  [ 1, 5 ]
S( 5 )=  [ 2, 3 ]
S( 6 )=  [ 2, 4 ]
S( 7 )=  [ 2, 5 ]
S( 8 )=  [ 3, 4 ]
>
```

5.4   Winternitz OTS (W-OTS)

Winternitz [30, 32] suggested to Merkle the idea, called W-OTS, of signing several bits simultaneously on the expense of more evaluations of the one-way function, thus trading time for space. Given a small positive integer w, the secret key is $SK := (x_1, \ldots, x_t)$ and public key is $PK := (y_1 || \ldots || y_t)$ where $y_i = f^{2^w - 1}(x_i)$. Here $t = \lceil \ell/w \rceil + \lceil \lceil \log 2^w \ell/w \rceil /w \rceil$ and f^k means the k-fold composition of f with itself. To sign an ℓ-bit M, split it into w-bit blocks (including a check sum), $d_1, \ldots, d_t$, then the signature is $(s_1, \ldots, s_t)$ where d_i is treated as an integer and $s_i = f^{d_i}(x_i)$ for $1 \leq i \leq t$. Verifying consists in forming the blocks as before, computing $y_i = f^{2^w - 1 - d_i}(s_i)$, and accepting if and only it verifies with PK.

5.5   Signing Many Messages

It is easy to transform our OTS scheme into a t-time signature scheme, using the same techniques as for schemes based on OWF. Indeed, the security game can be generalized as follows. Run the key generation algorithm t times to obtain $SK_1, \ldots, SK_t$ and $PK_1, \ldots, PK_t$. Give $\mathcal{A}$ the keys PK_i for $1 \leq i \leq t$ and signatures for at most t adaptive messages, one signature per key, where t is a polynomial in the security parameter. Let $Q = \{M_1, \ldots, M_t\}$ be the set of queries asked by $\mathcal{A}$ and $\{\sigma_1, \ldots, \sigma_t\}$ the corresponding signatures. Finally, $\mathcal{A}$ outputs (M, σ, i). The probability that $\mathcal{A}$ forges a signature is defined to be $\text{PROB}^{\text{uf-cma}}_{\text{DSS}, \mathcal{A}} = \Pr[(M \notin Q) \wedge \text{Ver}(PK_i, M, \sigma) = 1]$. Note that the signature now include a number i to indicate the PK with which to verify. However, the standard approach to sign many messages is to use Merkle tree.

References

1. Abdalla, M., Reyzin, L.: A new forward-secure digital signature scheme. In: Okamoto, T. (ed.) ASIACRYPT 2000. LNCS, vol. 1976, pp. 116–129. Springer, Heidelberg (2000)

2. Bellare, M., Shoup, S.: Two-tier signatures, strongly unforgeable signatures, and fiat-shamir without random oracles. In: Okamoto, T., Wang, X. (eds.) PKC 2007. LNCS, vol. 4450, pp. 201–216. Springer, Heidelberg (2007)
3. Bicakci, K., Tung, B., Tsudik, G.: How to construct optimal one-time signatures. J. Comput. Netw. **43**(3), 339–349 (2003)
4. Bleichenbacher, D., Maurer, U.M.: Directed acyclic graphs, one-way functions and digital signatures. In: Desmedt, Y.G. (ed.) CRYPTO 1994. LNCS, vol. 839, pp. 75–82. Springer, Heidelberg (1994)
5. Bleichenbacher, D., Maurer, U.M.: Optimal tree-based one-time digital signature schemes. In: Puech, C., Reischuk, R. (eds.) STACS 1996. LNCS, vol. 1046, pp. 361–374. Springer, Heidelberg (1996)
6. Bos, J.N.E., Chaum, D.: Provably unforgeable signatures. In: Brickell, E.F. (ed.) CRYPTO 1992. LNCS, vol. 740, pp. 1–14. Springer, Heidelberg (1993)
7. Buchmann, J., Dahmen, E., Ereth, S., Hülsing, A., Rückert, M.: On the security of the winternitz one-time signature scheme. In: Nitaj, A., Pointcheval, D. (eds.) AFRICACRYPT 2011. LNCS, vol. 6737, pp. 363–378. Springer, Heidelberg (2011)
8. Buchmann, J., Dahmen, E., Hülsing, A.: XMSS - a practical forward secure signature scheme based on minimal security assumptions. In: Yang, B.-Y. (ed.) PQCrypto 2011. LNCS, vol. 7071, pp. 117–129. Springer, Heidelberg (2011)
9. Buchmann, J., García, L.C.C., Dahmen, E., Döring, M., Klintsevich, E.: CMSS – an improved merkle signature scheme. In: Barua, R., Lange, T. (eds.) INDOCRYPT 2006. LNCS, vol. 4329, pp. 349–363. Springer, Heidelberg (2006)
10. Buchmann, J., Lindner, R., Rückert, M., Schneider, M.: Post-quantum cryptography: lattice signatures. Computing **85**(1–2), 105–125 (2009)
11. Cover, T.: Enumerative source encoding. IEEE Trans. Inf. Theor. **19**(1), 73–77 (1973)
12. Dang, Q.: Randomized Hashing for Digital Signatures. NIST Special Publication 800–106 (2009)
13. David, J.P., Kalach, K., Tittley, N.: Hardware complexity of modular multiplication and exponentiation. IEEE Trans. Comput. **56**(10), 1308–1319 (2007)
14. Diffie, W., Hellman, M.: New directions in cryptography. IEEE Trans. Inf. Theor. **22**(6), 644–654 (1976)
15. Dodis, Y., Katz, J.: Chosen-ciphertext security of multiple encryption. In: Kilian, J. (ed.) TCC 2005. LNCS, vol. 3378, pp. 188–209. Springer, Heidelberg (2005)
16. Dods, C., Smart, N.P., Stam, M.: Hash based digital signature schemes. In: Smart, N.P. (ed.) Cryptography and Coding 2005. LNCS, vol. 3796, pp. 96–115. Springer, Heidelberg (2005)
17. Barker, E., William Barker, W., Smid, M.: Recommendation for Key Management - Part 1: General (Revision 3), NIST Special Publication 800–57, July 2012
18. Elgamal, T.: A public key cryptosystem and a signature scheme based on discrete logarithms. IEEE Trans. Inf. Theor. **31**(4), 469–472 (1985)
19. Even, S., Goldreich, O., Micali, S.: On-line/off-line digital signatures. J. Cryptology **9**(1), 35–67 (1996)
20. Groth, J.: Simulation-sound NIZK proofs for a practical language and constant size group signatures. In: Lai, X., Chen, K. (eds.) ASIACRYPT 2006. LNCS, vol. 4284, pp. 444–459. Springer, Heidelberg (2006)
21. Grover, L.K.: Quantum mechanics helps in searching for a needle in a haystack. Phys. Rev. Lett. **79**(2), 325–328 (1997)
22. van Heyst, E., Pedersen, T.P.: How to make efficient fail-stop signatures. In: Rueppel, R.A. (ed.) EUROCRYPT 1992. LNCS, vol. 658, pp. 366–377. Springer, Heidelberg (1993)

23. Huang, Q., Wong, D.S., Zhao, Y.: Generic transformation to strongly unforgeable signatures. In: Katz, J., Yung, M. (eds.) ACNS 2007. LNCS, vol. 4521, pp. 1–17. Springer, Heidelberg (2007)
24. Johnson, D., Menezes, A., Vanstone, S.: The elliptic curve digital signature algorithm (ECDSA). Int. J. Inf. Secur. 1(1), 36–63 (2001)
25. Lamport, L.: Constructing digital signatures from a one-way function. Technical report, SRI International Computer Science Laboratory (1979)
26. Lyubashevsky, V., Micciancio, D.: Asymptotically Efficient Lattice-Based Digital Signatures. Cryptology ePrint Archive, Report 2013/746 (2013)
27. Lyubashevsky, V., Micciancio, D., Peikert, C., Rosen, A.: SWIFFT: a modest proposal for FFT hashing. In: Nyberg, K. (ed.) FSE 2008. LNCS, vol. 5086, pp. 54–72. Springer, Heidelberg (2008)
28. Mehta, M., Harn, L.: Efficient one-time proxy signatures. IEE Proc. Commun. 152(2), 129–133 (2005)
29. Menezes, A.J., van Oorschot, P.C., Vanstone, S.A.: Handbook of Applied Cryptography. CRC Press, Boca Raton (1996)
30. Merkle, R.C.: Secrecy, Authentication, and Public Key Systems. Ph.D. thesis, Stanford University (1979)
31. Merkle, R.C.: A digital signature based on a conventional encryption function. In: Pomerance, C. (ed.) CRYPTO 1987. LNCS, vol. 293, pp. 369–378. Springer, Heidelberg (1988)
32. Merkle, R.C.: A certified digital signature. In: Brassard, G. (ed.) CRYPTO 1989. LNCS, vol. 435, pp. 218–238. Springer, Heidelberg (1990)
33. Micciancio, D.: Personal communication
34. Micciancio, D.: Generalized Compact Knapsacks Cyclic Lattices and Efficient One-Way Functions. Computational Complexity 16(4), 365–411 (2007). preliminary version in FOCS 2002
35. Mohassel, P.: One-time signatures and chameleon hash functions. In: Biryukov, A., Gong, G., Stinson, D.R. (eds.) SAC 2010. LNCS, vol. 6544, pp. 302–319. Springer, Heidelberg (2011)
36. Naor, D., Shenhav, A., Wool, A.: One-Time Signatures Revisited: Have They Become Practical? Cryptology ePrint Archive, Report 2005/442 (2005)
37. Pedersen, T.P.: Non-interactive and information-theoretic secure verifiable secret sharing. In: Feigenbaum, J. (ed.) CRYPTO 1991. LNCS, vol. 576, pp. 129–140. Springer, Heidelberg (1992)
38. Perrig, A.: The BiBa one-time signature and broadcast authentication protocol. In: Proceedings of the 8th ACM Conference on Computer and Communications Security, CCS 2001, pp. 28–37 (2001)
39. Pieprzyk, J., Wang, H., Xing, C.: Multiple-time signature schemes against adaptive chosen message attacks. In: Matsui, M., Zuccherato, R.J. (eds.) SAC 2003. LNCS, vol. 3006, pp. 88–100. Springer, Heidelberg (2004)
40. Reyzin, L., Reyzin, N.: Better than BiBa: short one-time signatures with fast signing and verifying. In: Batten, L.M., Seberry, J. (eds.) ACISP 2002. LNCS, vol. 2384, pp. 144–153. Springer, Heidelberg (2002)
41. Rivest, R.L., Shamir, A., Adleman, L.: A method for obtaining digital signatures and public-key cryptosystems. Commun. ACM 21(2), 120–126 (1978)
42. Rohatgi, P.: A compact and fast hybrid signature scheme for multicast packet authentication. In: Proceedings of the 6th ACM Conference on Computer and Communications Security, CCS 1999, pp. 93–100 (1999)

43. Seys, S., Preneel, B.: Power consumption evaluation of efficient digital signature schemes for low power devices. In: IEEE International Conference on Wireless and Mobile Computing, Networking and Communications, vol. 1, pp. 79–86 (2005)
44. Shor, P.W.: Polynomial-time algorithms for prime factorization and discrete logarithms on a quantum computer. SIAM J. Comput. **26**(5), 1484–1509 (1997)
45. Song, F.: A note on quantum security for post-quantum cryptography. In: Mosca, M. (ed.) PQCrypto 2014. LNCS, vol. 8772, pp. 246–265. Springer, Heidelberg (2014)
46. Stinson, D.R.: Cryptography: Theory and Practice, 3rd edn. Chapman and Hall/CRC, Boca Raton (2005)
47. Vaudenay, S.: One-time identification with low memory. In: Camion, P., Charpin, P., Harari, S. (eds.) EUROCODE 1992. International Centre for Mechanical Sciences, CISM Courses and Lectures, vol. 339, pp. 217–228. Springer, Heidelberg (1992)
48. Zaverucha, G.M., Stinson, D.R.: Short one-time signatures. Adv. Math. Commun. **5**, 473–488 (2011)

Constructing Lightweight Optimal Diffusion Primitives with Feistel Structure

Zhiyuan Guo[1,2,3]($\boxtimes$), Wenling Wu[1,2,3], and Si Gao[1,2,3]

[1] TCA Laboratory, SKLCS, Institute of Software,
Chinese Academy of Sciences, Beijing, China
{guozhiyuan,wwl}@tca.iscas.ac.cn
[2] State Key Laboratory of Cryptology, P.O. Box 5159, Beijing 100878, China
[3] University of Chinese Academy of Sciences, Beijing, China

Abstract. As one of the core components in any SPN block cipher and hash function, diffusion layers are mainly introduced by matrices with maximal branch number. Surprisingly, the research on optimal binary matrices is rather limited compared with that on MDS matrices. Especially, not many general constructions for binary matrices are known that give the best possible branch number and guarantee the efficient software/hardware implementations as well. In this paper, we propose a new class of binary matrices constructed by Feistel structure with bit permutation as round functions. Through investigating bounds on the branch number our structure can achieve, we construct optimal binary matrices for a series of parameters with the lowest hardware cost up to now. Compared to the best known results, our optimal solutions for size 16×16 and 32×32 can save about 20 % and 33.3 % gate equivalents respectively. Without loss of hardware efficiency, a list of software-friendly optimal binary matrices can be constructed by Feistel structure with cyclic shift as round functions. The characteristics of this class of matrices are summarized and involutory optimal instances with commonly used dimensions are also provided. In the case of 8×8, we prove that optimal matrices from our structure can not be involutory. Finally, we extend the strategy to Generalized Feistel Structure and present some typical experimental results.

Keywords: Lightweight cryptography · Diffusion layer · Optimal binary matrix · Feistel structure · Multiple platforms

1 Introduction

As a central part of Substitution-Permutation Networks, diffusion layers are very important for the overall security and efficiency of cryptographic schemes. On the one hand, they play a role in spreading internal dependencies, which contributes to enhancing the resistance of statistical cryptanalysis. On the other hand, with the rapid development of lightweight cryptography, designing hardware-efficient diffusion layers has already been a hot research topic due to the increasing importance of ubiquitous computing.

© Springer International Publishing Switzerland 2016
O. Dunkelman and L. Keliher (Eds.): SAC 2015, LNCS 9566, pp. 352–372, 2016.
DOI: 10.1007/978-3-319-31301-6_21

The quality of a diffusion layer is connected to its branch number, whose cryptographic significance corresponds to the minimal number of active S-boxes in any two consecutive rounds. Obviously, the larger the branch number is, the better the diffusion effect will be, and simultaneously the cipher will not be vulnerable to unexpected attacks. Therefore, most designers chose to focus on diffusion layers with the best possible branch number to ensure a relatively strongest security.

From a coding theory perspective, Maximum Distance Separable (MDS) codes are quite good choices for the construction of diffusion layers since their branch numbers are maximum (known as the Singleton bound [1]). Not only are MDS matrices used in many block ciphers [2–4], but they promote generations of various related design strategies [5–7]. However, a problem with using MDS matrices is that they usually come at the price of a less efficient implementation. Due to Galois field multiplications, hardware implementations will often suffer from an important area requirement, with the result that MDS matrices are not suitable for the resource-constrained environments, such as RFID systems and sensor networks. Although this unfavorable situation is greatly improved with the advent of recursive MDS matrices [8–10], the temporary memory required (and hence hardware area) for the computation of matrices is still not reduced to a degree of satisfaction sometimes.

Another attractive type of diffusion layers is derived from Maximum Distance Binary Linear (MDBL) codes. The corresponding binary matrices are optimal in the sense that they achieve the largest possible branch number. Though the diffusion speed of optimal binary matrices can not keep pace with the one of MDS matrices, it is an overwhelming advantage that they involve no finite field multiplication, which is more propitious to a low-cost implementation. Typical examples are block ciphers E2 [11], Camellia [12] and ARIA [13], who get an excellent hardware efficiency and remarkable software performance on various platforms as well. It is accordingly our belief that, in many cases, it is easier to obtain an overall construction through using optimal binary matrix (or in general a matrix with branch number not meeting the Singleton bound), despite sacrificing the diffusion speed to a certain extent.

Compared with the study on constructions of MDS matrices, the research on designs of MDBL matrices is rather limited [14,15]. Early strategy from [16] (partially) guided the design of diffusion layers in E2 and Camellia, and unified method presented in [17] was conducive to summarizing the characteristics of 8×8 optimal binary matrices. For constructions of large dimensions (e.g. 16×16 and 32×32), designers in [18,19] considered combining small matrices into bigger ones, where each block matrix corresponds a finite field element. Indeed, in our opinion, the generalities of most previous constructions (focusing only on a few dimensions) are very weak, not to mention making them have efficient implementation. Here, one exception is the proposal of Dehnavi et al. [20], who recently investigated a special kind of binary linear layers for commonly used sizes with efficient implementation.

Feistel structure is one of the most prominently used structures in cryptography and accounts for substantial portion of data encrypted today. This was facilitated by the introduction of DES [21], which indicated the generation of modern block cipher. Not only is this classical structure used in large quantities of symmetric-key algorithms, but it inspires plenty of designs of cryptographic primitives. For example, S-boxes in [22–24] are constructed by 3-round Feistel structure, and linear layers in E2 and Camellia are implicitly implemented with 4-round Feistel structure.

Our Contributions. In this paper, we propose constructing diffusion layers over $\mathbb{F}_2$ with maximal branch number and efficient hardware implementation by use of Feistel structure with bit permutation as round functions. After introducing necessary notations and concepts in Sect. 2, we investigate the bounds on the branch number this construction can achieve, which will later help us judge whether hardware efficiency (hereby focus mainly on the area and latency) of the resulting matrix is optimal. Meanwhile, taking account of the improvement of software performance, we restrict the round function to cyclic shift and give the overall search strategy for "optimal solutions" in Sect. 3.

In order to demonstrate the generality of our construction, we provide typical optimal solutions for a series of feasible parameters (up to 32) in Sect. 4. To the best of our knowledge, the hardware cost of most proposals is the lowest compared with previous results. For cryptographic applications, our focus is further placed on involutory optimal solutions with commonly used dimensions in Sect. 5, and we prove that it is impossible to obtain an involutory 8×8 optimal diffusion layer from this structure.

Along similar lines, we present diffusion layers constructed by Generalized Feistel Structure in Sect. 6, improving their applicabilities on other platforms without loss of hardware efficiency. According to figures listed in Sect. 7, we afterwards show that optimal solutions for size 16×16 and 32×32 can save about 20 % and 33.3 % gate equivalents respectively, compared to the best known results. Finally, we conclude the paper in Sect. 8.

2   Preliminaries

In this section, we fix the basic notions and further more introduce several judgement methods of branch number. Since diffusion layers investigated in the present paper are linear transformations on the n-dimensional vector space over $\mathbb{F}_2$, we directly use an $n \times n$ binary matrix to represent a linear layer in the subsequent discussions.

2.1   Branch Number

Assume $\mathbf{v} = (v_1, v_2, \ldots, v_n)^T$ is a vector such that $v_i \in \mathbb{F}_2$, $1 \leq i \leq n$. Then the Hamming weight of $\mathbf{v}$, denoted by $w_b(\mathbf{v})$, is equal to the number of non-zero elements in $\mathbf{v}$.

Definition 1. *[3] The differential branch number of a diffusion layer D is given by*

$$\mathcal{B}_d(D) = \min_{\mathbf{v} \neq 0}\{w_b(\mathbf{v}) + w_b(D(\mathbf{v}))\}. \tag{1}$$

Analogously, we can define the linear branch number.

Definition 2. *[3] The linear branch number of a diffusion layer D is given by*

$$\mathcal{B}_l(D) = \min_{\mathbf{v} \neq 0}\{w_b(\mathbf{v}) + w_b(D^T(\mathbf{v}))\}, \tag{2}$$

where D^T is the transposition of D.

As the differential branch number of an $n \times n$ linear transformation is equal to the minimum distance of its associated $[2n, n]$ linear code, the maximal $\mathcal{B}_d$ of a binary matrix is known for small dimension according to [25]. A binary matrix is optimal if it achieves the maximal $\mathcal{B}_d$ and $\mathcal{B}_l$. Since each $n \times n$ (with the exception of $n = 32$) diffusion layer over $\mathbb{F}_2$ constructed in this article satisfies $\mathcal{B}_d = \mathcal{B}_l$, we omit linear branch number in the sequel.

Definition 3. *[14] Two matrices A, B are permutation homomorphic to each other if there exists a row permutation ρ and a column permutation γ satisfying*

$$\rho(\gamma(A)) = \gamma(\rho(A)) = B. \tag{3}$$

Proposition 1. *[14] If two matrices A, B are permutation homomorphic to each other, then A, B are of the same branch number.*

2.2 Judgement Methods

There is a one-to-one correspondence between an $n \times n$ linear transformation $\theta(x) = M \cdot x$ and a linear code $\mathcal{C}_\theta$ with the generator matrix $G_\theta = [I_{n \times n}|M]$, so we can use the following property to determine $\mathcal{B}_d(M)$.

Proposition 2. *[3] A linear code has minimum distance d if and only if every $d - 1$ columns of its parity check matrix are linearly independent and there exists some set of d columns that are linearly dependent.*

To deal with an $n \times n$ binary matrix with branch number s, it costs approximately $\sum_{i=1}^{s}\binom{2n}{i}$ Gaussian eliminations according to Proposition 2, while it needs to exhaust all possible non-zero input vectors based on Definition 1. After analyzing the characteristics of minimum-weight codewords among $\mathcal{C}_\theta$, we give Algorithm 2 (cf. Appendix A for more details) as main detection method, reducing the time complexity to $2\sum_{i=1}^{\lfloor s/2 \rfloor}\binom{n}{i}$.

3   On Properties of Proposed Diffusion Layers

3.1   The General Construction

Throughout this paper, we consider diffusion layers over $\mathbb{F}_2$ constructed by Feistel structure with bit permutation as round functions. Let $x = (x_L, x_R)$ and $y = (y_L, y_R)$ be the n-bit input and output respectively, then a diffusion layer shown in Fig. 1 can be characterized as

$$M = \begin{pmatrix} 0 & I \\ I & 0 \end{pmatrix} \begin{pmatrix} P_r & I \\ I & 0 \end{pmatrix} \cdots \begin{pmatrix} P_1 & I \\ I & 0 \end{pmatrix}, \tag{4}$$

where the size of each block matrix is $\frac{n}{2} \times \frac{n}{2}$. In the rest of this paper, we only extract the sequence of permutation matrices, namely, $[P_1, P_2, \ldots, P_r]$ to represent M for simplicity. According to [21], it holds that

$$M^{-1} = [P_r, P_{r-1}, \ldots, P_1], \tag{5}$$

which means the inverse matrix could be implemented with the same structure by simply reversing the order of round functions. This decided advantage guarantees the diffusion layer and its inverse require equal XOR's in terms of hardware implementation. In particular, the encryption and decryption can even use the exact same circuit in the case of involutory instances, i.e. round transformations appearing symmetrically (e.g. $[P_1, P_2, P_1]$ and $[P_1, P_2, P_2, P_1]$).

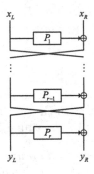

Fig. 1. A diffusion layer over $\mathbb{F}_2$ constructed with Feistel structure

As hardware efficiency can have very different meanings depending on the utilization scenario targeted by the designer, we hereby chose to focus on two classical metrics: silicon area and latency. Clearly, to make the perfect diffusion layer hardware-optimal under this construction, the number of iterations should be as small as possible on the premise of maximal branch number. Therefore it is necessary to investigate the bound on the branch number of resulting diffusion layer.

3.2   Upper Bound of the Branch Number

Before elaborating our main theorem, we need to introduce the following novel observation. Remember that in technical terms, the Fibonacci sequence $F(n)$ is defined by the recurrence relation $F(n+2) = F(n+1) + F(n)$, with seed values $F(0) = 1$, $F(1) = 1$.[1]

Lemma 1. *Assume $(0, \alpha)$ is the input of the structure shown in Fig. 1 such that $w_b(\alpha) = 1$. Then the Hamming weight of the output of round i, $w_b(y_i)$, is upper bounded by the i-th number of the Fibonacci sequence:*

$$w_b(y_i) \leq F(i). \tag{6}$$

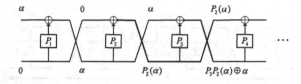

Fig. 2. Propagation of the Hamming weight on the input $(0, \alpha)$

Proof. Let us illustrate it by mathematical induction. According to the output of the first three rounds (see Fig. 2), we know that $w_b(y_1) = F(1)$, $w_b(y_2) = F(2)$ and $w_b(y_3) \leq F(3)$.

Now suppose the induction hypothesis is true for round i, $3 \leq i < r$. Then we only need to prove $w_b(y_{i+1}) \leq F(i+1)$. Notice that the transformation of round $i+1$ can be represented as

$$\begin{cases} y_{i+1,L} = P_{i+1}(y_{i,L}) \oplus y_{i,R} \\ y_{i+1,R} = y_{i,L} \end{cases}$$

and we always have $w_b(y_{i,L}) = w_b(P_{i+1}(y_{i,L}))$, which implies $w_b(y_{i+1,L}) \leq w_b(y_{i,L}) + w_b(y_{i,R})$.
Likewise, we obtain

$$w_b(y_{i+1,R}) = w_b(y_{i,L}) \leq w_b(y_{i-1,L}) + w_b(y_{i-1,R}).$$

Thus it holds

$$\begin{aligned} w_b(y_{i+1}) &= w_b(y_{i+1,L}) + w_b(y_{i+1,R}) \\ &\leq w_b(y_{i,L}) + w_b(y_{i,R}) + w_b(y_{i-1,L}) + w_b(y_{i-1,R}) \\ &\leq F(i) + F(i-1) \\ &= F(i+1), \end{aligned}$$

and we complete the proof. □

---

[1] Alternatively, the chosen starting points are fixed to $F(0) = 0$, $F(1) = 1$, which has no substantial impact on the global sequence.

Theorem 1. *The branch number of the diffusion layer constructed as in Fig. 1 satisfies*

$$\mathcal{B}_d^{(r)} \leq \begin{cases} 2F\left(\frac{r+1}{2}\right) & r \text{ is odd} \\ F\left(\frac{r}{2}\right) + F\left(\frac{r}{2}+1\right) & r \text{ is even,} \end{cases} \tag{7}$$

where the superscript is used to emphasize the number of rounds in the proposed construction.

Proof. Our strategy is similar to the start-from-the-middle technique [26]: start with a particular state value at the middle round and then propagate forward and backward to the output and input of the Feistel structure respectively. Note that middle round means different positions depending on whether the number of rounds is odd or even.

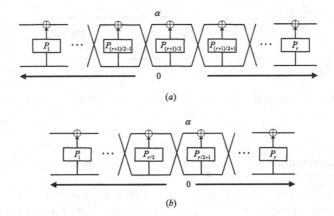

Fig. 3. Upper bound of the branch number of proposed diffusion layers

When r is odd, let $(0, \alpha)$ be the input of round $(r+1)/2$ such that $w_b(\alpha) = 1$. For $r = 1$, it is easy to see that $w_b(y_0) + w_b(y_1) = 2F(1)$. For $r \geq 3$, both the forward and backward propagations begin with the same initial value[2] $(\alpha, 0)$ and contain $(r+1)/2 - 1$ rounds (see Fig. 3.(a)). According to Lemma 1, we obtain one input/output pair whose Hamming weight satisfies

$$w_b(y_0) \leq F\left(\frac{r+1}{2}\right), \quad w_b(y_r) \leq F\left(\frac{r+1}{2}\right),$$

which implies the branch number of the resulting diffusion layer is at most $2F((r+1)/2)$.

When r is even, we change the target position to round $r/2 + 1$, with the result that each direction consists of $r/2$ rounds (see Fig. 3.(b)). Likewise it holds

$$w_b(y_0) + w_b(y_r) \leq F\left(\frac{r}{2}+1\right) + F\left(\frac{r}{2}\right),$$

---

[2] Notice that we consider the input of round $(r+1)/2 + 1$ as the forward starting point.

since the inputs of the backward and forward direction are $(\alpha, 0)$ and $(0, \alpha)$ respectively. Hence we complete the proof.   □

Remark 1. For an expected branch number, Theorem 1 gives insights on the lower bound on the number of rounds our construction should have, efficiently reducing a lot of unnecessary search works. As an illustration, we need at least 8 iterations in the Feistel structure to get a diffusion layer with $\mathcal{B}_d = 12$ due to $\mathcal{B}_d^{(7)} \leq 2F(4) = 10$.

3.3 Search Strategy for Software-Friendly Diffusion Layers

In this section, we will explain how to improve software performances of the proposed diffusion layers without loss of hardware efficiency. Compared with the bit permutation, cyclic shift is undoubtedly much more attractive as suitable rotation can be implemented as a single instruction on the corresponding processor. For example, while constructing a 16×16 binary matrix with cyclic shift as round transformations, all operations of each round are based on 32-bit words on condition that 4-bit S-boxes are used. As a result, instead of bit permutation, cyclic shift is our first choice and the round function is afterwards restricted to $P_i(x) = x <<< t_i, \, 0 \leq t_i < n/2$.

The pseudo-code of our basic search procedure is shown in Algorithm 1. The function BASICSEARCH $(n, r, T, \mathbb{G})$ returns all $n \times n$ binary matrices with $\mathcal{B}_d \geq T$ constructed by r-round Feistel structure. Here $\mathbb{G}$ denotes the set of transformation matrices that can be selected as round functions. On the basis of the above strategy, we initialize it to the set of matrices representing cyclic shift (which implies $|\mathbb{G}| = n/2$) and begin the first attempt with minimum possible r according to Theorem 1. If no optimal solution is found (i.e. $\mathbb{E} = \emptyset$), choose to increase r or relax restrictions on some round functions to continue searching, until suboptimal solutions are returned.

Algorithm 1. Search for optimal diffusion layers over $\mathbb{F}_2$
___
1: **function** BASICSEARCH$(n, r, T, \mathbb{G})$
2:     $\mathbb{E} \leftarrow \emptyset$
3:     **for all** $M \in \{ [P_1, P_2, \ldots, P_r] \mid P_i \in \mathbb{G}, 1 \leq i \leq r \}$ **do**
4:         **if** $\mathcal{B}_d(M) \geq T$ **then**
5:             $\mathbb{E} \leftarrow \mathbb{E} \cup \{M\}$
6:         **end if**
7:     **end for**
8:     **return** $\mathbb{E}$
9: **end function**
___

Remark 2. "Optimal solutions" here refer to binary matrices with maximal $\mathcal{B}_d$ constructed by the least possible number of cyclic shift operations. For instance, an 8×8 matrix with $\mathcal{B}_d = 5$ constructed by 4 cyclic shifts is optimal solution owing to $\mathcal{B}_d^{(3)} \leq 4$. Moreover, "suboptimal solutions" have different forms since bigger r results in higher cost in hardware implementation, while enlarged $\mathbb{G}$ (from the set of cyclic shift matrices to the one of permutation matrices) leads to the loss of advantages in software performance. Consequently, there are various trade-offs when we search for suboptimal solutions[3].

Next, we introduce the following statement to relate certain matrices that lead to the same branch number.

Theorem 2. *For any diffusion layer $M = [P_1, P_2, \ldots, P_r]$ constructed by r-round Feistel structure, there always exists a corresponding $M' = [I, P_2', \ldots, P_r']$ such that $\mathcal{B}_d(M') = \mathcal{B}_d(M)$.*

Proof. First of all, it is not difficult to see that we can place P_1 after the XOR operation in round 1 as shown in Fig. 4.(b), since

$$y_{1,L} = P_1(x_L) \oplus x_R = P_1(x_L \oplus P_1^{-1}(x_R)).$$

By using similar equivalent transforms, P_1 can be moved to the end of the structure (see Fig. 4.(c) and (d)), with each round function redefined as $P_i' = P_i \cdot P_1$ (i is even) or $P_i' = P_1^{-1} \cdot P_i$ (i is odd). Then depending on whether the number of rounds is even or odd, it holds

$$M = \begin{pmatrix} P_1 & 0 \\ 0 & I \end{pmatrix} \cdot M' \cdot \begin{pmatrix} I & 0 \\ 0 & P_1^{-1} \end{pmatrix} \; or \; M = \begin{pmatrix} I & 0 \\ 0 & P_1 \end{pmatrix} \cdot M' \cdot \begin{pmatrix} I & 0 \\ 0 & P_1^{-1} \end{pmatrix},$$

respectively, which means M and M' are permutation homomorphic to each other. Thus their branch numbers are equal according to Proposition 1 and we complete the proof. □

Note that all matrices constructed by r-round Feistel structure can be classified according to any P_i, $i \in \{1, \ldots r\}$, although we simply choose $P_1 = I$ in Theorem 2. In other words, each diffusion layer constructed with $P_1 = I$ is a representative of an equivalence class, from which one can obtain all diffusion layers in the same equivalence class through analogous transforms mentioned above. We will make use of this property to reduce the search space by one round in the subsequent experiment.

4   Constructing Optimal Diffusion Layers with Feistel Structure

In this section, we will provide the results on constructing diffusion layers for various parameters. According to the search strategy, cyclic shift is preferred choice and for convenience, we abuse the symbol R_i, $i = 0, \ldots, n/2 - 1$, to

---

[3] As we take maximal branch number as primary premise, the branch number of suboptimal solutions is equal to that of optimal solutions.

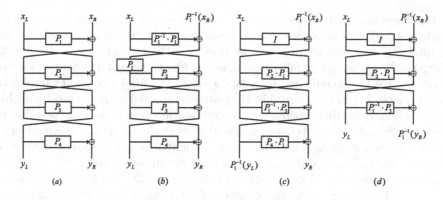

Fig. 4. Equivalence partitioning of the proposed diffusion layers

represent the matrix which corresponds to the transformation $L(x) = x <<< i$, where the size of x is $n/2$. As an example, an 8×8 matrix $M = [R_0, R_3, R_2, R_1]$ denotes the diffusion layer constructed by the following round functions:

$$P_1 = \begin{pmatrix} 1\,0\,0\,0 \\ 0\,1\,0\,0 \\ 0\,0\,1\,0 \\ 0\,0\,0\,1 \end{pmatrix}, \ P_2 = \begin{pmatrix} 0\,0\,0\,1 \\ 1\,0\,0\,0 \\ 0\,1\,0\,0 \\ 0\,0\,1\,0 \end{pmatrix}, \ P_3 = \begin{pmatrix} 0\,0\,1\,0 \\ 0\,0\,0\,1 \\ 1\,0\,0\,0 \\ 0\,1\,0\,0 \end{pmatrix}, \ P_4 = \begin{pmatrix} 0\,1\,0\,0 \\ 0\,0\,1\,0 \\ 0\,0\,0\,1 \\ 1\,0\,0\,0 \end{pmatrix}.$$

Clearly, R_0 is the representation of identity matrix and more R_0's implies more efficient software implementation.

4.1 Diffusion Layers for $n = 4, 8, 16$ and 32

First, we ran Algorithm 1 to search for optimal diffusion layers with commonly used dimensions in cryptography, that is, $n = 2^k$, where $k = 2, 3, 4, 5$. The total number of optimal solutions and typical instance of M we obtained are summarized in Table 1, accompanied by the cost in hardware implementation for each parameter. Notice that the branch number of 32×32 binary matrices we can find is 12, which is consistent with the best known value.

Table 1. Experimental results for optimal diffusion layers with $n = 4, 8, 16$, and 32

n	$\mathcal{B}_d$	Optimal solutions		XOR gates
		Total number	Example of M	
4	4	2	$[R_0, R_1, R_0]$	6
8	5	32	$[R_0, R_1, R_2, R_0]$	16
16	8	9760	$[R_0, R_1, R_1, R_2, R_2, R_0]$	48
32	$12^\dagger$	6272	$[R_0, R_1, R_1, R_{13}, R_{13}, R_0, R_8, R_6]$	128

The time complexity of the exhaustive search for optimal solutions with given length n is $(n/2)^r$, where r is the least possible number of rounds to achieve the maximal $\mathcal{B}_d$. Actually it took us less than 20 h to find all the best 32×32 binary matrices through parallel search, and with the help of Theorem 2, we can further reduce the search time by a factor of 2^4. Additionally, the examples in Table 1 are perfect in the sense that they are constructed with the most possible number of R_0's, cutting down on as many rotation instructions as possible in software implementation. Yet it is noteworthy that the technique of equivalence partitioning is not applicable in this case, as matrices in the same equivalence class are constructed by different sequences of round functions.

Moreover, we need to point out for any $n \times n$ diffusion layer constructed by r-round Feistel structure, the number of XOR gates required for hardware implementation is $nr/2$, which enjoys an overwhelming advantage even if we have not done any other optimization. For instance, although the total number of ones in any 32×32 binary matrix with $\mathcal{B}_d = 12$ is lower bounded by $11 \times 32 = 352$, each of our optimal solutions can be implemented just with 128 XOR's, which is the best result up to our knowledge.

4.2 Results for Other Parameters

To illustrate the generality of our proposal, we also search for diffusion layers with other sizes (n is even and $n < 20$), despite the fact that they are probably not often used. As shown in Table 2, we obtain optimal solutions for each given length, with the exception of $n = 12$. In other words, no 12×12 optimal binary matrix can be constructed by 6-round Feistel structure with cyclic shift as round functions. This means we need to consider searching for suboptimal solutions using the methods described in Sect. 3.3.

Laying particular stress on hardware efficiency, we hereby adopt the second strategy (i.e. enlarging $\mathbb{G}$ for only one round function) and find 120 suboptimal solutions. One of the best results is

$$M_{12\times12} = [R_5, P_1, R_4, R_1, R_1, R_0],$$

where P_1 is a permutation matrix. Due to the lack of space, we will give the concrete form of P_1 and $M_{12\times12}$ in Appendix B.

Table 2. Experimental results for diffusion layers with other interesting sizes

n	$\mathcal{B}_d$	Optimal solutions		XOR gates
		Total number	Example of M	
6	4	12	$[R_0, R_1, R_0]$	9
10	6	80	$[R_0, R_1, R_2, R_0, R_4]$	25
14	8	42	$[R_0, R_1, R_3, R_6, R_5, R_3]$	42
18	8	36720	$[R_0, R_1, R_1, R_2, R_2, R_0]$	54

4.3   Other Information from the Proposed Structure

Below we elaborate how to acquire a prior knowledge of the resulting matrix. First, it is clear that any matrix constructed by r-round Feistel structure can be characterized as

$$M^{(r)} = \begin{pmatrix} A_1 & A_2 \\ A_3 & A_4 \end{pmatrix}, \tag{8}$$

where each block matrix is an expression that consists of P_i, $1 \leq i \leq r$. For example, based on

$$M^{(3)} = \begin{pmatrix} P_2 P_1 \oplus I & P_2 \\ P_3 P_2 P_1 \oplus P_3 \oplus P_1 & P_3 P_2 \oplus I \end{pmatrix},$$

we get $A_4 = P_3 P_2 \oplus I$ in this case. Let $T(A_i)$ be the number of terms in A_i and through exploring the regularity on changes of $T(A_i)$, we have

Theorem 3. *The four block matrices constituting $M^{(r)}$ as shown in (8) satisfy*

$$T(A_1) = F(r-1), T(A_2) = F(r-2), T(A_3) = F(r), T(A_4) = F(r-1).$$

This observation is straightforward and we omit the proof here. It seems that Theorem 3 places major focus only on the expanded form, nevertheless, we will later see that it contributes to understanding the generic picture of optimal matrices.

As a matter of fact, each $\frac{n}{2} \times \frac{n}{2}$ block matrix in the resulting matrix $M_{n \times n}$ is a circulant matrix [27] for the optimal solution. To explain conveniently, we denote a $t \times t$ circulant matrix with i ones in the first row by $U_i^{(t)}$. Then each optimal solution in the case of $n = 4$ satisfies[4]

$$M_{4 \times 4} = \begin{pmatrix} U_2^{(2)} & U_1^{(2)} \\ U_1^{(2)} & U_2^{(2)} \end{pmatrix},$$

since the terms in A_i are always eliminated pairwise during calculating. Taking $n = 8$ as anther illustration, in the light of $T(A_1) = 3, T(A_2) = 2, T(A_3) = 5, T(A_4) = 3$ and the fact that each row in an optimal solution has at least 4 ones, we conclude

$$M_{8 \times 8} = \begin{pmatrix} U_3^{(4)} & U_2^{(4)} \\ U_3^{(4)} & U_3^{(4)} \end{pmatrix}, \quad or \quad \begin{pmatrix} U_3^{(4)} & U_2^{(4)} \\ U_1^{(4)} & U_3^{(4)} \end{pmatrix},$$

which is in accord with the experimental results. Specifically, there are only 8 instances with the latter form (among 32 optimal solutions) and one example will be introduced in Appendix C.

---

[4] The necessary condition implies the number of optimal solutions of $M_{4 \times 4}$ is at most 4 since $U_2^{(2)}$ is determined and $U_1^{(2)}$ has only two forms. As can be seen in our search result, solutions achieve the maximal branch number only if "$U_1^{(2)} = U_1^{(2)}$".

5   Searching for Involutory Optimal Diffusion Layers

In this section, we consider constructing involutory optimal matrix by Feistel structure with cyclic shift as round function, which enjoys an attractive advantage as it requires only one procedure to be implemented for the encryption and decryption. Notice that the search strategy is derived from Algorithm 1 and most instances come from the optimal solutions. For sizes that are often used in block ciphers, we list some examples in Table 3. Below are some explications of our experimental results:

(1) For parameter $n = 4$, the two optimal solutions we find themselves are involutory, however, the proportion ($= 24/9760$) is very low for $n = 16$.
(2) In the case of $n = 32$, the largest branch number of the involutory matrices constructed by 8-round Feistel structure is 11. We do not search further (for involutory matrices with $\mathcal{B}_d = 12$) since the results are lightweight enough to have promising applications.
(3) As for $n = 8$, we do not obtain even one involutory instance despite of many attempts on the number of rounds. Before jumping to the full explanation of this situation, we need the following lemma.

Table 3. Experimental results for involutory diffusion layers with $n = 4$, 16, and 32

n	$\mathcal{B}_d$	Total number	Example of M
4	4	2	$[R_1, R_0, R_1]$
16	8	24	$[R_0, R_1, R_2, R_2, R_1, R_0]$
32	11	640	$[R_0, R_1, R_6, R_{14}, R_{14}, R_6, R_1, R_0]$

Lemma 2. *Assume M is an 8×8 optimal matrix as shown in (8) where each A_i, $1 \leq i \leq 4$, is a circulant matrix. Then no A_i can be "0" or $U_4^{(4)}$.*

Proof. Suppose not, then two cases should be discussed:

(a) Without loss of generality, we let $A_1 = 0$, then $A_3 = U_4^{(4)}$ since every column in M contains at least 4 ones, which implies M is singular and hence is a contradiction.
(b) Similarly let $A_1 = U_4^{(4)}$. Then it needs $A_3 = U_3^{(4)}$ or $A_3 = U_1^{(4)}$ to make M invertible. Note that in these cases, we have $\mathcal{B}_d \leq 4$ as there exists a vector $\mathbf{v} = (1, 1, 0, \ldots, 0)^T$ with $w_b(\mathbf{v}) = 2$ such that $w_b(\mathbf{v}) + w_b(M \cdot \mathbf{v}) = 4$. This contradicts the optimality condition.

Consequently, there is no "0" or $U_4^{(4)}$ among the four block matrices and we complete the proof. □

In addition, the following statement, deduced from [17], is very useful for our illustration.

Proposition 3. *For any* 8×8 *binary matrix with* $\mathcal{B}_d = 5$, *if the rows have only two different Hamming weights and each contains half number of rows, then it must belong to one of the following cases:*

(1) the rows are of Hamming weight 4 and 5.
(2) the rows are of Hamming weight 5 and 6.

Theorem 4. *Any* 8×8 *optimal diffusion layer constructed by* r-*round Feistel structure with cyclic shift as round functions can not be involutory.*

Proof[5]. First, the conditions in Proposition 3 are always satisfied for every 8×8 optimal matrix constructed by the structure as shown in Fig. 1. Furthermore, the four block matrices are all circulant matrices as explained before and according to Lemma 2 and Proposition 3, it is easy to see that the form of any resulting matrix belongs to one of the following eight cases:[6]

$$\begin{pmatrix} U_3 & U_3 \\ U_3 & U_2 \end{pmatrix}, \begin{pmatrix} U_2 & U_3 \\ U_3 & U_3 \end{pmatrix}, \begin{pmatrix} U_1 & U_3 \\ U_3 & U_2 \end{pmatrix}, \begin{pmatrix} U_2 & U_3 \\ U_3 & U_1 \end{pmatrix}, \tag{9}$$

$$\begin{pmatrix} U_3 & U_3 \\ U_2 & U_3 \end{pmatrix}, \begin{pmatrix} U_3 & U_2 \\ U_3 & U_3 \end{pmatrix}, \begin{pmatrix} U_3 & U_2 \\ U_1 & U_3 \end{pmatrix}, \begin{pmatrix} U_3 & U_1 \\ U_2 & U_3 \end{pmatrix}. \tag{10}$$

For the sake of clarity, we simply denote the resulting matrix by

$$\tilde{M} = \begin{pmatrix} A & B \\ C & D \end{pmatrix},$$

and now suppose $\tilde{M}$ is involutory. Then bases on $\tilde{M}^2 = I$ and the property that multiplications here are commutative, we have

$$\begin{cases} A^2 \oplus BC = I & (11) \\ (A \oplus D)B = 0 & (12) \\ (A \oplus D)C = 0 & (13) \\ A^2 = D^2 & (14) \end{cases}$$

Next, we claim neither A nor D is equal to U_2. Otherwise, one is singular and the other is invertible, since there is only one U_2 among each of the eight matrices. This contradicts (14) and we thereby exclude all cases in (9).

---

[5] Throughout this proof, we omit the superscript in $U_i^{(4)}$ for simplicity.
[6] With the view of permutation homomorphic, these forms can be considered as two types on condition that the (row and column) permutation in Definition 2 is block-wise.

For cases in (10), we suppose $C = U_2$ without loss of generality. Then B is a nonsingular matrix and we have $B^{-1} = U_1$, or $B^{-1} = U_3$.

Due to $BC \neq 0$, we have $A^2 \neq I$ from (11). Furthermore, as $A = U_3$ and $A^2 = (U_4 \oplus U_1)(U_4 \oplus U_1) = U_1^2$, it holds

$$A^2 = \begin{pmatrix} 0\,0\,1\,0 \\ 0\,0\,0\,1 \\ 1\,0\,0\,0 \\ 0\,1\,0\,0 \end{pmatrix}.$$

Then according to (11), we can easily obtain

$$C = \begin{cases} (I \oplus A^2)U_1 & B^{-1} = U_1 \\ (I \oplus A^2)(U_4 \oplus U_1) = (I \oplus A^2)U_1 & B^{-1} = U_3, \end{cases}$$

which implies the first and third columns in C are the same (the remaining two columns are also the same). Therefore, there always exists a vector $\mathbf{v} = (1,0,1,0,\ldots,0)^T$ with $w_b(\mathbf{v}) = 2$ such that $w_b(\mathbf{v}) + w_b(C \cdot \mathbf{v}) = 4$. This is a contradiction and all cases in (10) are thus excluded.

In summary, no involutory 8×8 optimal binary matrix can be constructed by Feistel structure with cyclic shift as round functions and we complete the proof. $\square$

6   Diffusion Layers Constructed with Generalized Feistel Structure

As explained in Sect. 3.3, we restrict the round function to cyclic shift with the purpose of improving software performance. However, an unfavourable situation we are likely to face is that the length of $n/2$ words is longer than the word size of the processor. For example, in the case of 8-bit S-box, the software efficiency of our 16×16 optimal diffusion layer on 32-bit processor is weaken since the rotation on a 64-bit word becomes complicated.

To make up for the above shortcomings, an instinctive idea is to construct diffusion layers using r-round Type-II Generalized Feistel Structure (GFS, [29]), which be characterized as

$$M_{gfs} = \begin{pmatrix} 0\,0\,0\,I \\ I\,0\,0\,0 \\ 0\,I\,0\,0 \\ 0\,0\,I\,0 \end{pmatrix} \begin{pmatrix} P_{2r-1}\,I\,0\,0 \\ 0\,\,\,0\,I\,0 \\ 0\,\,\,0\,P_{2r}\,I \\ I\,\,\,0\,0\,0 \end{pmatrix} \cdots \begin{pmatrix} P_1\,I\,0\,0 \\ 0\,0\,I\,0 \\ 0\,0\,P_2\,I \\ I\,0\,0\,0 \end{pmatrix} \tag{15}$$

where round functions in round i, P_{2i-1} and P_{2i}, are cyclic left shifts. Also, we use $[P_1, P_2, \ldots, P_{2r}]$ to represent M_{gfs} for simplicity.

Owing to the slow diffusion property of Type-II GFS, one may take it for granted that the number of rounds will be increased compared with the Feistel structure on the premise of the same $\mathcal{B}_d$. However, that is not the case according

Table 4. Experimental results for diffusion layers constructed by Type-II GFS

n	$\mathcal{B}_d$	example of M_{gfs}
8	5	$[R_0, R_0, R_0, R_1,$ $R_1, R_1, R_0, R_1]$
16	8	$[R_0, R_0, R_0, R_0, R_1, R_1,$ $R_1, R_2, R_0, R_0, R_0, R_0]$
32	11	$[R_0, R_0, R_0, R_0, R_4, R_5, R_1, R_1,$ $R_1, R_0, R_0, R_0, R_1, R_3, R_3, R_4]$

to our search results listed in Table 4. Actually, the conclusion of Theorem 1 also holds for Type-II GFS and hence the solutions we find are optimal in terms of the number of rounds.

Compared to the Feistel structure, the cost of hardware implementation of each round in Type-II GFS remains unchanged, which means we can almost perfectly solve the problem introduced at the beginning of this section. Yet, it is to be noticed that the time complexity of r-round search becomes $(n/4)^{2r}$, far greater than $(n/2)^r$ when $n > 8$. For example, while searching for 32×32 binary matrix with $\mathcal{B}_d = 11$ constructed by Type-II GFS, the total number of matrices to be detected is 2^{48}. Despite the help of equivalence partitioning technique, the search space (i.e. 2^{42}) is still so huge that we need highly parallel computations to obtain all solutions.

7 Comparison with Known Results

In this section, we compare our Feistel-structure-based proposals with previous known results on hardware implementation. As can be seen in Table 5, solutions we found for $n = 32$ can save approximately 33.3 % gate equivalents compared to the best known result. Furthermore, while considering the constructions with $\mathcal{B}_d = 11$, this improvement shots up to a staggering 64.7 %.

For the size $n = 16$, a noteworthy comparison comes between the diffusion layer used in ARIA and ours. The area can be reduced by 36.8 % provided that the original linear layer is replaced by our optimal instances. Moreover, in the case of $n = 8$, the hardware cost of the design in [16] is equal to ours. The reason, which we have mentioned in the introduction, is that the examples given in [16] can be implicitly implemented by 4-round Feistel structure (while the last swap is not removed).

Here we omit comparisons on the software performance for two reasons. One is some of the previous constructions place the major focus on maximizing the branch number using algebraic methods, ignoring the estimate of implementation efficiency, and thus there is no need to make comparisons. The other is diffusion layers in cryptographic algorithms are often implemented together with S-boxes, which makes the comparisons complicated. Nevertheless, as explained

Table 5. Comparison of our diffusion layers with the known results

n	$\mathcal{B}_d$	XOR gates	Involutory	Reference
4	4	**6**	Yes	This paper
8	5	34	Yes	[18]
	5	16	No	[16]
	5	16	No	This paper
16	8	95	Yes	[18]
	8	76	Yes	[13]
	8	64	No	[20]
	8	60	Yes	[14]
	8	**48**	Yes	This paper
32	10	286	No	[15]
	10	**112**	Yes	This paper
	11	363	No	[19]
	11	**128**	Yes	This paper
	12	328	Yes	[19]
	12	192	No	[20]
	12	**128**	No	This paper

in Sects. 3.3 and 6, our proposals still have excellent software performance even without any optimization.

8   Conclusion

In this paper, we propose a new class of optimal diffusion layers over $\mathbb{F}_2$ by use of Feistel structure with bit permutation as round functions. Through investigating bounds on the branch number our structure can achieve, we construct optimal binary diffusion layers for a series of parameters (up to 32×32) with excellent software/hardware performances. As far as we know, the hardware cost of most proposals is the lowest compared to the previous results. Involutory optimal instances for the commonly used dimensions are also presented, with the exception of 8×8. Finally, we investigate optimal diffusion layers constructed by Type-II GFS and provide some typical solutions. Since the hardware cost of our results are extremely low, we expect our strategy will be useful for future construction of lightweight ciphers based on (involutory) binary diffusion components.

Acknowledgments. The authors would like to thank the anonymous referees for their valuable comments. We are also grateful to Shengbao Wu and Renzhang Liu for providing useful suggestions. This work is supported by the National Basic Research Program of China (No. 2013CB338002) and National Natural Science Foundation of China (No. 61272476, No. 61232009, No. 61202422).

A   Determination of Branch Number

For a $[2n, n, d]$ linear code $\mathcal{C}_\theta$, there exists at least one codeword $\mathbf{v} = (v_1, \ldots, v_{2n})^T$ such that $w_b(\mathbf{v}) = d$. Let

$$\mathbf{v_{left}} = (v_1, \ldots, v_n)^T, \quad \mathbf{v_{right}} = (v_{n+1}, \ldots, v_{2n})^T,$$

and then it must hold

$$w_b(\mathbf{v_{left}}) \leq d/2 \quad or \quad w_b(\mathbf{v_{right}}) \leq d/2,$$

which implies the number of input vectors to be computed according to (1) is at most $2^{d/2}$.

Specifically, as both $G_\theta = [I|M]$ and $G'_\theta = [M^{-1}|I]$ are the generator matrices of $\mathcal{C}_\theta$, we can determine $\mathcal{B}_d(M)$ by searching for the minimum value between $w_b(\mathbf{x}) + w_b(M \cdot \mathbf{x})$ and $w_b(\mathbf{x}) + w_b(M^{-1} \cdot \mathbf{x})$ among at most $2^{d/2}$ input vectors of low Hamming weights.

Algorithm 2. Determining the branch number of a binary matrix

---

```
1: function BINARYBRANCHNUMBER(M, dimension)
2:      InvM ← inverse matrix of M
3:      x ← (1, 0, ..., 0)^T
4:      miniweight ← 2 · dimension
5:      while w_b(x) < miniweight/2 do
6:          weight ← w_b(x) + w_b(M · x)
7:          if weight ≤ miniweight then
8:              miniweight ← weight
9:          end if
10:         weight ← w_b(x) + w_b(InvM · x)
11:         if weight ≤ miniweight then
12:             miniweight ← weight
13:         end if
14:         update x in the increasing order by Hamming weight
15:     end while
16:     return miniweight
17: end function
```

B $M_{12\times12}$ Constructed by $[R_5, P_1, R_4, R_1, R_1, R_0]$

$$P_1 = \begin{pmatrix} 0\,0\,0\,0\,0\,1 \\ 0\,0\,0\,0\,1\,0 \\ 1\,0\,0\,0\,0\,0 \\ 0\,0\,1\,0\,0\,0 \\ 0\,1\,0\,0\,0\,0 \\ 0\,0\,0\,1\,0\,0 \end{pmatrix} \qquad M_{12\times12} = \begin{pmatrix} 0\,0\,0\,1\,0\,1\,1\,1\,1\,0\,1\,1 \\ 1\,1\,1\,1\,0\,0\,1\,1\,0\,0\,1\,0 \\ 1\,0\,1\,0\,1\,1\,1\,0\,0\,0\,1\,1 \\ 1\,0\,1\,0\,0\,0\,0\,1\,1\,1\,1\,1 \\ 1\,0\,0\,1\,1\,1\,0\,1\,0\,1\,1\,0 \\ 0\,1\,1\,1\,0\,1\,0\,1\,1\,1\,0\,0 \\ 0\,0\,1\,1\,1\,0\,1\,0\,1\,1\,1\,0 \\ 0\,1\,1\,0\,1\,0\,1\,1\,1\,0\,0\,1 \\ 1\,1\,1\,1\,1\,0\,0\,0\,0\,1\,0\,1 \\ 1\,1\,0\,0\,0\,1\,1\,1\,0\,1\,0\,1 \\ 0\,1\,0\,0\,1\,1\,0\,0\,1\,1\,1\,1 \\ 1\,1\,0\,1\,1\,1\,1\,0\,1\,0\,0\,0 \end{pmatrix}$$

C $M_{8\times8}$ Constructed by $[R_0, R_2, R_1, R_1]$

$$M_{8\times8} = \begin{pmatrix} 1\,1\,0\,1\,1\,0\,0\,1 \\ 1\,1\,1\,0\,1\,1\,0\,0 \\ 0\,1\,1\,1\,0\,1\,1\,0 \\ 1\,0\,1\,1\,0\,0\,1\,1 \\ 0\,1\,0\,0\,1\,1\,1\,0 \\ 0\,0\,1\,0\,0\,1\,1\,1 \\ 0\,0\,0\,1\,1\,0\,1\,1 \\ 1\,0\,0\,0\,1\,1\,0\,1 \end{pmatrix}$$

References

1. Tilborg, H.C.A.: Coding Theory - A First Course. Lecture Notes on Error-Correcting Codes. Springer (1993)
2. Daemen, J., Knudsen, L.R., Rijmen, V.: The block cipher SQUARE. In: Biham, E. (ed.) FSE 1997. LNCS, vol. 1267, pp. 149–165. Springer, Heidelberg (1997)
3. Daemen, J., Rijmen, V.: The Design of Rijndael: AES - The Advanced Encryption Standard. Springer, Heidelberg (2002)
4. Shirai, T., Shibutani, K., Akishita, T., Moriai, S., Iwata, T.: The 128-bit blockcipher CLEFIA (extended abstract). In: Biryukov, A. (ed.) FSE 2007. LNCS, vol. 4593, pp. 181–195. Springer, Heidelberg (2007)
5. Sajadieh, M., Dakhilalian, M., Mala, H., Omoomi, B.: On construction of involutory MDS matrices from Vandermonde Matrices in $GF(2^q)$. Des. Codes Crypt. 64(3), 287–308 (2012)
6. Chand Gupta, K., Ghosh Ray, I.: On constructions of circulant MDS matrices for lightweight cryptography. In: Huang, X., Zhou, J. (eds.) ISPEC 2014. LNCS, vol. 8434, pp. 564–576. Springer, Heidelberg (2014)

7. Sim, S.M., Khoo, K., Oggier, F., Peyrin, T.: Lightweight MDS involution matrices. In: Leander, G. (ed.) FSE 2015. LNCS, vol. 9054, pp. 471–493. Springer, Heidelberg (2015)
8. Sajadieh, M., Dakhilalian, M., Mala, H., Sepehrdad, P.: Recursive diffusion layers for block ciphers and hash functions. In: Canteaut, A. (ed.) FSE 2012. LNCS, vol. 7549, pp. 385–401. Springer, Heidelberg (2012)
9. Wu, S., Wang, M., Wu, W.: Recursive diffusion layers for (lightweight) block ciphers and hash functions. In: Knudsen, L.R., Wu, H. (eds.) SAC 2012. LNCS, vol. 7707, pp. 355–371. Springer, Heidelberg (2013)
10. Augot, D., Finiasz, M.: Direct construction of recursive MDS diffusion layers using shortened BCH codes. In: Cid, C., Rechberger, C. (eds.) FSE 2014. LNCS, vol. 8540, pp. 3–17. Springer, Heidelberg (2015)
11. Kanda, M., Moriai, S., Aoki, K., Ueda, H., Takashima, Y., Ohta, K., Matsumoto, T.: E2 - a new 128-bit block cipher. IEICE Trans. Fundam. Electron. Commun. Comput. Sci. **E83**(A1), 48–59 (2000)
12. Aoki, K., Ichikawa, T., Kanda, M., Matsui, M., Moriai, S., Nakajima, J., Tokita, T.: *Camellia*: a 128-bit block cipher suitable for multiple platforms - design and analysis. In: Stinson, D.R., Tavares, S. (eds.) SAC 2000. LNCS, vol. 2012, pp. 39–56. Springer, Heidelberg (2001)
13. Kwon, D., Kim, J., Park, S., Sung, S., Sohn, Y., Song, J., Yeom, Y., Yoon, E., Lee, S., Lee, J., Chee, S., Han, D., Hong, J.: New block cipher: ARIA. In: Lim, J.-I., Lee, D.-H. (eds.) ICISC 2003. LNCS, vol. 2971, pp. 432–445. Springer, Heidelberg (2004)
14. Koo, B.-W., Jang, H.S., Song, J.H.: Constructing and cryptanalysis of a 16×16 binary matrix as a diffusion layer. In: Chae, K.-J., Yung, M. (eds.) WISA 2003. LNCS, vol. 2908, pp. 489–503. Springer, Heidelberg (2004)
15. Koo, B.-W., Jang, H.S., Song, J.H.: On constructing of a 32×32 binary matrix as a diffusion layer for a 256-bit block cipher. In: Rhee, M.S., Lee, B. (eds.) ICISC 2006. LNCS, vol. 4296, pp. 51–64. Springer, Heidelberg (2006)
16. Kanda, M., Takashima, Y., Matsumoto, T., Aoki, K., Ohta, K.: A strategy for constructing fast round functions with practical security against differential and linear cryptanalysis. In: Tavares, S., Meijer, H. (eds.) SAC 1998. LNCS, vol. 1556, pp. 264–279. Springer, Heidelberg (1999)
17. Gao, Y., Guo, G.: Unified approach to construct 8×8 binary matrices with branch number 5. In: First ACIS International Symposium on Cryptography, and Network Security, Data Mining and Knowledge Discovery, E-Commerce and Its Applications, and Embedded Systems, pp. 413–416 (2010)
18. Aslan, B., Sakalli, M.: Algebraic construction of cryptographically good binary linear transformations. Secur. Commun. Netw. **7**(1), 53–63 (2014)
19. Sakalli, M., Aslan, B.: On the algebraic construction of cryptographically good 32×32 binary linear transformations. J. Comput. Appl. Math. **259**, 485–494 (2014)
20. Dehnavi, S., Rishakani, A., Shamsabad, M.: Bitwise linear mappings with good cryptographic properties and efficient implementation. IACR Cryptology ePrint Archive, 225 (2015)
21. DAta Encryption Standard (1999). http://csrc.nist.gov/publications/fips/fips46-3/fips46-3.pdf
22. Lim, C.: CRYPTON: a new 128-bit block cipher. In: The First AES Candidate Conference. National Institute for Standards and Technology (1998)
23. Specification of the 3Gpp. Confidentiality, Integrity Algorithms 128-EEA3, 128-EIA3. Document 4: Design and Evaluation Report, version 1.3 (2011)

24. Li, Y., Wang, M.: Constructing S-boxes for lightweight cryptography with Feistel structure. In: Batina, L., Robshaw, M. (eds.) CHES 2014. LNCS, vol. 8731, pp. 127–146. Springer, Heidelberg (2014)
25. Code Tables: Bounds on the parameters of various types of codes. http://codetables.de
26. Gilbert, H., Peyrin, T.: Super-Sbox cryptanalysis: improved attacks for AES-like permutations. In: Hong, S., Iwata, T. (eds.) FSE 2010. LNCS, vol. 6147, pp. 365–383. Springer, Heidelberg (2010)
27. Davis, P.: Circulant Matrices, 2nd edn. American Mathematical Society, Providence (2012)
28. Sakalli, M., Akleylek, S., Aslan, B., Bulus, E., Sakalli, F.: On the construction of 20 × 20 and 24 × 24 binary matrices with good implementation properties for lightweight block ciphers and hash functions. Math. Prob. Eng. **2014**, 1–12 (2014). Article ID 540253
29. Zheng, Y., Matsumoto, T., Imai, H.: On the construction of block ciphers provably secure and not relying on any unproved hypotheses. In: Brassard, G. (ed.) CRYPTO 1989. LNCS, vol. 435, pp. 461–480. Springer, Heidelberg (1990)

Construction of Lightweight S-Boxes Using Feistel and MISTY Structures

Anne Canteaut[⊠], Sébastien Duval, and Gaëtan Leurent

Inria, project-team SECRET, Rocquencourt, France
{Anne.Canteaut,Sebastien.Duval,Gaetan.Leurent}@inria.fr

Abstract. The aim of this work is to find large S-Boxes, typically operating on 8 bits, having both good cryptographic properties and a low implementation cost. Such S-Boxes are suitable building-blocks in many lightweight block ciphers since they may achieve a better security level than designs based directly on smaller S-Boxes. We focus on S-Boxes corresponding to three rounds of a balanced Feistel and of a balanced MISTY structure, and generalize the recent results by Li and Wang on the best differential uniformity and linearity offered by such a construction. Most notably, we prove that Feistel networks supersede MISTY networks for the construction of 8-bit permutations. Based on these results, we also provide a particular instantiation of an 8-bit permutation with better properties than the S-Boxes used in several ciphers, including Robin, Fantomas or CRYPTON.

Keywords: S-Box · Feistel network · MISTY network · Lightweight block-cipher

1 Introduction

A secure block cipher must follow Shannon's criteria and provide confusion and diffusion [42]. In most cases, confusion is achieved with small substitution boxes (S-Boxes) operating on parts of the state (usually bytes) in parallel, and diffusion is achieved with linear operations mixing the state. The security of the cipher is then strongly dependent on the cryptographic properties of the S-Boxes. For instance, the AES uses an 8-bit S-Box based on the inversion in the finite field with 2^8 elements. This S-Box has the smallest known differential probability and linear correlation, and then allows the AES to be secure with a small number of rounds, and to reach very good performances. However, it is not always the best option for constrained environments. In software, an S-Box can be implemented with a look-up table in memory, but this takes 256 bytes for the AES S-Box, and there might be issues with cache-timing attacks [7]. In hardware, the best known implementation of the AES S-Box requires 115 gates [13]; this hardware description can also be used for a bit-sliced software implementation [25].

Partially supported by the French Agence Nationale de la Recherche through the BLOC project under Contract ANR-11-INS-011.

O. Dunkelman and L. Keliher (Eds.): SAC 2015, LNCS 9566, pp. 373–393, 2016.
DOI: 10.1007/978-3-319-31301-6_22

For some constrained environments, this cost might be too high. Therefore, the field of lightweight cryptography has produced many alternatives with a smaller footprint, such as TEA [47], CRYPTON [29,30], NOEKEON [16], PRESENT [11], KATAN [17], LBLOCK [48], PRINCE [12], TWINE [44], the LS-Designs [23], or PRIDE [2]. In particular, many of those lightweight ciphers use S-Boxes operating on 4-bit words, or even on a smaller alphabet like in [1]. But, reducing the number of variables increases the values of the optimal differential probability and linear correlation. Therefore, more rounds are required in order to achieve the same resistance against differential and linear attacks.

An alternative approach when constructing a lightweight cipher consists in using larger S-Boxes, typically operating on 8 bits like in the AES, but with a lower implementation cost. Then, we search for S-Boxes with better implementations than the AES S-Box, at the cost of suboptimal cryptographic properties. Finding 8-bit S-Boxes which offer such an interesting trade-off is a difficult problem: they cannot be classified like in the 4-bit case [18,27], and randomly chosen S-Boxes have a high implementation cost [46]. Therefore, we focus on constructions based on smaller S-Boxes and linear operations. This general approach has been used in several previous constructions: CRYPTON v0.5 [29] (3-round Feistel), CRYPTON v1.0 [30] (2-round SPN), WHIRLPOOL [5] (using five small S-Boxes), KHAZAD [4] (3-round SPN), ICEBERG [43] (3-round SPN), ZORRO [21] (4-round Feistel with mixing layer), and the LS-Designs [23] (3-round Feistel and MISTY network). As in [23], we here focus on constructions with a 3-round Feistel network, or a 3-round balanced MISTY network, because they use only 3 smaller S-Boxes, but can still provide good large S-Boxes. And we study the respective merits of these two constructions, since this comparison is raised as an open question in [23].

The Feistel and MISTY structures have been intensively studied in the context of block cipher design, and bounds are known for the maximum expected differential probability (MEDP) [3,32,38] and maximum expected linear potential (MELP) [3,37]. However, those results are not relevant for the construction of S-Boxes, because they only consider the average value over all the keys, while an S-Box is unkeyed. Therefore, the differential and linear properties of the Feistel and MISTY constructions need to be analyzed in the unkeyed setting. Such a study has been initiated recently by Li and Wang [28] in the case of 3 rounds of a Feistel network. In this work, we expand the results of Li and Wang, by giving some more general theoretical results for unkeyed Feistel and MISTY structures, with a particular focus on the construction of 8-bit permutations. Due to the page limitation, some of the results are not detailed here and are presented in the full version of this paper [15] only.

Our contributions. We first explain why the usual MEDP and MELP notions are meaningless in the unkeyed setting. In particular we exhibit a 3-round MISTY network where there exists a differential with probability higher than the MEDP *for any fixed key*, but this optimal differential depends on the key. Then, Sect. 3 gives some lower bounds on the differential uniformity and linearity of any 3-round balanced MISTY structure, which involve the properties of the three inner S-Boxes. Similar results on 3-round Feistel networks are also detailed in Sect. 5 of [15], which generalize the previous results from [28]. Section 4 then focuses on the construction of 8-bit permutations. Most notably, we show that 3 rounds of a Feistel network with appropriate inner S-Boxes provide better cryptographic properties than any 3-round MISTY network, explaining some experimental results reported in [23]. Section 5 then gives an instantiation of such an 8-bit permutation, which offers a very good trade-off between the cryptographic properties and the implementation cost. It can be implemented efficiently in hardware and for bit-sliced software, and has good properties for side-channel resistant implementations with masking. In particular, this S-Box supersedes the S-Boxes considered in many lightweight ciphers including CRYPTON, Robin and Fantomas.

2 From Keyed Constructions to Unkeyed S-Boxes

2.1 Main Cryptographic Properties for an S-Box

In this paper, we focus on S-Boxes having the same number of input and output bits. The resistance offered by an S-Box against differential [10] and linear [31] cryptanalysis is quantified by the highest value in its difference table (resp. table of linear biases, aka linear-approximation table). More precisely, these two major security parameters are defined as follows.

Definition 1 (Differential uniformity [36]). *Let F be a function from $\mathbb{F}_2^n$ into $\mathbb{F}_2^n$. For any pair of differences (a, b) in $\mathbb{F}_2^n$, we define the set*

$$D_F(a \rightarrow b) = \{x \in \mathbb{F}_2^n \mid F(x \oplus a) \oplus F(x) = b\}.$$

The entry at position (a, b) in the difference table of F then corresponds to the cardinality of $D_F(a \rightarrow b)$ and will be denoted by $\delta_F(a, b)$.

Moreover, the differential uniformity of F is

$$\delta(F) = \max_{a \neq 0, b} \delta_F(a, b).$$

Obviously, the differential uniformity of an S-Box is always even, implying that, for any F, $\delta(F) \geq 2$. The functions F for which equality holds arze named *almost perfect nonlinear (APN) functions.*

Similarly, the bias of the best linear approximation of an S-Box is measured by its linearity.

Definition 2 (Walsh transform of an S-Box). *Let F be a function from $\mathbb{F}_2^n$ into $\mathbb{F}_2^n$. The Walsh transform of F is the function*

$$\mathbb{F}_2^n \times \mathbb{F}_2^n \rightarrow \mathbb{Z}$$
$$(a, b) \mapsto \lambda_F(a, b) = \sum_{x \in \mathbb{F}_2^n} (-1)^{b \cdot F(x) + a \cdot x}.$$

Moreover, the linearity of F is

$$\mathcal{L}(F) = \max_{a, b \in \mathbb{F}_2^n, b \neq 0} |\lambda_F(a, b)|.$$

Indeed, up to a factor 2^n, the linearity corresponds to the bias of the best linear relation between the input and output of F:

$$\Pr_X[b \cdot F(X) + a \cdot X = 1] = \frac{1}{2^n}\left(2^{n-1} - \frac{1}{2}\sum_{x \in \mathbb{F}_2^n}(-1)^{b \cdot F(x) + a \cdot x}\right) = \frac{1}{2}\left(1 - \frac{\lambda_F(a,b)}{2^n}\right).$$

It is worth noticing that, for any fixed output mask $b \in \mathbb{F}_2^n$, the function $a \mapsto \lambda_F(a,b)$, corresponds to the Walsh transform of the n-variable Boolean component of $F\colon x \mapsto b \cdot F(x)$. In particular, it enjoys all properties of a discrete Fourier transform, for instance the Parseval relation.

2.2   Constructing S-Boxes from Smaller Ones

If this paper we focus on the construction of S-Boxes using several smaller S-Boxes. Indeed small S-Boxes are much cheaper to implement that large S-Boxes:

- for table-based software implementations, the tables are smaller;
- for hardware implementations, the gate count is lower;
- for bit-sliced software implementation, the instructions count is lower;
- for vectorized implementation, small S-Boxes can use vector permutations.

In many cases, implementing several small S-Boxes requires less resources than implementing a large one. Therefore, constructing S-Boxes from smaller ones can reduce the implementation cost.

The Feistel construction is a well-known construction to build a $2n$-bit permutation from smaller n-bit functions, introduced in 1971 for the design of Lucifer (which later became DES [34]). It is a good candidate for constructing large S-Boxes from smaller ones at a reasonable implementation cost. In particular, this construction has been used for the S-Boxes of CRYPTON v0.5 [29], ZUC [20] (for S_0), Robin [23] and iSCREAM [22]. The MISTY construction introduced by Matsui [32] uses a different structure, but offers a similar level of security. The main advantage of the MISTY network is that it can offer a reduced latency because the first two S-Boxes can be evaluated in parallel. Therefore it is a natural alternative to Feistel networks for the construction of lightweight S-Boxes, and it has been used in the design of Fantomas [23] and SCREAM [22]. In order to reduce the number of gates used for implementing the construction, we focus on *balanced* MISTY networks, while the MISTY block cipher proposed in [33] is unbalanced and combines an $(n-1)$-bit S-Box and an $(n+1)$-bit S-Box.

The two structures we study are depicted in Figs. 1 and 2. It is worth noticing a major difference between the two: the function resulting from the Feistel construction is always invertible (since one round is an involution, up to a permutation of the outputs), while the function resulting from the MISTY construction is invertible if and only if all the inner S-Boxes are invertible.

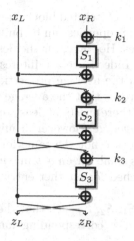

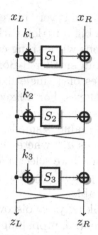

Fig. 1. 3-round MISTY network   **Fig. 2.** 3-round Feistel network

Analysis of Feistel and MISTY Structures. Since these two constructions have been used for the design of many block ciphers (in particular the DES [34] and MISTY [33], respectively), their security properties have been intensively studied. A natural way to measure the resistance of the resulting block cipher against differential and linear cryptanalysis is to study the probabilities of the differentials (respectively the potentials of the linear approximations) averaged over all keys.

Definition 3 (MEDP and MELP). *Let F_K be a family of function from $\mathbb{F}_2^n$ into $\mathbb{F}_2^n$. The MEDP is the maximum probability of a differential, averaged over all keys:*

$$\text{MEDP}(F_K) = \max_{a \neq 0, b} \frac{1}{2^k} \sum_{K \in \mathbb{F}_2^k} \frac{\delta_{F_K}(a, b)}{2^n}.$$

The MELP is the maximum potential of a linear approximation, averaged over all keys:

$$\text{MELP}(F_K) = \max_{a, b \neq 0} \frac{1}{2^k} \sum_{K \in \mathbb{F}_2^k} \left(\frac{\lambda_{F_K}(a, b)}{2^n} \right)^2.$$

The following theorem shows that the MEDP and MELP of a Feistel or MISTY network can be bounded.

Theorem 1 (Feistel or MISTY, averaged over all keys, [3,32,38]). *Given S_1, S_2 and S_3 three n-bit permutations, let $p = \max_i \delta(S_i)/2^n$ and $q = \max_i (\mathcal{L}(S_i)/2^n)^2$. Then the family of functions $(F_K)_{K=(K_1, K_2, K_3) \in \mathbb{F}_2^{3k}}$ defined by 3 rounds of a Feistel or of a MISTY network with S_i as inner functions verifies*

$$\text{MEDP}(F_K) \leq p^2 \text{ and } \text{MELP}(F_K) \leq q^2.$$

This theorem is very powerful for the construction of iterated block ciphers: it shows that a big function with strong cryptographic properties can be built from small functions with strong cryptographic properties. However, for the design of an S-Box from smaller S-Boxes, it is of little use. Indeed, we are interested in the properties of a single fixed S-Box, rather than the average properties of a family of S-Boxes. For a fixed (a, b) the theorem proves that the average values of $\delta_{F_K}(a, b)$ and $\lambda_{F_K}(a, b)$ are bounded, therefore there exists at least one key for which the value is smaller than or equal to the average. However, it might be that the values a, b where the maximum is reached are not the same for every key. Therefore if we select a key so that $\delta_{F_K}(a, b)$ is small for an a, b maximizing the average probability, the maximum can be reached for another entry of the differential table.

More strikingly, we discovered some choices of S_1, S_2, S_3 such that the maximum differential probability of the functions in the corresponding family is always higher than the MEDP.

Example 1. We consider a MISTY structure with three identical S-Boxes:

$$S_i = [\mathsf{A}, 7, 9, 6, 0, 1, 5, \mathsf{B}, 3, \mathsf{E}, 8, 2, \mathsf{C}, \mathsf{D}, 4, \mathsf{F}].$$

We have $\mathsf{MEDP}(F_K) \leq 16/256$ according to Theorem 1, because $\delta(S_i) = 4$. However, for *any* function in this family, there exists a differential with probability $32/256$. This is not a contradiction, because the differential reaching the maximum depends on the key.

The relevant property for the construction of an S-Box is the maximum differential probability (respectively maximum linear potential). Therefore, we could derive some information on this quantity for F_K for some fixed keys from the knowledge of the average value of the maximal differential probability, i.e., the EMDP (resp. EMLP), which may significantly differ from the MEDP (resp. MELP). We would like to point out that there is a confusion between the two notions in [33]: the definition corresponds to the expected maximum differential probability (respectively expected maximum linear potential), while the theorems apply to the MEDP and MELP.

Analysis of Feistel and MISTY Structures with Fixed Key. In order to study the properties of Feistel and MISTY structures for the construction of lightweight S-Boxes, we must study these structures with a fixed key. Equivalently, we can consider the structures without any key, because a structure with a fixed key is equivalent to an unkeyed one with different S-Boxes. Indeed, using an S-Box S_i with round key k_i is equivalent to using $S_i' : x \mapsto S_i(x + k_i)$ as an S-Box without any key. In the following, we always consider a key-less variant.

In a recent analysis of the fixed-key Feistel structure [28], Li and Wang derive the best differential uniformity and linearity which can be achieved by a 3-round Feistel cipher with a fixed key, and give examples reaching this bound. Their main results are as follows:

Theorem 2 (Feistel unkeyed, [28]). *Let S_1, S_2 and S_3 be three n-bit S-Boxes and F be the 2n-bit function defined by the corresponding 3-round Feistel network. Then, $\delta(F) \geq 2\delta(S_2)$. Moreover, if S_2 is not a permutation, $\delta(F) \geq 2^{n+1}$. If $n = 4$, F satisfies $\delta(F) \geq 8$. If equality holds, then $\mathcal{L}(F) \geq 64$.*

3   S-Boxes Obtained from 3 Rounds of MISTY or Feistel

3.1   Our Results

In this paper, we generalize the bounds of Li and Wang [28] on Feistel structures, and derive bounds for MISTY structures. The results are very similar for the two structures, but for a MISTY structure, optimal results are only achieved with non-invertible inner functions. Therefore, our work shows that Feistel structures allow better results than MISTY structures for the design of invertible 8-bit S-Boxes.

More precisely, we introduce two new S-Box properties $\delta_{\min}$ and $\mathcal{L}_{\min}$ in order to derive our bounds: $\mathcal{L}_{\min}$ is the smallest linearity we can have for a non-trivial component of the S-Box. Similarly, $\delta_{\min}$ is the smallest value we can have for the maximum $\max_b \delta(a, b)$ within a row in the difference table. In particular, for any 4-bit function S, $\delta_{\min}(S) \geq 2$ and $\mathcal{L}_{\min}(S) \geq 4$. Moreover, if S is a 4-bit permutation, then $\delta_{\min}(S) \geq 4$ and $\mathcal{L}_{\min}(S) \geq 8$.

We first present the general lower bounds we obtain on the differential uniformity and linearity of 3 rounds of a Feistel and of a MISTY construction.

1. For a Feistel network with inner S-Boxes S_1, S_2 and S_3:
   - $\delta(F) \geq \delta(S_2) \max(\delta_{\min}(S_1), \delta_{\min}(S_3))$
   - if S_2 is not a permutation, $\delta(F) \geq 2^{n+1}$.
   - if S_2 is a permutation, $\delta(F) \geq \max_{i \neq 2, j \neq i, 2} (\delta(S_i)\delta_{\min}(S_j), \delta(S_i)\delta_{\min}(S_2^{-1}))$.
   - $\mathcal{L}(F) \geq \mathcal{L}(S_2) \max(\mathcal{L}_{\min}(S_1), \mathcal{L}_{\min}(S_3))$
   - if S_2 is a permutation, $\mathcal{L}(F) \geq \max_{i \neq 2, j \neq i, 2} (\mathcal{L}(S_i)\mathcal{L}_{\min}(S_j), \mathcal{L}(S_i)\mathcal{L}_{\min}(S_2^{-1}))$.
2. For a MISTY network with inner S-Boxes S_1, S_2 and S_3:
   - $\delta(F) \geq \delta(S_1) \max(\delta_{\min}(S_2), \delta_{\min}(S_3))$
   - if S_1 is not a permutation, $\delta(F) \geq 2^{n+1}$.
   - if S_1 is a permutation, $\delta(F) \geq \max_{i \neq 1, j \neq 1, i} (\delta(S_i)\delta_{\min}(S_j), \delta(S_i)\delta_{\min}(S_1^{-1}))$;
   - $\mathcal{L}(F) \geq \max(\mathcal{L}(S_1)\mathcal{L}_{\min}(S_2), \mathcal{L}(S_2)\mathcal{L}_{\min}(S_1), \mathcal{L}(S_3)\mathcal{L}_{\min}(S_1))$;
   - if S_3 is a permutation, $\mathcal{L}(F) \geq \mathcal{L}(S_1)\mathcal{L}_{\min}(S_3^{-1})$.
   - if S_1 is a permutation, $\mathcal{L}(F) \geq \mathcal{L}(S_3)\mathcal{L}_{\min}(S_2)$.
   - if S_1 and S_3 are permutations, $\mathcal{L}(F) \geq \mathcal{L}(S_2)\mathcal{L}_{\min}(S_3^{-1})$.

If $n = 4$ this yields for both constructions:

$$\delta(F) \geq 8 \text{ and } \mathcal{L}(F) \geq 48.$$

Moreover, $\mathcal{L}(F) \geq 64$ unless $\delta(F) \geq 32$.

For the MISTY construction with $n = 4$, if F is a permutation, we obtain tighter bounds: $\delta(F) \geq 16$ and $\mathcal{L}(F) \geq 64$. This implies that the Feistel construction is more appropriate for constructing 8-bit permutations. We will also show that all these bounds for $n = 4$ are tight. We now detail the results in the case of the MISTY construction, while the results for the Feistel construction are presented in the full version [15].

3.2   Differential Uniformity of 3 Rounds of MISTY

Our lower bound on the differential uniformity of the 3-round MISTY relies on the evaluation of the number of solutions of some differentials for which the input difference of one of the 3 S-Boxes is canceled (see Fig. 3 in [15]).

Proposition 1. Let S_1, S_2 and S_3 be three n-bit S-Boxes and F be the $2n$-bit function defined by the corresponding 3-round MISTY network. Then, for all a, b and c in $\mathbb{F}_2^n$, we have:

(i)
$$\delta_F(0\|a, b\|c) = \delta_{S_1}(a, c) \times \delta_{S_3}(c, b \oplus c);$$

(ii) If S_1 is bijective,

$$\delta_F(a\|0, b\|c) = \delta_{S_2}(a, a \oplus c) \times \delta_{S_3}(a, b \oplus c);$$

(iii) $\delta_{S_1}(a, b) \times \delta_{S_2}(b, c) \leq \delta_F(b\|a, c\|c) \leq \sum_{d \in \mathbb{F}_2^n} \delta_{S_1}(a, b \oplus d) \times \delta_{S_2}(b, c \oplus d) \times \gamma_{S_3}(d)$

where $\gamma_{S_3}(d)$ is 0 if $\delta_{S_3}(d, 0) = 0$ and 1 otherwise. Most notably, if S_3 is bijective,
$$\delta_F(b\|a, c\|c) = \delta_{S_1}(a, b) \times \delta_{S_2}(b, c).$$

Proof. Let x be the input of the MISTY network, and let x_L and x_R be its left and right parts respectively.

(i) $x = (x_L, x_R)$ satisfies $F(x_L\|x_R) \oplus F(x_L\|(x_R \oplus a)) = b\|c$ if and only if

$$\begin{cases} S_3(S_1(x_R) \oplus x_L) \oplus S_3(S_1(x_R \oplus a) \oplus x_L) = b \oplus c, \\ S_2(x_L) \oplus S_1(x_R) \oplus x_L \oplus S_2(x_L) \oplus S_1(x_R \oplus a) \oplus x_L = c \end{cases}$$

$$\Leftrightarrow \begin{cases} S_3(S_1(x_R) \oplus x_L) \oplus S_3(S_1(x_R \oplus a) \oplus x_L) = b \oplus c, \\ S_1(x_R) \oplus S_1(x_R \oplus a) = c \end{cases}$$

or equivalently

$$x_R \in D_{S_1}(a \to c) \text{ and } x_L \in S_1(x_R) \oplus D_{S_3}(c \to b \oplus c).$$

Hence, we deduce that there are exactly $\delta_{S_1}(a, c)$ values of x_R, and for each of those, $\delta_{S_3}(c, b \oplus c)$ values of x_L, such that x verifies the differential.

(ii) $x = (x_L, x_R)$ satisfies $F(x_L \| x_R) \oplus F((x_L \oplus a) \| x_R) = b \| c$ if and only if

$$\begin{cases} S_3(S_1(x_R) \oplus x_L) \oplus S_3(S_1(x_R) \oplus x_L \oplus a) = b \oplus c, \\ S_2(x_L) \oplus S_1(x_R) \oplus x_L \oplus S_2(x_L \oplus a) \oplus S_1(x_R) \oplus x_L \oplus a = c \end{cases}$$

$$\Leftrightarrow \begin{cases} S_1(x_R) \oplus x_L \in D_{S_3}(a \to b \oplus c), \\ S_2(x_L) \oplus S_2(x_L \oplus a) = a \oplus c \end{cases}$$

or equivalently,

$$x_L \in D_{S_2}(a \to a \oplus c) \text{ and } S_1(x_R) \in x_L \oplus D_{S_3}(a \to b \oplus c).$$

If S_1 is invertible, for any fixed x_L, each one of the $\delta_{S_3}(a, b \oplus c)$ values defined by the second condition determines a unique value of x_R. Therefore, the number of (x_L, x_R) satisfying the differential is exactly $\delta_{S_2}(a, a \oplus c) \times \delta_{S_3}(a, b \oplus c)$.

(iii) (x_L, x_R) satisfies $F(x_L \| x_R) \oplus F((x_L \oplus b) \| (x_R \oplus a)) = c \| c$ if and only if

$$\begin{cases} S_3(S_1(x_R) \oplus x_L) \oplus S_3(S_1(x_R \oplus a) \oplus x_L \oplus b) = 0, \\ S_2(x_L) \oplus S_1(x_R) \oplus x_L \oplus S_2(x_L \oplus b) \oplus S_1(x_R \oplus a) \oplus x_L \oplus b = c \end{cases}$$

$$\Leftrightarrow \begin{cases} S_3(S_1(x_R) \oplus x_L) \oplus S_3(S_1(x_R \oplus a) \oplus x_L \oplus b) = 0, \\ S_2(x_L) \oplus S_1(x_R) \oplus S_2(x_L \oplus b) \oplus S_1(x_R \oplus a) = b \oplus c \end{cases}$$

This equivalently means that there exists some $d \in \mathbb{F}_2^n$ such that

$$\begin{cases} x_R \in D_{S_1}(a \to b \oplus d), \ x_L \in D_{S_2}(b \to c \oplus d), \\ S_3(S_1(x_R) \oplus x_L) \oplus S_3(S_1(x_R \oplus a) \oplus x_L \oplus b) = 0, \end{cases}$$

i.e.,

$$x_R \in D_{S_1}(a \to b \oplus d), \ x_L \in D_{S_2}(b \to c \oplus d) \text{ and } S_1(x_R) \oplus x_L \in D_{S_3}(d \to 0).$$

Then, for any fixed $d \in \mathbb{F}_2^n$ such that $\delta_{S_3}(d, 0) = 0$, no pair (x_L, x_R) satisfies the third condition. If $\delta_{S_3}(d, 0) > 0$, then some of the values (x_L, x_R) defined by the first two conditions may also satisfy the third one, and if $d = 0$, the third condition is always satisfied. It then follows that

$$\delta_{S_1}(a, b) \times \delta_{S_2}(b, c) \leq \delta_F(b \| a, c \| c) \leq \sum_{d \in \mathbb{F}_2^n} \delta_{S_1}(a, b \oplus d) \times \delta_{S_2}(b, c \oplus d) \times \gamma_{S_3}(d)$$

where $\gamma_{S_3}(d)$ is 0 if $\delta_{S_3}(d, 0) = 0$ and 1 otherwise. Moreover, if S_3 is bijective, $\delta_{S_3}(d, 0) > 0$ if and only if $d = 0$, implying that the two previous bounds are equal, i.e.,

$$\delta_F(b \| a, c \| c) = \delta_{S_1}(a, b) \times \delta_{S_2}(b, c).$$

$\square$

These three particular types of differentials provide us with the following lower bound on the differential uniformity of any 3-round MISTY network.

Theorem 3. *Let S_1, S_2 and S_3 be three n-bit S-Boxes and let F be the 2n-bit function defined by the corresponding 3-round MISTY network. Then,*

$$\delta(F) \geq \delta(S_1) \max\left(\delta_{\min}(S_2), \delta_{\min}(S_3)\right) \text{ where } \delta_{\min}(S) = \min_{a \neq 0} \max_b \delta_S(a, b).$$

Moreover,

– *if S_1 is a permutation,*

$$\delta(F) \geq \max_{i \neq 1, j \neq 1, i} \max\left(\delta(S_i)\delta_{\min}(S_j),\ \delta(S_i)\delta_{\min}(S_1^{-1})\right),$$

– *if S_1 is not a permutation, $\delta(F) \geq 2^{n+1}$.*

Proof. The result is a direct consequence of Proposition 1. We here derive the bounds from the first item in Proposition 1; the other cases can be similarly deduced from the other two items. Let us first consider a pair of differences (α, β) which achieves the differential uniformity of S_1, i.e., $\delta(S_1) = \delta_{S_1}(\alpha, \beta)$. Then, we choose $a = \alpha$ and $c = \beta$, and get that, for any $b \in \mathbb{F}_2^n$,

$$\delta_F(0\|\alpha, b\|\beta) = \delta(S_1) \times \delta_{S_3}(\beta, \beta \oplus b).$$

Then, we can choose for b the value which maximizes $\delta_{S_3}(\beta, \beta \oplus b)$. This value is always greater than or equal to $\delta_{\min}(S_3)$. Similarly, we can now consider a pair of differences (α, β) which achieves the differential uniformity of S_3, i.e., $\delta(S_3) = \delta_{S_3}(\alpha, \beta)$. In this case, we choose $c = \alpha$ and $b = \alpha \oplus \beta$, and get that, for any $a \in \mathbb{F}_2^n$,

$$\delta_F(0\|a, (\alpha \oplus \beta)\|\alpha) = \delta_{S_1}(a, \alpha) \times \delta(S_3).$$

We then choose for a the value which maximizes $\delta_{S_1}(a, \alpha)$ which is always greater than or equal to $\delta_{\min}(S_1^{-1})$ when S_1 is a permutation.

Let us now assume that S_1 is not bijective. This means that there exists some nonzero $a \in \mathbb{F}_2^n$ such that $\delta_{S_1}(a, 0) \geq 2$. Then, we deduce from the first item in Proposition 1, with $b = c = 0$, that $F(x_L\|x_R) \oplus F_K(x_L \oplus a\|x_R) = (0,0)$ has $\delta_{S_1}(a, 0) \times \delta_{S_3}(0, 0) \geq 2 \times 2^n = 2^{n+1}$ solutions in $\mathbb{F}_2^{2n}$.   □

3.3   Linearity of 3 Rounds of MISTY

The lower bound on the linearity of a three-round MISTY structure can be derived in a similar way. The proofs of the following results are given in [15].

Proposition 2. *Let S_1, S_2 and S_3 be three n-bit S-Boxes and F the 2n-bit function defined by the corresponding 3-round MISTY network. Then, for all a, b and c in $\mathbb{F}_2^n$, we have:*

(i) $\lambda_F(a\|b, 0\|c) = \lambda_{S_1}(b, c)\lambda_{S_2}(a \oplus c, c)$
(ii) $\lambda_F(a\|b, c\|c) = \lambda_{S_1}(b, a)\lambda_{S_3}(a, c)$
(iii) *If S_1 is bijective,* $\lambda_F(a\|0, b\|c) = \lambda_{S_2}(a, b \oplus c)\lambda_{S_3}(b \oplus c, b)$

As in the differential case, the previous three linear approximations provide us with a lower bound on the linearity of any 3-round MISTY network. This bound involves both the linearity of the constituent S-Boxes and another quantity denoted by $\mathcal{L}_{\min}$ computed from the table of linear biases as follows.

Definition 4 ($\mathcal{L}_{\min}$). *Let F be an n-bit S-Box. We define*

$$\mathcal{L}_{\min}(F) = \min_{b \in \mathbb{F}_2^n, b \neq 0} \max_{a \in \mathbb{F}_2^n} |\lambda_F(a,b)|.$$

Most notably, $\mathcal{L}_{\min}(F) \geq 2^{\frac{n}{2}}$ and this bound is not tight when F is bijective.

Proof. By definition, $\mathcal{L}_{\min}(F)$ is the smallest linearity achieved by a component $F_b : x \mapsto b \cdot F(x)$ of F, when b varies in $\mathbb{F}_2^n \backslash \{0\}$. Since any F_b is an n-variable Boolean function, its linearity is at least $2^{\frac{n}{2}}$ with equality if and only if F_b is bent [41]. Since bent functions are not balanced, none of the components of a permutation is bent, implying that $\mathcal{L}_{\min}(F) > 2^{\frac{n}{2}}$ when F is a permutation. $\square$

We then derive the following lower bound on the linearity of any 3-round MISTY network. The proof is given in [15].

Theorem 4. *Let S_1, S_2 and S_3 be three n-bit S-Boxes and let F be the $2n$-bit function defined by the corresponding 3-round MISTY network. Then,*

$$\mathcal{L}(F) \geq \max\left(\mathcal{L}(S_1)\mathcal{L}_{\min}(S_2), \ \mathcal{L}(S_2)\mathcal{L}_{\min}(S_1), \ \mathcal{L}(S_3)\mathcal{L}_{\min}(S_1)\right).$$

Moreover, if S_1 is a permutation, $\mathcal{L}(F) \geq \mathcal{L}(S_3)\mathcal{L}_{\min}(S_2)$; if S_3 is a permutation, $\mathcal{L}(F) \geq \mathcal{L}(S_1)\mathcal{L}_{\min}(S_3^{-1})$, and if both S_1 and S_3 are permutations, then $\mathcal{L}(F) \geq \mathcal{L}(S_2)\mathcal{L}_{\min}(S_3^{-1})$.

4   Application to 8-Bit S-Boxes

In this section, we investigate the cryptographic properties of 8-bit S-Boxes corresponding to a 3-round MISTY structure with 4-bit inner S-Boxes, with a particular focus on the case where the three inner S-Boxes are bijective, since it corresponds to the case where the resulting function is a permutation.

4.1   Differential Uniformity

The following bound on the differential uniformity of any 3-round MISTY network over $\mathbb{F}_2^8$ is a direct consequence of Theorem 3.

Corollary 1. *Any 8-bit function F corresponding to a 3-round MISTY network satisfies $\delta(F) \geq 8$.*

Proof. The bound clearly holds when S_1 is not bijective, since we known from Theorem 3 that $\delta(F) \geq 32$ in this case. If S_1 is bijective, then $\delta(S_1) \geq 4$ since APN permutations over $\mathbb{F}_2^4$ do not exist, as proved in [24, Theorem 2.3]. Obviously, any 4-bit S-Box S satisfies $\delta_{\min}(S) \geq 2$, implying that

$$\delta(F) \geq \delta(S_1)\delta_{\min}(S_2) \geq 8.$$

$\square$

Besides this general result, we can provide some necessary conditions on the constituent S-Boxes to achieve the previous lower bound.

Theorem 5. *Let S_1, S_2 and S_3 be three 4-bit S-Boxes and let F be the 8-bit function defined by the corresponding 3-round MISTY network. Then, $\delta(F) = 8$ implies that S_1 is a permutation with $\delta(S_1) = 4$ and S_2 and S_3 are two APN functions. Otherwise, $\delta(F) \geq 12$.*

Proof. Since $\delta(F) \geq 32$ when S_1 is not bijective, we only need to focus on the case where S_1 is a permutation. If any of the constituent S-Boxes S_i has differential uniformity strictly greater than 4, i.e., $\delta(S_i) \geq 6$, we deduce from Theorem 3 that $\delta(F) \geq \delta(S_i)\delta_{\min}(S_j) \geq 12$. Therefore, $\delta(F) = 8$ can be achieved only if $\delta(S_1) = 4$, $\delta(S_2) \leq 4$, and $\delta(S_3) \leq 4$. The fact that $\delta(F) \geq 16$ when at least one of the S-Boxes S_2 or S_3 has differential uniformity 4 is proved in Lemma 1 in the full version [15]. □

We can then prove that the lower bound in Corollary 1 is tight by exhibiting three 4-bit S-Boxes satisfying the previous conditions which lead to a 3-round MISTY network with differential uniformity 8.

Example 2. The following 4-bit S-Boxes yield an 8-bit S-Box with differential uniformity 8 and linearity 64 when used in a MISTY structure:

$$S_1 = [4, 0, 1, \mathsf{f}, 2, \mathsf{b}, 6, 7, 3, 9, \mathsf{a}, 5, \mathsf{c}, \mathsf{d}, \mathsf{e}, 8]$$
$$S_2 = [0, 0, 0, 1, 0, \mathsf{a}, 8, 3, 0, 8, 2, \mathsf{b}, 4, 6, \mathsf{e}, \mathsf{d}]$$
$$S_3 = [0, 7, \mathsf{b}, \mathsf{d}, 4, 1, \mathsf{b}, \mathsf{f}, 1, 2, \mathsf{c}, \mathsf{e}, \mathsf{d}, \mathsf{c}, 5, 5]$$

With Bijective Inner S-Boxes. We now focus on the case where the three inner S-Boxes are permutations since this guarantees that the resulting MISTY network is a permutation. We have proved that, in this case, the lowest possible differential uniformity we can obtain is 12. Here, we refine this result and show that the differential uniformity cannot be lower than 16. This improved bound exploits the following lemma on the difference tables of 4-bit permutations.

Lemma 1. *Let S_1, S_2 and S_3 be 4-bit permutations. Then, there exists a nonzero difference $\gamma \in \mathbb{F}_2^4 \backslash \{0\}$ such that at least one of the following statements holds:*

- *The difference table of S_1 has at least one value greater than or equal to 4 in Column γ and the difference table of S_2 has at least one value greater than or equal to 4 in Row γ;*
- *The difference table of S_1 has at least one value greater than or equal to 4 in Column γ and the difference table of S_3 has at least one value greater than or equal to 4 in Row γ,*
- *The difference table of S_2 has at least one value greater than or equal to 4 in Row γ and the difference table of S_3 has at least one value greater than or equal to 4 in Row γ.*

Proof. This result relies on an exhaustive search over the equivalence classes defined by composition on the left and on the right by an affine transformation, exactly as in the classification of optimal 4-bit S-Boxes in [18,27]. There are 302 equivalence classes for 4-bit permutations. From each of the classes we picked a representative, and checked that its difference table has at least six rows defined by some nonzero input difference a which contain a value greater than or equal to 4. Let $\mathcal{R}(S)$ denote the corresponding set (of size at least six):

$$\mathcal{R}(S) = \{a \in \mathbb{F}_2^4 \backslash \{0\} : \exists b \in \mathbb{F}_2^4 \backslash \{0\}, \delta_S(a,b) \geq 4\}.$$

Therefore, if there exists no difference $\gamma \in \mathbb{F}_2^4 \backslash \{0\}$ satisfying one of the three statements in the lemma, then this would mean that the three sets $\mathcal{R}(S_2)$, $\mathcal{R}(S_3)$ and $\mathcal{R}(S_1^{-1})$ are disjoint. In other words, we could find 18 distinct values amongst the 15 nonzero elements in $\mathbb{F}_2^4$, which is impossible.   □

We then deduce the following refined lower bound on the differential uniformity of a 3-round MISTY network over $\mathbb{F}_2^8$ with inner permutations.

Theorem 6. *Let S_1, S_2 and S_3 be three 4-bit permutations and let F be the 8-bit function defined by the corresponding 3-round MISTY network. Then, $\delta(F) \geq 16$.*

Proof. The result is a direct consequence of Proposition 1 combined with the previous lemma. Indeed, Lemma 1 guarantees the existence of a, b and c such that at least one of the three following properties holds:

- $\delta_{S_1}(a,c) \geq 4$ and $\delta_{S_3}(c, b \oplus c) \geq 4$,
- $\delta_{S_2}(a, a \oplus c) \geq 4$ and $\delta_{S_3}(a, b \oplus c) \geq 4$,
- $\delta_{S_1}(a,b) \geq 4$ and $\delta_{S_2}(b,c) \geq 4$.

In each of these three situations, Proposition 1 exhibits a differential (α, β) for F with $\delta_F(\alpha, \beta) = 16$.   □

4.2   Linearity

In order to apply Theorem 4 to the case of 8-bit MISTY network, we need to estimate the best linearity (and $\mathcal{L}_{\min}$) for 4-bit S-Boxes. It is well-known that the lowest linearity for a 4-bit permutation is 8. But, this result still holds if the S-Box is not bijective.

Lemma 2. *Any 4-bit S-Box S satisfies $\mathcal{L}(S) \geq 8$.*

Proof. Assume that there exists some S from $\mathbb{F}_2^4$ into $\mathbb{F}_2^4$ with $\mathcal{L}(S) < 8$, i.e., with $\mathcal{L}(S) \leq 6$. Then, all nonzero Boolean components of S, $S_c : x \mapsto c \cdot S(x)$ with $c \neq 0$, satisfy $\mathcal{L}(S_c) \leq 6$. From the classification of all Boolean functions of at most 5 variables by Berlekamp and Welch [6], we deduce that any S_c, $c \neq 0$, is affine equivalent either to $x_1 x_2 x_3 x_4 + x_1 x_2 + x_3 x_4$ or to $x_1 x_2 + x_3 x_4$, because these are the only classes of Boolean functions with linearity at most 6. Let L_1

(resp. L_2) denote the set of all nonzero $c \in \mathbb{F}_2^4$ such that S_c belongs to the first (resp. second) class. Since the degree is invariant under affine transformations, L_1 (resp. L_2) corresponds to the components with degree 4 (resp. with degree at most 2). The sum of two components of degree at most 2 has degree at most 2, implying that $L_2 \cup \{0\}$ is a linear subspace V of $\mathbb{F}_2^4$. It follows that the projection of S on V can be seen as a function from $\mathbb{F}_2^4$ into $\mathbb{F}_2^{\dim V}$ with linearity 4, i.e., a bent function. It has been shown by Nyberg [35] that, if a function F from $\mathbb{F}_2^n$ into $\mathbb{F}_2^m$ is bent, then $m \leq n/2$. Therefore, $\dim V \leq 2$. But, the sum of any two components S_c of degree 4 cannot have degree 4 since there is a single monomial of degree 4 of 4 variables. We deduce that, if L_1 contains t words of weight 1 (i.e., if S has t coordinates with linearity 6), then

$$\#L_2 \geq \binom{t}{2} + 2^{4-t} - 1 > 3,$$

for all $0 \leq t \leq 4$, a contradiction. $\square$

Combined with the previous lemma and with Definition 4, Theorem 4 provides the following lower bound on the linearity of a 3-round MISTY network over $\mathbb{F}_2^8$.

Corollary 2. *Any 8-bit function F corresponding to a 3-round MISTY network satisfies $\mathcal{L}(F) \geq 32$.*

This bound is of marginal interest since, up to our best knowledge, $\mathcal{L}(S) = 32$ is the lowest known linearity for an 8-bit S-Box. But, once again, the previous lower bound can be improved when focusing on permutations. Indeed, we can exploit that $\mathcal{L}_{\min}(S) \geq 8$ for any 4-bit permutation:

Lemma 3. *For any 4-bit permutation S, the table of linear biases of S has at least one value greater than or equal to 8 on every row and column.*

Proof. This result is obtained by an exhaustive search over all affine equivalence classes. The 302 representatives have been examined, and we could check the result for each of them. $\square$

Using that any 4-bit permutation S satisfies $\mathcal{L}(S) \geq 8$ and $\mathcal{L}_{\min}(S) \geq 8$, we directly deduce from Theorem 4 the following improved lower bound.

Proposition 3. *Let S_1, S_2 and S_3 be three 4-bit S-Boxes and let F be the 8-bit function defined by the corresponding 3-round MISTY network. If any of the three inner S-Boxes is a permutation, then $\mathcal{L}(F) \geq 64$. Most notably, if $\mathcal{L}(F) < 64$, then $\delta(F) \geq 32$.*

The last statement in the previous theorem is deduced from the first item in Theorem 3. While it shows that 3-round MISTY with $\mathcal{L}(F) < 64$ would be of little interest, we show in [15] the following theorem proving that their linearity is at least 48.

Theorem 7. *Let S_1, S_2 and S_3 be three 4-bit S-Boxes and let F be the 8-bit function defined by the corresponding 3-round MISTY network. Then $\mathcal{L}(F) \geq 48$.*

We conjecture that any MISTY network with 4-bit inner functions actually satisfies $\mathcal{L}(F) \geq 64$, but it seems hard to prove without a full classification of the 4-bit functions.

5 Constructions

We now use the previous results to design concrete 8-bit invertible S-Boxes optimized for lightweight implementations. We use Feistel and MISTY networks, and select 4-bit S-Boxes S_i's with a low-cost implementation that provide good cryptographic properties of the resulting 8-bit S-Box. Such S-Boxes have been considered as good candidates for many lightweight constructions (*e.g.* the LS-designs [23]), but their respective merits and their cryptographic properties remained open.

We focus on implementing functions with a low gate count for hardware implementations, and a low instruction count for bit-sliced implementations (for table-based implementations, the table size is independent of the concrete S-Boxes). Moreover, we focus on implementations with a small number of non-linear gates, because non-linear gates are much harder to implement than linear gates in some dedicated settings such as masking [40], multi-party computation, or homomorphic encryption [1]. Bit-slicing can be used as an implementation technique to take advantage of some platform characteristics (for instance, it yields the fastest known implementation of AES on some Intel processors [25]), but it can also be a design criterion. Indeed, using a bit-sliced S-Box allows compact implementations without tables, and good performances both in software and hardware. In addition, S-Boxes implemented in this way are easier to protect against side-channel attacks with masking. Therefore, this approach is used by many lightweight designs such as SERPENT [9], NOEKEON [16], KECCAK [8], ROBIN and FANTOMAS [23], PRIDE [2], PRØST [26], or ASCON [19]. This makes the construction of S-Boxes with a low gate count particularly relevant for lightweight cryptography.

Following the previous sections, the best results we can achieve for an 8-bit invertible S-Box are:

With a MISTY network: $\delta(F) = 16$ and $\mathcal{L}(F) = 64$.
With a Feistel network: $\delta(F) = 8$ and $\mathcal{L}(F) = 64$.

We can provide some examples fulfilling these bounds: Example 2 is optimal for the MISTY construction, and an example for the Feistel construction is now exhibited. It is worth noticing that these results explain the compared properties of the S-Boxes obtained by the simulations reported in [23]. Since Feistel networks can reach a better security, we will mostly consider this construction. In this case, the optimal differential uniformity can be reached only if S_1, S_3 are APN, and S_2 is a permutation with $\delta(S_2) = 4$, as proved in Th. 9 in the full version [15]. Note that, in some other contexts, the MISTY construction presents some advantages since it offers better performance in terms of throughput and latency because the first two S-Boxes can be evaluated in parallel.

5.1 Feistel Network with Low Gate Count and Instruction Count

Rather than choosing S-Boxes S_1, S_2 and S_3 with good properties first, and then searching for an efficient implementation of these S-Boxes (as in [13, 39] for instance), we take the opposite approach, following Ullrich *et al.* [45]. We build gate descriptions of S-Boxes, and we test their cryptographic properties until we find a good candidate. Indeed, we do not have to specify in advance the 4-bit S-Boxes S_1, S_2, S_3. Instead, we look for a good implementation of a permutation with $\delta(S) = 4$ for S_2, a good implementation of an APN function for S_1 and S_3, and we test the properties of the resulting Feistel structure. With good probability, this results in a Feistel network F with $\delta(F) = 8$ and $\mathcal{L}(F) = 64$.

Following Ullrich *et al.*, we run a search oriented towards bit-sliced implementations. We consider sequences of software instructions, with instructions AND, OR, XOR, NOT, and MOV, using at most 5 registers. This directly translates to a hardware representation: the MOV instruction becomes a branch while the other instructions represent the corresponding gates. There are 85 choices of instructions at each step, but we use an equivalence relation to restrict the search. For S_2, we can directly reuse the results of [45]: they give an optimal implementation of a 4-bit permutation with $\delta(S) = 4$. For S_1 and S_3, we implemented a version of their algorithm, and searched for APN functions. We found that there is no construction of an APN function with 9 or fewer instructions. There are solutions with 10 instructions, but they have at least 6 non-linear instructions (AND, OR), which is not efficient for a masked implementation. Finally, with 11 instructions, there are constructions of APN functions with 4 non-linear instructions, 5 XOR instructions, and 2 MOV (copy) instructions. This search requires about 6000 core-hours of computation. The branching factor of our search is close to 10, while Ullrich *et al.* report a branching factor of less than 7; this is because we do not restrict the search to permutations (indeed, 4-bit APN functions are not permutations). This results in a very efficient 8-bit S-Box with good cryptographic properties, using 12 nonlinear gates, and 26 XORs. According to Theorem 9 in [15] and to the following lemma, this is the optimal number of non-linear gates.

Lemma 4. *Let S be a 4-bit permutation with $\delta(S) \leq 4$ or a 4-bit APN function. Any implementation of S requires at least 4 non-linear gates.*

Proof. If S can be implemented with 3 non-linear gates or less, then the algebraic expression of the output variables is a linear combination of the input variables, and of the 3 polynomials corresponding to the output of the 3 non-linear gates. Therefore, there exists a linear combination of the input and output variables that sums to a constant, *i.e.* $\mathcal{L}(S) = 16$. According to the classification of 4-bit permutation in [18], any permutation with $\delta(S) = 4$ satisfies $\mathcal{L}(S) \leq 12$. Furthermore, the classification of 4-bit APN functions [14] shows that they satisfy $\mathcal{L}(S) = 8$, which proves the lemma. $\square$

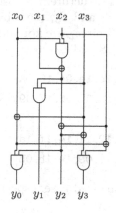

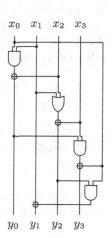

3.1. S_1, APN function with $\delta(S_1) = 2$.
$S_1 = [0, 0, 4, \mathsf{d}, \mathsf{c}, 0, 0, 5, 8, 0, 7, 6, 5, \mathsf{a}, 2, 4]$

3.2. S_2, permutation with $\delta(S_2) = 4$.
$S_2 = [0, 8, 6, \mathsf{d}, 5, \mathsf{f}, 7, \mathsf{c}, 4, \mathsf{e}, 2, 3, 9, 1, \mathsf{b}, \mathsf{a}]$

Fig. 3. Construction of a lightweight S-Box S with a three-round Feistel (S_1, S_2, S_1) satisfying $\delta(S) = 8$ and $\mathcal{L}(S) = 64$.

We give an example of such an implementation in Fig. 3, and we compare our results with previous designs in Table 1. In particular, we reach a better differential uniformity than the S-Boxes used in Robin and Fantomas [23], for a small number of extra gates. For comparing the respective merits of the S-Boxes considered in Table 1, we use the fact that, as a simple approximation, the number of rounds needed to reach a fixed security level against differential attacks is proportional to $1/\log(\delta(S)/256)$, and the implementation cost per round is proportional to the number of non-linear gates (for a bit-sliced software implementation with masking). This allows to derive a simple implementation cost metric for the S-Boxes presented in the last column, taking 1 for the AES, and considering only security against differential attacks.

5.2   Unbalanced MISTY Structure

Finally, we consider an alternative to MISTY structures as studied in this paper. Instead of dividing the input into two halves of equal size, we consider unbalanced networks. The idea is to split the 8 input bits in two unequal parts of 3 and 5 bits. Thus, the MISTY network will use only 3- and 5-bit S-Boxes. The advantage of 3- and 5-bit S-Boxes is that invertible S-Boxes with $\delta = 2$ exist, contrarily to the case of 4-bit S-Boxes. We managed to obtain 8-bit S-Boxes S with $\delta(S) = 8$ using unbalanced MISTY networks, which is better than the lower bound $\delta(S) \geq 16$ proved for balanced MISTY networks. However, this method uses 5-bit S-Boxes, which are more complicated to implement than 4-bit S-Boxes.

Example 3. We consider a 3-round unbalanced MISTY structure, with 5-bit permutations S_1, S_3 and a 3-bit permutation S_2. After S_1 and S_3, the 3-bit x_L is xored in the 3 MSB of x_R; after S_2 the 3 MSB of the 5-bit x_L are xored into x_L. The following S-Boxes define an 8-bit S-Box with $\delta = 8$ and $\mathcal{L} = 64$:

$$S_1 = [00, 01, 02, 04, 03, 08, 0d, 10, 05, 11, 1c, 1b, 1e, 0e, 18, 0a,$$
$$06, 13, 0b, 14, 1f, 1d, 0c, 15, 12, 1a, 0f, 19, 07, 16, 17, 09]$$
$$S_2 = [2, 5, 6, 4, 0, 1, 3, 7]$$
$$S_3 = [00, 01, 02, 04, 03, 08, 10, 1c, 05, 0a, 1a, 12, 11, 14, 1f, 1d,$$
$$06, 15, 18, 0c, 16, 0f, 19, 07, 0e, 13, 0d, 17, 09, 1e, 1b, 0b]$$

This shows that generalizing our results to the unbalanced case, especially for the MISTY construction, may be of interest.

Table 1. Comparison of some 8-bit S-Boxes. δ and $\mathcal{L}$ respectively denote the differential uniformity and the linearity of the S-Box (see Sect. 2.1 for the definitions), the last column presents the relative overall implementation cost (taking 1 for the AES).

S-Box	Construction	Implementation		Properties		
		AND/OR	XOR	$\mathcal{L}$	δ	cost
AES [13]	Inversion in $\mathbb{F}_{2^8}$ + affine	32	83	32	4	1
Whirlpool [5]	Lai-Massey	36	58	56	8	1.35
CRYPTON [29]	3-round Feistel	49	12	64	8	1.83
Robin [23]	3-round Feistel	12	24	64	16	0.56
Fantomas [23]	3-round MISTY (3/5 bits)	11	25	64	16	0.51
Unnamed [23]	Whirlpool-like	16	41	64	10	0.64
New	3-round Feistel	12	26	64	8	0.45

6 Conclusion

Our results give a better understanding of the cryptographic properties of light-weight S-Boxes built from smaller S-Boxes. We give a precise description of the best security achievable with a 3-round balanced Feistel or MISTY structure for an 8-bit S-Box, and necessary conditions to reach the bound. Interestingly, the MISTY network cannot offer the same security as the Feistel network for constructing an invertible 8-bit S-Box. Using those results, we describe an 8-bit S-Box S using only 12 non-linear gates and 26 XOR gates, with $\delta(S) = 8$ and $\mathcal{L}(S) = 64$. This is the best security that can be achieved with a 3-round Feistel or MISTY structure, and our construction uses the minimal number of non-linear gates to reach this security. This is an improvement over previous proposals, including the S-Boxes used in CRYPTON, Fantomas and Robin, but further work is required to determine whether different structures can provide better S-Boxes.

References

1. Albrecht, M.R., Rechberger, C., Schneider, T., Tiessen, T., Zohner, M.: Ciphers for MPC and FHE. In: Oswald, E., Fischlin, M. (eds.) EUROCRYPT 2015. LNCS, vol. 9056, pp. 430–454. Springer, Heidelberg (2015)
2. Albrecht, M.R., Driessen, B., Kavun, E.B., Leander, G., Paar, C., Yalçın, T.: Block ciphers – focus on the linear layer (feat. PRIDE). In: Garay, J.A., Gennaro, R. (eds.) CRYPTO 2014, Part I. LNCS, vol. 8616, pp. 57–76. Springer, Heidelberg (2014)
3. Aoki, K., Ohta, K.: Strict evaluation of the maximum average of differential probability and the maximum average of linear probability. IEICE Trans. Fundam. Electron. Commun. Comput. Sci. **E80A**(1), 2–8 (1997)
4. Barreto, P.S., Rijmen, V.: The KHAZAD Legacy-Level Block Cipher. NESSIE submission
5. Barreto, P.S., Rijmen, V.: The WHIRLPOOL Hashing Function. NESSIE submission
6. Berlekamp, E.R., Welch, L.R.: Weight distributions of the cosets of the (32, 6) Reed-Muller code. IEEE Trans. Inf. Theor. **18**(1), 203–207 (1972)
7. Bernstein, D.J.: Cache-timing on AES. In: Symmetric-Key Encryption Workshop - SKEW 2005 (2005). http://cr.yp.to/antiforgery/cachetiming-20050414.pdf
8. Bertoni, G., Daemen, J., Peeters, M., Assche, G.V.: The Keccak reference, January 2011. http://keccak.noekeon.org/
9. Biham, E., Anderson, R., Knudsen, L.R.: Serpent: a new block cipher proposal. In: Vaudenay, S. (ed.) FSE 1998. LNCS, vol. 1372, pp. 222–238. Springer, Heidelberg (1998)
10. Biham, E., Shamir, A.: Differential cryptanalysis of DES-like cryptosystems. J. Crypt. **4**(1), 3–72 (1991)
11. Bogdanov, A.A., Knudsen, L.R., Leander, G., Paar, C., Poschmann, A., Robshaw, M., Seurin, Y., Vikkelsoe, C.: PRESENT: an ultra-lightweight block cipher. In: Paillier, P., Verbauwhede, I. (eds.) CHES 2007. LNCS, vol. 4727, pp. 450–466. Springer, Heidelberg (2007)
12. Borghoff, J., Canteaut, A., Güneysu, T., Kavun, E.B., Knezevic, M., Knudsen, L.R., Leander, G., Nikov, V., Paar, C., Rechberger, C., Rombouts, P., Thomsen, S.S., Yalçın, T.: PRINCE – a low-latency block cipher for pervasive computing applications. In: Wang, X., Sako, K. (eds.) ASIACRYPT 2012. LNCS, vol. 7658, pp. 208–225. Springer, Heidelberg (2012)
13. Boyar, J., Peralta, R.: A new combinational logic minimization technique with applications to cryptology. In: Festa, P. (ed.) SEA 2010. LNCS, vol. 6049, pp. 178–189. Springer, Heidelberg (2010)

14. Brinkmann, M., Leander, G.: On the classification of APN functions up to dimension five. Des. Codes Crypt. **49**(1–3), 273–288 (2008)
15. Canteaut, A., Duval, S., Leurent, G.: Construction of Lightweight S-Boxes using Feistel and MISTY structures (Full Version). IACR eprint report 2015/711, Jul 2015. http://eprint.iacr.org/2015/711
16. Daemen, J., Peeters, M., Assche, G.V., Rijmen, V.: Nessie proposal. In: NOEKEON (2000)
17. De Cannière, C., Dunkelman, O., Knežević, M.: KATAN and KTANTAN — a family of small and efficient hardware-oriented block ciphers. In: Clavier, C., Gaj, K. (eds.) CHES 2009. LNCS, vol. 5747, pp. 272–288. Springer, Heidelberg (2009)
18. De Cannière, C.: Analysis and Design of Symmetric Encryption Algorithms. Ph.D. thesis, KU Leuven (2007)
19. Dobraunig, C., Eichlseder, M., Mendel, F., Schläffer, M.: Ascon v1. In: CAESAR Competition (2014). http://competitions.cr.yp.to/round1/asconv1.pdf
20. ETSI/SAGE: specification of the 3GPP confidentiality and integrity algorithms 128-EEA3 & 128-EIA3. document 4: design and evaluation report. Technical report (2011)
21. Gérard, B., Grosso, V., Naya-Plasencia, M., Standaert, F.-X.: Block ciphers that are easier to mask: how far can we go? In: Bertoni, G., Coron, J.-S. (eds.) CHES 2013. LNCS, vol. 8086, pp. 383–399. Springer, Heidelberg (2013)
22. Grosso, V., Leurent, G., Standaert, F.X., Varici, K., Durvaux, F., Gaspar, L., Kerckhof, S.: SCREAM & iSCREAM side-channel resistant authenticated encryption with masking. In: CAESAR Competition (2014). http://competitions.cr.yp.to/round1/screamv1.pdf
23. Grosso, V., Leurent, G., Standaert, F.-X., Varıcı, K.: LS-designs: bitslice encryption for efficient masked software implementations. In: Cid, C., Rechberger, C. (eds.) FSE 2014. LNCS, vol. 8540, pp. 18–37. Springer, Heidelberg (2015)
24. Hou, X.D.: Affinity of permutations of $\mathbb{F}_2^n$. Discrete Appl. Math. **154**(2), 313–325 (2006)
25. Käsper, E., Schwabe, P.: Faster and timing-attack resistant AES-GCM. In: Clavier, C., Gaj, K. (eds.) CHES 2009. LNCS, vol. 5747, pp. 1–17. Springer, Heidelberg (2009)
26. Kavun, E.B., Lauridsen, M.M., Leander, G., Rechberger, C., Schwabe, P., Yalçın, T.: Prøst. CAESAR Proposal (2014). http://proest.compute.dtu.dk
27. Leander, G., Poschmann, A.: On the classification of 4 bit S-Boxes. In: Carlet, C., Sunar, B. (eds.) WAIFI 2007. LNCS, vol. 4547, pp. 159–176. Springer, Heidelberg (2007)
28. Li, Y., Wang, M.: Constructing S-Boxes for lightweight cryptography with Feistel structure. In: Batina, L., Robshaw, M. (eds.) CHES 2014. LNCS, vol. 8731, pp. 127–146. Springer, Heidelberg (2014)
29. Lim, C.H.: CRYPTON: A new 128-bit block cipher. AES submission (1998)
30. Lim, C.H.: A revised version of CRYPTON - CRYPTON V1.0. In: Knudsen, L.R. (ed.) FSE 1999. LNCS, vol. 1636, pp. 31–45. Springer, Heidelberg (1999)
31. Matsui, M.: Linear cryptanalysis method for DES cipher. In: Helleseth, T. (ed.) EUROCRYPT 1993. LNCS, vol. 765, pp. 386–397. Springer, Heidelberg (1994)
32. Matsui, M.: New structure of block ciphers with provable security against differential. In: Gollmann, Dieter (ed.) FSE 1996. LNCS, vol. 1039, pp. 205–218. Springer, Heidelberg (1996)
33. Matsui, M.: New block encryption algorithm MISTY. In: Biham, E. (ed.) FSE 1997. LNCS, vol. 1267, pp. 54–68. Springer, Heidelberg (1997)

34. National Institute of Standards and Technology: Data Encryption Standard, FIPS Publication 46-2, December 1993
35. Nyberg, K.: Perfect nonlinear S-Boxes. In: Davies, D.W. (ed.) EUROCRYPT 1991. LNCS, vol. 547, pp. 378–386. Springer, Heidelberg (1991)
36. Nyberg, K.: Differentially uniform mappings for cryptography. In: Helleseth, T. (ed.) EUROCRYPT 1993. LNCS, vol. 765, pp. 55–64. Springer, Heidelberg (1994)
37. Nyberg, K.: Linear approximation of block ciphers. In: De Santis, A. (ed.) EURO-CRYPT 1994. LNCS, vol. 950, pp. 439–444. Springer, Heidelberg (1995)
38. Nyberg, K., Knudsen, L.R.: Provable security against a differential attack. J. Crypt. 8(1), 27–37 (1995)
39. Osvik, D.A.: Speeding up Serpent. In: AES Candidate Conference, pp. 317–329 (2000)
40. Rivain, M., Prouff, E.: Provably secure higher-order masking of AES. In: Mangard, S., Standaert, F.-X. (eds.) CHES 2010. LNCS, vol. 6225, pp. 413–427. Springer, Heidelberg (2010)
41. Rothaus, O.S.: On "bent" functions. J. Comb. Theor. Ser. A 20(3), 300–305 (1976)
42. Shannon, C.E.: Communication theory of secrecy systems. Bell Syst. Tech. J. 28(4), 656–715 (1949)
43. Standaert, F.-X., Piret, G., Rouvroy, G., Quisquater, J.-J., Legat, J.-D.: ICEBERG: an involutional cipher efficient for block encryption in reconfigurable hardware. In: Roy, B., Meier, W. (eds.) FSE 2004. LNCS, vol. 3017, pp. 279–299. Springer, Heidelberg (2004)
44. Suzaki, T., Minematsu, K., Morioka, S., Kobayashi, E.: TWINE: a lightweight block cipher for multiple platforms. In: Knudsen, L.R., Wu, H. (eds.) SAC 2012. LNCS, vol. 7707, pp. 339–354. Springer, Heidelberg (2013)
45. Ullrich, M., De Cannière, C., Indesteege, S., Küçük, Ö., Mouha, N., Preneel, B.: Finding optimal bitsliced implementations of 4×4-bit S-Boxes. In: SKEW 2011 Symmetric Key Encryption Workshop, Copenhagen, Denmark, pp. 16–17 (2011)
46. Wegener, I.: The Complexity of Boolean Functions. Wiley-Teubner, New York (1987)
47. Wheeler, D.J., Needham, R.M.: TEA, a Tiny Encryption Algorithm. In: Preneel, B. (ed.) FSE 1994. LNCS, vol. 1008, pp. 363–366. Springer, Heidelberg (1994)
48. Wu, W., Zhang, L.: LBlock: a lightweight block cipher. In: Lopez, J., Tsudik, G. (eds.) ACNS 2011. LNCS, vol. 6715, pp. 327–344. Springer, Heidelberg (2011)

Authenticated Encryption

A New Mode of Operation for Incremental Authenticated Encryption with Associated Data

Yu Sasaki[1,2] and Kan Yasuda[1(✉)]

[1] NTT Secure Platform Laboratories, Tokyo, Japan
{sasaki.yu,yasuda.kan}@lab.ntt.co.jp
[2] Nanyang Technological University, Singapore, Singapore

Abstract. We propose a new mode of operation for authenticated encryption with associated data (AEAD) that achieves incrementality and competitive performance with many designs in CAESAR. The incrementality of a function $F(\cdot)$ means the property that once $F(x)$ is stored, the value $F(x')$ with a slightly modified input x' can be updated from $F(x)$ much faster than computing $F(x')$ from scratch. It turns out that the security of incremental AEAD needs to be treated carefully. Incremental operations leak more information than ordinary ones. Moreover, if the scheme is nonce-based, nonce repetition must be taken into account. We discuss which structures are (un)suitable for incrementality. Parallelizability is a minimum requirement, but there are many other requirements, often subtle and complex, especially for combining associated data A with a message M. For example, using PMAC for both A and M would spoil incrementality. We go through 57 designs submitted to CAESAR and show that none of them meets all the requirements. It turns out that options are quite limited and the design of an incremental AEAD mode is almost uniquely determined. We propose a new construction providing incremental operations such as update, append and chop. Interestingly, our encryption part is composed of the XE construction instead of XEX, hence optimizing the efficiency.

Keywords: Incremental cryptography · AEAD · PMAC · OCB · CAESAR · Enc-then-MAC · XE

1 Introduction

Recently, many AEAD schemes have been proposed especially for CAESAR [10], which determines a new portfolio of AEAD. AEAD schemes are evaluated in many criteria including the performance in various platforms, security of integrity/confidentiality in nonce-respect/nonce-repeat models, and additional features such as parallelizability, provable security, length optimality, efficient processing for repeated associated data, variable tag size, a small masking cost against side-channel analysis, *etc.* Incrementality is one of such additional features.

The preliminary version was presented at Dagstuhl Seminar 14021 in 2014.

© Springer International Publishing Switzerland 2016
O. Dunkelman and L. Keliher (Eds.): SAC 2015, LNCS 9566, pp. 397–416, 2016.
DOI: 10.1007/978-3-319-31301-6_23

Incremental cryptography was introduced by Bellare et al. [6]. Suppose that an input x was once processed by a function $F(\cdot)$, and the output was stored. Then, a fraction of x is modified to x'. The incremental cryptography allows to quickly update $F(x)$ to $F(x')$ in time proportional to the amount of modification made to x, which is faster than recomputing $F(x')$ from scratch.

Incremental cryptography in early days focused on hashing, signature, and MAC rather than encryption [6,8]. A work in [7] initiated the discussion for incremental encryption/decryption. The security of incremental encryption schemes is always worse than the ordinary schemes. [7] introduced the security policy as hiding all the information other than the amount of difference for incremental operations. The proposed schemes provide most of practical operations i.e. update, delete, and insert. As a drawback, their scheme requires extra randomness and the randomness must be sent to a verifier, which causes a tag expansion. Moreover the construction needs to make the plaintext size at least twice of the original plaintext so that the modification history for all message blocks are recorded in the extended blocks.

More closely related work is the one by Buonanno et al. [15] about incremental AE (without AD). Buonanno et al. study a generic composition of probabilistic encryptions and incremental-MACs (instantiated with randomized ECB and the XOR scheme in [7]) to achieve a provable security with respect to unforgeability. One of the proposed schemes, RPC-mode, provides insert and delete functionalities. As a side-effect, the encryption cost becomes 4 times more than the ordinary scheme, which is unlikely to be accepted in the CAESAR competition.

There are a few more researches on incremental cryptography, e.g. byte-wise incremental encryption [4,5] and incremental MAC with SHA-3 [18]. We omit their details in this paper.

Achieving the incrementality for both of the efficient encryption and MAC at the same time is not an easy task. Considering the basic requirements of the incrementality, the following two conditions are inevitable: (1) parallelizability for encryption and decryption and (2) the same nonce must be iterated multiple times, thus the scheme must provide some robustness against nonce repetition. The GCM mode [28] can meet the first condition. However, under the nonce-repeating model, the key stream for all blocks are leaked, which does not meet the second condition. The OCB mode [35] can meet the both conditions for encryption/decryption. However, its authentication (tag generation) is the encryption of the message checksum, which can be broken easily under the nonce-repeating model. The COPA mode [2,3] is parallelizable and nonce-misuse resistant. However, its onlineness, i.e. modifying 1 message block affects all the subsequent blocks, cannot allow the incremental operation.

Our Contributions. This paper investigates incremental AEAD. Our final goal is developing a new mode of operation for incremental AEAD, which is also secure and practically efficient; as efficient as CAESAR candidates. Therefore, we avoid the ciphertext expansion, tag expansion, and extra randomness. We make the scheme simple and minimize the cost of each operation. We begin with formalizing nonce-based incremental AEAD. Then, we investigate various operations used in the CAESAR candidates with respect to (un)suitability for the incremental operation. Finally we propose a new mode of operation with security proof. Contributions of this paper are further detailed as follows.

1. FORMALIZATION. As pointed out by Bellare et al. [7], in principle it is inevitable that incremental operations, especially encrypting, should leak more information than ordinary, non-incremental ones. In particular, for nonce-based schemes [34], one needs to carefully consider the problem of security under "nonce repetition." Moreover, in nonce-based AEAD [32], the combination of various components raises subtle and complex issues. In this work, a framework of incremental AEAD is formalized, rigorously defining the syntax of incremental operations, the model of adversaries, and the notion of security.

2. DESIGN REQUIREMENTS. The presence of associated data A causes potential problems, which need to be carefully addressed. For instance, consider Enc-then-MAC approach [9]. Suppose that A is processed by PMAC [13] and that its output and C are again processed by PMAC. We call this construction *layered PMAC* and illustrate it in Fig. 1. Here one would expect the parallelizability of PMAC to be useful for incremental operations. However, one would soon notice that incremental operations cannot be performed on A, since the two PMACs are "stacked." This motivates us to slightly change the processing method for A from PMAC to the one used in OCB3 [26]. The resulting scheme is illustrated in Fig. 2. This scheme now allows incremental update of A. In this way we identify suitable and unsuitable constructions for designing an incremental AEAD mode of operation.

3. NEW MODE OF OPERATION. It turns out that there are a number of requirements essential for a secure incremental AEAD mode of operation. Indeed, we perform a survey on the CAESAR candidates and confirm that none of them fulfills all the requirements. This motivates us to build a new mode that satisfies all these conditions. Interestingly, options are quite limited, and the overall structure of the mode is almost uniquely determined according to the design conditions. Fortunately there *is* one that satisfies all the conditions. This work presents one concrete, optimized example. The new AEAD mode achieves greater efficiency and enables incremental operations at a minimal cost. Unlike previous proposals [7], the new mode does not need ciphertext expansion, tag expansion, or additional randomness. The computational cost of incremental operations is only linear to the number of blocks to be updated. The new mode is capable of executing *update*, *append* and *chop* operations.

The overall design of the new AEAD mode is inspired by PMAC [13] which was already incremental to a certain degree. Specifically, we utilize XE construction [33] within Enc-then-MAC paradigm [9]. Unlike OCB [35], the XE construction, rather than XEX [33], turns out to be sufficient for realizing the encryption part, owing to the Enc-then-MAC composition and "verify-then-decrypt" functionality [1]. This reduces the number of masks and the cost of an XOR operation per block.

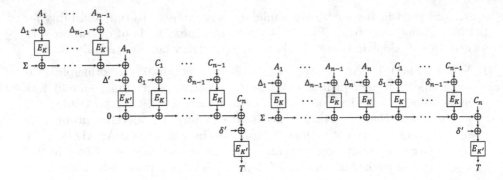

Fig. 1. Bad example: layered PMAC.   **Fig. 2.** Good example: OCB3 + PMAC.

2 Our Goals

We begin with stating properties that we want an incremental AEAD scheme to have. At the same time we mention properties that we do not—that is, the properties that ordinary AEAD schemes possess but seem "incompatible" with incrementality and inessential in the practical use.

Let Π be an AEAD scheme. Let N denote the nonce. Let $A = A_1 A_2 \cdots A_a$ be associated data divided into n-bit blocks A_i. Let $M = M_1 M_2 \cdots M_m$ be a message and $C = C_1 C_2 \cdots C_m$ the ciphertext. Let T denote the tag. In the following we consider the case of incremental authenticated *encryption*; the case of verification and decryption can be treated in the same manner.

Types of Incremental Operations. We demand Π to be capable of *update*, *append* and *chop* operations on A and M, since these appear to be feasible by basing on a PMAC-like internal structure. Here, the update operation either recomputes new T' given new A_i' and the original T and A_i or recomputes new T' and C_i' given the new M_i' and the original T, C_i and M_i. The append operation either recomputes new T' given an additional data block A_{a+1} and the original T or recomputes new T' and C_{m+1}' given an additional message block M_{m+1} and the original T. The chop operation is similarly defined. It should be noted that in many of the existing nonce-based AEAD schemes [32] the ciphertext C is not affected by A, which is a desirable property for incremental AEAD.

Efficiency. We demand the incremental operations to be efficient. Specifically, the computational cost should increase only linearly to the number of blocks to be modified.

Nonce-Based. We follow the well-established nonce-based framework, which was formalized by Rogaway [34]. The crucial problem here is that incremental operations by definition imply the use of the same nonce value N multiple times, which contradicts the security notions of nonce-based framework. We formalize the security notions under such circumstances.

No Randomization or Expansion. Since we follow the nonce-based framework, we do not allow Π to be randomized. In general randomization would require the use of pseudo-random number generator, which is costly. Moreover, randomization would expand the output size of the encryption algorithm. We would like to keep the original size of the ciphertext and the tag of an ordinary nonce-based AEAD scheme.

3   Formalization

In this section we formalize the framework of incremental AEAD by defining its algorithmic syntax, adversarial model, and security notions.

3.1   Syntax

An incremental AEAD scheme (Π, Δ) consists of an ordinary AEAD algorithms $\Pi = (\mathcal{E}_K, \mathcal{D}_K)$ and the set of incremental operations Δ which consists of $(\mathbb{U}_K^A, \mathbb{A}_K^A, \mathbb{C}_K^A)$ for associated data and $(\mathbb{U}_K^M, \mathbb{A}_K^M, \mathbb{C}_K^M)$ for the message part. The syntax of these algorithms are as follows.

- The encryption algorithm $\mathcal{E}_K$ takes as input a nonce $N \in \mathcal{N}$, associated data $A \in \mathcal{A}$ and a message $M \in \mathcal{M}$ and outputs a pair (C, T) of ciphertext $C \in \mathcal{C}$ and tag $T \in \{0,1\}^n$, so that we have $(C, T) \leftarrow \mathcal{E}_K(N, A, M)$. For simplicity we assume that the length of associated data A, of a message M, or of ciphertext C is always a multiple of n bits.
- The decryption algorithm $\mathcal{D}_K$ takes as input a nonce N, associated data A, a ciphertext C and a tag T and outputs either the reject symbol $\perp$ or a message M, so that we have $\perp$ or $M \leftarrow \mathcal{D}_K(N, A, C, T)$.
- We demand that $\mathcal{E}_K$ and $\mathcal{D}_K$ should be consistent; that is, whenever $(C, T) \leftarrow \mathcal{E}_K(N, A, M)$, we must have $M = \mathcal{D}_K(N, A, C, T)$.
- The update operation $\mathbb{U}_K^A$ takes as input a nonce N, the position i to be updated, the original A_i, the new A_i' and the original T and outputs a new tag value T', so that $T' \leftarrow \mathbb{U}_K^A(N, i, A_i, A_i', T)$. Similarly, the update function $\mathbb{U}_K^M$ takes as input a nonce N, the position i to be updated, the original C_i, the new M_i' and the original T and outputs a new ciphertext block C_i' and a new tag value T', so that we have $(C_i', T') \leftarrow \mathbb{U}_K^M(N, i, C_i, M_i', T)$.
- The append operation $\mathbb{A}_K^A$ takes as input a nonce N, the current number of blocks a of A, a new block A_{a+1} to be appended, and the original T and outputs a new tag value T', as $T' \leftarrow \mathbb{A}_K^A(N, a, A_{a+1}, T)$. Similarly, the append function $\mathbb{A}_K^M$ takes as input a nonce N, the current number m of ciphertext blocks, the new M_{m+1} and the original T and outputs a new ciphertext block C_{a+1} and a new tag value T', so that we have $(C_{m+1}, T') \leftarrow \mathbb{A}_K^M(N, m, M_{m+1}, T)$.

- The chop operation $\mathbb{C}_K^A$ takes as input a nonce N, the current number a of blocks in A, the original A_a, and the original T and outputs a new tag value T', so we can write $T' \leftarrow \mathbb{C}_K^A(N, a, A_a, T)$. Similarly, the chop function $\mathbb{C}_K^M$ takes as input a nonce N, the current number m of ciphertext blocks, the original C_m, and the original T and outputs a new tag value T', and so we simply have $T' \leftarrow \mathbb{C}_K^M(N, m, C_m, T)$.
- We demand that the incremental operations are consistent with each other; that is, if the same value of a block A_i or M_j is "restored" after performing a sequence of incremental operations, then we must have the same value for C_j and T.
- We demand that the incremental operations are consistent with the encryption and decryption algorithms; that is, if (N, A', C', T') is a result after performing incremental operations, we must have $M' = \mathcal{D}_K(N, A', C', T')$.

3.2   Adversarial Model

An adversary is an oracle machine. In the conventional setting of AEAD, an adversary A is given access to the encryption oracle $\mathcal{E}_K(\cdot, \cdot, \cdot)$ and the decryption oracle $\mathcal{D}_K(\cdot, \cdot, \cdot, \cdot)$. We follow the framework of nonce-based symmetric-key encryption and demand that A should be *nonce-respecting* to its encryption oracle. That is, adversary A is not allowed to use the same value of N in making queries to its encryption oracle $\mathcal{E}_K$.

Now in the current setting of incremental AEAD we give A access also to the incremental oracles, $\mathbb{U}_K^A$, $\mathbb{A}_K^A$, $\mathbb{C}_K^A$, $\mathbb{U}_K^M$, $\mathbb{A}_K^M$ and $\mathbb{C}_K^M$. We forbid A from making queries to these incremental oracles with an "unused" nonce N; that is, these values of N which have not been used in the encryption oracle $\mathcal{E}_K$. We write $\Delta = (\mathbb{U}_K^A, \mathbb{A}_K^A, \ldots, \mathbb{C}_K^M)$ to gather the six oracles.

Since we assume that adversary A is nonce-respecting and does not make incremental queries with an unused nonce, from the value of N it should be clear which (A, C, T) is relevant when A makes a query to one of its incremental oracles. We allow A to interleave its queries to the encryption, decryption and incremental oracles. A can make its queries to these oracles in any combination.

Suppose that A has access to its oracles $\mathcal{O}_1, \mathcal{O}_2, \ldots$ We write $A^{\mathcal{O}_1, \mathcal{O}_2, \cdots}$ to denote the value returned by A after interacting with its oracles.

3.3   Security Notions

In the conventional AEAD, the key notions of security were privacy and authenticity. We follow these notions. We formalize the notions of security and authenticity when adversaries have additional access to the incremental oracles.

Privacy. In the conventional AEAD setting the notion of privacy was defined in terms of indistinguishability between the real encryption oracle $\mathcal{E}_K$ and the ideal oracles \$ where \$ is an oracle having the same interface with $\mathcal{E}_K$ but returns a random string of expected length upon a query (N, A, M). Here of course the adversary is assumed to be nonce-respecting.

We do similarly for the incremental setting. Specifically, we consider random oracles having the same interfaces with the six incremental operations. In the ideal world adversary has access to this random oracle, which we write $\& = (\$, \$, \$, \$, \$, \$)$ where each oracle is an independent random oracle having the corresponding interface. We define

$$\text{Adv}_{\Pi, \Delta}^{\text{priv}}(A) := \Pr[A^{\mathcal{E}_K, \Delta} = 1] - \Pr[A^{\$, \&} = 1],$$

where the probabilities are taken over the choice of key K and randomness used in the oracles and in the adversary A. This convention of probability definition apply to the rest of the paper. Here we demand that A be nonce-respecting. Without loss of generality we also assume that A never makes incremental queries that lead to a trivial win. These are incremental queries which A can check the consistency conditions of the incremental oracles in the real world.

Authenticity. In the conventional AEAD the authenticity was defined in terms of unforgeability under chosen-message attacks. Here adversary A is given access to both the encryption oracle $\mathcal{E}_K$ and the decryption oracle $\mathcal{D}_K$. By a forgery we mean the event that A makes a non-trivial query (N, A, C, T) to its decryption oracle where the oracle would return something other than the reject symbol $\perp$.

This notion of authenticity is extended to our incremental setting in a straightforward way. We simply give A additional access to the incremental oracles Δ and consider a forgery event. That is, we define

$$\text{Adv}_{\Pi, \Delta}^{\text{auth}}(A) := \Pr[A^{\mathcal{E}_K, \mathcal{D}_K, \Delta} \text{ forges}],$$

where by "forges" we mean the event that $\mathcal{D}_K$ would return something other than $\perp$. We assume that A is nonce-respecting and does not make a trivial-win query to its decryption oracle $\mathcal{D}_K$.

4 Design Requirements

Parallelizability. An obvious requirement to be incremental is the parallelizability for encryption, decryption and tag generation. Ideally, those computations should be block-wise, but they can be processed with neighboring blocks as long as the scheme is parallelizable. This requirement excludes several design principles, e.g. online ciphers with up-to-prefix privacy and offline ciphers. Serial designs, e.g. SpongeWrap [11], is also excluded.

Encryption for Combined Data. For the MAC generation, intermediate values computed in parallel must be combined efficiently. The combined data is sensitive, thus some secure finalization computation must be applied to the combined data so that adversaries cannot recover it from the tag. Moreover, when a part of blocks are modified for incremental operations, the combined data also needs to be updated. Ideally, the design allows to obtain the combined data without touching any unmodified blocks. The possible solution is recovering the combined data from the tag with the knowledge of the key. Those constraints lead to the usage of encryption for the combined data.

Security Against Nonce Reuse. During the incremental operations, unmodified blocks need to stay unchanged, which requires the reuse of the same nonce. Thus, the scheme must provide some robustness against nonce-repeating adversaries. As a result, the schemes using keystream for encryption and decryption like GCM is not suitable. Indeed, trivial attacks can be mounted against GCM with incremental operation. The adversary first makes query of (N, A, M) to the encryption oracle and obtains the corresponding (N, A, C, T). Key stream can be recovered with $M \oplus C$. Then, the adversary incrementally modifies a fraction of M to M'. The nonce N is repeated and the key stream never changes, which allows to predict corresponding $C' = M' \oplus \texttt{keystream}$. Similarly, integrity must be ensured against the nonce repetition to avoid trivial distinguishers. Encrypting the check-sum of M as OCB is unsuitable for incremental AEAD.

No Tag Truncation. Truncating a state to generate tag is incompatible with incremental operations. Let us explain an example of incremental operation for PMAC, which computes a tag T for a message $M = M_1 \| M_2 \| \cdots \| M_n$ as follows.

$$\Sigma \leftarrow 0, \quad \Sigma \leftarrow \Sigma \oplus E_K(M_i \oplus L_i) \text{ for } i = 1, \ldots, n-1, \quad T \leftarrow E_K(\Sigma \oplus M_n \oplus L'),$$

where L_i and L' are the masks. With the tag truncation, $\hat{T} \leftarrow trunc(T)$ is output as a tag, where $trunc$ truncates the input value to the tag size. To incrementally update tag, the user first needs to recover Σ by inverting the last block cipher call. If the tag truncation is applied, T cannot be recovered from $\hat{T}$ (no clue for the user to guess the discarded bits of T).

No Layered PMAC. Associated data A needs to be incorporated carefully. An unsuitable choice is a layered PMAC in Minalpher [36], which computes a tag by PMAC(PMAC(A)$\|C$). Because PMAC(A) cannot be recovered from T, the user cannot update A incrementally. A slightly modified procedure in OCB3 [26] can avoid this problem. Let OCB3^A be the function to process A in OCB3, which initializes Σ to 0 and updates it as $\Sigma \leftarrow \Sigma \oplus E_K(A_i \oplus L_i)$ for $i = 1, 2, \ldots, n$. In short, the special treatment of the last block of PMAC is removed. The original motivation of replacing PMAC with OCB3^A was improving hardware implementation, but it also makes the scheme suitable for incremental AEAD. Namely, if the tag is computed as PMAC(OCB3^A(A)$\|C$), both of A and C can be incrementally updated (see Figs. 1 and 2).

No Ciphertext Translation. Another unsuitable example is the ciphertext translation proposed by [32] as a generic way to incorporate A in nonce-based

AE schemes. It processes A by a universal hash function f^A, say OCB3^A, and generates the tag by XORing $f^A(A)$ and a part of C. Intuitively C is a random string as long as the nonce is respected, and it takes a role of hiding $f^A(A)$. It is also possible to generate the tag as $f^A(A) \oplus f^C(C)$ with a deterministic MAC f^C, say PMAC. Using a deterministic MAC for f^C is crucial because incremental AEAD reuses the same nonce. Here the combination of universal hash f^A and deterministic MAC f^C spoils incremental AEAD. The construction with OCB3^A for A and PMAC for f^C is given in Fig. 3. In order to incrementally update C, the last block-cipher call in PMAC needs to be inverted. Without the knowledge of OCB3$^A(A)$, the user cannot recover the block-cipher's output. Note that the same applies when f^A is PMAC and f^C is OCB3^A (then A cannot be updated incrementally).

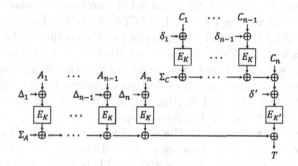

Fig. 3. Ciphertext translation assuming f^A is OCB3^A and f^C is PMAC.

Parallel Mask Generation. The mask for each block needs to be generated in parallel. Using serial pseudo-random number generates for generating masks is unsuitable for incremental AEAD.

5   Survey of Existing Schemes

We investigate the modes of operation for 57 algorithms submitted to CAESAR with respect to the suitability for incremental AEAD. We list the algorithms providing block-wise encryption and discuss their suitability for incremental AEAD. GCM is also discussed for comparison.

CPFB [30] basically follows the CTR mode. In each block, the keystream is generated with XOR of the counter and the previous plaintext block. The serial structure of the decryption and the reproduction of the same keystream in the nonce-repeating model make CPFB unsuitable for incremental AEAD. The tag generation is suitable for incremental AEAD. In short, it initializes $\Sigma \leftarrow 0$ and generates a tag T for $A_1, \ldots, A_m$ and $M_1, \ldots, M_n$ as $\Sigma \leftarrow$

$\Sigma \oplus E_K(A_i\|i)$ for $i = 1,\ldots, m$, $\Sigma \leftarrow \Sigma \oplus E_{K_j}((M_j\|j) \oplus K)$ for $j = 1,\ldots, n$, and $T \leftarrow E_{K'}(\Sigma \oplus L)$, where K, K_j, K', and L are key and mask values. Each block can be incrementally updated after $D_{K'}(T) \oplus L$ is computed.

Enchilada [20] processes M with the XEX construction and computes tag with GHASH(A, C), which seems suitable for incremental AEAD. However the mask values for the XEX construction are generated by ChaCha stream-cipher, which breaks parallelizability.

GCM [28] adopts the CTR mode. The reproduction of the same keystream in the nonce-repeating model raises a security concern for incremental AEAD.

iFeed [37] processes A with PMAC and processes M with the inversion of the CBC mode. The tag is generated by CBC-MAC$(M) \oplus$ PMAC(A) with the ciphertext translation. Decryption of iFeed is not parallelizable and the ciphertext translation spoils incremental operations.

Minalpher [36] processes M with the XEX construction and computes tag with layered-PMAC, i.e. PMAC(PMAC(A)$\|C$). The layered PMAC spoils incremental AEAD. The block-cipher is built with the double-block-length permutation with the Even-Mansour construction which causes the tag truncation at the end. The serial mask generation is also a drawback.

OCB [27] is also used in some other designs including CBA [21], Deoxys [22], Joltik [23], KIASU [24], and SCREAM [19]. The only but important concern for OCB is the integrity in the nonce-repeating model.

OTR [29] which is also used in PRØST-OTR [25] is similar to OCB except that encryption and decryption are processed in every 2 blocks with 2-round Feistel network. Interestingly, security of OTR is worse than OCB for incremental AEAD. In addition to the modified block position, it leaks how the 2 blocks are modified. Suppose that M_{2i} and M_{2i-1} are encrypted as

$$C_{2i-1} = E_K(2^{i-1}L \oplus M_{2i-1}) \oplus M_{2i},$$
$$C_{2i} = E_K(2^{i-1}L \oplus \delta \oplus C_{2i-1}) \oplus M_{2i-1},$$

where L and δ are mask values. Then, incrementally updating M_{2i} only simply affects to C_{2i-1}, which does not provide the ideal confidentiality.

PAEQ [12] works like the CTR mode with a big permutation. In each block, a part of permutation output is used as keystream which raises the problem against nonce-repetition. The overall structure for tag generation, which takes XOR of results of each message block and each associated data block then performs the permutation at the end, is suitable for incremental AEAD. However, the last permutation output is always truncated due to the big permutation size, which prevents incremental operations.

π-Cipher [17] has the similar structure as PAEQ. Besides the security issue about the reproduction of keystream, the tag for A and M are lastly XORed for the ciphertext translation, which is unsuitable for incremental AEAD.

Silver [31] is a tweakable block-cipher (TBC) based design. The computation structure basically follows the one in OCB in a TBC level. Similarly to OCB, the integrity in the nonce-repeating model is the main drawback.

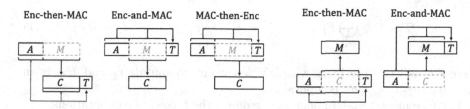

Fig. 4. Incremental encryption.

Fig. 5. Incremental decryption. MtE is omitted.

YAES [14] processes A with a PMAC-like structure. The encryption is similar to the CTR mode. The PMAC-like structure is again used by taking intermediate value of each block as input. Finally, the results of two PMAC-like computations are XORed for ciphertext translation. Reproduction of keystream and ciphertext translation are unsuitable for incremental AEAD.

Summary is available in Table 1 in Appendix. None of the existing schemes achieves incremental AEAD. In the next section, we will propose a new construction suitable for incremental AEAD.

6 A Concrete Example

6.1 Choice of Generic Constructions

In general, there are three approaches to construct authenticated encryptions: Enc-then-MAC (EtM), Enc-and-MAC, (E&M), and MAC-then-Enc (MtE). Here, we compare the cost of incremental operations with those three approaches.

Encryption Cost. Encryption with those approaches are illustrated in Fig. 4. For incremental AEAD, $(C, T) \leftarrow (K, N, A, M)$ is once computed and (K, N, A, C, T) is stored for decryption. An important remark is that M disappears in the 5-tuple value passed to the decryption side. Then, for modified values (A', M'), new values (C', T') are updated from (C, T, A', M').

As illustrated in Fig. 4, E&M and MtE are similar. With MtE, modifying 1 bit of A can affect to C, which is not desired for the incrementality. Hereafter, we exclude MtE from our choices and show the comparison between EtM and E&M.

Several conditions are required to achieve incrementality. First, we suppose that the conversion from M to C (or C to M in decryption) is processed in block-wise, i.e. 1 block change in M only requires 1-block computation to obtain new C'. For integrity, we need to avoid simple forgery attack in the nonce-repeating model. Let X be C in EtM and X be M in E&M. We assume that each block in A and X is processed by some computation, and moreover the sum of their computations is processed again. Namely, for a given $(A_1\|A_2\|\cdots\|A_n, X_1\|X_2\|\cdots\|X_l)$, T is computed as follows.

$$\Sigma \leftarrow 0, \quad \Sigma \leftarrow \Sigma \oplus \bigoplus_{i=1}^{n} F_K^i(A_i), \quad \Sigma \leftarrow \Sigma \oplus \bigoplus_{j=1}^{l} G_K^j(X_j), \quad T \leftarrow H_K(\Sigma), \quad (1)$$

where F_K^i and G_K^j are block-wise key-dependent computations and H_K is an invertible key-dependent computation.

The incremental update operation requires the following computations.

1. Recover the intermediate value Σ from T.
2. Compute the impact from old (A, X) to Σ for the modified blocks.
3. (a) If M is modified to M', compute new C' for the modified blocks. (skip if only A is modified)
   (b) Compute the impact from new (A', M') to Σ' for the modified blocks.
4. Compute new T' from Σ'.

Let us consider the cost of the update operation in EtM and E&M when s blocks are modified. In Step 1, both approaches require 1-block computation of H_K^{-1}. Step 2 yields a difference. In EtM, X is C which is stored in (K, N, A, C, T), thus it requires s-block computations. While in E&M, X is M however M is not stored. M needs to be recovered from C and then the impact from M to Σ is computed, which requires $2s$-block computations. In Step 3, both approaches require between $2s$- and s-block computations depending on how many blocks of A and M are modified. In 4, both approaches require 1-block computation. In the end, Step 2 yields a difference and it brings an advantage for EtM.

Regarding the append operation, Step 2 can be removed from the above procedure, because the append operations only adds the new impact to Σ. Hence, the advantage of EtM disappears. Regarding the chop operation, Step 3 is removed from the above procedure, because nothing new is added to Σ. Hence, EtM keeps the advantage over E&M. All in all, we conclude that EtM is better for incremental encryption.

Decryption Cost. Decryption with EtM and E&M are illustrated in Fig. 5. $(M, T) \leftarrow (K, N, A, C, T)$ is once computed and (K, N, A, M, T) is stored for incremental computations. C does not have to be stored. There are two ways to utilize incrementality of decryption. In the first way, the tag verification fails and later only a fraction of (A, C) is modified as correction. In the second way, after decrypting (K, N, A, C, T), it is expected to decrypt the similar tuples of (A', C'). Actually, authentication of several messages in which only the header part is different was the original motivation of the incremental cryptography [6].

As it can be seen in Figs. 4 and 5, encryption of EtM and encryption of E&M have duality against decryption of E&M and decryption of EtM, respectively. The procedure of the incremental operations is basically the same as Step 1 to Step 4. When the impact to Σ from old C (for EtM) or old M (for E&M) is computed, EtM requires $2s$-block computations while E&M requires only s-block computation. Thus, E&M has an advantage compared to EtM.

In summary, due to the duality, there is no choice with an overwhelming advantage. EtM and E&M should be chosen depending on which of encryption and decryption is the main target.

Favoring Enc-then-MAC. Given the above discussion, it seems reasonable to choose EtM for incremental encryption and E&M for incremental decryption. However, we argue that EtM can be also an attractive option for incremental decryption.

The reason is that one can use a "weak" (IND-CPA) algorithm for the encryption part in EtM composition [9], whereas in E&M one needs relatively stronger encryption algorithm such as IND-CCA. In particular, for block-wise encryption the XE construction, rather than XEX, is sufficient for building the encryption part if the scheme is based on the EtM composition. As a result, within the EtM paradigm one can construct a mode of operation that is efficient for both incremental encryption and decryption, as we shall show in the following section.

6.2 New Mode of Operation

In this section we present a new mode of operation for incremental AEAD. We first explain the method of mask generations. Consider the set $\{0,1\}^n$ as the finite field of 2^n elements with respect to some primitive polynomial $f(x)$, so that the degree of f is n and the element x generates the whole multiplicative group. Let Γ_i denote the i-th Gray code, for $i \geq 1$. We define $\gamma_i \leftarrow \Gamma_i L_0$, $\Delta_i \leftarrow \Gamma_i L_1$, $\delta_i \leftarrow \Gamma_i L_2$ and $\delta' \leftarrow L_3$ where $L_j \leftarrow E_K(N\|(j)_2)$ for $j = 0, 1, 2, 3$ with $(j)_2$ being a 2-bit representation of integer j. Here E_K is the underlying n-bit block cipher with a secret key K and $N \in \{0,1\}^{n-2}$ is a nonce. The choice of Gray code rather than doubling comes from the fact that we need to compute i-th mask from scratch when one of the incremental operations is performed.

For simplicity we assume that $T \in \{0,1\}^n$ and the lengths of A, M and C are always a multiple of n bits. In practice A and M might need to be padded, so that their lengths become a multiple of n bits. In principle this would not affect the feasibility of our construction, as in most of the cases the padding affect only the last block (as in the standard 10* padding), and the last block can be always handles by the update, append and chop operations.

Now the new mode of operation iterates an n-bit block cipher E_K and is based on XE construction given by Rogaway [33]. Its pseudo-code is given in Figs. 6 and 7. The encryption algorithm is also illustrated in Fig. 8.

6.3 Security of the New Mode

We prove the security of the incremental AEAD scheme specified above. We follow the adversarial model and the security notions defined in Sect. 3.

The first step is to replace the block cipher calls with the secret masks with independent random functions (not permutations). This is possible owing to the result of XE construction given by Rogaway. The result says that the block

```
1: function 𝒟_K(N, A, C, T)
2:     A_1A_2 ⋯ A_a ← A
3:     C_1C_2 ⋯ C_m ← C
4:     Σ ← 0
5:     for i = 1 to a do
6:         Σ ← Σ ⊕ E_K(A_i ⊕ Δ_i)
7:     end for
8:     for i = 1 to m do
9:         M_i ← E_K^{-1}(C_i)γ_i
10:        Σ ← Σ ⊕ E_K(C_i ⊕ δ_i)
11:    end for
12:    M ← M_1M_2 ⋯ M_m
13:    T' ← E_K(Σ ⊕ δ')
14:    if T = T' then
15:        return M
16:    else
17:        return ⊥
18:    end if
19: end function
```

```
1: function ℰ_K(N, A, M)
2:     A_1A_2 ⋯ A_a ← A
3:     M_1M_2 ⋯ M_m ← M
4:     Σ ← 0
5:     for i = 1 to a do
6:         Σ ← Σ ⊕ E_K(A_i ⊕ Δ_i)
7:     end for
8:     for i = 1 to m do
9:         C_i ← E_K(M_i ⊕ γ_i)
10:        Σ ← Σ ⊕ E_K(C_i ⊕ δ_i)
11:    end for
12:    C ← C_1C_2 ⋯ C_m
13:    T ← E_K(Σ ⊕ δ')
14:    return (C, T)
15: end function
```

Fig. 6. Pseudo-code of the new mode of operation, the ordinary AEAD part.

```
1: function 𝕌_K^M(N, i, C_i, M'_i, T)
2:     Σ* ← E_K^{-1}(T)
3:     C'_i ← E_K(M'_i ⊕ γ_i)
4:     Σ* ← Σ* ⊕ E_K(C_i ⊕ δ_i) ⊕ E_K(C'_i ⊕ δ_i)
5:     T' ← E_K(Σ*)
6:     return (C'_i, T')
7: end function
8:
9: function 𝔸_K^M(N, m, M_{m+1}, T)
10:    Σ* ← E_K^{-1}(T)
11:    C_{m+1} ← E_K(M_{m+1} ⊕ γ_{m+1})
12:    Σ* ← Σ* ⊕ E_K(C_{m+1} ⊕ δ_{m+1})
13:    T' ← E_K(Σ*)
14:    return (C_{m+1}, T')
15: end function
16:
17: function ℂ_K^M(N, m, C_m, T)
18:    Σ* ← E_K^{-1}(T)
19:    Σ* ← Σ* ⊕ E_K(C_m ⊕ δ_m)
20:    T' ← E_K(Σ*)
21:    return T'
22: end function
```

```
1: function 𝕌_K^A(N, i, A_i, A'_i, T)
2:     Σ* ← E_K^{-1}(T)
3:     Σ* ← Σ* ⊕ E_K(A_i ⊕ Δ_i) ⊕ E_K(A'_i ⊕ Δ_i)
4:     T' ← E_K(Σ*)
5:     return T'
6: end function
7:
8: function 𝔸_K^A(N, a, A_{a+1}, T)
9:     Σ* ← E_K^{-1}(T)
10:    Σ* ← Σ* ⊕ E_K(A_{a+1} ⊕ Δ_{a+1})
11:    T' ← E_K(Σ*)
12:    return T'
13: end function
14:
15: function ℂ_K^A(N, a, A_a, T)
16:    Σ* ← E_K^{-1}(T)
17:    Σ* ← Σ* ⊕ E_K(A_a ⊕ Δ_a)
18:    T' ← E_K(Σ*)
19:    return T'
20: end function
```

Fig. 7. Pseudo-code of the new mode of operation, the incremental operations.

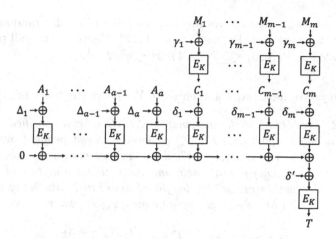

Fig. 8. An illustration of the encryption algorithm of the new mode of operation

cipher calls can be considered as secure tweakable block ciphers, with tweak space (N, i). Moreover, the independence among the masks γ, Δ, δ and δ' are ensured. So now we replace these block cipher calls with independent random functions and prove the security of the mode under such assumptions.

Privacy. We start with privacy. We prove the following theorem.

Theorem 1. *Let (Π, Δ) be the incremental AEAD scheme specified above, with the underlying block cipher calls $E(\cdot \oplus \xi)$ replaced with independent random functions φ_ξ, where $\xi \in \{\gamma_i, \Delta_i, \delta_i, \delta'\}$. Then for any adversary A which makes at most q queries, we have*

$$\mathrm{Adv}^{\mathrm{priv}}_{(\Pi,\Delta)}(A) \leq \frac{q^2}{2^{n+1}}.$$

Proof. Since A is assumed to be nonce-respecting and not making a trivial-win query, all ciphertext blocks C_i, whether output by the encryption oracle $\mathcal{E}_K$ or output by the incremental operations, are truly random strings. So we only need to show that tag T should behave like a random string. There is an event that the value T is not an independent random string. Namely, whenever we have a collision of the input value Σ or Σ^* to the tag-generation function $\varphi_{\delta'}$, the output tag T does not become a freshly random string.

So let bad be the event that the values Σ or Σ^* collide. Since the scheme (Π, Δ) behaves like the ideal object as long as the event bad does not occur, we have $\mathrm{Adv}^{\mathrm{priv}}_{(\Pi,\Delta)}(A) \leq \Pr[A \text{ sets bad}]$.

For $i = 1, 2, \ldots, q$ let bad_i denote the event that the event bad occurs at the i-th query for the first time; i.e. the event bad has not occurred for the queries $j = 1, 2, \ldots, i-1$ so far. Clearly $\Pr[\mathsf{bad}] \leq \sum_{i=0}^{q} \Pr[\mathsf{bad}_i]$.

We bound $\Pr[\mathsf{bad}_i]$. Suppose bad has not occurred so far and now A makes the i-th query. Observe that by construction of the scheme, the computation of

Σ or Σ^* at the i-th query always involves an invocation of a random function $\varphi_\cdot$; therefore, we must have $\Pr[\mathsf{bad}_i] \leq (i-1)/2^n$. Hence the overall probability can be bounded as $\Pr[\mathsf{bad}] \leq \sum_i^q (i-1)/2^n \leq q^2/2^{n+1}$. $\square$

Authenticity. We next prove authenticity. We prove the following theorem.

Theorem 2. *Let (Π, Δ) be the incremental AEAD scheme specified above, with the underlying block cipher calls $E(\cdot \oplus \xi)$ replaced with independent random functions φ_ξ, where $\beta \in \{\gamma_i, \Delta_i, \delta_i, \delta'\}$. Then for any adversary $\mathbf{A}$ which makes at most q queries to its encryption oracle and incremental oracles and at most q' queries to its decryption oracle, the length of associated data being at most ℓ_A blocks and the length of a message being at most ℓ_M blocks, we have*

$$\mathrm{Adv}^{\mathrm{auth}}_{(\Pi, \Delta)}(\mathbf{A}) \leq \frac{q^2}{2^{n+1}} + \frac{q'(\ell_A \ell_M q + 3)}{2^n}.$$

Proof. Let bad be the event as defined in the privacy proof, neglecting queries to the decryption oracle $\mathcal{D}_K$. Namely, queries to the $\mathcal{D}_K$ oracle has no effect on setting the event bad. Let forge denote the event that $\mathbf{A}$ forges. We have $\mathrm{Adv}^{\mathrm{auth}}_{(\Pi, \Delta)}(A) = \Pr[\mathsf{forge}] \leq \Pr[\mathsf{forge} \mid \neg\mathsf{bad}] + [\mathsf{bad}]$. Similarly to the privacy proof, we can show that $\Pr[\mathsf{bad}] \leq q^2/2^{n+1}$.

Now we bound $\Pr[\mathsf{forge} \mid \neg\mathsf{bad}]$. For this, we slightly change the game and consider a game where $\mathbf{A}$ is allowed to make only a single query to its decryption oracle. Let forge' be the event that $\mathbf{A}$ forges in such a game. Then we have $\Pr[\mathsf{forge} \mid \neg\mathsf{bad}] \leq q' \Pr[\mathsf{forge'} \mid \neg\mathsf{bad}]$ where with abuse of notation each bad denotes a bad event in the corresponding game.

So it remains to bound $\Pr[\mathsf{forge'} \mid \neg\mathsf{bad}]$. Let $(N^\circ, A^\circ, C^\circ, T^\circ)$ denote the query to the decryption oracle.

- *Case N° is new.* In this case the computation of a tag from $(N^\circ, A^\circ, C^\circ)$ involves an invocation of a random function $\varphi_\cdot$. Therefore, for any value T°, we must have $\Pr[\mathsf{forge'} \mid \neg\mathsf{bad}] \leq 1/2^n$.
- *Case N° is old but either A° or C° contains a new block relative to N°.* Again, in this case the new block in A° or C° would cause an invocation of the random function φ to compute a tag value from $(N^\circ, A^\circ, C^\circ)$. Therefore, we have $\Pr[\mathsf{forge'} \mid \neg\mathsf{bad}] \leq 1/2^n$.
- *Case N° is old and all block of A° and C° are also old relative to N°.* This is a tricky case. There is no new invocation of the functions φ when computing a tag value from $(N^\circ, A^\circ, C^\circ)$. Let Σ° denote the input value to the tag-generation function φ'_δ, computed from the triplet $(N^\circ, A^\circ, C^\circ)$. We emphasize that in computing Σ° there is no fresh invocation of random functions φ. Let A^* and C^* denote the associated data and the ciphertext corresponding to the N°-query to the encryption oracle $\mathcal{E}_K$. Since A° and C° must be chopped forms of A^* and C^*, respectively, the number of possible values of Σ° are at most $\ell_A \cdot \ell_M$.
  - *Case Σ° is a new input to φ'_δ.* We have $\Pr[\mathsf{forge'} \mid \neg\mathsf{bad}] \leq 1/2^n$.

- *Case Σ° is an old input to φ'_{δ}* Since there are at most q possible values of Σ and Σ^{*}, under the event $\neg\mathsf{bad}$, the event forge' implies the event that the value Σ° gets hit by one of these q values. Therefore, we must have $\Pr[\mathsf{forge}' \mid \neg\mathsf{bad}] \le \ell_A \ell_M q / 2^n$.

Hence the overall probability becomes

$$\Pr[\mathsf{forge}' \mid \neg\mathsf{bad}] \le \frac{1}{2^n} + \frac{1}{2^n} + \frac{1}{2^n} + \frac{\ell_A \ell_M q}{2^n} = \frac{\ell_A \ell_M q + 3}{2^n}.$$

$\square$

7 Discussion

In this section, we would discuss several topics on incremental authenticated encryption with associated data.

- Incremental AEAD loses its efficiency when the number of modified blocks is big. Suppose that there is a non-incremental AEAD scheme that requires only 1 block-cipher call per block (twice faster than our construction). When more than a half of the input blocks are modified, recalculating the result from scratch with such a fast non-incremental is better.
- In many AEAD schemes especially for 1-pass schemes, the ciphertext is computed block by block and sent to the receiver as soon as each block is computed. The receiver also starts to decrypt the received ciphertext block by block. To apply the incremental decryption, the ciphertext must be stored even after the decryption is finished. In practice, such a scenario will occur in storage encryption; storing the encrypted data is the main purpose.
- Under the assumption that the computation results, (C, T) for encryption and M for decryption, are not disclosed, incrementality might be achieved with the AEAD schemes requiring a single use of nonce. Abed et al. [16] summarized the CAESAR candidates in this line.

Acknowledgments. We would like to thank anonymous reviewers and the attendees of Dagstuhl Seminar 14021 on Symmetric Cryptography in 2014 for their helpful comments. We are very grateful to Liting Zhang for pointing out an issue with the previous version of our scheme in the pre-proceedings and for suggesting a possible solution.

A Summary of Existing Schemes

Table 1 lists the unsuitable factors for incremental AEAD adopted in each design. Each column represents "block-wise encryption," "block-wise decryption," "integrity for nonce repeat," "PRP for nonce repeat," "tag-truncation free," "ciphertext translation free," "layered PMAC for processing A," and "parallel generation of mask (or counter)," respectively. "✓" represents that the

Table 1. Survey of CAESAR candidates and GCM for incremental AEAD.

Scheme	BW-Enc	BW-Dec	INT-NR	PRP-NR	TT-free	CT-free	lPMAC	mask/ctr
CPFB	✓	–	✓	–	✓	✓	✓	✓
Enchilada	✓	✓	✓	✓	✓	✓	✓	–
GCM	✓	✓	✓	–	✓	✓	✓	✓
iFeed	✓	–	✓	–	✓	–	✓	✓
Minalpher	✓	✓	✓	✓	–	✓	–	–
OCB-like	✓	✓	–	✓	✓	✓	✓	✓
OTR	✓	✓	–	–	✓	✓	✓	✓
PAEQ	✓	✓	✓	–	–	✓	✓	✓
π-Cipher	✓	✓	✓	–	✓	–	✓	✓
Silver	✓	✓	–	✓	✓	✓	✓	✓
YAES	✓	✓	✓	–	✓	–	✓	✓

scheme is suitable for incremental AEAD while "–" represents that the scheme is unsuitable.

Parallelizability is a crucial requirement for incrementality. Without having parallelizability, the scheme cannot be incremental. However for other requirements, the scheme can be incremental with paying additional cost or with additional restrictions in applications even if they are not satisfied. If the incremental update is applied prior to the release of the encryption/decryption result to a public domain, *security issues in the nonce-repeating model* do not arise. Besides, security issues can be solved by combining other cryptographic techniques as long as they can be processed in block-wise. The problems with *tag truncation*, *ciphertext translation* and *layered PMAC* can be practically addressed by safely storing the internal state value, which is of a relatively small cost, e.g. only with additional one block information Also, the *serial mask generation* can also be avoided if masks are precomputed and stored offline.

References

1. Andreeva, E., Bogdanov, A., Luykx, A., Mennink, B., Mouha, N., Yasuda, K.: How to securely release unverified plaintext in authenticated encryption. IACR Cryptology ePrint Archive 144 (2014)
2. Andreeva, E., Bogdanov, A., Luykx, A., Mennink, B., Tischhauser, E., Yasuda, K.: Parallelizable and authenticated online ciphers. In: Sako, K., Sarkar, P. (eds.) ASIACRYPT 2013, Part I. LNCS, vol. 8269, pp. 424–443. Springer, Heidelberg (2013)
3. Andreeva, E., Bogdanov, A., Luykx, A., Mennink, B., Tischhauser, E., Yasuda, K.: AES-COPA v. 1. Submission to CAESAR (2014)
4. Atighehchi, K.: Space-efficient, byte-wise incremental and perfectly private encryption schemes. Cryptology ePrint Archive, Report 2014/104 (2014). http://eprint.iacr.org/

5. Atighehchi, K., Muntean, T.: Towards fully incremental cryptographic schemes. In: Chen, K., Xie, Q., Qiu, W., Li, N., Tzeng, W. (eds.) ASIA CCS 2013, pp. 505–510. ACM (2013)
6. Bellare, M., Goldreich, O., Goldwasser, S.: Incremental cryptography: the case of hashing and signing. In: Desmedt, Y.G. (ed.) CRYPTO 1994. LNCS, vol. 839, pp. 216–233. Springer, Heidelberg (1994)
7. Bellare, M., Goldreich, O., Goldwasser, S.: Incremental cryptography and application to virus protection. In: Leighton, F.T., Borodin, A. (eds.) STOC 1995, pp. 45–56. ACM (1995)
8. Bellare, M., Micciancio, D.: A new paradigm for collision-free hashing: incrementality at reduced cost. In: Fumy, W. (ed.) EUROCRYPT 1997. LNCS, vol. 1233, pp. 163–192. Springer, Heidelberg (1997)
9. Bellare, M., Namprempre, C.: Authenticated encryption: relations among notions and analysis of the generic composition paradigm. In: Okamoto, T. (ed.) ASIACRYPT 2000. LNCS, vol. 1976, pp. 531–545. Springer, Heidelberg (2000)
10. Bernstein, D.: CAESAR Competition (2013). http://competitions.cr.yp.to/caesar.html
11. Bertoni, G., Daemen, J., Peeters, M., Van Assche, G.: Duplexing the sponge: single-pass authenticated encryption and other applications. In: Miri, A., Vaudenay, S. (eds.) SAC 2011. LNCS, vol. 7118, pp. 320–337. Springer, Heidelberg (2012)
12. Biryukov, A., Khovratovich, D.: PAEQ v1. Submission to CAESAR (2014)
13. Black, J.A., Rogaway, P.: A block-cipher mode of operation for parallelizable message authentication. In: Knudsen, L.R. (ed.) EUROCRYPT 2002. LNCS, vol. 2332, pp. 384–397. Springer, Heidelberg (2002)
14. Bosselaers, A., Vercauteren, F.: YAES v1. Submission to CAESAR (2014)
15. Buonanno, E., Katz, J., Yung, M.: Incremental unforgeable encryption. In: Matsui, M. (ed.) FSE 2001. LNCS, vol. 2355, p. 109. Springer, Heidelberg (2002)
16. Abed, F., Christian Forler, S.L.: General overview of the first-round CAESAR candidates for authenticated ecryption. Cryptology ePrint Archive, Report 2014/792 (2014)
17. Gligoroski, D., Mihajloska, H., Samardjiska, S., Jacobsen, H., El-Hadedy, M., Jensen, R.E.: π-Cipher v1. Submission to CAESAR (2014)
18. Gligoroski, D., Samardjiska, S.: iSHAKE: Incremental hashing with SHAKE128 and SHAKE256 for the zettabyte era. Presented at NIST SHA-3 Workshop 2014 (2014). http://csrc.nist.gov/groups/ST/hash/sha-3/Aug2014/
19. Grosso, V., Leurent, G., Standaert, F.X., Varici, K., Durvaux, F., Gaspar, L., Kerckhof, S.: SCREAM & iSCREAM side-channel resistant authenticated encryption with masking. Submission to CAESAR (2014)
20. Harris, S.: The Enchilada authenticated ciphers, v1. Submission to CAESAR (2014)
21. Hosseini, H., Khazaei, S.: CBA mode (v1). Submission to CAESAR (2014)
22. Jean, J., Nikolić, I., Peyrin, T.: Deoxys v1. Submission to CAESAR (2014)
23. Jean, J., Nikolić, I., Peyrin, T.: Joltik v1. Submission to CAESAR (2014)
24. Jean, J., Nikolić, I., Peyrin, T.: KIASU v1. Submission to CAESAR (2014)
25. Kavun, E.B., Lauridsen, M.M., Leander, G., Rechberger, C., Schwabe, P., Yalçın, T.: Prøst v1. Submission to CAESAR (2014)
26. Krovetz, T., Rogaway, P.: The software performance of authenticated-encryption modes. In: Joux, A. (ed.) FSE 2011. LNCS, vol. 6733, pp. 306–327. Springer, Heidelberg (2011)
27. Krovetz, T., Rogaway, P.: OCB (v1). Submission to CAESAR (2014)

28. McGrew, D.A., Viega, J.: The security and performance of the Galois/Counter Mode (GCM) of operation. In: Canteaut, A., Viswanathan, K. (eds.) INDOCRYPT 2004. LNCS, vol. 3348, pp. 343–355. Springer, Heidelberg (2004)
29. Minematsu, K.: AES-OTR v1. Submission to CAESAR (2014)
30. Montes, M., Penazzi, D.: AES-CPFB v1. Submission to CAESAR (2014)
31. Penazzi, D., Montes, M.: Silver V. 1. Submission to CAESAR (2014)
32. Rogaway, P.: Authenticated-encryption with associated-data. In: Atluri, V. (ed.) ACM CCS 2002, pp. 98–107. ACM (2002)
33. Rogaway, P.: Efficient instantiations of tweakable blockciphers and refinements to modes OCB and PMAC. In: Lee, P.J. (ed.) ASIACRYPT 2004. LNCS, vol. 3329, pp. 16–31. Springer, Heidelberg (2004)
34. Rogaway, P.: Nonce-based symmetric encryption. In: Roy, B., Meier, W. (eds.) FSE 2004. LNCS, vol. 3017, pp. 348–359. Springer, Heidelberg (2004)
35. Rogaway, P., Bellare, M., Black, J., Krovetz, T.: OCB: a block-cipher mode of operation for efficient authenticated encryption. In: Reiter, M.K., Samarati, P. (eds.) ACM CCS 2001, pp. 196–205. ACM (2001)
36. Sasaki, Y., Todo, Y., Aoki, K., Naito, Y., Sugawara, T., Murakami, Y., Matsui, M., Hirose, S.: Minalpher v1. Submission to CAESAR (2014)
37. Zhang, L., Wu, W., Sui, H., Wang, P.: iFeed[aes] v1. Submission to CAESAR (2014)

SCOPE: On the Side Channel Vulnerability of Releasing Unverified Plaintexts

Dhiman Saha[(✉)] and Dipanwita Roy Chowdhury

Crypto Research Lab, Department of Computer Science and Engineering,
IIT Kharagpur, Kharagpur, India
{dhimans,drc}@cse.iitkgp.ernet.in

Abstract. In Asiacrypt 2014, Andreeva et al. proposed an interesting idea of intermittently releasing plaintexts before verifying the tag which was inspired from various practical applications and constraints. In this work we try to asses the idea of releasing unverified plaintexts in the light of side channel attacks like fault attacks. In particular we show that this opens up new avenues of attacking the decryption module. We further show a case-study on the APE authenticated encryption scheme and reduce its key space from 2^{160} to 2^{50} using 12 faults and to 2^{24} using 16 faults on the decryption module. These results are of particular interest since attacking the decryption enables the attacker to completely bypass the nonce constraint imposed by the encryption. Finally, at the outset this work also addresses a related problem of fault attacks with partial state information.

Keywords: Authenticated encryption · Releasing unverified plaintexts · APE · Differential fault analysis

1 Introduction

In conventional security notions of Authenticated Encryption (AE), release of decrypted plaintext is subject to successful verification. In their pioneering paper in Asiacrypt 2014, Andreeva et al. challenged this model by introducing and formalizing the idea of *releasing unverified plaintexts* (RUP) [5,6]. The idea was motivated by a lot of practical problems faced by the classical approach like insufficient memory in constrained environments, real-time usage requirements and inefficiency issues. The basic idea is to separate the plaintext computation and verification during AE decryption, so that the plaintexts are always released irrespective of the status of the verification process. In order to assess the security under RUP and to bridge the gap with the classical approach, the authors have introduced two new definitions: INT-RUP (for integrity) and *plaintext awareness* or PA for privacy (in combination with IND-CPA).

In this work, we try to answer the question pertaining to RUP that arises from a side-channel view-point: *Can the ability to observe unverified plaintexts serve as a source of side-channel information?* Our research reveals that the answer is

© Springer International Publishing Switzerland 2016
O. Dunkelman and L. Keliher (Eds.): SAC 2015, LNCS 9566, pp. 417–438, 2016.
DOI: 10.1007/978-3-319-31301-6_24

affirmative with respect to differential fault analysis (DFA) [8,10–14,16] which is known to be one of the most effective side-channel attacks on symmetric-key constructions. The basic requirement of any form of fault analysis is the ability to induce a fault in the intermediate state of the cipher and consequently observe the faulty output. Our first observation is that in the classical approach where successful verification precedes release of plaintexts, fault attacks are infeasible. This is attributed to the fact that if the attacker induces a fault, the probability of the faulty plaintext to pass the verification is negligible, thereby denying the ability to observe the faulty output. This scenario changes in the presence of unverified plaintexts. So the first *scope* that RUP provides at the hands of the attacker is the ability to observe *faulty unverified plaintexts*. Our second observation is in terms of the *nonce* constraint. In Indocrypt 2014, Saha et al. studied the impact of the nonce constraint in their EscApe fault attack [15] on the authenticated cipher APE [3]. The authors showcased the restriction that the uniqueness of nonces imposes on the *replaying criterion*[1] of fault analysis and demonstrated the idea of *faulty collisions* to overcome it. In this work we argue that ability to attack the decryption, provided by RUP, gives the additional benefit of totally *bypassing* the nonce constraint. This follows from the very definition of AE decryption which allows an attacker to make multiple queries to the decryption oracle with the same nonce. Thus prospect of *nonce bypass* makes fault analysis highly feasible.

Following these observations, we mount Scope: a differential fault attack on the decryption of APE which is also one of the submissions to the on-going CAESAR [1] competition. The choice of APE is motivated by the fact that according to PA classification of schemes provided by Andreeva et al. in [5,6], *APE which has* **offline decryption***, is one of the CAESAR submissions that supports RUP*. Authenticated Permutation-based Encryption [4] or APE was introduced by Andreeva et al. in FSE 2014 and later reintroduced in CAESAR along with GIBBON and HANUMAN and an indigenous permutation called PRIMATE as part of the authenticated encryption family PRIMATEs [2,3]. We studied the fault diffusion in the inverse of the internal permutation PRIMATE of APE using random uni-word fault injections in the penultimate round. We capitalize on properties arising out of the non-square nature of the internal state and also the knowledge of the fault-free unverified plaintext. Our analysis shows average key-space reduction from 2^{160} to 2^{50} using 12 faults and to 2^{24} using 16 faults. Finally, this work identifies and addresses a broader problem in differential fault analysis: *Fault analysis with partial state information*. Since, only part of the state is observable, the fault analysis presented here deviates from the classical DFA [8,10–14,16] which generally assumes availability of the entire state at the output. Here we showcase that even knowledge of just one-fifth (the size of a plaintext block) of the state can be used to reconstruct the differential state and finally reduce the key-space. Moreover, close similarity between PRIMATE permutation and AES [9], automatically amplifies the scope

---

[1] The replaying criterion in differential fault analysis states that the attacker must be able to induce faults while replaying a previous fault free run of the algorithm.

of the results presented here. The contributions of this work can be summarized as below:

- Scrutinizing the recently introduced RUP model in the light of fault attacks.
- Showing that unverified plaintext can be an important source of side-channel information.
- Showing the feasibility of fault induction using nonce bypass.
- For the first time attacking the decryption of an AE scheme using DFA.
- Presenting SCOPE attack exploiting: fault diffusion in the last two rounds of the Inverse PRIMATE permutation and the ability to observe faulty unverified plaintexts.
- Finally, achieving a key space reduction from 2^{160} to 2^{50} with 12 faults and 2^{24} with 16 faults using the random word fault model.
- Moreover, this work also brings into focus the idea of fault analysis of AES based constructions with partial state information.

The rest of the work is organized as follows: Sect. 2 gives a brief description of the PRIMATE permutation and its inverse and introduces the notations used in this work. Section 3 looks at the RUP and classical models in the light of side-channel analysis. Some properties of APE decryption that become relevant in the presence of faults are discussed in Sect. 4. The proposed SCOPE attack is introduced in Sect. 5. Section 6 furnishes the experimental results with a brief discussion while Sect. 7 gives the concluding remarks.

2 Preliminaries

2.1 The Design of PRIMATE

PRIMATE has two variants in terms of size: PRIMATE-80 (200-bit permutation) and PRIMATE-120 (280-bit) which operate on states of (5×8) and (7×8) 5-bit elements respectively. The family consists of four permutations p_1, p_2, p_3, p_4 which differ in the round constants used and the number of rounds. All notations introduced in this section are with reference to PRIMATEs-80 with the APE mode of operation.

Definition 1 (Word). *Let* $\mathbb{T} = \mathbb{F}[x]/(x^5 + x^2 + 1)$ *be the field* $\mathbb{F}_{2^5}$ *used in the PRIMATE MixColumn operation. Then a **word** is defined as an element of* $\mathbb{T}$.

Definition 2 (State). *Let* $\mathbb{S} = (\mathbb{T}^5)^8$ *be the set of* (5×8)-*word matrices. Then the internal **state** of the PRIMATE-80 permutation family is defined as an element of* $\mathbb{S}$. *We denote a state* $s \in \mathbb{S}$ *with elements* $s_{i,j}$ *as* $[s_{i,j}]_{5,8}$.

$$s = [s_{i,j}]_{5,8}, \ where \begin{cases} s_{i,j} \in \mathbb{T} \\ 0 \leq i \leq 4, \ 0 \leq j \leq 7 \end{cases} \tag{1}$$

In the rest of the paper, for simplicity, we omit the dimensions in $[s_{i,j}]_{5,8}$ and use $[s_{i,j}]$ as the default notation for the 5×8 state. We denote a column of $[s_{i,j}]$ as $s_{*,j}$ while a row is referred to as $s_{i,*}$. We now describe in brief the design of PRIMATE permutation. In this work we also deal with the inverse of the PRIMATE permutation. APE instantiates p_1 which is a composition of 12 round functions. The inverse permutation p_1^{-1} applies the round functions in the reverse order with each component operations itself being inverted. For the sake of consistency, in the rest of the work rounds of p^{-1} will be denoted w.r.t to the corresponding rounds of p. For instance, the last round of p^{-1} will be referred to as $\mathcal{R}_1^{-1}$ since functionally it is the inverse of the first round of p.

$$p_1, p_1^{-1} : \mathbb{S} \longrightarrow \mathbb{S}, \qquad p_1 = \mathcal{R}_{12} \circ \mathcal{R}_{11} \circ \cdots \circ \mathcal{R}_1 \quad \bigg| \quad p_1^{-1} = \mathcal{R}_1^{-1} \circ \mathcal{R}_2^{-1} \circ \cdots \circ \mathcal{R}_{12}^{-1}$$
$$\mathcal{R}_r, \mathcal{R}_r^{-1} : \mathbb{S} \longrightarrow \mathbb{S}, \qquad \mathcal{R}_r = \alpha_r \circ \mu_r \circ \rho_r \circ \beta_r \quad \bigg| \quad \mathcal{R}_r^{-1} = \beta_r^{-1} \circ \rho_r^{-1} \circ \mu_r^{-1} \circ \alpha_r^{-1}$$

where $\mathcal{R}_r$ is a composition of four bijective functions on $\mathbb{S}$ while $\mathcal{R}_r^{-1}$ denotes the inverse round function. The index r denotes the rth round and may be dropped if the context is obvious. Here, the component function β represents the non-linear transformation SubBytes which constitutes word-wise substitution of the state according to predefined S-box. The definitions extend analogously for the inverse.

$$\beta_r, \beta_r^{-1} : \mathbb{S} \longrightarrow \mathbb{S}, \qquad s = [s_{i,j}] \overset{\beta}{\mapsto} [S(s_{i,j})], \qquad\qquad s = [s_{i,j}] \overset{\beta^{-1}}{\mapsto} [S^{-1}(s_{i,j})]$$

where $S : \mathbb{T} \longrightarrow \mathbb{T}$ is the S-box given in Table 1. The transformation ρ corresponds to ShiftRows which cyclically shifts each row of the state based on a set of offsets. The same applies to ρ^{-1} with only the direction of shift being reversed.

$$\rho_r, \rho^{-1} : \mathbb{S} \longrightarrow \mathbb{S}, \qquad s = [s_{i,j}] \overset{\rho}{\mapsto} [s_{i,(j-\sigma(i)) \bmod 8}], \qquad s = [s_{i,j}] \overset{\rho^{-1}}{\mapsto} [s_{i,(j+\sigma(i)) \bmod 8}]$$

where, $\sigma = \{0, 1, 2, 4, 7\}$ is the ShiftRow offset vector and $\sigma(i)$ defines shift-offset for the ith row. The MixColumn operation, denoted by μ, operates on the state column-wise. μ is actually a left-multiplication by a 5×5 matrix (M_μ) over the finite field $\mathbb{T}$. For the InverseMixColumn (μ^{-1}), the multiplication is carried out using (M_μ^{-1}).

$$\mu_r : \mathbb{S} \longrightarrow \mathbb{S}, \qquad s = [s_{i,j}] \longmapsto s' = [s'_{i,j}], \qquad s'_{*,j} = M_\mu \times s_{*,j}$$

The last operation of the round function is α which corresponds to the round constant addition. The constants are the output $\{\mathcal{B}_1, \mathcal{B}_2, \cdots, \mathcal{B}_{12}\}$ of a 5-bit LFSR and are xored to the word $s_{1,1}$ of the state $[s_{i,j}]$. α is involutory implying $\alpha = \alpha^{-1}$.

$$\alpha_r : \mathbb{S} \longrightarrow \mathbb{S}, \qquad [s_{i,j}] \longmapsto [s'_{i,j}], \qquad s'_{i,j} = \begin{cases} s_{i,j} \oplus \mathcal{B}_r \text{ if } i, j = 1 \\ s_{i,j}, \text{ Otherwise} \end{cases}$$

The APE mode of operation is depicted in Fig. 1. Here, $N[\cdot]$ represents a Nonce block while $A[\cdot]$ and $M[\cdot]$ denote blocks of associated data and message respectively. The IVs shown in Fig. 1 are predefined and vary according to the nature

Table 1. The PRIMATE 5-bit S-box

x	0	1	2	3	4	5	6	7	8	9	10	11	12	13	14	15
$S(x)$	1	0	25	26	17	29	21	27	20	5	4	23	14	18	2	28
x	16	17	18	19	20	21	22	23	24	25	26	27	28	29	30	31
$S(x)$	15	8	6	3	13	7	24	16	30	9	31	10	22	12	11	19

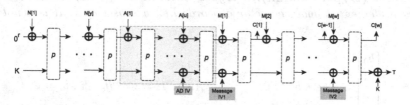

(a) APE Encryption

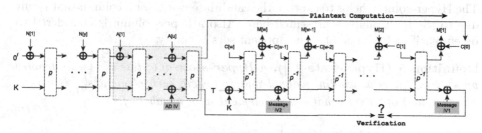

(b) APE Decryption

Fig. 1. The APE mode of operation

of the length of message and associated data. Figure 1a and b show the encryption and decryption modules of APE respectively. In is evident from Fig. 1b that the decryption starts from the last ciphertext block and proceeds in the reverse direction which implies that APE decryption is *offline*.

2.2 Notations

Definition 3 (Differential state). *A **differential state** is defined as the element-wise XOR between a state $[s_{i,j}]$ and the corresponding faulty state $[s'_{i,j}]$.*

$$s'_{i,j} = s_{i,j} \oplus \delta_{i,j}, \ \forall \, i, j \tag{2}$$

δ fully captures the initial fault as well as the dispersion of the fault in the state. In this work we assume induction of random faults in some word of a state. Thus, if the initial fault occurs in word $s_{I,J} \in s$, the differential state is of the following form:

$$\delta_{i,j} = \begin{cases} f : f \xleftarrow{R} \mathbb{T} \setminus \{0\}, & \text{if } (i = I, j = J) \\ 0, & \text{Otherwise} \end{cases} \tag{3}$$

If $\exists j : \delta_{i,j} = 0 \ \forall i$ then $\delta_{*,j}$ is called a *pure* column, otherwise $\delta_{*,j}$ is referred to as a *faulty* column.

Definition 4 (Hyper-column). *A **Hyper-column** is a (5×1) column vector where each element is again a vector of words i.e., a subset of $\mathbb{T}$. It is denoted by $\mathcal{H}$.*

$$\mathcal{H} = \begin{bmatrix} b_0 \\ b_1 \\ \vdots \\ b_4 \end{bmatrix} \quad \text{where } b_j \subset \mathbb{T}, \qquad \text{Also, } \mathcal{H} = \varnothing \quad \text{if} \quad \exists i : b_i = \varnothing$$

The Hyper-column helps to capture the candidate words for a column that result due to the fault analysis presented here. Also a hyper-column is considered to be empty if at least one of its component sets is empty.

Definition 5 (Hyper-state [15]). *A **Hyper-state** of a state $s = [s_{i,j}]$, denoted by $s^h = [s^h_{i,j}]$, is a two-dimensional matrix, where each element $s^h_{i,j}$ is a non-empty subset of $\mathbb{T}$, such that s is an element-wise member of s^h.*

$$s^h = \begin{bmatrix} s^h_{00} & s^h_{01} & \cdots & s^h_{07} \\ s^h_{10} & s^h_{11} & \cdots & s^h_{17} \\ \vdots & \vdots & \ddots & \vdots \\ s^h_{40} & s^h_{41} & \cdots & s^h_{47} \end{bmatrix} \quad \text{where} \quad \begin{cases} s^h_{i,j} \subset \mathbb{T}, \ s^h_{i,j} \neq \varnothing \\ s_{i,j} \in s^h_{i,j} \ \forall i, j \end{cases} \tag{4}$$

The significance of a hyper-state s^h is that the state s is in a way 'hidden' inside it. This means that if we create all possible states taking one word from each element of s^h, then one of them will be exactly equal to s.

The hyper-state has some interesting properties with respect to the component transformations of the PRIMATE permutation and consequently its inverse. For instance all the inverse operations like InverseShiftRow(ρ^{-1}), InverseSubByte(β^{-1}), InverseAddRoundConstant(α^{-1}) can be applied on a hyper-state with little technical changes. This is possible since all these operations work word-wise and thus can be applied as a whole to each element-set of a hyper-state too with an equivalent effect. We define the analogs of these operations on a hyper-state as hyper-state-<operation>: $(\rho^{-1})', (\beta^{-1})', (\alpha_r^{-1})'$. The formal definitions are provided in Appendix A. Another observation of particular interest is that hyper-state-<operation>$(s^h) = ($<operation>$(s))^h$.

Definition 6 (Kernel [15]). *If s^h is a hyper-state of s, then **Kernel** of a column $s^h_{*,j} \in s^h$, denoted by $\mathcal{K}^{s^h_{*,j}}$, is defined as the cross-product of $s^h_{0,j}, s^h_{1,j}, \cdots, s^h_{4,j}$.*

$$\mathcal{K}^{s^h_{*,j}} = \left\{ w_k : w_k^T \in \bigtimes_{i=0}^{4} s^h_{i,j},\ 1 \le k \le \prod_{i=0}^{4} |s^h_{i,j}| \right\}$$

Subsequently, Kernel of the entire hyper-state is the set of the Kernels of all of its columns: $\mathcal{K}^{s^h} = \{\mathcal{K}^{s^h_{*,0}}, \mathcal{K}^{s^h_{*,1}}, \cdots, \mathcal{K}^{s^h_{*,7}}\}$

Here, w_k^T represents the transpose of w_k, thereby implying that w_k is a column vector. One should note that $s_{*,j} \in \mathcal{K}^{s^h_{*,j}} \forall j$. Thus each column of s is contained in each element of $\mathcal{K}^{s^h}$. We now define an operation $(\mu^{-1})'$ over the Kernel of a hyper-state which is equivalent to μ^{-1} that operates on a state.

Definition 7 (Kernel-InverseMixColumn).
*The **Kernel-InverseMixColumn** transformation denoted by $(\mu^{-1})'$ is the left multiplication of M_μ^{-1} to each element of each $\mathcal{K}^{s^h_{*,j}} \in \mathcal{K}^{s^h}$.*

$$(\mu^{-1})'(\mathcal{K}^{s^h_{*,j}}) = \{M_\mu^{-1} \times w_i,\ \forall w_i \in \mathcal{K}^{s^h_{*,j}}\}$$
$$(\mu^{-1})'(\mathcal{K}^{s^h}) = \{(\mu^{-1})''(\mathcal{K}^{s^h_{*,0}}), (\mu^{-1})'(\mathcal{K}^{s^h_{*,1}}), \cdots, (\mu^{-1})'(\mathcal{K}^{s^h_{*,7}})\}$$

An important implication is that $(\mu^{-1})'(\mathcal{K}^{s^h}) = \mathcal{K}^{(\mu^{-1}(s))^h}$. The notion of **Hyper-state** and **Kernel** will be used in the OUTBOUND phase of SCOPE detailed in Subsect. 5.3.

3   RUP in the Light of Side-Channels

RUP which has been argued to be a very desirable property can be a major source for side channel information. In this work we try to study how RUP stands out in the light of fault attacks. Our research reveals that RUP opens up an exploitable opportunity with respect to fault analysis which would not have been possible if verification would precede release of the plaintexts. Moreover, attacking the decryption also allows the attacker to bypass the nonce constraint imposed by the encryption. It has been shown that nonce based encryption has an automatic protection against DFA and hence ability to bypass the nonce constraint exposes the AE scheme to fault attacks. In the rest of the paper we refer to the classical model that does not allow RUP as **RVP (Release of Verified Plaintexts)**. We now argue why RVP has an implicit protection against fault attacks which makes attacking the decryption infeasible.

In order to understand the significance of the *scope* that the RUP model puts at the hands of the attacker, one has to be aware of why fault analysis in the RVP model is infeasible. According to the classical RVP model, the decryption oracle returns the entire plaintext if the verification passes and $\perp$ otherwise.

Now we define the term *faulty forgery* as the ability of the attacker to produce a plaintext $(p* \neq p)$ after inducing faults such that the verification passes. Now, in standard fault analysis it is assumed that *the attacker can induce random faults in the state but the value of the fault is unknown.* Under this scenario, the probability of the attacker to produce a faulty forgery is negligible (Fig. 2).

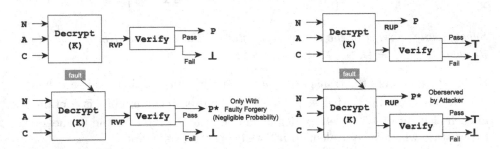

Fig. 2. RVP vs RUP from the perspective of fault analysis

On the contrary RUP gives the attacker the scope of inducing random faults while decrypting any chosen or known ciphertext and **unconditionally** observe the corresponding faulty plaintexts (which would never have passed verification in the RVP model). This power opens up the side-channel for fault analysis and is the basis of the differential fault attack presented in this work. Moreover, the ability to attack the decryption has the additional and important advantage of bypassing the nonce constraint that is imposed while making encryption queries. This magnifies the feasibility of mounting fault attacks.

In the next section, we look at some of the features of APE decryption and the inverse PRIMATE permutation p^{-1} that gain importance from a fault attack perspective. Finally, building upon these observations we introduce the SCOPE attack where for the first time we show how the decryption can also be attacked under RUP to retrieve the entire internal state of p^{-1} leading to recovery of the key with practical complexities.

4   Analyzing APE Decryption in the Presence of Faults

In this section we look at certain properties of APE decryption that become relevant in the context of RUP and from the prospect of fault induction. We first look at a property which by itself is of no threat to the security of APE but becomes exploitable in the presence of faults in the RUP scenario.

4.1   The Block Inversion Property

The Block Inversion Property is purely attributed to the APE mode of operation. This property allows the attacker to retrieve partial information about the contents of the state matrix after the last round InverseMixColumn operation.

Property 1. *If the state after μ_1^{-1} in $\mathcal{R}_1^{-1}$ (the last round of p^{-1}) be represented as $t = [t_{i,j}]$ and the released plaintext block and next ciphertext block be p and c respectively, then $t_{0,*}$ is public by the following expression:*

$$t_{0,i} = S(v_i) \quad where, \begin{cases} v_i \in (p \oplus c), \\ S \rightarrow \texttt{PRIMATE Sbox} \quad (\text{Table 1}) \end{cases}$$

Analysis: By virtue of the APE mode of operation and the SPONGE [7] construction it follows, the *rate* part (top row of the state) after $\mathcal{R}_1^{-1}$ of p^{-1} is released after XORing with the next ciphertext block as the plaintext block (which can be observed unconditionally under RUP). If the state after $\mathcal{R}_1^{-1}$ be $s = [s_{i,j}]$ then $p \oplus c$ gives back $s_{0,*}$. We can now invert this block to get inside $\mathcal{R}_1^{-1}$ despite partial knowledge of the state. This becomes possible since β operates word-wise and ρ operates row-wise. Moreover, ρ can be ignored for it has no effect on top row as the shift-offset is zero. Thus applying β on $s_{0,*}$ we get the value of $t_{0,*}$. However, the inversion stops here since μ operates column-wise and only word of each column is known. ∎

Later in this work we show how the SCOPE attack can exploit the Block Inversion Property along with RUP and use both faulty and fault-free plaintexts to reconstruct differential state after μ_2^{-1} in $\mathcal{R}_2^{-1}$. We now study the fault induction and diffusion in the state of p^{-1} which is vital to understanding of the attack presented here.

4.2 Fault Diffusion in the Inverse PRIMATE Permutation

In this section we describe the induction and diffusion of faults in the inverse (p^{-1}) of the PRIMATE permutation during APE decryption. In fact, our intention is to study the fault diffusion in the *differential state* of p^{-1} which we exploit to formulate a fault attack on APE Decryption. The fault induction and subsequent differential plaintext block formation are illustrated in Fig. 3. One can see from Fig. 3 that the fault is induced in the input of the penultimate round $\mathcal{R}_2^{-1}$ of p^{-1}. The logic behind this will be clear from the following important property of fault diffusion in the internal state of p^{-1}.

Property 2. *If a single column is faulty at the start of $\mathcal{R}_{r+1}^{-1}$ then there are exactly three fault-free words in each row of the differential state after $\mathcal{R}_r^{-1}$.*

Analysis: This property surfaces because in two rounds the fault does not spread to the entire state matrix. This is primarily attributed to the fact that the state matrix is non-square. To visualize this we need to first look at fault diffusion in the $\mathcal{R}_{r+1}^{-1}$ round. Let us denote the differential state at the input of $\mathcal{R}_{r+1}^{-1}$ as $s = [s_{i,j}]$. This analysis takes into account the structural dispersion of the fault and is independent of the actual value of s. At the beginning of $\mathcal{R}_{r+1}^{-1}$ only one column $s_{*,j}$ is faulty. The operation α^{-1} is omitted from analysis since round-constant addition has no effect on the differential state.

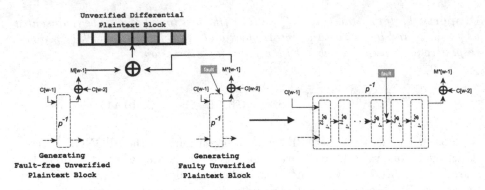

Fig. 3. Fault induction in APE decryption before releasing unverified plaintext and the unverified differential plaintext block

– Fault diffusion in $\mathcal{R}_{r+1}^{-1}$
  - μ_{r+1}^{-1} : Intra-column diffusion. Fault spreads to entire column $s_{*,j}$.
  - ρ_{r+1}^{-1} : No diffusion, fault shifts to the words $\{s_{i,(j+\sigma(i)) \bmod 8} : 0 \leq i < |\sigma|\}$.
  - β_{r+1}^{-1} : No diffusion, fault limited to the same words as after ρ_{r+1}^{-1}.

$$s_{*,j} \xrightarrow{\mu_{r+1}^{-1}} s_{*,j} \xrightarrow{\beta_{r+1}^{-1} \circ \rho_{r+1}^{-1}} \{s_{i,(j+\sigma(i)) \bmod 8}\} \qquad (5)$$

– Fault diffusion in $\mathcal{R}_r$
  - μ_r^{-1} : Fault spreads to each column $s_{*,(j+\sigma(i)) \bmod 8}$.
  - ρ_r^{-1} : No diffusion, fault shifts to the words $\{s_{i,(j+\sigma(i)+\sigma(k)) \bmod 8} : 0 \leq i,k < |\sigma|\}$.
  - β_r^{-1} : No diffusion, fault limited to the same words as after ρ_r^{-1}.

$$\{s_{i,(j+\sigma(i)) \bmod 8}\} \xrightarrow{\mu_r^{-1}} s_{*,(j+\sigma(i)) \bmod 8} \xrightarrow{\beta_r^{-1} \circ \rho_r^{-1}} \{s_{i,(j+\sigma(i)+\sigma(k)) \bmod 8}\}$$
$$(6)$$

From (5) and (6) we have the following relation between the faulty column $s_{*,j}$ at the start $\mathcal{R}_{r+1}^{-1}$ and the faulty words after $\mathcal{R}_r^{-1}$.

$$s_{*,j} \xrightarrow{\mathcal{R}_r^{-1} \circ \mathcal{R}_{r-1}^{-1}} \{s_{i,(j+\sigma(i)+\sigma(k)) \bmod 8} : 0 \leq i,k < |\sigma|\} \qquad (7)$$

For PRIMATE-80, $\sigma = \{0,1,2,4,7\}$, implying that $|\sigma| = 5$. From (7), we have $|\{s_{i,(j+\sigma(i)+\sigma(k)) \bmod 8}\}| = 25$. Thus a single faulty column before $\mathcal{R}_{r+1}^{-1}$ results in 25 faulty words at the end of $\mathcal{R}_r^{-1}$. Moreover, for each value of i we have $|\{s_{i,(j+\sigma(i)+\sigma(k)) \bmod 8} : 0 \leq k < 5\}| = 5$ implying that each row has 5 faulty words and respectively $8 - 5 = 3$ fault-free words at the end of $\mathcal{R}_r^{-1}$. An illustration of the above analysis with the source fault in column $s_{*,3}$ is depicted in Fig. 4. ∎

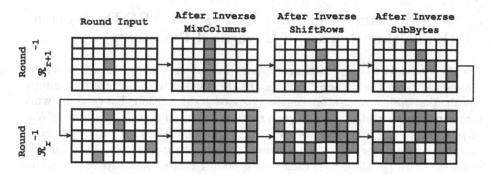

Fig. 4. 2-round fault diffusion with a uni-word fault in column $s_{*,3}$

4.3   The Bijection Lemma

This lemma stems out of the property mentioned above and is pivotal in increasing the efficiency of the SCOPE attack. Again it is a direct consequence of the non-square nature of the internal state of p^{-1}.

Lemma 1. *If fault is induced in the jth column of the state at the input of $\mathcal{R}_{r+1}^{-1}$, then the fault-free words in the differential plaintext block released after $\mathcal{R}_r^{-1}$ are $((j+3),(j+5),(j+6))$ mod 8.*

Proof. This directly follows from relation (7). One can recall that for APE decryption under RUP, the first row of the state is released after XORing with next ciphertext block. However, since we are considering a differential here, the effect of the ciphertext block is nullified. Now, for $i = 0$, from relation (7) we have $\{s_{0,(j+\sigma(0)+\sigma(k)) \bmod 8} : 0 \leq k < 5\} = \{s_{0,j}, s_{0,j+1}, s_{0,j+2}, s_{0,j+4}, s_{0,j+7}\}$ which signifies the set of faulty words in the differential plaintext block. Hence, the complement of this set w.r.t the set of all the words in the plaintext block is $\{s_{0,j+3}, s_{0,j+5}, s_{0,j+6}\}$, which signify the fault-free words.   ∎

The implication of this lemma is that there exists a bijection between the positions of the fault-free words in the differential plaintext block released after $\mathcal{R}_r^{-1}$ and position of the column in which the fault was induced before $\mathcal{R}_{r+1}^{-1}$. This is vital to the analysis presented in this work and shows that by looking at the unverified differential plaintext block the attacker can ascertain the column position of the fault. This makes the attack 8 times faster. However, this information is not sufficient to guess the row position since all faults in the same column will produce the same pattern for the fault-free words.

In case of $p^{-1}, r = 1$ and the Bijection Lemma implies that by looking at the unverified differential block (Fig. 3) released after $\mathcal{R}_1^{-1}$, the attacker can ascertain in which column the fault was induced before $\mathcal{R}_2^{-1}$. With knowledge of all these characteristics of the APE mode of operation as well as p^{-1}, we are now in a place to finally introduce the differential fault attack developed in this work: SCOPE.

5 SCOPE: Differential Fault Analysis of APE Decryption (Exploiting Release of Unverified Plaintexts)

The first task is to run APE decryption and observe the released unverified plaintexts. Next the attacker queries the decryption with same set of inputs. Recall, that *nonce constraint can be bypassed* by definition. Every time, while replaying the decryption, he induces a random uni-word fault at the input of $\mathcal{R}_2^{-1}$ of p^{-1} during the processing of the same ciphertext block. *By RUP principle, the attacker can observe the corresponding faulty plaintext blocks.* The fault-free plaintext block (p) along with each corresponding faulty plaintexts block (p_i') are stored. Now using the Bijection Lemma every differential plaintext block $(p \oplus p_i')$ is analyzed to get the faulty column before $\mathcal{R}_2^{-1}$. The information is stored in the fault count vector $(\mathcal{F})$ which is an array keeping count of the number of faults traced back to each column before $\mathcal{R}_2^{-1}$. For each unverified faulty plaintext, the INBOUND phase is initiated to get back a set of hyper-columns. The process is detailed in the next subsection.

5.1 The INBOUND Phase

The main aim of this phase is to reduce the number of candidate words for the column to which the fault was traced back. Let the state after μ_1^{-1} for the fault-free case be $s = [s_{i,j}]$ and for the faulty case be $s' = [s_{i,j}']$. Now, by virtue of the Block Inversion Property, $s_{0,*}$ and $s_{0,*}'$ are known to the attacker. He now exploits the relation between the differential state before and after μ_1^{-1} that arises from the fault diffusion to reconstruct the entire differential state after $\mathcal{R}_1^{-1}$. To be more precise, the attacker is interested in the nature of the differential block $(s_{0,*} \oplus s_{0,*}')$. Due to the InverseMixColumn operation every non-zero word $(s_{0,*} \oplus s_{0,*}')$ is a multiple of the non-zero word in the corresponding column before μ_1^{-1} and the relation is governed by the InverseMixColumns matrix. Thus if the source fault is in column 4, $(s_{0,*} \oplus s_{0,*}')$ is of the following form: $\{0, 0, x_1 \times F_5, x_2 \times F_1, x_3 \times F_2, x_4 \times F_3, 0, x_5 \times F_4\}$. Now to get back each F_i from the differential row, the attacker makes use of the Factor Matrix given in Table 2. As one can notice the Factor Matrix is a circulant matrix. The ith row corresponds to the factors to be used if the source fault is in the ith column. The '*' represents the positions of the zero values of the corresponding differential row. So, the attacker retrieves each F_i by word-wise Galois Field division of the differential row $(s_{0,*} \oplus s_{0,*}')$ by using the appropriate row from the Factor Matrix. The method of generating the Factor Matrix is detailed in Appendix D.

The attacker now has the entire differential state after $\mathcal{R}_2^{-1}$. He cannot invert further deterministically since β is nonlinear. However, as ρ and β are commutative, he can apply ρ before β. By virtue of the fault diffusion described in Property (2), the differential state after β_2^{-1} has only one non-zero column and it is the same column where the fault was induced. The attacker now solves differential equations involving the same column at the input of β_2^{-1} which arise due the InverseMixColumns of $\mathcal{R}_2^{-1}$. However, these equations are characterized by

Table 2. The Factor Matrix

6	22	31	*	1	*	*	15
15	6	22	31	*	1	*	*
*	15	6	22	31	*	1	*
*	*	15	6	22	31	*	1
1	*	*	15	6	22	31	*
*	1	*	*	15	6	22	31
31	*	1	*	*	15	6	22
22	31	*	1	*	*	15	6

the row in which the initial fault was induced. One can recall that from Lemma 1, that the information available is not sufficient to ascertain the exact row. For instance, the fault invariants[2] for different rows of column 4 is shown in Fig. 5. So, the attacker solves the five sets of equations assuming all the possibilities. Out of these one set corresponds to the actual row that was affected. Solving the equations results in significant reduction in column space. The candidate words that satisfy the equations are stored into hyper-columns (Definition 4). So each row guess results in a different hyper-column and hence there can be maximum of 5 hyper-columns. However, one may encounter a lot of wrong candidate words getting accepted as they satisfy the wrong set of equations arising from the incorrect row guess. We refer to all accepted words other than the legitimate ones as NOISE. Thus one run of the INBOUND Phase returns a set of hyper-columns with a maximum cardinality of 5. The phase, is repeated for each faulty unverified plaintext and corresponding set of hyper-columns appropriately stored in a set of sets of hyper-columns: $\mathbb{H}$. After all faulty plaintexts have been processed, the set $\mathbb{H}$ along with the fault count vector $\mathcal{F}$ are passed on to the NOISE handling phase.

5.2   NOISE Handling

Here the attacker takes the advantage of the fact that while he induces random uni-word faults in input of $\mathcal{R}_2^{-1}$, there is a high probability that some faults get induced in the same column. Thus he will have multiple sets of hyper-columns from the INBOUND phase that reduced the column space for the same column before $\mathcal{R}_2^{-1}$. On the contrary, it might so happen that only one fault gets induced for a particular column. The worst-case scenario occurs if none of the induced faults affects some specific column. The former cases are dealt with in the next subsections while for the later case the attacker is left with exhaustive search implying that NOISE handling phase will return a hyper-column that spans the entire column space.

---

[2] A discussion on the generation and nature of the fault invariants is furnished in Appendix C.

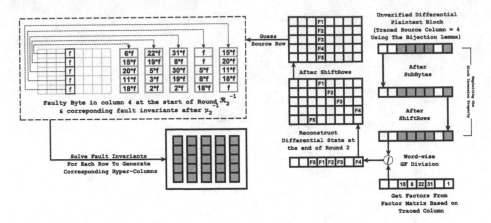

Fig. 5. Generation of hyper-columns using a word-fault at the beginning of $\mathcal{R}_2^{-1}$

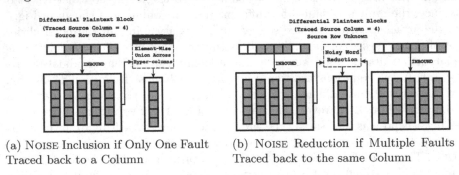

(a) NOISE Inclusion if Only One Fault Traced back to a Column

(b) NOISE Reduction if Multiple Faults Traced back to the same Column

Fig. 6. The NOISE handling phase

NOISE **Inclusion.** When the attacker traces only one fault back to a column, he faces an ambiguity regarding the source row. In this scenario, he has no other option but to include all the hyper-columns for the next phase of the attack. So he includes all the NOISE in the final step. So NOISE Inclusion corresponds to word-wise union of all hyper-columns as depicted in Fig. 6a. NOISE Inclusion, definitely, increases the column-space, however, computer simulations show that the final cardinality is still much better that brute force.

NOISE **Reduction.** When the attacker traces multiple faults to the same column, he can significantly reduce the column space by eliminating Noisy hyper-columns. For e.g. if two faults are traced back to column x, then the attacker has two sets of hyper-columns. He now takes the cross-product of these two sets. Every element of the cross-product is a pair of hyper-columns. He now takes the set intersection between each such pair. The result is again a hyper-column with the cardinality of its component sets highly reduced. However, if the hyper-column turns out to be empty[3], it is discarded. Experiments show that most of the elements from the cross-product get eliminated due to this and the attacker is left with a single final hyper-column. In case multiple hyper-columns remain, a element-wise union is taken to form the final hyper-column.

This NOISE handling phase is repeated for all the columns and returns a set of eight hyper-columns for the last phase of the attack.

1: **procedure** HANDLENOISE($\mathcal{F}, \mathbb{H}$) ▷ $\left| \begin{array}{l} \mathcal{F} \to \text{Fault count vector} \\ \mathbb{H} \to \text{Set of all sets of hyper-columns} \end{array} \right.$

2:    $\mathcal{H}_U = \{b_0, b_1, b_2, b_3, b_4\}^T$, where $b_i = \{0, 1, \cdots, 31\}$
      ▷ $\mathcal{H}_U \to$ Exhaustive Hyper-column

3:    **for** $i = 0 : 7$ **do**

4:        **if** $\mathcal{F}(i) = 0$ **then**   ▷ If no fault traced to column i

5:            $\mathcal{H}_i = \mathcal{H}_U$ ▷ Set hyper-column to be exhaustive

6:        **else if** $\mathcal{F}(i) = 1$ **then**   ▷ $\left\{ \begin{array}{l} \text{If only one fault traced back} \\ \text{to column } i: \text{NOISE Inclusion} \end{array} \right.$

7:            $\mathcal{H}_i = \bigcup \mathbb{H}_{i,1}$ ▷ Take union over the hyper-column set

8:        **else if** $\mathcal{F}(i) > 1$ **then**   ▷ $\left\{ \begin{array}{l} \text{If multiple faults traced back} \\ \text{to column } i: \text{NOISE Reduction} \end{array} \right.$

9:            $\mathcal{C} = \mathbb{H}_{i,1} \times \mathbb{H}_{i,2} \times \cdots \times \mathbb{H}_{i,\mathcal{F}(i)}$ ▷ $\left\{ \begin{array}{l} \text{Take cross-product over} \\ \text{all sets of hyper-columns} \end{array} \right.$

10:           $j = 0$

11:           **for** $\forall d \in \mathcal{C}$ **do**   ▷ $\left\{ \begin{array}{l} \text{Each element } d \in \mathcal{C} \text{ is a set of} \\ \text{hyper-columns and } |d| = \mathcal{F}(i) \end{array} \right.$

12:               $\mathcal{H}^j_{temp} = \bigcap d$ ▷ $\left\{ \begin{array}{l} \text{Take intersection over} \\ \text{each set of hyper-columns} \end{array} \right.$

13:               **if** $\mathcal{H}^j_{temp} \neq \varnothing$ **then** ▷ $\left\{ \begin{array}{l} \text{Recall (Definition 4)} \\ \mathcal{H} = \varnothing \text{ if } \exists i : b_i = \varnothing, b_i \in \mathcal{H} \end{array} \right.$

14:                   $j = j + 1$

15:               **end if**

16:           **end for**

17:           $\mathcal{H}_i = \bigcup \mathcal{H}^j_{temp}$ ▷ $\left\{ \begin{array}{l} \text{Take union over the set of} \\ \text{all non-empty hyper-columns} \end{array} \right.$

18:        **end if**

19:    **end for**

20:    **return** $\{\mathcal{H}_0, \mathcal{H}_1, \cdots, \mathcal{H}_7\}$
           ▷ Each $\mathcal{H}_i$ has candidate words for the ith column in the state after μ_2^{-1}.

21: **end procedure**

---

[3] Recall, that by Definition 4, a hyper-column is empty if any of its components is empty.

5.3   The OUTBOUND Phase

The OUTBOUND phase of SCOPE is inspired from the OUTBOUND phase of the EsCAPE [15] attack proposed by Saha et al. in Indocrypt 2014 and closely follows it. It borrows the idea of a *Hyper-state* and *Kernel* from there. The input to this phase is the set of eight hyper-columns. Since none of the hyper-columns are empty, they can easily be combined structurally to form the hyper-state of the state after μ_2^{-1}. Let us denote the state by $s = [s_{i,i}]$ and then the hyper-state is s^h. This hyper-state s^h captures the reduced state-space for the state s that has been generated using the last two phases. In this phase we want to further reduce the state-space using knowledge of the fault-free plaintext block by again employing the Block Inversion property. This phase is called OUTBOUND since it tries to move outward from μ_2^{-1}. We start by propagating further into $\mathcal{R}_2^{-1}$ and then move into $\mathcal{R}_1^{-1}$ by applying some hyper-state-<operations> on s^h. The steps of the OUTBOUND phase are enlisted below.

1. The attacker starts the OUTBOUND phase by applying Hyper-state Inverse-ShiftRow transformation (Definition 8) on s^h followed by Hyper-state Inverse-SubByte (Definition 9) on s^h. This completes $\mathcal{R}_2^{-1}$ propagation.

$$s^h \xrightarrow{(\rho^{-1})'} (\rho_2^{-1}(s))^h \xrightarrow{(\beta^{-1})'} (\beta_2^{-1}(\rho_2^{-1}(s)))^h \to v^h (say)$$

2. We now move forward into the last round of $p^{-1} : \mathcal{R}_1^{-1}$. Let us denote the state $\beta_2^{-1}(\rho_2^{-1}(s))$ as v. We now apply Hyper-state InverseAddRoundConstant (Definition 10): $(\alpha_1^{-1})'$ on the hyper-state v^h. The next step is to compute the Kernel for $(\alpha_1^{-1}(v))^h : \mathcal{K}^{(\alpha_1^{-1}(v))^h}$.

$$v^h \xrightarrow{(\alpha_1^{-1})'} (\alpha_1^{-1}(v))^h \xrightarrow{\text{Compute Kernel}} \mathcal{K}^{(\alpha_1^{-1}(v))^h}$$

3. Then the attacker applies the Kernel-InverseMixColumn transformation on the Kernel $\mathcal{K}^{(\alpha_1^{-1}(v))^h}$

$$\mathcal{K}^{(\alpha_1^{-1}(v))^h} \xrightarrow{(\mu^{-1})'} \mathcal{K}^{(\mu_1^{-1}(\alpha_1^{-1}(v)))^h}$$

4. Next comes the reduction step. It can be noted that $\mathcal{K}^{(\mu_1^{-1}(\alpha_1^{-1}(v)))^h}$ represents the kernel for the hyper-state of $(\mu_1^{-1}(\alpha_1^{-1}(v)))$. i.e., the state just before the application of ρ_1^{-1}. Now let $t = (\mu_1^{-1}(\alpha_1^{-1}(v)))$. Then by the Block Inversion property, the actual value of $t_{0,*}$ is known. This knowledge is used to reduce the size of each $\mathcal{K}^{t_{*,j}^h} \in \mathcal{K}^{t^h}$. This reduction algorithm is almost similar to REDUCEKERNEL given in [15] and is restated in Appendix B for easy reference.

A pictorial description of the OUTBOUND phase is furnished in Fig. 7. Thus, after the OUTBOUND phase we get a reduced Kernel for the state at the end of μ_1^{-1}. Every element of the cross-product of Kernels of each column is a candidate state. Finally, applying ρ_1^{-1} and β_1^{-1} on each candidate state produces the reduced state-space at the end of $\mathcal{R}_1^{-1}$ of p^{-1}. This reduced state-space directly corresponds to the key-space of the state since recovering the internal state implies recovery of the key. The overall SCOPE attack is summarized by the following algorithm:

```
1: procedure SCOPE(p, c, {p'_1, p'_2, ··· , p'_n})
        ⎧ p → Fault-free unverified plaintext ⎫
        ⎪   c → Next ciphertext block        ⎪
        ⎨  p'_i → Faulty unverified plaintext ⎬
        ⎩    n → # of faulty outputs         ⎭
2:      for i = 0 : 7 do
3:          F(i) = 0                                    ▷ Initialize fault count vector
4:      end for
5:      for i = 1 : n do
6:          col ⟵^{Lemma (1)} (p ⊕ p'_i)          ▷ ⎰ Get faulty column using
                                                       ⎱ the Bijection Lemma
7:          F(col) = F(col) + 1                         ▷ Update fault count vector
8:          H_{col,F(col)} ← INBOUND(p, c, p'_i)   ▷ ⎰ Get set of hyper-columns
                                                       ⎱ for column col (Fig 5)
9:      end for
10:     {H_0, H_1, ··· , H_7} ← HANDLENOISE(F, H)
                                                  ▷ Final set of 8 hyper-columns
11:     s^h ← {H_0, H_1, ··· , H_7}               ▷ Construct Hyper-State
12:     K^{t^h}_{red} ←OUTBOUND(s^h, p ⊕ c)   ▷ ⎰ Get Reduced Kernel using the
                                                  ⎱ Block Inversion Property (Fig. 7)
13:     S = ∅                                     ▷ Initialze final state-space set
14:     for all w ∈ (×^{7}_{j=0} K^{t^h_{*,j}}_{red}) do   ▷ Unroll Kernel to generate state-space
15:         s = β_1^{-1}(ρ_1^{-1}(w))
16:         S = S ∪ {s}
17:     end for
18:     return S                                 ▷ Return final state-space after R_1^{-1}
19: end procedure
```

6 Experimental Results and Discussion

SCOPE was verified by extensive computer simulations. The experimental results confirm large scale reduction in the state-space and consequently the key-space. Average case analysis reveals that with 12 random uni-word faults at the input of $\mathcal{R}_2^{-1}$, the state-space at the end of $\mathcal{R}_1^{-1}$ reduces from 2^{160} to 2^{50} while 16 faults give a reduced state-space of 2^{24}. It is interesting to note that the fault distribution had a direct impact on state(key)-space reduction. To highlight the impact we look at two different fault distributions with 12 faults. Let the fault count vectors be $\mathcal{F}_1 = \{1, 2, 3, 0, 2, 2, 1, 1\}$ and $\mathcal{F}_2 = \{2, 2, 2, 0, 2, 2, 1, 1\}$.

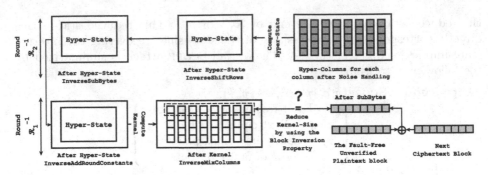

Fig. 7. Final reduction in state-space using fault-free unverified plaintext block

The average reduction with these distributions are 2^{45} and 2^{28} respectively. This extreme variance in the reduced key-spaces is attributed firstly to the fact that $\mathcal{F}_2$ is a more uniform distribution. Secondly, $\mathcal{F}_1$ has three columns which get just one fault. Thus, NOISE reduction cannot be applied to them. While for $\mathcal{F}_2$ such cases are two which leads to a better NOISE reduction in the NOISE handling phase and hence the better reduction in overall key-space. To conclude, it can be said that best results are obtained when fault distribution is such that maximum number of columns receive at least two faults.

It might be argued that in comparison to EscApe attack by Saha et al. SCOPE requires more faults. However, it must be kept in mind that SCOPE works with only partial state information while EscApe has the full state at its disposal. Moreover, since SCOPE attacks APE decryption it can bypass the nonce constraint and hence also avoid the need of *faulty collisions* which are inevitable for EscApe. Overall, SCOPE shows an interesting case-study where an AES-like construction is analyzed using faults with *partial* state information available to the attacker.

7   Conclusion

In this work we explore the *scope* provided by the RUP model with regards to fault analysis. We argue that ability to observe unverified plaintext opens up the fault side channel to attackers which is otherwise unavailable or available with negligible probability. In this work for the first time we show how the decryption of APE, an AE scheme that supports RUP, becomes vulnerable to DFA. Experiments reveal that using the random word fault model the key-space can be reduced from 2^{160} to 2^{50} using 12 faults while 16 faults reduce it to 2^{24}. An important implication of the ability to attack the decryption using RUP is that the attacker can totally bypass the nonce constraint imposed by the encryption. Finally, this work shows that though RUP is a desirable property addressing a lot of practical problems, it provides a unique scope to the attacker for mounting the SCOPE fault attack.

A   Some More Definitions

Definition 8 (Hyper-state InverseShiftRow). *This transformation, denoted by $(\rho^{-1})'$, corresponds to cyclically **right** shifting each row of s^h based on the predefined set of offsets σ.*

$$(\rho^{-1})' : \bigtimes_{j=0}^{7} s_{i,j}^h \longrightarrow \bigtimes_{j=0}^{7} s_{i,(j+\sigma(i)) \bmod 8}^h \quad \forall i$$

$$s_{i,*}^h \longmapsto s_{i,(*+\sigma(i)) \bmod 8}^h \quad \forall i$$

It is interesting to note that, every word in the state $\rho^{-1}(s)$ will be a member of the corresponding element of $(\rho^{-1})'(s^h)$, thereby implying that $(\rho^{-1})'(s^h) = (\rho^{-1}(s))^h$.

Definition 9 (Hyper-state InverseSubByte). *This transformation, denoted by $(\beta^{-1})'$, corresponds to word-wise substitution of each word of each element-set of s^h based on the inverse of the PRIMATE Sbox.*

$$(\beta^{-1})' : w \longmapsto S^{-1}(w), \quad \forall w \in s_{i,j}^h, \quad \forall i,j$$

Definition 10 (Hyper-state InverseAddRoundConstant). *This transformation, denoted by $(\alpha_r^{-1})'$, corresponds to appropriate round constant addition to all words of element-set $s_{1,1}^h$ of the hyper-state s^h.*

$$(\alpha_r^{-1})' : w \longmapsto w \oplus \mathcal{B}_r, \quad \forall w \in s_{1,1}^h$$

B   REDUCEKERNEL Algorithm [15]

```
1: procedure REDUCEKERNEL(𝒦^{t^h}, t_{0,*})
2:     for j = 0 : 7 do
3:         for all {e_0, e_1, e_2, e_3, e_4}^T ∈ 𝒦^{t^h}_{*,j} do
4:             if e_0 ≠ t_{0,j} then
5:                 𝒦^{t^h}_{*,j} = 𝒦^{t^h}_{*,j} − {e_0, e_1, e_2, e_3, e_4}^T
6:             end if
7:         end for
8:     end for
9:     𝒦^{t^h}_{red} = 𝒦^{t^h}
10:    return 𝒦^{t^h}_{red}
11: end procedure
```

C   A Discussion on the Fault Invariants After μ_2^{-1} in $\mathcal{R}_2^{-1}$

The formation of the fault invariants due to a uni-word fault in the input of $\mathcal{R}_2^{-1}$ is completely governed by the InverseMixColumn Matrix (M_μ^{-1}). It is worth mentioning that the fault-invariants are unique w.r.t to the row in which the fault was induced in the state before $\mathcal{R}_2^{-1}$ and are independent of the column. This implies that fault-invariants are fixed for a particular row irrespective of which column the faulty word belonged to. Let the differential state before μ_2^{-1} be $s = [s_{i,j}]$ and the one after μ_2^{-1} be $t = [t_{i,j}]$. Also let the differential fault value be f. Figure 8 depicts an example of fault invariant formation in $t_{*,7}$ for a word-fault in $s_{4,7}$. It can be noted that **value** of the invariant given in $t_{*,7}$ is same for any word-fault in $s_{4,*}$ i.e. the fifth row of s. However, the position of the invariant is due the column position of the fault which is the eighth column of s. Table 3 gives the exhaustive list of fault invariants exploited in SCOPE and their relation to the row location of the induced word fault.

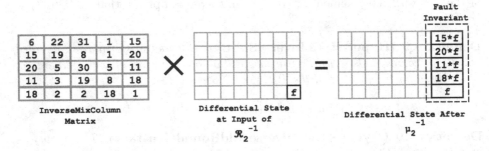

Fig. 8. Formation of fault invariant. Value of the fault invariant determined by M_μ^{-1} and the **row** position of the word-fault before $\mathcal{R}_2^{-1}$. Position of fault invariant determined by **column** position of the word-fault. Formation of fault invariant. Value of the fault invariant determined by M_μ^{-1} and the **row** position of the word-fault before $\mathcal{R}_2^{-1}$. Position of fault invariant determined by **column** position of the word-fault.

Table 3. Fault invariants exploited in SCOPE attack and the rows they correspond to.

Row position of fault	0	1	2	3	4
Fault invariant value	$\begin{bmatrix} 6 \times f \\ 15 \times f \\ 20 \times f \\ 11 \times f \\ 18 \times f \end{bmatrix}$	$\begin{bmatrix} 22 \times f \\ 19 \times f \\ 5 \times f \\ 3 \times f \\ 2 \times f \end{bmatrix}$	$\begin{bmatrix} 31 \times f \\ 8 \times f \\ 30 \times f \\ 19 \times f \\ 2 \times f \end{bmatrix}$	$\begin{bmatrix} f \\ f \\ 5 \times f \\ 8 \times f \\ 18 \times f \end{bmatrix}$	$\begin{bmatrix} 15 \times f \\ 20 \times f \\ 11 \times f \\ 18 \times f \\ f \end{bmatrix}$

D   Constructing the Factor State

The factor state as stated earlier gives us the factors with which the differential row computed using the Block Inversion property needs to be divided to reconstruct the state at the end of $\mathcal{R}_2^{-1}$. The pseudo-code for generating the factor matrix is given below:

```
 1: procedure GENERATEFACTORMATRIX
 2:     Take state s = [s_{i,j}]
 3:     for col = 0 : 7 do
 4:         s_{i,j} = 0, ∀ i, j
 5:         s_{*,col} = [1, 1, 1, 1, 1]^T
 6:         s = μ^{-1}(ρ^{-1}(s))
 7:         factorMat_{col,*} = s_{1,*}
 8:     end for
 9:     return factorMat
10: end procedure
```

References

1. CAESAR: Competition for Authenticated Encryption: Security, Applicability, and Robustness. http://competitions.cr.yp.to/caesar.html
2. Andreeva, E., Bilgin, B., Bogdanov, A., Luykx, A., Mennink, B., Mouha, N., Wang, Q., Yasuda, K.: PRIMATEs v1 (2014). http://competitions.cr.yp.to/round1/primatesv1.pdf
3. Andreeva, E., Bilgin, B., Bogdanov, A., Luykx, A., Mennink, B., Mouha, N., Wang, Q., Yasuda, K.: PRIMATEs v1.01 (2014). http://primates.ae/wp-content/uploads/primatesv1.01.pdf
4. Andreeva, E., Bilgin, B., Bogdanov, A., Luykx, A., Mennink, B., Mouha, N., Yasuda, K.: APE: Authenticated Permutation-Based Encryption for Lightweight Cryptography. In: Cid, C., Rechberger, C. (eds.) FSE 2014. LNCS, vol. 8540, pp. 168–186. Springer, Heidelberg (2015). https://lirias.kuleuven.be/handle/123456789/450105
5. Andreeva, E., Bogdanov, A., Luykx, A., Mennink, B., Mouha, N., Yasuda, K.: How to Securely Release Unverified Plaintext in Authenticated Encryption. In: Sarkar, P., Iwata, T. (eds.) ASIACRYPT 2014. LNCS, vol. 8873, pp. 105–125. Springer, Heidelberg (2014)
6. Andreeva, E., Bogdanov, A., Luykx, A., Mennink, B., Mouha, N., Yasuda, K.: How to securely release unverified plaintext in authenticated encryption. Cryptology ePrint Archive, Report 2014/144 (2014). http://eprint.iacr.org/
7. Bertoni, G., Daemen, J., Peeters, M., Assche, G.V.: Cryptographic sponge functions. http://sponge.noekeon.org/CSF-0.1.pdf
8. Biham, E., Shamir, A.: Differential Fault Analysis of Secret Key Cryptosystems. In: Kaliski Jr., B.S. (ed.) CRYPTO 1997. LNCS, vol. 1294, pp. 513–525. Springer, Heidelberg (1997)
9. Daemen, J., Rijmen, V.: The Design of Rijndael: AES - The Advanced Encryption Standard. Information Security and Cryptography. Springer, Heidelberg (2002)
10. Dusart, P., Letourneux, G., Vivolo, O.: Differential Fault Analysis on A.E.S. IACR Cryptology ePrint Archive 2003, 10 (2003). http://eprint.iacr.org/2003/010
11. Giraud, C.: DFA on AES. In: Dobbertin, H., Rijmen, V., Sowa, A. (eds.) AES 2005. LNCS, vol. 3373, pp. 27–41. Springer, Heidelberg (2005)

12. Moradi, A., Shalmani, M.T.M., Salmasizadeh, M.: A Generalized Method of Differential Fault Attack Against AES Cryptosystem. In: Goubin, L., Matsui, M. (eds.) CHES 2006. LNCS, vol. 4249, pp. 91–100. Springer, Heidelberg (2006)
13. Mukhopadhyay, D.: An Improved Fault Based Attack of the Advanced Encryption Standard. In: Preneel, B. (ed.) AFRICACRYPT 2009. LNCS, vol. 5580, pp. 421–434. Springer, Heidelberg (2009)
14. Piret, G., Quisquater, J.-J.: A Differential Fault Attack Technique Against SPN Structures, with Application to the AES and KHAZAD. In: Walter, C.D., Koç, Ç.K., Paar, C. (eds.) CHES 2003. LNCS, vol. 2779, pp. 77–88. Springer, Heidelberg (2003)
15. Saha, D., Kuila, S., Chowdhury, D.R.: EscApe: Diagonal Fault Analysis of APE. In: Meier, W., Mukhopadhyay, D. (eds.) INDOCRYPT 2014. LNCS, vol. 8885, pp. 197–216. Springer, Heidelberg (2014)
16. Saha, D., Mukhopadhyay, D., RoyChowdhury, D.: A Diagonal Fault Attack on the Advanced Encryption Standard. Cryptology ePrint Archive, Report 2009/581 (2009). http://eprint.iacr.org/

On the Hardness of Mathematical Problems

Bit Security of the CDH Problems
over Finite Fields

Mingqiang Wang[1], Tao Zhan[1], and Haibin Zhang[2]([✉])

[1] Shandong University, Jinan, China
wangmingqiang@sdu.edu.cn, zhantao@moe.edu.cn
[2] University of North Carolina, Chapel Hill, Chapel Hill, USA
haibin@cs.unc.edu

Abstract. It is a long-standing open problem to prove the existence of (deterministic) hard-core predicates for the Computational Diffie-Hellman (CDH) problem over finite fields, without resorting to the *generic* approaches for any one-way functions (*e.g.,* the Goldreich-Levin hard-core predicates). Fazio *et al.* (FGPS, Crypto '13) made important progress on this problem by defining a *weaker* Computational Diffie-Hellman problem over $\mathbb{F}_{p^2}$, *i.e.,* Partial-CDH problem, and proving, when allowing changing field representations, the unpredictability of every single bit of one of the coordinates of the secret Diffie-Hellman value. In this paper, we show that *all* the individual bits of the CDH problem over $\mathbb{F}_{p^2}$ and *almost all* the individual bits of the CDH problem over $\mathbb{F}_{p^t}$ for $t > 2$ are hard-core.

Keywords: CDH · Diffie-Hellman problem · d-th CDH problem · Finite fields · Hard-core bits · List decoding · Multiplication code · Noisy oracle · Partial-CDH problem

1 Introduction

Hard-core predicates [4,14] are central to cryptography. Of particular interest is the hard-core predicate for the CDH problem, which is essential to establishing the security for Diffie-Hellman (DH) key exchange protocol [7] and ElGamal encryption scheme [9] without having to make a (potentially) much stronger DH assumption—the Decisional Diffie-Hellman (DDH) assumption.

However, despite the generic approaches for *randomized* predicates working for any computationally hard problems [13,19], showing the existence of *deterministic* and *specific* hard-core predicates for the CDH problem over *finite fields* has proven elusive. This is in contrast to other conjectured hard problems such as discrete logs, RSA, and Rabin, whose deterministic hard-core predicates were discovered roughly three decades ago [2,4]. Recently, Fazio, Gennaro, Perera, and Skeith (FGPS) [10] made a significant breakthrough by introducing a relaxed variant of the CDH problem over finite fields $\mathbb{F}_{p^2}$, *i.e.,* the Partial-CDH problem and proving the unpredictability for a large class of predicates.

PARTIAL-CDH PROBLEM. Given a prime p, there are many different fields $\mathbb{F}_{p^2}$ which are all isomorphic to each other. Let $h(x) = x^2 + h_1 x + h_0$ be a monic

© Springer International Publishing Switzerland 2016
O. Dunkelman and L. Keliher (Eds.): SAC 2015, LNCS 9566, pp. 441–461, 2016.
DOI: 10.1007/978-3-319-31301-6_25

irreducible polynomial of degree 2 in $\mathbb{F}_p$. We know that $\mathbb{F}_{p^2}$ is isomorphic to the field $\mathbb{F}_p[x]/(h)$, where $(h(x))$ is a principal ideal in the polynomial ring $\mathbb{F}_p[x]$ and elements of $\mathbb{F}_{p^2}$ can be written as linear polynomials. Namely, if $g \in \mathbb{F}_{p^2}$ then $g = g_1 x + g_0$ and addition and multiplication are performed as polynomial operations modulo h. Given $g \in \mathbb{F}_{p^2}$ we denote by $[g]_i$ the coefficient of the degree-i term.

Let g denote a random generator of the multiplicative group of $\mathbb{F}_{p^2}$. FGPS defined the following Partial-CDH problem over $\mathbb{F}_{p^2}$ [10]: the Partial-CDH problem is hard over $\mathbb{F}_{p^2}$ if given random inputs $g, A = g^a$, $B = g^b \in \mathbb{F}_{p^2}$, it is computationally hard to output $K = [g^{ab}]_1 \in \mathbb{F}_p$ (*i.e.*, the coefficient of the degree 1 term of g^{ab}), for any representation of $\mathbb{F}_{p^2}$.

Assuming the hardness of the Partial-CDH problem, FGPS developed the idea of randomizing the problem representation originally suggested by Boneh and Shparlinski [5] and proved a large class of hard-core predicates over a *random representation* of the finite field $\mathbb{F}_{p^2}$. Namely, given an oracle that predicts any bit of $K = \left[g^{ab}\right]_1$ over a random representation of $\mathbb{F}_{p^2}$ with non-negligible advantage, one can recover K with non-negligible probability.

However, the Partial-CDH problem is clearly weaker than the regular CDH problem. Given a CDH oracle, one can easily solve the Partial-CDH problem. Note that the reason why we need hard-core predicates is exactly that we do not want to make stronger assumptions. Without characterizing the hardness of the Partial-CDH problem, the FGPS result can hardly be based on a firm foundation. Thus, studying the hardness of the Partial-CDH problem is left by FGPS as an important open problem [10, Sect. 6].

THE d-TH CDH PROBLEMS. It is natural to generalize the Partial-CDH problem over $\mathbb{F}_{p^2}$ to define the d-th CDH problems over $\mathbb{F}_{p^t}$ for $t > 1$ (history and related work coming shortly). For a prime p and an integer $t > 1$, there are many different fields $\mathbb{F}_{p^t}$, but they are all isomorphic to each other. Let $h(x)$ be a monic irreducible polynomial of degree t in $\mathbb{F}_p$. It is well known that $\mathbb{F}_{p^t}$ is isomorphic to the field $\mathbb{F}_p[x]/(h)$, where $(h(x))$ is a principal ideal in the polynomial ring $\mathbb{F}_p[x]$ and elements of $\mathbb{F}_{p^t}$ can be written as polynomials of degree $t-1$. Namely, if $g \in \mathbb{F}_{p^t}$ then $g = g_{t-1} x^{t-1} + g_{t-2} x^{t-2} + \cdots + g_1 x + g_0$. Addition and multiplication of the elements in $\mathbb{F}_{p^t}$ are performed as polynomial operations modulo h. In the following, given $g \in \mathbb{F}_{p^t}$ we denote by $[g]_i$ the coefficient of the degree-i term, *i.e.*, $g_i = [g]_i$.

Let g be a random generator of the multiplicative group of $\mathbb{F}_{p^t}$ and d be an integer such that $0 \le d \le t - 1$. Informally we say that the d-th CDH problem is hard in $\mathbb{F}_{p^t}$ if given $g, g^a, g^b \in \mathbb{F}_{p^t}$, it is computationally hard to compute $[g^{ab}]_d$, for any representations of $\mathbb{F}_{p^t}$.

PRIOR WORK ON HARDNESS OF d-TH CDH PROBLEMS: NOT YET PERFECT. FGPS and an earlier version of this paper did not realize that the hardness of d-th CDH problem had already been studied in [20,22]. Verheul [22, Theorem 21] showed that given a *perfect d-th CDH problem* oracle (which always returns correct answers), one can solve the CDH problem over the same fields. Concretely, given a CDH instance $(g^x, g^y) \in (\mathbb{F}_{p^t})^2$, Verheul's algorithm needs to run the

d-th CDH problem oracle on $(g^x, g^y \cdot g^r)$ for at least $poly(t)$ times, with the same g^x and g^y, yet uniformly chosen $r \xleftarrow{\$} \mathbb{Z}_{p^t - 1}$. For some d, say, $d = \lceil t/2 \rceil$, Verheul's algorithm even has to run the d-th CDH oracle for at least 2^t times such that the algorithm can have exponential running time in t.

Shparlinski [20] generalized Verheul's result to handle the case of *noisy* oracles (which return correct answers with some probabilities). Shparlinski's reduction uses a strategy that is the same as Verheul's to limit the behavior of the oracle. Namely, the queries given to the d-th CDH oracle have the form of $(g^x, g^y \cdot g^r)$ with uniformly chosen r. In this case, it is not guaranteed that the noisy d-th CDH oracle would answer this type of queries correctly. It might well be the case that a malicious d-th CDH problem oracle (adversary) simply always returns incorrect answers for any query of the form $(X, \cdot)$, if it has previously been given a query with the same X. Hence, Shparlinski's reduction is problematic in the sense it failed to prove what's claimed in the presence of noisy oracles. (Note that Verheul's reduction does not suffer from the same problem, as the answers returned by the perfect oracles are always correct.)

1.1 Our Contributions

In this paper, we show that *all* the individual bits of the CDH problem over $\mathbb{F}_{p^2}$ and *almost all* the individual bits of the CDH problem over $\mathbb{F}_{p^t}$ for $t > 2$ are hard-core. Let's explain our main contributions in a bit more detail.

THE HARDNESS OF d-TH CDH PROBLEM. In order to characterize the hardness of d-th CDH problem, we consider a case of noisy oracles which is more general than those of Verheul [22] and Shparlinski [20]. In our model, to compute the secret CDH value, we just require that the d-th CDH oracle return correct answers at some probability. Given a CDH instance $(g^x, g^y) \in (\mathbb{F}_{p^t})^2$, we need to run the d-th CDH oracle on inputs $(g^x \cdot g^r, g^y \cdot g^s)$ with uniformly chosen r and s. The analysis for general t turns out to take some work.

With this model, we show that the 1-th CDH problem (i.e., the Partial-CDH problem) and 0-th CDH problem (which we call Dual-Partial-CDH problem) over finite fields $\mathbb{F}_{p^2}$ are strictly as hard as the regular CDH problem over the same fields. Regarding general extension fields, we are able to prove that all the d-th CDH problems over a random representation of finite fields $\mathbb{F}_{p^t}$ (with $t > 1$) are as hard as the regular CDH problem over the same fields; in particular, the 0-th CDH problem and $(t - 1)$-th CDH problem given *any* field representation are as hard as the CDH problem. We comment that applying our approach to the case of perfect oracles, our reduction leads to no security loss, which is in contrast to Verheul's, where for many d's, the algorithm can easily have exponential running time in t.

THE CASE OF $\mathbb{F}_{p^2}$. At the heart of the FGPS result is the *list decoding* approach for hard-core predicates, which was developed by Akavia, Goldwasser and Safra [1], and extended by Morillo and Ràfols [18] and Duc and Jetchev [8]. Up to now, the list decoding approach has only been proven successful for multiplicative codes [1,8,18]. It is unclear if the approach can work more generally. In this paper, we will work *directly* on a non-multiplicative code. *Still* assuming the hardness of the Partial-CDH problem, we are able to prove the unpredictability of every single bit of the *other* coordinate (*i.e.*, the coefficient of the lower degree term) of the secret CDH value, by using a careful analysis of the Fourier coefficients of the function. To the best of our knowledge, this is the first positive result that the list decoding approach can be applied to a non-multiplicative code, a result of independent interest.

Combining all the above-mentioned results, we are able to prove our main result for the regular CDH problem over $\mathbb{F}_{p^2}$: given an oracle $\mathcal{O}$ that predicts *any* bit of the CDH value over a random representation of the field $\mathbb{F}_{p^2}$ with non-negligible advantage, we can solve the *regular* CDH problem over $\mathbb{F}_{p^2}$ with non-negligible probability.

THE CASE OF $\mathbb{F}_{p^t}$. We go on to prove that assuming the hardness of the d-th CDH problem, every single bit of the d-th CDH coordinate for $d \neq 0$ is hard-to-compute. FGPS [10, Sect. 6] found that their technique was not powerful enough to solve the generalized problem. To overcome the difficulty, we identify a *general* yet *simplified* class of isomorphisms. The isomorphisms identified generalize those of finite field $\mathbb{F}_{p^2}$ in FGPS to the case of general finite fields $\mathbb{F}_{p^t}$ for any $t > 1$. More importantly, they simplify those of FGPS by adopting a more restrictive class of isomorphisms. We comment that it is the simplicity that is essential to overcoming the original technical difficulty and establishing the bit security for general finite fields. To achieve this result, we also use another idea of Boneh and Shparlinski [5] using d-th residues modulo p.

Together with the equivalence result between all the d-th CDH problems over $\mathbb{F}_{p^t}$ (with $t > 1$) and the regular CDH problem, we obtain another main result of the paper: all bits except the bits of the degree-0 term of the usual CDH problem over a random representation of the finite field $\mathbb{F}_{p^t}$ are hard-core.

1.2   Further History and Discussion

An earlier version was put online [23]. Galbraith and Shani [11] extended our work to obtain an essentially stronger hard-core result that works for every individual bit for any finite fields $\mathbb{F}_{p^t}$ with *any* t. Thus, as claimed by the authors, this improvement can allow us to consider "the case of large t, and in particular the case of fields with small characteristic" [11, p. 264]. We certainly agree with this point of view, but one may not understand that our reduction approach is inherently defective. The security loss in our reduction only comes from the loss in proving the equivalence between the d-th CDH problem and the conventional CDH problem. If one can find a way to prove their equivalence with no security loss, as what we did for the case of perfect oracles, our result can be equally expressive.

As commented by Galbraith and Shani [11, Remark 25], their approach does not work for the popular polynomial basis, while our approach deals with this case, and therefore our result will be useful when one desires a hard-core bit in its polynomial basis.

Another reason that makes our paper worth attending to is that as discussed earlier, we point out the "problem" of studying the hardness of the d-th CDH problem in prior work by Shparlinski [20]. We regard identifying the problem and providing a more general and correct proof as an important contribution of the paper. However, one may not really deem Verheul's result [22] as being "faulty" or "flawed"; rather, it is that our result provides a stronger result for the problem.

By the same token, with Verheul's result, one may regard that FGPS is actually the first (though they did not notice this) to solve the open problem whether there exists "specific" hard-core bits over finite fields: half of the individual bits of the secret CDH value over $\mathbb{F}_{p^2}$ are unpredictable. If one is uncomfortable about their restricted reduction from CDH to d-th CDH, our result for the case of both perfects oracles and noisy oracles can then come into use.

2 Preliminaries

2.1 Notation

We use the standard symbols $\mathbb{N}$, $\mathbb{Z}$, $\mathbb{R}$ and $\mathbb{C}$ to denote the natural numbers, the integers, the real numbers and the complex numbers, respectively. Let $\mathbb{Z}_+$ and $\mathbb{R}_+$ stand for the positive integers and reals, respectively. A function $\nu(l)\colon \mathbb{N} \to \mathbb{R}$ is *negligible* if for every constant $c \in \mathbb{R}_+$ there exists $l_c \in \mathbb{N}$ such that $\nu(l) < l^{-c}$ for all $l > l_c$. A function $\rho(l)\colon \mathbb{N} \to \mathbb{R}$ is *non-negligible* if there exists a constant $c \in \mathbb{R}_+$ and $l_c \in \mathbb{N}$ such that $\rho(l) > l^{-c}$ for all $l > l_c$. For a Boolean function $f\colon \mathcal{D} \to \{\pm 1\}$ over an arbitrary domain $\mathcal{D}$, denote by $\mathsf{maj}_f = \max_{\{b=\pm 1\}} \Pr_{\alpha \in \mathcal{D}}[f(\alpha) = b]$ the *bias* of f toward its majority value.

2.2 Fourier Transform

Let $\mathbb{G}$ be a finite abelian group. For any two functions $f, g\colon \mathbb{G} \to \mathbb{C}$, their *inner product* is defined as $\langle f, g \rangle = 1/|\mathbb{G}| \sum_{x \in G} \overline{f(x)}g(x)$. The l_2-norm of f on the vector space $\mathbb{C}(\mathbb{G})$ is defined as $\|f\|_2 = \sqrt{\langle f, f \rangle}$. A *character* of $\mathbb{G}$ is a homomorphism $\chi\colon \mathbb{G} \to \mathbb{C}^*$, i.e., $\chi(x+y) = \chi(x)\chi(y)$ for all $x, y \in \mathbb{G}$. The set of all characters of $\mathbb{G}$ forms a *character group* $\widehat{\mathbb{G}}$, whose elements form an orthogonal basis (the *Fourier basis*) for the vector space $\mathbb{C}(\mathbb{G})$. One can then describe any function $f \in \mathbb{C}(\mathbb{G})$ via its *Fourier expansion* $\sum_{\chi \in \widehat{\mathbb{G}}} \widehat{f}(\chi)\chi$, where $\widehat{f}\colon \widehat{\mathbb{G}} \to \mathbb{C}$ is the *Fourier transform* of f and we have $\widehat{f}(\chi) = \langle f, \chi \rangle$. The coefficients $\widehat{f}(\chi)$ in the Fourier basis $\{\chi\}_{\chi \in \widehat{\mathbb{G}}}$ are the *Fourier coefficients* of f. The *weight* of a Fourier coefficient is denoted by $|\widehat{f}(\chi)|^2$. When $\mathbb{G} = \mathbb{Z}_n$ (i.e., the additive group of integers modulo n) and $\widehat{\mathbb{G}} = \widehat{\mathbb{Z}}_n$, for each $\alpha \in \mathbb{Z}_n$, the α-character is defined as a function $\chi_\alpha\colon \mathbb{Z}_n \to \mathbb{C}$ such that $\chi_\alpha(x) = \omega_n^{\alpha x}$, where $\omega_n = e^{2\pi i/n}$. If Γ is a subset of $\mathbb{Z}_n$ then it is natural to consider the projection of f in set Γ, i.e., $f_{|\Gamma} = \sum_{\alpha \in \Gamma} \widehat{f}(\alpha)\chi_\alpha$, where $\widehat{f}(\alpha) = \langle f, \chi_\alpha \rangle$. Since the characters are orthogonal, we have $\|f\|_2^2 = \sum_{\alpha \in \mathbb{Z}_n} |\widehat{f}(\alpha)|^2$ and $\|f_{|\Gamma}\|_2^2 = \sum_{\alpha \in \Gamma} |\widehat{f}(\alpha)|^2$.

Definition 1 (Fourier concentrated function [1]). *A function $f\colon \mathbb{Z}_n \to \mathbb{C}$ is Fourier ϵ-concentrated if there exists a set $\Gamma \subseteq \mathbb{Z}_n$ consisting of $poly(\log n, 1/\epsilon)$ characters, so that*

$$\|f - f_{|\Gamma}\|_2^2 = \sum_{\alpha \notin \Gamma} |\widehat{f}(\alpha)|^2 \le \epsilon.$$

A function f is called Fourier concentrated if it is Fourier ϵ-concentrated for every $\epsilon > 0$.

This and subsequent definitions can be readily made *asymptotic* by requiring that ϵ depend on the security parameter.

Definition 2 (τ-heavy characters [1]). *Given a threshold $\tau > 0$ and an arbitrary function $f\colon \mathbb{Z}_n \to \mathbb{C}$, we say that a character χ_α is τ-heavy if the weight of its corresponding Fourier coefficient is at least τ. The set of all τ-heavy characters is denoted by*

$$\mathsf{Heavy}_\tau(f) = \{\chi_\alpha \colon |\widehat{f}(\alpha)|^2 \ge \tau\}.$$

2.3 Error Correcting Codes: Definitions and Properties

Error correcting codes can encode messages into codewords by adding redundant data such that the message can be recovered even in the presence of noise. The code to be discussed here encodes each element $\alpha \in \mathbb{Z}_n$ into a codeword C_α of length n. Each codeword C_α can be represented by a function $C_\alpha \colon \mathbb{Z}_n \to \{\pm 1\}$. We now recall a number of definitions and lemmata [1,8] about codes over $\mathbb{Z}_n$.

Definition 3 (Fourier concentrated code). *A code $\mathcal{C} = \{C_\alpha \colon \mathbb{Z}_n \to \{\pm 1\}\}$ is concentrated if each of its codewords C_α is Fourier concentrated.*

Definition 4 (Recoverable code). *A code $\mathcal{C} = \{C_\alpha \colon \mathbb{Z}_n \to \{\pm 1\}\}$ is recoverable, if there exists a recovery algorithm that, given a character $\chi \in \widehat{\mathbb{Z}}_n$ and a threshold τ, returns in time $poly(\log n, 1/\tau)$ a list of all elements α associated with codewords C_α for which χ is a τ-heavy coefficient (i.e., $\{\alpha \in \mathbb{Z}_n \colon \chi \in \mathsf{Heavy}_\tau(C_\alpha)\}$).*

Lemma 1 below shows that in a concentrated code $\mathcal{C}$, any corrupted ("noisy") versions $\widetilde{C}_\alpha$ of codeword C_α share at least one heavy coefficient with C_α. Lemma 2 shows that when given query access to any function f one can efficiently learn all its heavy characters.

Lemma 1 ([1, Lemma 1]). *Let $f, g\colon \mathbb{Z}_n \to \{\pm 1\}$ such that f is concentrated and for some $\epsilon > 0$,*

$$\Pr_{\alpha \in \mathbb{Z}_n} [f(\alpha) = g(\alpha)] \ge \mathsf{maj}_f + \epsilon.$$

There exists a threshold τ such that $1/\tau \in poly(1/\epsilon, \log n)$, and there exists a nontrivial character $\chi \ne 0$ which is heavy for f and g: $\chi \in \mathsf{Heavy}_\tau(f) \cap \mathsf{Heavy}_\tau(g)$.

Lemma 2 ([1, Theorem 6]). *There is a probabilistic algorithm that, given query access to* $w\colon \mathbb{Z}_n \to \{\pm 1\}$, $\tau > 0$ *and* $0 < \delta < 1$, *outputs a list* L *of* $O(1/\tau)$ *characters containing* $\mathsf{Heavy}_\tau(w)$ *with probability at least* $1 - \delta$, *whose running time is* $\tilde{O}\left(\log(n) \cdot \ln^2 \frac{(1/\delta)}{\tau^{5.5}} \right)$.

2.4   Review of List Decoding Approach for Hard-Core Predicates

Informally, a cryptographic one-way function $f\colon \mathcal{D} \to \mathcal{R}$ is a function which is easy to compute but hard to invert. Given a one-way function f and a predicate π, we say π is hard-core if there is an efficient probabilistic polynomial-time (PPT) algorithm that given $\alpha \in \mathcal{D}$ computes $\pi(\alpha)$, but there is no PPT algorithm $\mathcal{A}$ that given $f(\alpha) \in \mathcal{R}$ predicts $\pi(\alpha)$ with probability $\mathsf{maj}_\pi + \epsilon$ for a non-negligible ϵ.

Goldreich and Levin [13] showed hard-core predicates for general one-way functions by providing a general list decoding algorithm for Hadamard code. Akavia, Goldwasser, and Safra (AGS) [1] formalized the list decoding methodology and applied it to a broad family of conjectured one-way functions. In particular, they proved the unpredictability of *segment predicates* [1] for any one-way function f with the following *homomorphic* property: given $f(\alpha)$ and λ, one can efficiently compute $f(\lambda\alpha)$. This includes discrete logarithms in finite fields and elliptic curves, RSA, and Rabin. Morillo and Ràfols [18] extended the AGS result to prove the unpredictability of every individual bit for these functions. Duc and Jetchev [8] showed how to extend to elliptic curve-based one-way functions which do not necessarily enjoy the homomorphic property. Their result instead requires introducing a random description of the curve, an idea originally developed by Boneh and Shparlinski [5]. In their paper, Boneh and Shparlinski proved for the elliptic curve Diffie-Hellman problem that the least significant bit of each coordinate of the secret CDH value is hard-core over a random representation of the curve. Recently, FGPS extended the Boneh and Shparlinski idea to prove every individual bit (not merely the least significant bit) of the elliptic curve Diffie-Hellman problem is hard-core. By extending the same idea to the case of finite fields $\mathbb{F}_{p^2}$, FGPS also proved for a weak CDH problem (*i.e.* Partial-CDH problem) the unpredictability of every single bit of one of the coordinates of the secret CDH value.

LIST DECODING APPROACH OVERVIEW. Given a one-way function $f\colon \mathcal{D} \to \mathcal{R}$ and a predicate π, one would have to identify an error-correcting code $\mathcal{C}^\pi = \{C_\alpha \colon \mathcal{D} \to \{\pm 1\}\}_{\alpha \in \mathcal{D}}$ such that every input α of the one-way function is associated with a codeword C_α. The code needs to satisfy the following properties:

(1) *Accessibility.* One should be able to obtain a corrupted ("noisy") version $\tilde{C}_\alpha$ of the original codeword C_α. Such a corrupted codeword must be close to the original codeword, *i.e.*, $\Pr_\lambda[C_\alpha(\lambda) = \tilde{C}_\alpha(\lambda)] > \mathsf{maj}_\pi + \epsilon$ for a non-negligible ϵ.

(2) *Concentration.* Each codeword C_α should be a Fourier concentrated function, *i.e.*, each codeword can be approximated by a small number of heavy coefficients in the Fourier representation.

(3) *Recoverability.* There exists a $poly(\log n, \tau^{-1})$ algorithm that on input a Fourier character χ and a threshold τ outputs a short list L_χ which contains all the values $\alpha \in \mathcal{D}$ such that χ is τ-heavy for the codeword C_α.

Roughly speaking, accessibility is related to both the code and the oracle, while concentration and recoverability concern the code itself. We now show how to invert $y = f(\alpha)$ with the prediction oracle Ω. Querying Ω will allow one to have access to a corrupted codeword $\widetilde{C}_\alpha$ that is close to C_α. According to Lemma 1, we know that there should exist a threshold τ and at least one Fourier character that is τ-heavy for both $\widetilde{C}_\alpha$ and C_α. Applying the learning algorithm in Lemma 2, we can find the set of all τ-heavy characters for $\widetilde{C}_\alpha$. Due to the recovery property, we are able to produce for each heavy character a polynomial size list containing possible α. Note that one can identify the correct α since f is efficiently computable.

LIST DECODING VIA MULTIPLICATION CODE. The crux of list decoding approach is to identify the "right" code which is accessible, concentrated, and recoverable. To this end, AGS and subsequent work either define a multiplication code, or transform the original code to an equivalent multiplication code. (Such a multiplication code is of the form $C_\alpha(\lambda) = \pi(\lambda\alpha)$.) Indeed, as argued in [1,8], this is at the basis of their proofs: multiplication codes can be proven to satisfy concentration and recoverability.

In Sect. 3, we will directly work on a code that is *not* multiplicative. Not surprisingly, this makes it hard to prove code concentration and recoverability. To our knowledge, we are the first to apply the list decoding approach to the case of a non-multiplicative code.

3   All Bits Security of the CDH Problems over $\mathbb{F}_{p^2}$

In this section, we show the following three results: (1) we show that over finite fields $\mathbb{F}_{p^2}$ the Partial-CDH problem [10] is as hard as the regular CDH problem. (2) assuming the hardness of the Partial-CDH problem over $\mathbb{F}_{p^2}$, we prove the unpredictability of every single bit of the *other* coordinate of the secret CDH value; (3) we go on to prove the unpredictability of *every* single bit of the secret CDH value for the regular CDH problem over $\mathbb{F}_{p^2}$.

THE PARTIAL-CDH ASSUMPTION IS EQUIVALENT TO THE CDH ASSUMPTION OVER $\mathbb{F}_{p^2}$. Throughout the paper we fix a security parameter l. We consider an instance generator $\mathcal{G}$ which takes as input 1^l and outputs an l-bit prime p. Let g be a random generator of the multiplicative group of $\mathbb{F}_{p^2}$. The Partial-CDH problem over $\mathbb{F}_{p^2}$ is a relaxed variant of the conventional CDH problem over $\mathbb{F}_{p^2}$, which we formally state as follows:

Assumption 1 (The CDH assumption over $\mathbb{F}_{p^2}$). *We say that the CDH problem is hard in $\mathbb{F}_{p^2}$ if for any PPT adversary $\mathcal{A}$, his CDH advantage*

$$\mathbf{Adv}^{\mathrm{cdh}}_{\mathcal{A}, \mathbb{F}_{p^2}} := \Pr\left[\mathcal{A}(p, g, g^a, g^b) = g^{ab} | p \xleftarrow{\$} \mathcal{G}(1^l); a, b \xleftarrow{\$} \{1, \cdots, p^2 - 1\}\right]$$

is negligible in l.

Let $I_2(p)$ be the set of monic irreducible polynomials of degree 2 in $\mathbb{F}_p$. Informally we say that the *Partial-CDH* problem [10] is hard in $\mathbb{F}_{p^2}$ if for all $h \in I_2(p)$ no efficient algorithm given $g, A = g^a, B = g^b \in \mathbb{F}_{p^2}$ can output $\left[g^{ab}\right]_1 \in \mathbb{F}_p$. Formally we consider the following assumption:

Assumption 2 (The Partial-CDH assumption over $\mathbb{F}_{p^2}$ [10]). *We say that the Partial-CDH problem is hard in $\mathbb{F}_{p^2}$ if for any PPT adversary $\mathcal{A}$, his Partial-CDH advantage for all $h \in I_2(p)$*

$$\mathbf{Adv}^{\text{pcdh}}_{\mathcal{A},h,\mathbb{F}_{p^2}} := \Pr\left[\mathcal{A}(p,h,g,g^a,g^b) = \left[g^{ab}\right]_1 \middle| p \xleftarrow{\$} \mathcal{G}(1^l); a, b \xleftarrow{\$} \{1,\cdots,p^2-1\}\right]$$

is negligible in l.

It is easy to see that the Partial-CDH problem is weaker than the regular CDH problem over $\mathbb{F}_{p^2}$. The following theorem shows that in the case of noisy oracles, the regular CDH problem can be also reduced to the Partial-CDH problem in $\mathbb{F}_{p^2}$.

Theorem 1. *Suppose $\mathcal{A}$ is a Partial-CDH adversary that runs in time at most φ and achieves advantage $\mathbf{Adv}^{\text{pcdh}}_{\mathcal{A},h,\mathbb{F}_{p^2}}$ for any $h \in I_2(p)$. Then there exists a CDH adversary $\mathcal{B}$, constructed from $\mathcal{A}$ in a blackbox manner, that runs in time at most 2φ plus the time to perform a small constant number of group operations and achieves advantage $\mathbf{Adv}^{\text{cdh}}_{\mathcal{B},h,\mathbb{F}_{p^2}} \geq (1 - \frac{1}{p}) \cdot (\mathbf{Adv}^{\text{pcdh}}_{\mathcal{A},h,\mathbb{F}_{p^2}})^2$.*

Proof: Our CDH adversary $\mathcal{B}$ works as follows, given input a random instance of the CDH problem $(g^a, g^b) \in (\mathbb{F}_{p^2})^2$ and given a Partial-CDH adversary $\mathcal{A}$ under the representation determined by any given polynomial $h(x) = x^2 + h_1 x + h_0 \in I_2(p)$.

First, adversary $\mathcal{B}$ chooses two random integers $r, s \xleftarrow{\$} \mathbb{Z}_{p^2-1}$, and computes (g^{a+r}, g^{b+s}). For brevity, let $A = a + r$ and $B = b + s$. Adversary $\mathcal{B}$ then runs the Partial-CDH adversary $\mathcal{A}$ on the generated instance (g^A, g^B) to obtain $\left[g^{AB}\right]_1$. Let $C = as + br + rs$. As $g^{AB} = g^{ab}g^C \bmod h(x)$, we have the following equation

$$\left(\left[g^C\right]_0 - \left[g^C\right]_1 h_1\right)\left[g^{ab}\right]_1 + \left[g^C\right]_1\left[g^{ab}\right]_0 = \left[g^{AB}\right]_1$$

Repeating the above process, $\mathcal{B}$ chooses two random integers $r', s' \xleftarrow{\$} \mathbb{Z}_{p^2-1}$ and gets the following equation

$$\left(\left[g^{C'}\right]_0 - \left[g^{C'}\right]_1 h_1\right)\left[g^{ab}\right]_1 + \left[g^{C'}\right]_1\left[g^{ab}\right]_0 = \left[g^{A'B'}\right]_1,$$

where $A' = a + r', B' = b + s'$, and $C' = as' + br' + r's'$.

Combining the above two equations, we obtain a linear equation set with the unknowns $\left[g^{ab}\right]_1$ and $\left[g^{ab}\right]_0$. If the coefficient matrix of the equation set has full rank then adversary $\mathcal{B}$ can solve the equation set and obtain g^{ab}. The coefficient matrix is of full rank if and only if its determinant is not zero, *i.e.*,

$$\left(\left[g^C\right]_0 - \left[g^C\right]_1 h_1\right)\left[g^{C'}\right]_1 - \left(\left[g^{C'}\right]_0 - \left[g^{C'}\right]_1 h_1\right)\left[g^C\right]_1 \neq 0.$$

Note that $[g^C]_i$ and $[g^{C'}]_i$ $(i = 0,1)$ in the above equation are independently and uniformly distributed at random from $\mathbb{F}_p$. Hence, the probability that the matrix is of full rank is $1 - 1/p$. This completes the proof of this theorem. ∎

We can define a *dual* variant of the Partial-CDH problem over $\mathbb{F}_{p^2}$: We say that the *Dual-Partial-CDH* problem is hard in $\mathbb{F}_{p^2}$ if for all $h \in I_2(p)$ no efficient algorithm given $g, A = g^a, B = g^b \in \mathbb{F}_{p^2}$ can output $[g^{ab}]_0 \in \mathbb{F}_p$. We can show that the Dual-Partial-CDH problem is also as hard as the conventional CDH problem. The formal definition and the proof can be found in our full paper [23]. Therefore, both the Partial-CDH and Dual-Partial CDH problems are as hard as the conventional CDH problem over $\mathbb{F}_{p^2}$.

BIT SECURITY FOR THE OTHER COORDINATE. Let $B_k : \mathbb{F}_p \to \{\pm 1\}$ denote the k-th bit predicate (with a 0 bit being encoded as $+1$). Let β_k be the bias of B_k. For all $h, \widehat{h} \in I_2(p)$ there exists an easily computable isomorphism $\phi_{h,\widehat{h}} : \mathbb{F}_p[x]/(h) \to \mathbb{F}_p[x]/(\widehat{h})$. Informally we show that when given an oracle $\mathcal{O}$ that predicts the k-th bit of the degree 0 coefficient of the CDH value with non-negligible advantage, and the representation of the field, then we can break the Partial-CDH assumption with non-negligible advantage.

Theorem 2. *Under the Partial-CDH assumption over $\mathbb{F}_{p^2}$ (i.e., Assumption 2), for any PPT adversary $\mathcal{O}$, we have that for all $h \in I_2(p)$ the following quantity is negligible in l:*

$$\left| \Pr\left[\mathcal{O}(h, \widehat{h}, g, g^a, g^b) = B_k\left([\phi_{h,\widehat{h}}(g^{ab})]_0 \right) \big| \widehat{h} \xleftarrow{\$} I_2(p); a, b \xleftarrow{\$} \{1, \cdots, p^2 - 1\} \right] - \beta_k \right|.$$

We first give an informal intuition of the proof of the theorem. We aim at constructing a code similar to those of FGPS and Duc and Jetchev [8]. For an element $\alpha \in \mathbb{F}_{p^2}$ and a monic irreducible polynomial $h \in I_2(p)$, we would define the following codeword:

$$C_\alpha(\widehat{h}) = B_k([\phi_{h,\widehat{h}}(\alpha)]_0).$$

Similar to the code defined in FGPS, the above code is accessible using $\mathcal{O}$. However, the predicate B_k is evaluated on the *other* coordinate of $\phi_{h,\widehat{h}}(\alpha)$. In this case, it holds that $[\phi_{h,\widehat{h}}(\alpha)]_0 = \eta[\alpha]_1 + [\alpha]_0$ for some $\eta \in \mathbb{F}_p$, according to FGPS [10, Lemma 5.3] (recalled in Lemma 3 below).

Lemma 3 ([10, Lemma 5.3]). *For any $h \in I_2(p)$, there exists a unique function $L_h : \mathbb{F}_p \times \mathbb{F}_p^* \to I_2(p)$ which takes a pair (η, λ) to the polynomial $\widehat{h} = L_h(\eta, \lambda)$ such that the matrix $\left(\begin{smallmatrix} 1 & \eta \\ 0 & \lambda \end{smallmatrix} \right)$ defines an isomorphism from $\mathbb{F}_p[x]/(h)$ to $\mathbb{F}_p[x]/(\widehat{h})$ that sends $[\alpha]_1 x + [\alpha]_0 \mapsto \lambda[\alpha]_1 x + \eta[\alpha]_1 + [\alpha]_0$.*

Intuitively, one would consider the following code: for $\alpha \in \mathbb{F}_{p^2}$ and for $\eta \in \mathbb{F}_p$ (and $\lambda \in \mathbb{F}_p^*$), set

$$C_\alpha(\eta) = B_k(\eta[\alpha]_1 + [\alpha]_0). \tag{1}$$

Unfortunately, the above code in (1) is not *multiplicative*. In particular, this makes it hard to prove concentration and recoverability. This is why FGPS considered defining the Partial-CDH problem over $\mathbb{F}_{p^2}$ as outputting the coefficient of the degree 1 term of g^{ab}, instead of the coefficient of the degree 0 term. More generally, the list decoding approach has only been proven successful for multiplicative codes so far [1, 8, 18]. One natural question is if it is (even) possible to apply list decoding approach to the case of non-multiplicative codes.

With a careful analysis, we are still able to show that the code in (1) is concentrated and recoverable. Concentration will follow from the key observation that the Fourier transform of the code in (1) is equal to that of a multiplication code (to be defined shortly) up to a factor of a character. This follows from a (well-known) scaling property of the Fourier transform, as shown in Lemma 4 below. Hence, the l_2-norm of the Fourier transform of the code is equal to that of the multiplication code. That is, the code in (1) is concentrated if and only if the multiplication code is. Note that it is easy to argue that the multiplication code is concentrated.

The goal of recoverability is to recover the secret value from the heavy characters of the code C_α. We find that a character χ_β is heavy for C_α if and only if χ_β is heavy for a multiplicative code C'_α. The associated constant of a heavy character χ_β for the multiplicative code C'_α equals the product of the secret value and an (easily determined) factor. Therefore, one can recover the secret value with a heavy character.

We first describe the scaling property of the Fourier transform.

Lemma 4. *Let F_1, F_2 be functions mapping $\mathbb{Z}_n$ to $\mathbb{C}$. If for any y, $F_2(y) = F_1(y - \sigma)$, where σ is a constant in $\mathbb{Z}_n$, then we have for $\alpha \in \mathbb{Z}_n$, $\widehat{F_2}(\alpha) = \chi_\alpha(\sigma)\widehat{F_1}(\alpha)$.*

Proof of Theorem 2: Suppose that there exists an oracle $\mathcal{O}$ such that

$$\left| \Pr_{\eta, a, b} \left[\mathcal{O}(h, \widehat{h}, g, g^a, g^b) = B_k \left([\phi_{h, \widehat{h}}(g^{ab})]_0 \right) \right] - \beta_k \right|$$

is larger than a non-negligible quantity ϵ. We construct another oracle $\mathcal{O}'$ that takes as input a base representation $h \in I_2(p)$, a Diffie-Hellman triple $g, g^a, g^b \in \mathbb{F}_{p^2}$, and an element of $\eta \in \mathbb{F}_p$ (instead of $\widehat{h} \in I_2(p)$). The new oracle selects $\lambda \xleftarrow{\$} \mathbb{F}_p^*$, constructs an isomorphism $\widetilde{h}$ from the matrix $\left(\begin{smallmatrix} 1 & \eta \\ 0 & \lambda \end{smallmatrix} \right)$ as described in Lemma 3, and returns $\mathcal{O}(h, \widetilde{h}, g, g^a, g^b)$. One can then show that

$$\left| \Pr_{\eta, a, b} \left[\mathcal{O}'(h, \eta, g, g^a, g^b) = B_k \left(\eta[g^{ab}]_1 + [g^{ab}]_0 \right) \right] - \beta_k \right|$$

is also larger than a non-negligible quantity.

For any element $\alpha \in \mathbb{F}_{p^2}$, we construct the following encoding of $\eta[\alpha]_1 + [\alpha]_0$ in its polynomial representation for $\mathbb{F}_p[x]/(h)$:

$$C_\alpha \colon \mathbb{F}_p \to \{\pm 1\} \quad \text{such that} \quad C_\alpha(\eta) = B_k(\eta[\alpha]_1 + [\alpha]_0),$$

where, above, $[\alpha]_1$ and $[\alpha]_0$ are under the representation determined by h.

Accessibility. Accessibility proof is the same as that of FGPS. In particular, the oracle $\mathcal{O}'$ allows us to have access to a corrupted codeword $\widetilde{C}_\alpha$ of the above codeword defined as $\widetilde{C}_\alpha = \mathcal{O}'(h, \eta, g, g^a, g^b)$. The code $C_\alpha(\eta)$ is conceptually the same as the code $C_\alpha(\widehat{h})$. Therefore, if the oracle $\mathcal{O}$ has advantage ϵ then we have $|\Pr_\eta[C_\alpha(\eta) = \widetilde{C}_\alpha(\eta)]| \geq \beta_k + \epsilon$. Accessibility of the code C_α follows.

Concentration. We now prove that the codeword C_α is a Fourier concentrated code. To prove so, we define the following related code:

$$C'_\alpha(\eta) = B_k(\eta[\alpha]_1).$$

It is easy to see that $C'_\alpha(\eta) = C_\alpha(\eta - [\alpha]_1^{-1}[\alpha]_0)$. According to Lemma 4, we can obtain

$$\chi_\beta([\alpha]_1^{-1}[\alpha]_0)\widehat{C_\alpha}(\beta) = \widehat{C'_\alpha}(\beta).$$

This immediately implies $|\widehat{C_\alpha}(\beta)| = |\widehat{C'_\alpha}(\beta)|$. Therefore, the code $C_\alpha(\eta)$ is concentrated if and only if the code $C'_\alpha(\eta)$ is. Note that it is easy to argue that $C'_\alpha(\eta)$ is a multiplication code. The proof for concentration of the code $C'_\alpha(\eta)$ is similar to those of [10,18], and now we describe our proof in some detail.

For $\beta \in \mathbb{F}_p$, if $C'_\alpha(\eta)$ is ϵ-concentrated in $\Gamma_\alpha = \{\chi_\beta\}$ then $B_k(\eta[\alpha]_1)$ is ϵ-concentrated in the set $\{\chi_\eta \colon \eta = \beta[\alpha]_1^{-1}\}$. Thus, we just need to prove the Fourier concentration of $B_k(\eta[\alpha]_1)$. We would need to analyze the Fourier coefficients of $B_k \colon \mathbb{F}_p \to \{\pm 1\}$.

We define $g(x)$ as

$$g(x) = \frac{B_k(x) + B_k(x + 2^k)}{2}.$$

Morillo and Ràfols [18] notice that the Fourier transform of $B_k(x)$ and the Fourier transform of $g(x)$ can be related with the following equation:

$$\widehat{g}(\eta) = \frac{\omega_p^{2^k \eta} + 1}{2}\widehat{B_k}(\eta),$$

where $\eta \in \mathbb{F}_p$ and $\omega_p = e^{2\pi i/p}$.

In particular, assuming $\eta \in [-\frac{p-1}{2}, \frac{p-1}{2}]$, they consider the following two cases for η:

1. $\eta \geq 0$, consider $\delta_{\eta,k} := 2^k \eta - (p-1)/2 \bmod p$ and let $\lambda_{\eta,k} \in [0, 2^{k-1} - 1]$ be the unique integer for which $2^k \eta = (p-1)/2 + \delta_{\eta,k} + p\lambda_{\eta,k}$.
2. $\eta < 0$, consider $\delta_{\eta,k} := 2^k \eta + (p+1)/2 \bmod p$ and let $\lambda_{\eta,k} \in [0, 2^{k-1} - 1]$ be the unique integer for which $2^k \eta = -(p+1)/2 + \delta_{\eta,k} + p\lambda_{\eta,k}$.

For both cases, there are unique integers $\mu_{\eta,k} \in [0, r]$, where r is the largest integer less than $p/2^{k+1}$ and $r_{\eta,k} \in [0, 2^k - 1]$ such that $a_p(2^k \eta - (p-1)/2) = \mu_{\eta,k} 2^k + r_{\eta,k}$, where $a_p(x) = \min\{x \bmod p, p - x \bmod p\}$ for $x \bmod p$ being taken in $[0, p-1]$. The definition of Γ_τ in Sect. 3 is as follows

$$\Gamma_\tau = \{\eta \colon (\lambda_{\eta,k}, \mu_{\eta,k}) \in [0, 1/\tau] \times [0, 1/\tau]\}.$$

Here we select τ such that $1/\tau = poly(\log p)$. Morillo and Ràfols [18] obtain the following upper bound of $\widehat{B_k}(\eta)$:

$$|\widehat{B_k}(\eta)|^2 < O(\frac{1}{\lambda^2_{\eta,k}\mu^2_{\eta,k}}).$$

Now one can conclude that $B_k(\eta[\alpha]_1)$ is Fourier concentrated.

A character χ_β is τ-heavy for C_α if and only if χ_β is τ-heavy for C'_α. Therefore, according to the discussion in FGPS, for a threshold $\tau > 0$, the τ-heavy characters of C_α belong to the set

$$\Gamma_{\alpha,\tau} = \{\chi_\beta : \beta = \eta[\alpha]_1 \text{ for } \eta \in \Gamma_\tau\},$$

where Γ_τ is a set containing the τ-heavy coefficients of the function B_k. For each $\eta \in \Gamma_\tau$, there exists a unique integer pair $(\xi_\eta, \varsigma_\eta) \in [0, 1/\tau] \times [0, 1/\tau]$. Note that by [18, Lemma 9], the size of Γ_τ is at most $4\tau^{-2}$.

Recoverability. The proof for recoverability is similar to those of [10,18]. According to Lemma 1, we know that there exists a threshold τ which is polynomial in the non-negligible quantity ϵ and at least one τ-heavy Fourier character $\chi \neq 0$ for C_α and $\widetilde{C}_\alpha$ such that $\chi \in \mathsf{Heavy}_\tau(C_\alpha) \cap \mathsf{Heavy}_\tau(\widetilde{C}_\alpha)$.

Given a polynomial $h(x) \in I_2(p)$, on input $g, g^a, g^b \in \mathbb{F}_{p^2}$, the following algorithm that has access to $\mathcal{O}$ produces a polynomial size list of elements in $\mathbb{F}_{p^2}$ which contains g^{ab} with probability $1 - \delta$.

Let τ be the threshold determined by Lemma 1. We write $\alpha = [\alpha]_1 x + [\alpha]_0$ to denote $g^{ab} \in \mathbb{F}_{p^2}$. Using the learning algorithm of AGS [1] (i.e., the algorithm in Lemma 2), we obtain a polynomial size list L_α of all the τ-heavy Fourier characters for $\widetilde{C}_\alpha$. If χ_β is a non-trivial τ-heavy character for C_α, we have $[\alpha]_1 = \eta^{-1}\beta$. Given $\chi_\beta \in L_\alpha$, we define $L_\beta = \{[\alpha]_1 : [\alpha]_1 = \eta^{-1}\beta \text{ for } \eta \in \Gamma_\tau\}$.

Let $L = \bigcup_{\chi_\beta \in L_\alpha} L_\beta$. Note that L is of polynomial size and $\alpha \in L$ with probability $1 - \delta$. Since this is a polynomial size set, we can guess a result for $[\alpha]_1$ and hence get $[g^{ab}]_1$. The theorem now follows.   ∎

HARD-CORE PREDICATES FOR THE CDH PROBLEM OVER $\mathbb{F}_{p^2}$. Note that for a given $h \in I_2(p)$, any element $\alpha \in \mathbb{F}_{p^2}$ of length $2l$ can be written as $[\alpha]_1 x + [\alpha]_0$, i.e., $[\alpha]_1$ and $[\alpha]_0$ are the leftmost and rightmost l bits value of α, respectively. Let $\widehat{B}_k : \mathbb{F}_{p^2} \to \{\pm 1\}$ denote the k-th bit predicate (where $1 \leq k \leq 2l$) and let β_k be the bias of $\widehat{B}_k$. In the following, we prove that given an oracle $\mathcal{O}$ that predicts the k-th bit of the CDH value over a random representation of the field $\mathbb{F}_{p^2}$ with non-negligible advantage, we can solve the *regular* CDH problem over $\mathbb{F}_{p^2}$ with non-negligible probability.

Theorem 3. *Under the CDH assumption over $\mathbb{F}_{p^2}$ (i.e., Assumption 1), for any PPT adversary $\mathcal{O}$, we have that for all $h \in I_2(p)$ the following quantity is negligible in l:*

$$\left| \Pr\left[\mathcal{O}(h, \widehat{h}, g, g^a, g^b) = \widetilde{B}_k(\phi_{h,\widehat{h}}(g^{ab})) \middle| \widehat{h} \xleftarrow{\$} I_2(p); a, b \xleftarrow{\$} \{1, \cdots, p^2 - 1\}\right] - \beta_k \right|.$$

Proof Sketch: For an element $\alpha \in \mathbb{F}_{p^2}$ and a given $h \in I_2(p)$, we define a codeword as follows: $C_\alpha(\widehat{h}) = \widetilde{B}_k(\phi_{h,\widehat{h}}(\alpha))$. If $k \leq l$, we have $\widetilde{B}_k(\phi_{h,\widehat{h}}(\alpha)) = B_k([\phi_{h,\widehat{h}}(\alpha)]_0)$. Otherwise if $k > l$, we have $\widetilde{B}_k(\phi_{h,\widehat{h}}(\alpha)) = B_{k-l}([\phi_{h,\widehat{h}}(\alpha)]_1)$. Along the same lines as the proofs of [10, Theorem 5.2] and Theorem 2, predicting any individual bit of the secret CDH value defined above can break the Partial-CDH assumption over $\mathbb{F}_{p^2}$, and hence break the CDH assumption over $\mathbb{F}_{p^2}$, as shown in Theorem 1. ∎

4 Almost All Bits Security of the CDH Problems over $\mathbb{F}_{p^t}$ for $t > 1$

4.1 Hardness of the d-th CDH Assumption over $\mathbb{F}_{p^t}$

We begin with the definition of the d-th CDH problem over $\mathbb{F}_{p^t}$. For a given prime p, there are many different fields $\mathbb{F}_{p^t}$, but they are all isomorphic to each other. Let $h(x) = x^t + h_{t-1}x^{t-1} + \cdots + h_1 x + h_0$ be a monic irreducible polynomial of degree t in $\mathbb{F}_p$. It is well known that $\mathbb{F}_{p^t}$ is isomorphic to the field $\mathbb{F}_p[x]/(h)$, where $(h(x))$ is a principal ideal in the polynomial ring $\mathbb{F}_p[x]$ and therefore elements of $\mathbb{F}_{p^t}$ can be written as polynomials of degree $t - 1$, *i.e.*, if $g \in \mathbb{F}_{p^t}$ then $g = g_{t-1}x^{t-1} + g_{t-2}x^{t-2} + \cdots + g_1 x + g_0$ and addition and multiplication are performed as polynomial operations modulo h. In the following, given $g \in \mathbb{F}_{p^t}$ we denote by $[g]_i$ the coefficient of the degree-i term, *i.e.*, $g_i = [g]_i$. Let $I_t(p)$ be the set of monic irreducible polynomials of degree t in $\mathbb{F}_p$, and let g be a generator of the multiplicative group of $\mathbb{F}_{p^t}$. First, the CDH problem can be easily extended to the case of finite fields $\mathbb{F}_{p^t}$ for $t > 1$.

Assumption 3 (The CDH assumption over $\mathbb{F}_{p^t}$). *We say that the CDH problem is hard in $\mathbb{F}_{p^t}$ for $t > 1$ if for any PPT adversary $\mathcal{A}$, his CDH advantage*

$$\mathbf{Adv}^{cdh}_{\mathcal{A},\mathbb{F}_{p^t}} := \Pr\left[\mathcal{A}(p,g,g^a,g^b) = g^{ab} \middle| p \xleftarrow{\$} \mathcal{G}(1^l); a,b \xleftarrow{\$} \{1,\cdots,p^t - 1\}\right]$$

is negligible in l.

We say that the d-th CDH problem (where $0 \leq d \leq t - 1$) is hard in $\mathbb{F}_{p^t}$ if for all $h \in I_t(p)$ no efficient algorithm given $g, A = g^a, B = g^b \in \mathbb{F}_{p^t}$ can output $[g^{ab}]_d \in \mathbb{F}_p$. Formally we consider the following assumption:

Assumption 4 (The d-th CDH assumption over $\mathbb{F}_{p^t}$). *We say that the d-th CDH problem (where $0 \leq d \leq t - 1$) is hard in $\mathbb{F}_{p^t}$ (for $t > 1$) if for any PPT adversary $\mathcal{A}$, his d-th CDH advantage for all $h \in I_t(p)$*

$$\mathbf{Adv}^{dcdh}_{\mathcal{A},h,\mathbb{F}_{p^t}} := \Pr\left[\mathcal{A}(p,h,g,g^a,g^b) = [g^{ab}]_d \middle| p \xleftarrow{\$} \mathcal{G}(1^l); a,b \xleftarrow{\$} \{1,\cdots,p^t - 1\}\right]$$

is negligible in l.

It is well known that the probability of a random polynomial $h \in \mathbb{F}_p[X]$ of degree t being irreducible is at least $\frac{1}{2t}$. The following theorem asserts that the regular CDH problem over $\mathbb{F}_{p^t}$ with $t > 1$ can be reduced to *any* d-th CDH problem $(0 \leq d \leq t-1)$ over a random representation of $\mathbb{F}_{p^t}$. Therefore, all the d-th CDH problems over a random representation of finite fields $\mathbb{F}_{p^t}$ for $t > 1$ are as hard as the regular CDH problem over the same fields.

Theorem 4. *Let $\mathbb{F}_{p^t}$ be a finite field of size l and $t > 1$. Suppose $\mathcal{A}$ is a d-th CDH adversary that runs in time at most φ and achieves advantage $\mathbf{Adv}_{\mathcal{A},h,\mathbb{F}_{p^t}}^{\text{dcdh}}$ for a monic polynomial $h \xleftarrow{\$} \mathbb{F}_p[X]$ of degree t and $h \in I_t(p)$. Then there exists a CDH adversary $\mathcal{B}$, constructed from $\mathcal{A}$ in a blackbox manner, that runs in time at most $t\varphi$ plus the time to perform $poly(l)$ group operations and achieves advantage $\mathbf{Adv}_{\mathcal{B},\mathbb{F}_{p^t}}^{\text{cdh}} \geq (1 - \frac{1}{p})^t \cdot e^{-\frac{2}{p-1}} \cdot (\mathbf{Adv}_{\mathcal{A},h,\mathbb{F}_{p^t}}^{\text{dcdh}})^t$.*

Before proceeding to the proof, we introduce a useful lemma, which claims that if all the entries in a square matrix are independently and uniformly chosen at random over a large finite field $\mathbb{F}_p$ then there is a good chance that the matrix is nonsingular. Note that we require that the probability depends only on the size of the finite field p, but not on the size of the matrix m. The proof of the lemma is fairly easy and can be found in our full paper [23].

Lemma 5. *Let M be an $m \times m$ square matrix over the finite field $\mathbb{F}_p$. If every element of the matrix is chosen independently and uniformly at random, then the probability that M is nonsingular is at least $e^{-\frac{2}{p-1}}$.*

Proof of Theorem 4: Let $h(x) = x^t + h_{t-1}x^{t-1} + \cdots + h_x + h_0$ be the irreducible polynomial of degree t over $\mathbb{F}_p$, where its coefficients being uniformly and independently selected at random.

Given a challenge instance $(g^a, g^b) \in (\mathbb{F}_{p^t})^2$ of the CDH problem, our CDH adversary $\mathcal{B}$ works as follows. First, adversary $\mathcal{B}$ chooses t pairs of integers $(r_\iota, s_\iota) \xleftarrow{\$} (\mathbb{Z}_{p^t-1})^2$ $(\iota = 0, 1, \cdots, t-1)$, and computes $(g^{a+r_\iota}, g^{b+s_\iota})$. For brevity, let $A_\iota = a + r_\iota$ and $B_\iota = b + s_\iota$ for $\iota = 0, 1, \cdots, t-1$. Adversary $\mathcal{B}$ runs the d-th CDH problem under the representation determined by $h(x)$ on each $(g^{A_\iota}, g^{B_\iota})$ to get the d-th coordinate of the CDH value $[g^{A_\iota B_\iota}]_d$ $(\iota = 0, 1, \cdots, t-1)$.

Adversary $\mathcal{B}$ computes $g^{as_\iota + br_\iota + r_\iota s_\iota} = (g^a)^{s_\iota}(g^b)^{r_\iota}g^{r_\iota s_\iota}$. Let $C_\iota = as_\iota + br_\iota + r_\iota s_\iota$. It is easy to see that $g^{A_\iota B_\iota} = g^{ab}g^{C_\iota} \bmod h(x)$, i.e.,

$$\sum_{k=0}^{t-1}[g^{A_\iota B_\iota}]_k x^k \equiv \left(\sum_{i=0}^{t-1}[g^{ab}]_i x^i\right)\left(\sum_{j=0}^{t-1}[g^{C_\iota}]_j x^j\right) \bmod h(x).$$

Therefore $[g^{A_\iota B_\iota}]_d$ can be written as a linear expression with the coordinates of g^{ab} being variables and with some known coefficients $e_{\iota\nu} \in \mathbb{F}_p$ $(0 \leq \iota, \nu \leq t-1)$ such that

$$[g^{A_\iota B_\iota}]_d = \sum_{\nu=0}^{t-1} e_{\iota\nu}[g^{ab}]_\nu, \quad \iota = 0, 1, \cdots, t-1.$$

If the coefficient matrix $(e_{\iota\nu})_{t\times t}$ for the above equation set has full rank, adversary $\mathcal{B}$ can use Gaussian elimination to compute the unknowns and therefore obtain g^{ab}, in polynomial time of l.

Indeed, we can show (with the proof in our full paper [23]) that the probability of every element of the coefficient matrix $(e_{\iota\nu})_{t\times t}$ being chosen independently and uniformly at random is at least $(1-\frac{1}{p})^t$, and then according to Lemma 5 we know that the probability of the coefficient matrix being nonsingular is at least $(1-\frac{1}{p})^t \cdot e^{-\frac{2}{p-1}}$.

Therefore, running adversary $\mathcal{A}$ for t times and solving the equation set obtained, adversary $\mathcal{B}$ can compute the desired CDH value, that runs in time at most $t\varphi$ plus the time to perform $poly(l)$ group operations with a non-negligible advantage $(1-\frac{1}{p})^t \cdot e^{-\frac{2}{p-1}} \cdot (\mathbf{Adv}^{\mathrm{dcdh}}_{\mathcal{A},h,\mathbb{F}_{p^t}})^t$. The theorem now follows. ∎

We comment that if an adversary $\mathcal{A}$ can solve the d-th CDH problem over $\mathbb{F}_{p^t}$ with respect to a monic polynomial $h \xleftarrow{\$} \mathbb{F}_p[X]$ of degree t and $h \in I_t(p)$ then we can construct an adversary $\mathcal{B}$ that solves all the d-CDH problems over $\mathbb{F}_{p^t}$ for $0 \leq d \leq t-1$ regarding any $h' \in I_t(p)$. To see this, for $h, h' \in I_t(p)$, we know that there exists an easily computable isomorphism $\phi_{h,h'}: \mathbb{F}_p[x]/(h) \to \mathbb{F}_p[x]/(h')$. When adversary $\mathcal{B}$ learns the CDH value with respect to h, it can easily compute all the d-th coordinates under any representation h'.

Theorem 4 proves a slightly weaker result than that of Theorem 1. In Theorem 1, the reduction works for any $h \in I_2(p)$, but in Theorem 4, it works for a random $h \xleftarrow{\$} \mathbb{F}_p[X]$ of degree t and $h \in I_t(p)$. (It could be the case that there exists some $h \in I_t(p)$ such that some d-th CDH problem might not be equivalent to the CDH problem over $\mathbb{F}_{p^t}$, although we conjecture that these two problems are equivalent with respect to any $h \in I_t(p)$.) However, we are able to prove that the 0-th CDH problem and the $(t-1)$-th CDH problem are both strictly equivalent to the CDH problem with respect to any $h \in I_t(p)$, and we have the following theorem:

Theorem 5. *Let $\mathbb{F}_{p^t}$ be a finite field of size l and $t > 1$. Suppose $\mathcal{A}$ is a 0-th (resp., $(t-1)$-th) CDH adversary that runs in time at most φ and achieves advantage $\mathbf{Adv}^{\mathrm{0cdh}}_{\mathcal{A},h,\mathbb{F}_{p^t}}$ (resp., $\mathbf{Adv}^{(t-1)\mathrm{cdh}}_{\mathcal{A},h,\mathbb{F}_{p^t}}$) for any $h \in I_t(p)$. Then there exists a CDH adversary $\mathcal{B}$, constructed from $\mathcal{A}$ in a blackbox manner, that runs in time at most $t\varphi$ plus the time to perform $poly(l)$ group operations and achieves advantage $\mathbf{Adv}^{\mathrm{cdh}}_{\mathcal{B},\mathbb{F}_{p^t}} \geq e^{-\frac{2}{p-1}} \cdot (\mathbf{Adv}^{\mathrm{0cdh}}_{\mathcal{A},h,\mathbb{F}_{p^t}})^t$ (resp., $\mathbf{Adv}^{\mathrm{cdh}}_{\mathcal{B},\mathbb{F}_{p^t}} \geq e^{-\frac{2}{p-1}} \cdot (\mathbf{Adv}^{(t-1)\mathrm{cdh}}_{\mathcal{A},h,\mathbb{F}_{p^t}})^t$).*

THE CASE OF PERFECT ORACLES. Applying our approach to the case of perfect oracles, our reduction leads to no security loss and a strict equivalence result. This is in contrast to Verheul's [22], where for many d's, the algorithm can easily have exponential running time in t.

4.2   Bit Security of the CDH Problem over $\mathbb{F}_{p^t}$

We now show the following result: assuming the hardness of the d-th CDH problem over $\mathbb{F}_{p^t}$ with $t > 1$, if $d \neq 0$, we prove the unpredictability of every single bit of the degree-d coordinate of the secret CDH value. Together with the equivalence result, this implies that for the conventional CDH problems over $\mathbb{F}_{p^t}$ for an l-bit prime p and an integer $t > 1$, $(t-1)l$ out of tl secret CDH bits—including every individual bit except that of the degree 0 coordinate—are hard-core.

We begin with the definition of d-th residues modulo p. Let p be a prime and d be an integer. We say that an element $\alpha \in \mathbb{F}_p^*$ is a d-th residue modulo p, if there exists an element $x \in \mathbb{F}_p$ such that $x^d \equiv \alpha \bmod p$. Let $\mathbb{F}_p^d$ denote the set of the d-th residues modulo p. The following lemma provides a well-known result on d-th residues modulo p:

Lemma 6. *Let p be a prime and $d \in \mathbb{Z}_+$. The number of the d-th residues modulo p is $(p-1)/(d, p-1)$.*

We present a lemma that gives a characterization of the isomorphisms between two representations of the fields $\mathbb{F}_{p^t}$. The isomorphisms generalize that of finite fields $\mathbb{F}_{p^2}$ in FGPS to the case of general finite fields $\mathbb{F}_{p^t}$ for any $t > 1$. More importantly, they simplify that of FGPS in the sense we identify a more restrictive class of isomorphisms. This simplicity turns out to be essential to establishing the bit security for general finite fields.

Lemma 7. *For any $h(x) \in I_t(p)$, there exists a unique function $L_h \colon \mathbb{F}_p^* \to I_t(p)$ which takes λ to the polynomial $\widehat{h}_\lambda = L_h(\lambda) = \frac{h(\lambda x)}{\lambda^t}$ such that λ defines an isomorphism from $\mathbb{F}_p[x]/(h)$ to $\mathbb{F}_p[x]/(\widehat{h}_\lambda)$ that sends*

$$\sum_{i=0}^{t}[\alpha]_i x^i \mapsto \sum_{i=0}^{t} \lambda^i [\alpha]_i x^i.$$

Proof: For any $\lambda \in \mathbb{F}_p^*$, let $\widehat{h}_\lambda(x) = \frac{h(\lambda x)}{\lambda^t}$. It is easy to see that $\widehat{h}_\lambda(x)$ is a monic irreducible polynomial over $\mathbb{F}_p$, i.e., $\widehat{h}_\lambda(x) \in I_t(p)$. Hence, there is an isomorphism from $\mathbb{F}_p[x]/(h)$ to $\mathbb{F}_p[x]/(\widehat{h}_\lambda)$. In order to specify a homomorphism ψ from $\mathbb{F}_p[x]/(h)$ to another field J of characteristic p, it is both necessary and sufficient to choose $\psi(x) = y \in J$ such that $h(y) = 0$ in J. The definition of $\widehat{h}_\lambda$ implies that x sends to λx. The lemma now follows.   □

Theorem 6. *Under the d-th CDH assumption over $\mathbb{F}_{p^t}$ for $t > 1$ (i.e., Assumption 4), for any PPT adversary $\mathcal{O}$, if $d \neq 0$, we have that for all $h \in I_t(p)$ the following quantity is negligible:*

$$\left| \Pr\left[\mathcal{O}(h, \lambda, g, g^a, g^b) = B_k\left(\left[\phi_{h, \widehat{h}_\lambda}(g^{ab})\right]_d\right) \mid \lambda \xleftarrow{\$} \mathbb{F}_p^*; a, b \xleftarrow{\$} \{1, \cdots, p^2 - 1\}\right] - \beta_k \right|.$$

Proof: For an element $\alpha \in \mathbb{F}_{p^t}$ and a monic irreducible polynomial $h \in I_t(p)$, $\lambda \xleftarrow{\$} \mathbb{F}_p^*$, the prediction oracle $\mathcal{O}$ gives noisy access to the codeword $B_k(\lambda^d[\alpha]_d)$. Note that when $d \neq 1$ the above code is not *multiplicative*. Again, this would make it hard to prove concentration and recoverability. In order to apply the techniques of [1], we would need noisy access to the multiplication code

$$C_\alpha : \mathbb{F}_p \mapsto \{\pm 1\}, \quad \text{defined as} \quad C_\alpha(\lambda) = B_k(\lambda[\alpha]_d) \quad (\text{extended by } C_\alpha(0) = -1).$$

We construct another oracle $\mathcal{O}'$ that takes as input a base representation $h \in I_t(p)$, a Diffie-Hellman triple $g, g^a, g^b \in \mathbb{F}_{p^t}$, and $\lambda \xleftarrow{\$} \mathbb{F}_p^*$, and returns $\mathcal{O}(h, r_\lambda, g, g^a, g^b)$ if λ is a d-th residue modulo p, where $r_\lambda^d \equiv \lambda \pmod{p}$, otherwise tosses a β_k-biased coin.

Suppose that there exists an oracle $\mathcal{O}$ such that

$$\left| \Pr_{\lambda,a,b} \left[\mathcal{O}(h, \lambda, g, g^a, g^b) = B_k \big(\big[\phi_{h,\widehat{h}_\lambda}(g^{ab}) \big]_d \big) \right] - \beta_k \right| \geq \epsilon \tag{2}$$

where ϵ is a non-negligible quantity. Following the technique in Boneh and Shparlinski [5], we now show that

$$\left| \Pr_{\lambda,a,b} \left[\mathcal{O}'(h, \lambda, g, g^a, g^b) = B_k \big(\lambda \big[g^{ab} \big]_d \big) \right] - \beta_k \right| \geq \epsilon/d.$$

Let $E_{g^{ab}}$ be the event that $\mathcal{O}'(h, \lambda, g, g^a, g^b) = B_k(\lambda[g^{ab}]_d)$. Note that if λ is uniform in $\mathbb{F}_p^d \backslash \{0\}$ then r_λ is uniform in $\mathbb{F}_p^*$. Therefore, we have

$$\Pr[E_{g^{ab}}] = \frac{1}{(d, p-1)} \Pr[E_{g^{ab}} | \lambda \in \mathbb{F}_p^d] + \left(1 - \frac{1}{(d, p-1)}\right) \Pr[E_{g^{ab}} | \lambda \notin \mathbb{F}_p^d] \quad (\text{Lemma 6})$$

$$\geq \frac{1}{(d, p-1)} (\beta_k + \epsilon) + \left(1 - \frac{1}{(d, p-1)}\right) \beta_k \quad (\text{condition (2)})$$

$$= \beta_k + \frac{\epsilon}{(d, p-1)} \geq \beta_k + \frac{\epsilon}{d}.$$

Note that $t > d$ and therefore the above quantity is non-negligible.

Accessibility. The oracle $\mathcal{O}'$ allows us to have access to a corrupted codeword $\widetilde{C}_\alpha$ of the above codeword defined as $\widetilde{C}_\alpha = \mathcal{O}'(h, \lambda, g, g^a, g^b)$. Therefore, if the oracle $\mathcal{O}$ has advantage ϵ then we have $|\Pr[C_\alpha(\lambda) = \widetilde{C}_\alpha(\lambda)]| \geq \beta_k + \epsilon/d$. Accessibility of the code C_α follows.

Concentration. The proof is similar to that of Theorem 2. For a threshold $\tau > 0$, the τ-heavy characters of C_α belong to the set

$$\Gamma_{\alpha,\tau} = \{\chi_\beta : \beta = \lambda[\alpha]_d \text{ for } \lambda \in \Gamma_\tau\},$$

where Γ_τ is a set containing the τ-heavy coefficients of the function B_k. For each $\lambda \in \Gamma_\tau$, there exists a unique integer pair $(\xi_\lambda, \varsigma_\lambda) \in [0, 1/\tau] \times [0, 1/\tau]$. As in Theorem 2, the proof for concentration of the code $C_\alpha(\lambda)$ is now similar to those of [10, 18].

Recoverability. First, by Lemma 1 we know that there exists a threshold τ which is polynomial in the non-negligible quantity ϵ and at least one τ-heavy Fourier character $\chi \neq 0$ for C_α and $\widetilde{C}_\alpha$ such that $\chi \in \mathsf{Heavy}_\tau(C_\alpha) \cap \mathsf{Heavy}_\tau(\widetilde{C}_\alpha)$.

Given a polynomial $h(x) \in I_t(p)$, on input $g, g^a, g^b \in \mathbb{F}_{p^t}$, the following algorithm that has access to $\mathcal{O}$ produces a polynomial size list of elements in $\mathbb{F}_{p^t}$ which contains g^{ab} with probability $1 - \delta$.

Let τ be the threshold determined by Lemma 1. We write $\alpha = \sum_{i=0}^{t-1} [\alpha]_i x^i$ to denote $g^{ab} \in \mathbb{F}_{p^t}$. Again using the learning algorithm of AGS [1], we obtain a polynomial size list L_α of all the τ-heavy Fourier characters for $\widetilde{C}_\alpha$. If χ_β is a non-trivial τ-heavy character for C_α, we have $[\alpha]_d = \lambda^{-1}\beta$. Given $\chi_\beta \in L_\alpha$, we define $L_\beta = \{[\alpha]_d : [\alpha]_d = \lambda^{-1}\beta$ for $\lambda \in \Gamma_\tau\}$.

Let $L = \bigcup_{\chi_\beta \in L_\alpha} L_\beta$, which is a set of polynomial size. Also we have $\alpha \in L$ with probability $1 - \delta$. We can guess a result for $[\alpha]_d$ and hence get $[g^{ab}]_d$. The theorem now follows. $\blacksquare$

DISCUSSION. It is worth mentioning that Theorem 6 proves what is slightly different in concept from those of FGPS and Theorem 2. In FGPS and Theorem 2, it is shown that any bit prediction oracle must have negligible success probability ranging over *all representations*, whereas Theorem 6 shows that the success probability must be negligible ranging over a restricted class. However, in any application, participants would agree upon some representation that they want to use, and therefore our result does not limit its applicability and it is in fact simpler.

Following from Theorems 4 and 6, we obtain the following result: almost all individual bits of the CDH value of the traditional CDH problem over finite fields $\mathbb{F}_{p^t}$ for $t > 1$ are hard-core. We require that the underlying field representation h be chosen uniformly at random (just as the generator g). Formally we have the following theorem:

Theorem 7. *Under the CDH assumption over $\mathbb{F}_{p^t}$ for $t > 1$ (i.e., Assumption 3), for any PPT adversary $\mathcal{O}$, if $d \neq 0$, the following quantity is negligible:*

$$\left| \Pr\left[\mathcal{O}(h, \lambda, g, g^a, g^b) = B_k\left([\phi_{h, \widehat{h}_\lambda}(g^{ab})]_d \right) \middle| h \xleftarrow{\$} \mathbb{F}_p[x] \text{ and } h \in I_t(p); \lambda \xleftarrow{\$} \mathbb{F}_p^*; \right. \right.$$
$$\left. \left. a, b \xleftarrow{\$} \{1, \cdots, p^t - 1\} \right] - \beta_k \right|.$$

Following from Theorems 5 and 6, we have the following theorem which holds for an arbitrary field representation:

Theorem 8. *Under the CDH assumption over $\mathbb{F}_{p^t}$ for $t > 1$ (i.e., Assumption 3), for any PPT adversary $\mathcal{O}$ and any $h \in I_t(p)$; the following quantity is negligible:*

$$\left| \Pr\left[\mathcal{O}(h, \lambda, g, g^a, g^b) = B_k\left([\phi_{h, \widehat{h}_\lambda}(g^{ab})]_{t-1} \right) \middle| \lambda \xleftarrow{\$} \mathbb{F}_p^*; a, b \xleftarrow{\$} \{1, \cdots, p^t - 1\} \right] - \beta_k \right|.$$

5   Conclusion

In this paper, we revisited the d-th CDH problem for any $0 \leq d \leq t - 1$ over finite fields $\mathbb{F}_{p^t}$ for $t > 1$ [20,22]. In contrast to prior work, we considered the most general case of noisy oracles. We proved that all the d-th CDH problems over a random representation of finite fields $\mathbb{F}_{p^t}$ for $t > 1$ are as hard as the regular CDH problem over the same fields. In particular, the 0-th CDH problem and $(t-1)$-th CDH problem given *any* field representation are as hard as the CDH problem. This latter claim applies to the special case of the Partial-CDH and the Dual-Partial CDH problems over $\mathbb{F}_{p^2}$.

We advanced the list decoding approach, and for the first time, we applied it to the case of a non-multiplicative code. We proved that the Partial-CDH problem also admits the hard-core predicates for every individual bit of the other coordinate of the secret CDH value over a random representation of the finite field $\mathbb{F}_{p^2}$. By combining all these, we obtained one of our main results: given an oracle $\mathcal{O}$ that predicts any bit of the CDH value over a random representation of the field $\mathbb{F}_{p^2}$ with non-negligible advantage, we can solve the *regular* CDH problem over $\mathbb{F}_{p^2}$ with non-negligible probability.

We continued to prove that over finite fields $\mathbb{F}_{p^t}$ for any $t > 1$, each d-th CDH problem except $d \neq 0$ admits a large class of hard-core predicates, including every individual bit of d-th coordinate. Hence we proved that almost all bits of the CDH value of the traditional CDH problem over finite fields $\mathbb{F}_{p^t}$ for $t > 1$ are hard-core.

Acknowledgments. Mingqiang Wang is supported by National 973 Grant 2013CB834205 and NSFC Grant 61272035. Haibin Zhang is supported by NSF CNS 1228828. We are greatly indebted to SAC15 and past PC members for their valuable comments and corrections. Many thanks to William E. Skeith for kindly verifying the proofs and for providing insightful corrections, comments, and suggestions. We thank Kai-Min Chung, Alexandre Duc, Matt Franklin, Dimitar Jetchev, Phil Rogaway, Xiaoyun Wang, and Haiyang Xue for helpful comments and discussion.

References

1. Akavia, A., Goldwasser, S., Safra, S.: Proving hard-core predicates using list decoding. In: FOCS, pp. 146–157. IEEE Computer Society (2003)
2. Alexi, W., Chor, B., Goldreich, O., Schnorr, C.: RSA and Rabin functions: certain parts are as hard as the whole. SIAM J. Comput. **17**(2), 194–209 (1988)
3. Ben-Or, M.: Probabilistic algorithms in finite fields. In: FOCS 1981, vol. 11, pp. 394–398 (1981)
4. Blum, M., Micali, S.: How to generate cryptographically strong sequences of pseudorandom bits. SIAM J. Comput. **13**(4), 850–864 (1984)
5. Boneh, D., Shparlinski, I.E.: On the unpredictability of bits of the elliptic curve Diffie-Hellman scheme. In: Kilian, J. (ed.) CRYPTO 2001. LNCS, vol. 2139, pp. 201–212. Springer, Heidelberg (2001)
6. Boneh, D., Venkatesan, R.: Hardness of computing the most significant bits of secret keys in Diffie-Hellman and related schemes. In: Koblitz, N. (ed.) CRYPTO 1996. LNCS, vol. 1109, pp. 129–142. Springer, Heidelberg (1996)
7. Diffie, W., Hellman, M.: New directions in cryptography. IEEE Trans. Inf. Theor. **22**(6), 644–654 (1976)

8. Duc, A., Jetchev, D.: Hardness of computing individual bits for one-way functions on elliptic curves. In: Safavi-Naini, R., Canetti, R. (eds.) CRYPTO 2012. LNCS, vol. 7417, pp. 832–849. Springer, Heidelberg (2012)
9. ElGamal, T.: A public-key cryptosystem, a signature scheme based on discrete logarithms. IEEE Trans. Inf. Theor. **IT–31**(4), 469–472 (1985)
10. Fazio, N., Gennaro, R., Perera, I.M., Skeith III, W.E.: Hard-core predicates for a Diffie-Hellman problem over finite fields. In: Canetti, R., Garay, J.A. (eds.) CRYPTO 2013, Part II. LNCS, vol. 8043, pp. 148–165. Springer, Heidelberg (2013)
11. Galbraith, S.D., Shani, B.: The multivariate hidden number problem. In: Lehmann, A., Wolf, S. (eds.) ICITS 2015. LNCS, vol. 9063, pp. 250–268. Springer, Heidelberg (2015)
12. von Zur Gathen, J., Gerhard, J.: Modern Computer Algebra. Cambridge University Press, Cambridge (1999)
13. Goldreich, O., Levin, L.A.: A hard-core predicate for all one-way functions. In: STOC, pp. 25–32. ACM Press (1989)
14. Goldwasser, S., Micali, S.: Probabilistic encryption. JCSS **28**(2), 270–299 (1984)
15. Håstad, J., Näslund, M.: The security of individual RSA bits. In: FOCS, pp. 510–521 (1998)
16. Joux, A.: A new index calculus algorithm with complexity $L(1/4 + o(1))$ in small characteristic. In: Lange, T., Lauter, K., Lisoněk, P. (eds.) SAC 2013. LNCS, vol. 8282, pp. 355–380. Springer, Heidelberg (2014)
17. Lidl, R., Niederreiter, H.: Finite Fields. Addison-Wesley, Reading (1983)
18. Morillo, P., Ràfols, C.: The security of all bits using list decoding. In: Jarecki, S., Tsudik, G. (eds.) PKC 2009. LNCS, vol. 5443, pp. 15–33. Springer, Heidelberg (2009)
19. Näslund, M.: All bits in $ax + b$ mod p are hard. In: Koblitz, N. (ed.) CRYPTO 1996. LNCS, vol. 1109, pp. 114–128. Springer, Heidelberg (1996)
20. Shparlinski, I.E.: Security of polynomial transformations of the Diffie-Hellman key. Finite Fields Appl. **10**(1), 123–131 (2014)
21. Slinko, A.: A generalization of Komlós's theorem on random matrices. N. Z. J. Math. **30**(1), 81–86 (2001)
22. Verheul, E.R.: Certificates of recoverability with scalable recovery agent security. In: Imai, H., Zheng, Y. (eds.) PKC 2000. LNCS, vol. 1751, pp. 258–275. Springer, Heidelberg (2000)
23. Wang, M., Zhan, T., Zhang, H.: Bit security of the CDH problems over finite fields. Full version, Cryptology ePrint Archive: Report 2014/685. http://eprint.iacr.org

Towards Optimal Bounds for Implicit Factorization Problem

Yao Lu[1,2], Liqiang Peng[1], Rui Zhang[1(✉)], Lei Hu[1], and Dongdai Lin[1]

[1] State Key Laboratory of Information Security (SKLOIS),
Institute of Information Engineering, Chinese Academy of Sciences,
Beijing 100093, China
lywhhit@gmail.com, {pengliqiang,r-zhang,hulei,ddlin}@iie.ac.cn
[2] The University of Tokyo, Tokyo, Japan

Abstract. We propose a new algorithm to solve the Implicit Factorization Problem, which was introduced by May and Ritzenhofen at PKC'09. In 2011, Sarkar and Maitra (IEEE TIT 57(6): 4002–4013, 2011) improved May and Ritzenhofen's results by making use of the technique for solving multivariate approximate common divisors problem. In this paper, based on the observation that the desired root of the equations that derived by Sarkar and Maitra contains large prime factors, which are already determined by some known integers, we develop new techniques to acquire better bounds. We show that our attack is the best among all known attacks, and give experimental results to verify the correctness. Additionally, for the first time, we can experimentally handle the implicit factorization for the case of balanced RSA moduli.

Keywords: Lattices · Implicit factorization problem · Coppersmith's method · LLL algorithm

1 Introduction

The RSA cryptosystem is the most widely used public-key cryptosystem in practice, and its security is closely related to the difficulty of Integer Factorization Problem (IFP): if IFP is solved then RSA is broken. It is conjectured that factoring cannot be solved in polynomial-time without quantum computers.

In Eurocrypt'85, Rivest and Shamir [20] first studied the factoring with known bits problem. They showed that $N = pq$ (p, q is of the same bit size) can be factored given $\frac{2}{3}$-fraction of the bits of p. In 1996, Coppersmith [2] improved [20]'s bound to $\frac{1}{2}$. Note that for the above results, the unknown bits are within one consecutive block. The case of n blocks was later considered in [7,15].

Motivated by the cold boot attack [4], in Crypto'09, Heninger and Shacham [6] considered the case of known bits are uniformly spread over the factors p and q, they presented a polynomial-time attack that works whenever a 0.59-fraction of the bits of p and q is given. As a follow-up work, Henecka et al. [5] focused on the attack scenario that allowed for error correction of secret factors,

© Springer International Publishing Switzerland 2016
O. Dunkelman and L. Keliher (Eds.): SAC 2015, LNCS 9566, pp. 462–476, 2016.
DOI: 10.1007/978-3-319-31301-6_26

which called Noisy Factoring Problem. Later, Kunihiro et al. [12] discussed secret key recovery from noisy secret key sequences with both errors and erasures. Recently, Kunihiro and Honda [11] discussed how to recover RSA secret keys from noisy analog data.

1.1 Implicit Factorization Problem (IFP)

The above works require the knowledge of explicitly knowing bits of secret factor. In PKC'09, May and Ritzenhofen [18] introduced a new factoring problem with implicit information, called Implicit Factorization Problem (IFP). Consider that $N_1 = p_1 q_1, \ldots, N_k = p_k q_k$ be n-bit RSA moduli, where $q_1, \ldots, q_k$ are $\alpha n (\alpha \in (0, 1))$-bit primes: Given the implicit information that $p_1, \ldots, p_k$ share certain portions of bit pattern, under what condition is it possible to factorize $N_1, \ldots, N_k$ efficiently? This problem can be applied in the area of malicious generation of RSA moduli, i.e. the construction of backdoored RSA moduli. Besides, it also helps to understand the complexity of the underlying factorization problem better.

Since then, there have been many cryptanalysis results for this problem [3,14,18,19,21–23]. Sarkar and Maitra [22] developed a new approach, they used the idea of [10], which is for the approximate common divisor problem (ACDP), to solve the IFP, and managed to improve the previous bounds significantly.

We now give a brief review of their method. Suppose that primes $p_1, \ldots, p_k$ share certain amount of most significant bits (MSBs). First, they notice that

$$\gcd(N_1, N_2 + (p_1 - p_2)q_2, \ldots, N_k + (p_1 - p_k)q_k) = p_1$$

Then they try to solve the simultaneous modular univariate linear equations

$$\begin{cases} N_2 + u_2 \equiv 0 \mod p_1 \\ \quad \vdots \\ N_k + u_k \equiv 0 \mod p_1 \end{cases} \tag{1}$$

for some unknown divisor p_1 of known modulus N_1. Note that if the root $(u_2^{(0)}, \ldots, u_k^{(0)}) = ((p_1 - p_2)q_2, \ldots, (p_1 - p_k)q_k)$ is small enough, we can extract them efficiently. In [22], Sarkar and Maitra proposed an algorithm to find the small root of Eq. (1). Recently, Lu et al. [14] performed a more effective analysis by making use of Cohn and Heninger's algorithm [1].

1.2 Our Contributions

In this paper, we present a new algorithm to obtain better bounds for solving the IFP. As far as we are aware, our attack is the best among all known attacks.

Technically, our algorithm is also to find a small root of Eq. (1). Concretely, our improvement is based on the observation that for $2 \le i \le k$, $u_i^{(0)}$ contains a large prime q_i, which is already determined by N_i.

Table 1. Comparison of our generalized bounds against previous bounds

	[18]	[3]	[22]	[14]	[19]	This paper
βn-bit LSBs case ($\beta > \cdot$)	$\frac{k}{k-1}\alpha$	-	$F(\alpha,k)$	$H(\alpha,k)$	$G(\alpha,k)$	$T(\alpha,k)$
γn-bit MSBs case ($\gamma > \cdot$)	-	$\frac{k}{k-1}\alpha + \frac{6}{n}$	$F(\alpha,k)$	$H(\alpha,k)$	$G(\alpha,k)$	$T(\alpha,k)$
γn-bit MSBs and βn-bit LSBs together case ($\gamma + \beta > \cdot$)	-	-	$F(\alpha,k)$	$H(\alpha,k)$	$G(\alpha,k)$	$T(\alpha,k)$
δn-bit in the Middle case ($\delta > \cdot$)	-	$\frac{2k}{k-1}\alpha + \frac{7}{n}$	-	-	-	-

[1] $F(\alpha,k) = \frac{\alpha k^2 - (2\alpha+1)k + 1 + \sqrt{k^2 + 2\alpha^2 k - \alpha^2 k^2 - 2k + 1}}{k^2 - 3k + 2}$

[2] $H(\alpha,k) = 1 - (1-\alpha)^{\frac{k}{k-1}}$

[3] $G(\alpha,k) = \frac{k}{k-1}\left(\alpha - 1 + (1-\alpha)^{\frac{k+1}{k}} + (k+1)(1-(1-\alpha)^{\frac{1}{k}})(1-\alpha)\right)$

[4] $T(\alpha,k) = k(1-\alpha)\left(1 - (1-\alpha)^{\frac{1}{k-1}}\right)$

[5] The symbol "-" means that this corresponding case has not been considered.

Therefore, we separate u_i into two unknown variables x_i and y_i i.e. $u_i = x_i y_i$. Consider the following equations

$$\begin{cases} N_2 + x_2 y_2 \equiv 0 \mod p_1 \\ \quad\vdots \\ N_k + x_k y_k \equiv 0 \mod p_1 \end{cases}$$

with the root $(x_2^{(0)}, \ldots, x_k^{(0)}, y_2^{(0)}, \ldots, y_k^{(0)}) = (q_2, \ldots, q_k, p_1 - p_2, \ldots, p_1 - p_k)$. Then we introduce $k-1$ new variables z_i for the prime factor p_i ($2 \leq i \leq k$), and use the equation $x_i z_i = N_i$ to decrease the determinant of the desired lattice. That is the key reason why we get better results than [22].

In Fig. 1, we give the comparison with previous bounds for the case $k = 2$. In Table 1, we list the comparisons between our generalized bounds and the previous bounds.

Recently in [19], Peng et al. proposed another method for the IFP. Instead of applying Coppersmith's technique directly to the ACDP, Peng et al. utilized the lattice proposed by May and Ritzenhofen [18], and tried to find the coordinate of the desired vector which is not included in the reduced basis, namely they introduced a method to deal with the case when the number of shared bits is not large enough to satisfy the bound in [18].

In this paper, we also investigate Peng et al.'s method [19]. Surprisingly, we get the same result with a different method. In Sect. 5, we give the experimental data for our two methods.

We organize the rest of the paper as follows. In Sect. 2, we review the necessary background for our approaches. In Sect. 3, based on new observations, we present our new analysis on the IFP. In Sect. 4, we revisit Peng et al.'s method [19]. Finally, in Sect. 5, we give the experimental data for the comparison with previous methods.

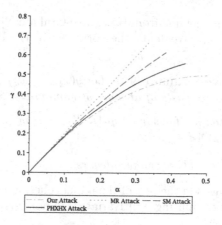

Fig. 1. Comparison with previous bounds on γ with respect to α: $k = 2$. MR Attack denotes May and Ritzenhofen's attack [18], SM Attack denotes Sarkar and Maitra's attack [22], PHXHX Attack denotes Peng et al.'s attack [19].

2 Preliminaries

2.1 Notations

Let $N_1 = p_1q_1, \ldots, N_k = p_kq_k$ be n-bit RSA moduli, where $q_1, \ldots, q_k$ are $\alpha n (\alpha \in (0, 1))$-bit primes. Three cases are considered in this paper, we list them below:

- $p_1, \ldots, p_k$ share βn LSBs where $\beta \in (0, 1)$;
- $p_1, \ldots, p_k$ share γn MSBs where $\gamma \in (0, 1)$;
- $p_1, \ldots, p_k$ share γn MSBs and βn LSBs together where $\gamma \in (0, 1)$ and $\beta \in (0, 1)$;

For simplicity, here we consider αn, βn and γn as integers.

2.2 Lattice

Consider a set of linearly independent vectors $u_1, \ldots, u_w \in \mathbb{Z}^n$, with $w \leqslant n$. The lattice $\mathcal{L}$, spanned by $\{u_1, \ldots, u_w\}$, is the set of all integer linear combinations of the vectors $u_1, \ldots, u_w$. The number w of vectors is the dimension of the lattice. The set $u_1, \ldots, u_w$ is called a basis of $\mathcal{L}$. In lattices with large dimension, finding the shortest vector is a very hard problem, however, approximations of a shortest vector can be obtained in polynomial-time by applying the well-known LLL basis reduction algorithm [13].

Lemma 1 (LLL [13]). *Let $\mathcal{L}$ be a lattice of dimension w. In polynomial-time, the LLL algorithm outputs reduced basis vectors v_i, $1 \leqslant i \leqslant w$ that satisfy*

$$\| v_1 \| \leqslant \| v_2 \| \leqslant \cdots \leqslant \| v_i \| \leqslant 2^{\frac{w(w-1)}{4(w+1-i)}} \det(\mathcal{L})^{\frac{1}{w+1-i}}.$$

We also state a useful lemma from Howgrave-Graham [9]. Let $g(x_1, \ldots, x_k) = \sum_{i_1, \ldots, i_k} a_{i_1, \ldots, i_k} x_1^{i_1} \cdots x_k^{i_k}$. We define the norm of g by the Euclidean norm of its coefficient vector: $\|g\|^2 = \sum_{i_1, \ldots, i_k} a_{i_1, \ldots, i_k}^2$.

Lemma 2 (Howgrave-Graham [9]). *Let* $g(x_1, \ldots, x_k) \in \mathbb{Z}[x_1, \ldots, x_k]$ *be an integer polynomial that consists of at most w monomials. Suppose that*

1. $g(y_1, \ldots, y_k) = 0 \bmod p^m$ *for some* $| y_1 | \leqslant X_1, \ldots, | y_k | \leqslant X_k$ *and*
2. $\| g(x_1 X_1, \ldots, x_k X_k) \| < \frac{p^m}{\sqrt{w}}$

Then $g(y_1, \ldots, y_k) = 0$ *holds over the integers.*

The approach we used in the rest of the paper relies on the following heuristic assumption [7,17] for computing multivariate polynomials.

Assumption 1. *The lattice-based construction in this work yields algebraically independent polynomials, this common roots of these polynomials can be computed using techniques like calculation of the resultants or finding a Gröbner basis.*

Gaussian Heuristic. For a random n-dimensional lattice $\mathcal{L}$ in $\mathbb{R}^n$ [8], the length of the shortest vector λ_1 is expected to be approximately

$$\sqrt{\frac{n}{2\pi e}} \det(\mathcal{L})^{\frac{1}{n}}.$$

In our attack, the low-dimensional lattice we constructed is not a random lattice, however, according to our practical experiments, the length of the first vector of the lattice basis outputted from the L^3 algorithm to that specific lattice is indeed asymptotically close to the Gaussian heuristic, similarly as the assumption says for random lattices. Moreover, the lengths of other vectors in the basis are also asymptotically close to the Gaussian heuristic. Hence, we can roughly estimate the sizes of the unknown coordinate of desired vector in the reduced basis.

3 Our New Analysis for Implicit Factorization

As described in the previous section, we will use the fact the desired common root of the target equations contains large prime factors q_i ($2 \leq i \leq k$) which are already determined by N_i to improve Sarkar-Maitra's results.

3.1 Analysis for Two RSA Moduli: The MSBs Case

Theorem 1. *Let* $N_1 = p_1 q_1, N_2 = p_2 q_2$ *be two different n-bit RSA moduli with αn-bit q_1, q_2 where $\alpha \in (0, 1)$. Suppose that p_1, p_2 share γn MSBs where $\gamma \in (0, 1)$. Then under Assumption 1, N_1 and N_2 can be factored in polynomial-time if*

$$\gamma > 2\alpha(1 - \alpha)$$

Proof. Let $\widetilde{p_2} = p_1 - p_2$. We have $N_1 = p_1 q_1$, $N_2 = p_2 q_2 = p_1 q_2 - \widetilde{p_2} q_2$, and $\gcd(N_1, N_2 + \widetilde{p_2} q_2) = p_1$. Then we want to recover $q_2, \widetilde{p_2}$ from N_1, N_2. We focus on a bivariate polynomial $f(x,y) = N_2 + xy$ with the root $(x^{(0)}, y^{(0)}) = (q_2, \widetilde{p_2})$ modulo p_1. Let $X = N^\alpha, Y = N^{1-\alpha-\gamma}, Z = N^{1-\alpha}$ be the upper bounds of $q_2, \widetilde{p_2}, p_2$. In the following we will use the fact that the small root q_2 is already determined by N_2 to improve Sarkar-Maitra's results.

First let us introduce a new variable z for p_2. We multiply the polynomial $f(x,y)$ by a power z^s for some s that has to be optimized. Additionally, we can replace every occurence of the monomial xz by N_2. Define two integers m and t, let us look at the following collection of trivariate polynomials that all have the root (x_0, y_0) modulo p_1^t.

$$g_k(x, y, z) = z^s f^k N_1^{\max\{t-k, 0\}} \quad \text{for } k = 0, \dots, m$$

For $g_k(x, y, z)$, we replace every occurrence of the monomial xz by N_2 because $N_2 = p_2 q_2$. Therefore, every monomial $x^k y^k z^s (k \geq s)$ with coefficient a_k is transformed into a monomial $x^{k-s} y^k$ with coefficient $a_k N_2^s$. And every monomial $x^k y^k z^s (k < s)$ with coefficient a_k is transformed into a monomial $y^k z^{s-k}$ with coefficient $a_k N_2^k$.

To keep the lattice determinant as small as possible, we try to eliminate the factor of N_2^i in the coefficient of diagonal entry. Since $\gcd(N_1, N_2) = 1$, we only need to multiply the corresponding polynomial with the inverse of N_2^i modulo N_1^t.

Compare to Sarkar-Maitra's lattice, the coefficient vectors $g_k(xX, yY, zZ)$ of our lattice contain less powers of X, which decreases the determinant of the lattice spanned by these vectors, however, on the other hand, the coefficient vectors contain powers of Z, which in turn increases the determinant. Hence, there is a trade-off and one has to optimize the parameter s subject to a minimization of the lattice determinant. That is the key reason why we can get better result than Sarkar-Maitra's results.

We have to find two short vectors in lattice $\mathcal{L}$. Suppose that these two vectors are the coefficient vectors of two trivariate polynomial $f_1(xX, yY, zZ)$ and $f_2(xX, yY, zZ)$. These two polynomials have the root $(q_2, \widetilde{p_2}, p_2)$ over the integers. Then we can eliminate the variable z from these polynomials by setting $z = \frac{N_2}{x}$. Finally, we can extract the desired root $(q_2, \widetilde{p_2})$ from the new two polynomials if these polynomials are algebraically independent. Therefore, our attack relies on Assumption 1.

We are able to confirm Assumption 1 by various experiments later. This shows that our attack works very well in practice.

Now we give the details of the condition for which we can find two sufficiently short vectors in the lattice $\mathcal{L}$. The determinant of the lattice $\mathcal{L}$ is

$$\det(\mathcal{L}) = N_1^{\frac{t(t+1)}{2}} X^{\frac{(m-s)(m-s+1)}{2}} Y^{\frac{m(m+1)}{2}} Z^{\frac{s(s+1)}{2}}$$

The dimension of the lattice is $w = m + 1$.

To get two polynomials sharing the root $q_2, \widetilde{p_2}, p_2$, we get the condition

$$2^{\frac{w(w-1)}{4w}} \det(\mathcal{L})^{\frac{1}{w}} < \frac{p_1^t}{\sqrt{w}}$$

Substituting the values of the $\det(\mathcal{L})$ and neglecting low-order terms, we obtain the new condition

$$\frac{t^2}{2} + \alpha\frac{(m-s)^2}{2} + (1-\alpha-\gamma)\frac{m^2}{2} + (1-\alpha)\frac{s^2}{2} < (1-\alpha)tm$$

Let $t = \tau m, s = \sigma m$. The optimized values of parameters τ and σ are given by

$$\tau = 1 - \alpha \qquad \sigma = \alpha$$

Plugging in this values, we finally end up with the condition

$$\gamma > 2\alpha(1-\alpha)$$

One can refer to Fig. 1 for the comparison with previous theoretical results.

3.2 Extension to k RSA Moduli

In this section, we give an analysis for k ($k > 2$) RSA moduli.

Theorem 2. *Let $N_1 = p_1q_1, \ldots, N_k = p_kq_k$ be k different n-bit RSA moduli with αn-bit $q_1, \ldots, q_k$ where $\alpha \in (0,1)$. Suppose that $p_1, \ldots, p_k$ share γn MSBs where $\gamma \in (0,1)$. Then under Assumption 1, $N_1, \ldots, N_k$ can be factored in polynomial-time if*

$$\gamma > k(1-\alpha)\left(1 - (1-\alpha)^{\frac{1}{k-1}}\right)$$

Proof. Let $\widetilde{p_i} = p_1 - p_i$. We have $N_1 = p_1q_1$ and $N_i = p_iq_i = p_1q_i - \widetilde{p_i}q_i$ ($2 \le i \le k$). We have $\gcd(N_1, N_2 + \widetilde{p_2}q_2, \ldots, N_k + \widetilde{p_k}q_k) = p_1$. Then we want to recover $q_i, \widetilde{p_i}$ ($2 \le i \le k$) from $N_1, \ldots, N_k$. We construct a system of $k-1$ polynomials

$$\begin{cases} f_2(x_2, y_2) = N_2 + x_2y_2 \\ \quad\vdots \\ f_k(x_k, y_k) = N_k + x_ky_k \end{cases}$$

with the root $(x_2^{(0)}, y_2^{(0)}, \ldots, x_k^{(0)}, y_k^{(0)}) = (q_2, \widetilde{p_2}, \ldots, q_k, \widetilde{p_k})$ modulo p_1. Using a technique similar to that of Theorem 1, and introducing $k-1$ new variables z_i for p_i ($2 \le i \le k$), we define the following collection of trivariate polynomials.

$$g_{i_2, \ldots, i_k}(x_2, \ldots, x_k, y_2, \ldots, y_k, z_2, \ldots, z_k) = (z_2 \cdots z_k)^s f_2^{i_2} \cdots f_k^{i_k} N_1^{\max\{t-i_2-\cdots-i_k, 0\}}$$

with $0 \le i_2 + \cdots + i_k \le m$ (Because of the symmetric nature of the unknown variables $x_2, \ldots, x_k$, i.e., all the $x_2, \ldots, x_k$ have the same size, we use the same parameter s).

For $g_{i_2,\ldots,i_k}$, we replace every occurrence of the monomial $x_i z_i$ by N_i. We can eliminate the factor of $N_2^{j_2} \cdots N_k^{j_k}$ in the coefficient of diagonal entry. The determinant of the lattice $\mathcal{L}$ is

$$\det(\mathcal{L}) = N_1^{s_N} \prod_{i=2}^{k} X_i^{s_{X_i}} Y_i^{s_{Y_i}} Z_i^{s_{Z_i}}$$

where

$$s_N = \sum_{j=0}^{t} j \binom{t-j+k-2}{k-2} = \binom{t+k-1}{k-1} \frac{t}{k}$$

$$s_{X_2} = \cdots = s_{X_k} = \sum_{j=0}^{m-s} j \binom{m-s-j+k-2}{k-2} = \binom{m-s+k-1}{k-1} \frac{m-s}{k}$$

$$s_{Y_2} = \cdots = s_{Y_k} = \sum_{j=0}^{m} j \binom{m-j+k-2}{k-2} = \binom{m+k-1}{k-1} \frac{m}{k}$$

$$s_{Z_2} = \cdots = s_{Z_k} = \sum_{j=0}^{s} j \binom{m-s+j+k-2}{k-2}$$

$$= \binom{m+k-1}{k} \frac{ks-m}{m} + \binom{m-s-1+k-1}{k} \frac{k+m-s-1}{m-s-1}$$

Here $X_i = N^{\alpha}, Y_i = N^{1-\alpha-\gamma}, Z_i = N^{1-\alpha}$ are the upper bounds of $q_i, \tilde{p}_i, p_i$. The dimension of the lattice is

$$w = \dim(\mathcal{L}) = \sum_{j=0}^{m} \binom{j+k-2}{j} = \binom{m+k-1}{m}$$

To get $2k-2$ polynomials sharing the root $q_2, \tilde{p}_2, p_2$, we get the condition

$$2^{\frac{w(w-1)}{4(w+4-2k)}} \det(\mathcal{L})^{\frac{1}{w+4-2k}} < \frac{p_1^t}{\sqrt{w}}$$

Substituting the values of the $\det(\mathcal{L})$ and neglecting low-order terms, we obtain the new condition

$$\binom{t+k-1}{k-1} \frac{t}{k} + (k-1)\alpha \binom{m-s+k-1}{k-1} \frac{m-s}{k}$$

$$+ (k-1)(1-\alpha-\gamma) \binom{m+k-1}{k-1} \frac{m}{k} + (k-1)(1-\alpha) \binom{m+k-1}{k} \frac{ks-m}{m}$$

$$+ (k-1)(1-\alpha) \binom{m-s-1+k-1}{k} \frac{k+m-s-1}{m-s-1}$$

$$< (1-\alpha)t \binom{m+k-1}{m}$$

Let $t = \tau m, s = \sigma m$, the optimized values of parameters τ and σ were given by

$$\tau = (1-\alpha)^{\frac{1}{k-1}} \qquad \sigma = 1 - (1-\alpha)^{\frac{1}{k-1}}$$

Plugging in this values, we finally end up with the condition

$$\gamma > k(1-\alpha)\left(1 - (1-\alpha)^{\frac{1}{k-1}}\right)$$

One can refer to Table 1 for the comparison with previous theoretical results.

3.3 Extension to the LSBs Case.

In the following, we show a similar result in the case of $p_1, \ldots, p_k$ share some MSBs and LSBs together. This also takes care of the case when only LSBs are shared.

Theorem 3. *Let $N_1 = p_1 q_1, \ldots, N_k = p_k q_k$ be k different n-bit RSA moduli with αn-bit q_i ($\alpha \in (0,1)$). Suppose that $p_1, \cdots, p_k$ share γn MSBs ($\gamma \in (0,1)$) and βn LSBs ($\beta \in (0,1)$) together. Then under Assumption 1, $N_1, \cdots, N_k$ can be factored in polynomial-time if*

$$\gamma + \beta > k(1-\alpha)\left(1 - (1-\alpha)^{\frac{1}{k-1}}\right)$$

Proof. Suppose that $p_1, \ldots, p_k$ share γn MSBs and βn LSBs together. Then we have the following equations:

$$\begin{cases} p_2 = p_1 + 2^{\beta n}\tilde{p}_2 \\ \quad\vdots \\ p_k = p_1 + 2^{\beta n}\tilde{p}_k \end{cases}$$

We can write as follows

$$N_i q_1 - N_1 q_i = 2^{\beta n}\tilde{p}_i q_1 q_i \quad \text{for } 2 \leq i \leq k$$

Then we get

$$(2^{\beta n})^{-1} N_i q_1 - \tilde{p}_i q_1 q_i \equiv 0 \mod N_1 \quad \text{for } 2 \leq i \leq k$$

Let $A_i \equiv (2^{\beta n})^{-1} N_i \bmod N_1$ for $2 \leq i \leq k$. Thus, we have

$$\begin{cases} A_2 - q_2\tilde{p}_2 \equiv 0 \mod p_1 \\ \quad\vdots \\ A_k - q_k\tilde{p}_k \equiv 0 \mod p_1 \end{cases}$$

Then we can construct a system of $k-1$ polynomials

$$\begin{cases} f_2(x_2, \cdots, x_k) = A_2 + x_2 y_2 \\ \quad\vdots \\ f_k(x_2, \cdots, x_k) = A_k + x_k y_k \end{cases}$$

with the root $(x_2^{(0)}, y_2^{(0)}, \ldots, x_k^{(0)}, y_k^{(0)}) = (q_2, \tilde{p}_2, \ldots, q_k, \tilde{p}_k)$ modulo p_1. The rest of the proof follows s similar technique as in the proof of Theorem 2. We omit the details here.

4 Revisiting Peng et al.'s Method [19]

In [19], Peng et al. gave a new idea for IFP. In this section, we revisit Peng et al.'s method and modify the construction of lattice which is used to solve the homogeneous linear modulo equation. Therefore, a further improved bound on the shared LSBs and MSBs is obtained.

Recall the method proposed by May and Ritzenhofen in [18], the lower bound on the number of shared LSBs has been determined, which can ensure the vector $(q_1, \cdots, q_k)$ is shortest in the lattice, namely the desired factorization can be obtained by lattice basis reduction algorithm.

Peng et al. took into consideration the lattice introduced in [18] and discussed a method which can deal with the case when the number of shared LSBs is not enough to ensure that the desired factorization can be solved by applying reduction algorithms to the lattice. More narrowly, since $(q_1, \cdots, q_k)$ is in the lattice, it can be represented as a linear combination of reduced lattice basis. Hence the problem of finding $(q_1, \cdots, q_k)$ is transformed into solving a homogeneous linear equation with unknown moduli. Peng et al. utilized the result from Herrmann and May [7] to solve the linear modulo equation and obtain a better result.

Firstly, we recall the case of primes shared LSBs. Assume that there are k different n-bit RSA moduli $N_1 = p_1q_1, \cdots, N_k = p_kq_k$, where $p_1, \cdots, p_k$ share γn LSBs and $q_1, \cdots, q_k$ are αn-bit primes. The moduli can be represented as

$$\begin{cases} N_1 = (p + 2^{\gamma n}\widetilde{p_1})q_1 \\ \quad \vdots \\ N_k = (p + 2^{\gamma n}\widetilde{p_k})q_k \end{cases}$$

Furthermore, we can get following modular equations

$$\begin{cases} N_1^{-1}N_2q_1 - q_2 \equiv 0 \mod 2^{\gamma n} \\ \quad \vdots \\ N_1^{-1}N_kq_1 - q_k \equiv 0 \mod 2^{\gamma n} \end{cases} \tag{2}$$

In [18], May and Ritzenhofen introduced a k-dimensional lattice $\mathcal{L}_1$ which is generated by the row vectors of following matrix

$$\begin{pmatrix} 1 & N_1^{-1}N_2 & N_1^{-1}N_3 & \cdots & N_1^{-1}N_k \\ 0 & 2^{\gamma n} & 0 & \cdots & 0 \\ 0 & 0 & 2^{\gamma n} & \cdots & 0 \\ \vdots & \vdots & \vdots & \ddots & \vdots \\ 0 & 0 & 0 & \cdots & 2^{\gamma n} \end{pmatrix}.$$

Since (2) holds, the vector $(q_1, \cdots, q_k)$ is the shortest vector in $\mathcal{L}_1$ with a good probability when $\gamma \geq \frac{k}{k-1}\alpha$. Then by applying the LLL reduction algorithm to the lattice, the vector $(q_1, \cdots, q_k)$ can be solved. Conversely, when $\gamma < \frac{k}{k-1}\alpha$ the reduced basis $(\lambda_1, \cdots, \lambda_k)$ doesn't contain vector $(q_1, \cdots, q_k)$, nevertheless, we can represent the vector $(q_1, \cdots, q_k)$ as a linear combination of reduced basis. Namely, there exist integers $x_1, x_2, \cdots, x_k$ such that $(q_1, \cdots, q_k) = x_1\lambda_1 + \cdots + x_k\lambda_k$. Moreover, the following system of modular equations can be obtained,

$$\begin{cases} x_1l_{11} + x_2l_{21} + \cdots + x_kl_{k1} = q_1 \equiv 0 \mod q_1 \\ \qquad\qquad \vdots \\ x_1l_{1k} + x_2l_{2k} + \cdots + x_kl_{kk} = q_k \equiv 0 \mod q_k \end{cases} \tag{3}$$

where $\lambda_i = (l_{i1}, l_{i2}, \cdots, l_{ik})$, $i = 1, 2, \cdots, k$.

Based on the experiments, the size of the reduced basis can be roughly estimated as Gaussian heuristic. We estimate the length of λ_i and the size of l_{ij} as $\det(L_2)^{\frac{1}{k}} = 2^{\frac{nt(k-1)}{k}}$, hence the solution of (3) is $|x_i| \approx \frac{q_i}{kl_{ij}} \approx 2^{\alpha n - \frac{nt(k-1)}{k} - \log_2 k} \leq 2^{\alpha n - \frac{nt(k-1)}{k}}$.

Then using the Chinese Remainder Theorem, from (3) we can get the following homogeneous modular equation

$$a_1x_1 + a_2x_2 + \cdots + a_kx_k \equiv 0 \mod q_1q_2\cdots q_k \tag{4}$$

where a_i is an integer satisfying $a_i \equiv l_{ij} \mod N_j$ for $1 \leq j \leq k$ and it can be calculated from the l_{ij} and N_j.

For this linear modular equation, Peng et al. directly utilized the method of Herrmann and May [7] to solve it and obtain that when

$$\gamma \geq \frac{k}{k-1}(\alpha - 1 + (1-\alpha)^{\frac{k+1}{k}} + (k+1)(1 - (1-\alpha)^{\frac{1}{k}})(1-\alpha)$$

the desired solution can be solved.

In this paper, we notice that the linear modular equation is homogeneous which is a variant of Herrmann and May's equation, hence we utilize the following theorem which is proposed by Lu et al. in [16] to modify the construction of lattice used in [19].

Theorem 4. *Let N be a sufficiently large composite integer (of unknown factorization) with a divisor p ($p \geq N^\beta$). Furthermore, let $f(x_1, \ldots, x_n) \in \mathbb{Z}[x_1, \ldots, x_n]$ be a homogenous linear polynomial in $n(n \geq 2)$ variables. Under Assumption 1, we can find all the solutions $(y_1, \ldots, y_n)$ of the equation $f(x_1, \ldots, x_n) = 0 \pmod{p}$ with $\gcd(y_1, \ldots, y_n) = 1$, $|y_1| \leq N^{\gamma_1}, \ldots |y_n| \leq N^{\gamma_n}$ if*

$$\sum_{i=1}^n \gamma_i < \left(1 - (1-\beta)^{\frac{n}{n-1}} - n(1-\beta)\left(1 - \sqrt[n-1]{1-\beta}\right)\right)$$

The running time of the algorithm is polynomial in $\log N$ but exponential in n.

For this homogeneous linear Eq. (4) in k variables modulo $q_1q_2\cdots q_k \approx (N_1N_2\cdots N_k)^\alpha$, by Theorem 4 with the variables $x_i < (N_1N_2\cdots N_k)^{\delta_i} \approx 2^{k\delta_i n}$, $i = 1, 2, \cdots, k$, we can solve the variables when

$$\sum_{i=1}^k \delta_i \approx k\delta_i \leq 1 - (1-\alpha)^{\frac{k}{k-1}} - k(1-\alpha)\left(1 - (1-\alpha)^{\frac{1}{k-1}}\right)$$

where $\delta_1 \approx \delta_2 \approx \cdots \approx \delta_k$.

Hence, when

$$\alpha - \frac{\gamma(k-1)}{k} \leq 1 - (1-\alpha)^{\frac{k}{k-1}} - k(1-\alpha)\left(1 - (1-\alpha)^{\frac{1}{k-1}}\right)$$

Namely,

$$\gamma \geq \frac{k}{k-1}\left(\alpha - 1 + (1-\alpha)^{\frac{k}{k-1}} + k(1-(1-\alpha)^{\frac{1}{k-1}})(1-\alpha)\right)$$
$$= k(1-\alpha)\left(1 - (1-\alpha)^{\frac{1}{k-1}}\right)$$

the desired vector can be found out.

The above result can be easily extend to MSBs case using the technique in [19]. Surprisingly we get the same result as Theorem 2 by modifying Peng et al.'s technique.

5 Experimental Results

We implemented our analysis in Magma 2.20 computer algebra system on a PC with Intel(R) Core(TM) Duo CPU(2.80 GHz, 2.16 GB RAM Windows 7). Note that for the first time, we can experimentally handle the IFP for the case of balanced RSA moduli. The column theo. denotes the asymptotic bound of shared bits when the dimension is infinite and the column expt. denotes the best experimental results for a fixed dimension of our constructed lattice. Since the method of [22] can not deal with the case of balanced RSA moduli, we use '-' to fill the Table 2. Moreover, [19] showed that they can obtain an theoretical bound when p and q are balanced, however, they failed to obtain the experimental results, thus we also use '-' to fill the Table 3. All of the running time of the experiments are measured in seconds.

Table 2. Theoretical and Experimental data of the number of shared MSBs in [22] and shared MSBs in Our Method in Sect. 3

k	Bitsize of (p_i, q_i), i.e., $((1-\alpha)\log_2 N_i, \alpha\log_2 N_i)$	No. of shared MSBs in p_i [22]				No. of shared MSBs in p_i (Sect.3)				
		Theo.	Expt.	Dim	Time of L^3	Theo.	Expt.	(m,t,s)	Dim	Time of L^3
2	(874,150)	278	289	16	1.38	257	265	(45,38,6)	46	2822.152
2	(824,200)	361	372	16	1.51	322	330	(45,36,9)	46	2075.406
2	(774,250)	439	453	16	1.78	378	390	(45,34,11)	46	1655.873
2	(724,300)	513	527	16	2.14	425	435	(45,32,13)	46	1282.422
3	(774,250)	352	375	56	51.04	304	335	(13,11,1)	105	11626.084
3	(724,300)	417	441	56	70.55	346	375	(13,11,2)	105	10060.380
3	(674,350)	480	505	56	87.18	382	420	(13,11,2)	105	14614.033
3	(624,400)	540	569	56	117.14	411	435	(13,10,3)	105	5368.806
3	(512,512)	-	-	-	-	450	460	(13,9,4)	105	2012.803

We present some numerical values for comparisons between our method of Sect. 3 and [22]'s method in Table 2. The running time of LLL algorithm depends on the lattice dimension and bit-size of the entries in lattice, and the largest coefficient of entries in lattice has a bit-size of at most $t \log(N_1)$. Thus the running time is decided by parameters m and t, that explains why the time is reduced as p and q get more balanced. For the case $k = 2$, when the bitlength of q increases, namely α increases, the optimal value of t decreases. Thus, the running time of LLL algorithm is reduced when α increased which means p and q get more balanced.

Table 3. Theoretical and Experimental data of the number of shared MSBs in [19] and shared MSBs in Our Method in Sect. 4

k	Bitsize of (p_i, q_i), i.e., $((1-\alpha)\log_2 N_i, \alpha\log_2 N_i)$	No. of shared MSBs in p_i [19]				No. of shared MSBs in p_i (Sect. 4)				
		Theo.	Expt.	dim	Time of L^3	theo.	Expt.	(m,t)	Dim	Time of L^3
2	(874,150)	267	278	190	1880.10	257	265	(45,7)	46	410.095
2	(824,200)	340	357	190	1899.21	322	335	(45,9)	46	470.827
2	(774,250)	405	412	190	2814.84	378	390	(45,11)	46	918.269
2	(724,300)	461	470	190	2964.74	425	440	(45,13)	46	1175.046
3	(774,250)	311	343	220	6773.48	304	335	(13,2)	105	4539.301
3	(724,300)	356	395	220	7510.86	346	380	(13,2)	105	8685.777
3	(674,350)	395	442	220	8403.91	382	420	(13,2)	105	10133.233
3	(624,400)	428	483	220	9244.42	410	435	(13,3)	105	22733.589
3	(512,512)	476	-	-	-	450	490	(13,4)	105	49424.252

Note that in the practical experiments, we always found many integer equations which share desired roots over the integers when the numbers of shared bits is greater than the listed results. It means that in the reduced basis, there are several vectors that satisfy Howgrave-Graham's bound. Moreover, the more integer equations corresponding to the vectors we choose, the less time calculating Gröbner basis. For an instance, when $k = 3$, $(m,t,s) = (13,9,4)$ and the bitlengths of p and q are both 512-bits, we constructed a 105-dimensional lattice and by applying the L^3 algorithm to the lattice, we successfully collected 74 polynomial equations which share desired roots over the integers when q_1, q_2, q_3 shared 460 MSBs. When we chose all of integer equations, the calculation of Gröbner basis took 12.839 s.

Meanwhile our method of Sect. 4 is based on an improved method of [19], we present some numerical values for comparison with these two methods in Table 3. As it is shown, by using an improved method to solve the homogeneous equations, we obtained an improved bound on the numbers of shared bits and the experiments also showed this improvement. For a fixed dimension of lattice, similarly since entries of our constructed lattice is decided by m and t, the running time of LLL algorithm increases when t increases.

Note that the running time of the method of Sect. 3 is faster than the method of Sect. 4 when p and q get more balanced, especially for balanced moduli. For the unbalanced case, the method of Sect. 4 is faster.

Acknowledgments. We would like to thank the anonymous reviewers for helpful comments. Y. Lu was supported by CREST, JST. Part of this work was also supported by Strategic Priority Research Program of the Chinese Academy of Sciences (No. XDA06010703, No. XDA06010701 and No. XDA06010702), the National Key Basic Research Project of China (No. 2011CB302400 and No. 2013CB834203), and National Science Foundation of China (No. 61379139 and No. 61472417).

References

1. Cohn, H., Heninger, N.: Approximate common divisors via lattices. In: ANTS X (2012)
2. Coppersmith, D.: Small solutions to polynomial equations, and low exponent RSA vulnerabilities. J. Cryptology **10**(4), 233–260 (1997)
3. Faugère, J.-C., Marinier, R., Renault, G.: Implicit factoring with shared most significant and middle bits. In: Nguyen, P.Q., Pointcheval, D. (eds.) PKC 2010. LNCS, vol. 6056, pp. 70–87. Springer, Heidelberg (2010)
4. Halderman, J.A., Schoen, S.D., Heninger, N., Clarkson, W., Paul, W., Calandrino, J.A., Feldman, A.J., Appelbaum, J., Felten, E.W.: Lest we remember: cold-boot attacks on encryption keys. Commun. ACM **52**(5), 91–98 (2009)
5. Henecka, W., May, A., Meurer, A.: Correcting errors in RSA private keys. In: Rabin, T. (ed.) CRYPTO 2010. LNCS, vol. 6223, pp. 351–369. Springer, Heidelberg (2010)
6. Heninger, N., Shacham, H.: Reconstructing RSA private keys from random key bits. In: Halevi, S. (ed.) CRYPTO 2009. LNCS, vol. 5677, pp. 1–17. Springer, Heidelberg (2009)
7. Herrmann, M., May, A.: Solving linear equations modulo divisors: on factoring given any bits. In: Pieprzyk, J. (ed.) ASIACRYPT 2008. LNCS, vol. 5350, pp. 406–424. Springer, Heidelberg (2008)
8. Hoffstein, J., Pipher, J., Silverman, J.H.: An Introduction to Mathematical Cryptography. Springer, New York (2008)
9. Howgrave-Graham, N.: Finding small roots of univariate modular equations revisited. In: Darnell, M.J. (ed.) Cryptography and Coding 1997. LNCS, vol. 1355, pp. 131–142. Springer, Heidelberg (1997)
10. Howgrave-Graham, N.: Approximate integer common divisors. In: Silverman, J.H. (ed.) CaLC 2001. LNCS, vol. 2146, p. 51. Springer, Heidelberg (2001)
11. Kunihiro, N., Honda, J.: RSA meets DPA: recovering RSA secret keys from noisy analog data. In: Batina, L., Robshaw, M. (eds.) CHES 2014. LNCS, vol. 8731, pp. 261–278. Springer, Heidelberg (2014)
12. Kunihiro, N., Shinohara, N., Izu, T.: Recovering RSA secret keys from noisy key bits with erasures and errors. In: Kurosawa, K., Hanaoka, G. (eds.) PKC 2013. LNCS, vol. 7778, pp. 180–197. Springer, Heidelberg (2013)
13. Lenstra, A.K., Lenstra, H.W., Lovász, L.: Factoring polynomials with rational coefficients. Math. Ann. **261**(4), 515–534 (1982)
14. Lu, Y., Zhang, R., Lin, D.: Improved bounds for the implicit factorization problem. Adv. Math. Comm. **7**(3), 243–251 (2013)
15. Lu, Y., Zhang, R., Lin, D.: Factoring multi-power RSA modulus $N = p^r q$ with partial known bits. In: Boyd, C., Simpson, L. (eds.) ACISP. LNCS, vol. 7959, pp. 57–71. Springer, Heidelberg (2013)

16. Lu, Y., Zhang, R., Peng, L., Lin, D.: Solving linear equations modulo unknown divisors: revisited. In: Iwata, T., Cheon, J.H. (eds.) ASIACRYPT 2015. LNCS, vol. 9452, pp. 189–213. Springer, Heidelberg (2015). doi:10.1007/978-3-662-48797-6_9
17. A. May: New RSA vulnerabilities using lattice reduction methods. Ph.D. thesis (2003)
18. May, A., Ritzenhofen, M.: Implicit factoring: on polynomial time factoring given only an implicit hint. In: Jarecki, S., Tsudik, G. (eds.) PKC 2009. LNCS, vol. 5443, pp. 1–14. Springer, Heidelberg (2009)
19. Peng, L., Hu, L., Xu, J., Huang, Z., Xie, Y.: Further improvement of factoring RSA moduli with implicit hint. In: Pointcheval, D., Vergnaud, D. (eds.) AFRICACRYPT. LNCS, vol. 8469, pp. 165–177. Springer, Heidelberg (2014)
20. Rivest, R.L., Shamir, A.: Efficient factoring based on partial information. In: Pichler, F. (ed.) EUROCRYPT 1985. LNCS, vol. 219, pp. 31–34. Springer, Heidelberg (1986)
21. Sarkar, S., Maitra, S.: Further results on implicit factoring in polynomial time. Adv. in Math. of Comm. 3(2), 205–217 (2009)
22. Sarkar, S., Maitra, S.: Approximate integer common divisor problem relates to implicit factorization. IEEE Trans. Inf. Theo. 57(6), 4002–4013 (2011)
23. Sarkar, S., Maitra, S.: Some applications of lattice based root finding techniques. Adv. in Math. of Comm. 4(4), 519–531 (2010)

Cryptanalysis of Authenticated Encryption Schemes

Forgery Attacks on Round-Reduced ICEPOLE-128

Christoph Dobraunig[✉], Maria Eichlseder, and Florian Mendel

IAIK, Graz University of Technology, Graz, Austria
christoph.dobraunig@iaik.tugraz.at

Abstract. ICEPOLE is a family of authenticated encryptions schemes submitted to the ongoing CAESAR competition and in addition presented at CHES 2014. To justify the use of ICEPOLE, or to point out potential weaknesses, third-party cryptanalysis is needed. In this work, we evaluate the resistance of ICEPOLE-128 against forgery attacks. By using differential cryptanalysis, we are able to create forgeries from a known ciphertext-tag pair with a probability of $2^{-60.3}$ for a round-reduced version of ICEPOLE-128, where the last permutation is reduced to 4 (out of 6) rounds. This is a noticeable advantage compared to simply guessing the right tag, which works with a probability of 2^{-128}. As far as we know, this is the first published attack in a nonce-respecting setting on round-reduced versions of ICEPOLE-128.

Keywords: CAESAR · ICEPOLE · Forgery · Differential cryptanalysis

1 Introduction

ICEPOLE is a family of authenticated encryption schemes, which has been presented at CHES 2014 [17] and submitted to CAESAR [16]. CAESAR [18] is an open cryptographic competition aiming to find a suitable portfolio of authenticated encryption algorithms for many use cases. For the first round, 57 candidates have been submitted. Due to the open nature of CAESAR, those candidates have different design goals ranging from high-speed software designs to designs suitable for compact hardware implementations. This makes comparison of the submitted ciphers difficult, which is nevertheless necessary to determine the ciphers for the next rounds. However, all designs have one goal in common: security. Thus, as much security analysis as possible is needed to sort out weak CAESAR candidates and get insight in the security of the others.

The goal of authenticated encryption is to provide confidentiality and authenticity. Our attacks focus solely on the authenticity in a forgery attack. The goal is to manipulate known ciphertext-tag pairs in a way such that they are valid with a certain probability. For ICEPOLE-128, the intended number of bits of security with respect to authenticity is 128. Therefore we consider only attacks with a success probability above the generic 2^{-128} applicable.

© Springer International Publishing Switzerland 2016
O. Dunkelman and L. Keliher (Eds.): SAC 2015, LNCS 9566, pp. 479–492, 2016.
DOI: 10.1007/978-3-319-31301-6_27

The method of choice for our attacks is differential cryptanalysis [7]. To create forgeries, we need differential characteristics which hold with a high probability and fulfill certain constraints explained later. By using this technique, we are able to attack versions of ICEPOLE-128 where the last permutation is reduced to 4 out of 6 rounds. As far as we know, no forgery attacks have been performed on round-reduced versions of ICEPOLE. In addition, we have analyzed the main building block of ICEPOLE, its permutation. We were able to improve the results of the designers regarding high-probability characteristics for the ICEPOLE permutation without any additional constrains. Our results are summarized in Table 1. Note that we have verified the forgery for 3 rounds practically by using the reference implementation of ICEPOLE-128 submitted to CAESAR.

Table 1. Results for ICEPOLE.

	Type	Rounds	Probability
ICEPOLE-128	Forgery	3/6	$2^{-14.8}$
		4/6	$2^{-60.3}$
ICEPOLE permutation	Differential characteristic	5	$2^{-104.5}$
		6	$2^{-258.3}$

Related Work. In the submission document [16], the designers bounded the minimum number of active S-boxes in a differential characteristic for the ICEPOLE permutation. They are able to show that for 3 rounds, the minimum number of active S-boxes is 9, and that there are no characteristics with 13 or fewer active S-boxes for 4 rounds. In addition, Morawiecki et al. [16] heuristically searched for differential characteristics. For 5 rounds, their best published differential characteristic has a probability of $2^{-186.2}$, and for 6 rounds $2^{-555.3}$.

Recently, Huang et al. [13] presented state-recovery attacks on ICEPOLE in a nonce-misuse scenario. They show that in this scenario, the internal state of ICEPOLE-128 and ICEPOLE-128a can be recovered with complexity 2^{46}, and the internal state of ICEPOLE-256a with complexity 2^{60}.

Outline. The remainder of the paper is organized as follows. We describe the design of ICEPOLE in Sect. 2. Afterwards, we give a high-level overview about the techniques used to find suitable differential characteristics in Sect. 3, followed by our attacks on round-reduced versions of ICEPOLE in Sect. 4. Finally, we conclude in Sect. 5.

2   Description of ICEPOLE

In this section, we give a short description of ICEPOLE-128 as it is specified in the CAESAR design document [16]. For more details about ICEPOLE-128 and the other members of the family, we refer to the CAESAR design document [16]. In case of disagreement between the specifications of ICEPOLE-128 in the CHES and CAESAR documents, we always stick to the version submitted to CAESAR and the available reference implementations of this version.

2.1 Mode of Operation

ICEPOLE uses a duplex-like [5] mode of operation, which operates on an internal state of 1280 bits. This state S is represented as 20 64-bit words $S[0..3][0..4]$. Bits on the same position of all 20 words are called slice. The encryption (as well as the decryption) can be split into the three subsequent phases: initialization, processing of data, and finalization (tag generation), and is shown in Fig. 1.

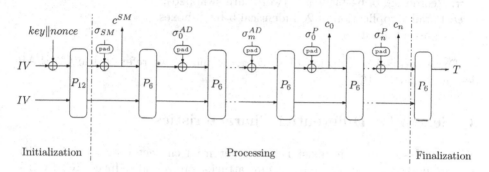

Fig. 1. Encryption of ICEPOLE-128.

Initialization. First, the state S is initialized with a constant value. Afterwards, the 128-bit key and the 128-bit nonce are xored to the internal state. Then, the 12-round variant P_{12} of the ICEPOLE permutation is applied.

Processing of Data. For processing, the associated data and the plaintext are split into 1024-bit blocks, with possibly smaller last blocks. Each of these blocks is padded to 1026 bits. The padding rule is to append a frame bit, followed by a single 1 and zeros until 1026 bits are reached.

After the initialization, the padded secret message number σ_{SM} is xored to the internal state and c_{SM} is extracted. Then, 6 rounds of the ICEPOLE permutation P_6 are applied. After the processing of the secret message number, the associated data blocks σ_i^{AD} are padded and injected, separated by the 6-round ICEPOLE permutation P_6. The plaintext blocks σ_i^P are processed in a similar way, except that ciphertext blocks c_i are extracted. For easier comparison with other sponge-based [3,4] primitives, we move the last permutation call P_6 (after the last plaintext block) to the finalization.

Finalization. Since we moved the last permutation call of the processing to the finalization, the finalization starts with calling P_6. Afterwards, the 128-bit tag T is extracted from the state:

$$T = S[0][0] \| S[1][0].$$

2.2 Permutation

Two variants of the ICEPOLE permutation are used: One with 6 rounds, P_6, and one with 12 rounds, P_{12}. Each round R consists of five steps, $R = \kappa \circ \psi \circ \pi \circ \rho \circ \mu$.

- μ: Mixing of every 20-bit slice through an MDS matrix over $GF(2^5)$.
- ρ: Rotation within all 64-bit words.
- π: Reordering of 64-bit words (words are swapped).
- ψ: Parallel application of 256 identical 5-bit S-boxes.
- κ: Constant addition.

For a detailed description of κ, ψ, π, ρ, and μ, we refer to the CAESAR design document [16].

3   Search for Differential Characteristics

As we will see later, the existence of differential characteristics holding with a high probability is crucial for our attacks on round-reduced ICEPOLE-128. Since ICEPOLE-128 is a bit-oriented construction, automatic search tools are helpful for finding complex differential characteristics with a high probability. ICEPOLE-128 has a rather big internal state of 1280 bits involving many operations per permutation round. Therefore, we have decided to use the guess-and-determine techniques already used for several attacks on hash functions [12,14,15] together with a greedy strategy, which has already been used to find differential characteristics with a high probability for SipHash [10].

We first describe the used concepts for representing differences within the used automatic search tool and propagating them in Sect. 3.1. Then, we give a high-level overview of our search strategy in Sect. 3.2.

3.1   Generalized Conditions and Propagation

To represent differential characteristics within the search tool, we have chosen generalized conditions [8]. These conditions are suitable for guess-and-determine-based searches, since they have a very high level of granularity. For instance, with a '?', it can be represented that no restrictions are given at some point of a search, or the value of a pair of bits can be completely determined, for instance by '1' denoting that both bits of the pair have to have the value 1. The complete set of all 16 generalized conditions can be found in Table 2.

Table 2. Generalized conditions [8].

x,x'	0,0	1,0	0,1	1,1		x,x'	0,0	1,0	0,1	1,1
?	✓	✓	✓	✓		3	✓	✓	–	–
-	✓	–	–	✓		5	✓	–	✓	–
x	–	✓	✓	–		7	✓	✓	✓	–
0	✓	–	–	–		A	–	✓	–	✓
u	–	✓	–	–		B	✓	✓	–	✓
n	–	–	✓	–		C	–	–	✓	✓
1	–	–	–	✓		D	✓	–	✓	✓
#	–	–	–	–		E	–	✓	✓	✓

Apart from the representation, the propagation of differences (or in this case of the generalized conditions) through the components of the ICEPOLE permutation has to be modeled. Here, we make the distinction between the linear part of one round, consisting of the application of μ, ρ, and π, and the nonlinear part ψ, which is the application of 256 5-bit S-boxes. The propagation for each S-box is done by exhaustively calculating all possible solutions for given input and output differences (basically look-ups in the difference distribution table). The propagation of the linear part of each round is modeled by techniques described in [11].

3.2 Search Strategy

On a high level, our search strategy can be split into the following two phases:

1. Search for a valid characteristic with a low number of active S-boxes.
2. Optimize the probability of the characteristic.

The first phase primarily serves to narrow the search space for the second phase. In this first phase, we search for truncated differentials with as few differentially active S-boxes as possible. In this context, an S-box is called active if there are differences on its inputs and outputs. The number of active S-boxes sets an upper bound on the best possible probability of a characteristic, since the maximum differential probability of the ICEPOLE S-box is 2^{-2}.

In the second phase, we search for the actual characteristic. In fact, just using the truncated differential and searching for the best assignment does not give us the best overall result. As we will see later, we search for characteristics having a special form, where a low number of active S-boxes does not necessarily give the best characteristic. Thus, we only fix the truncated differential for one or two rounds, leaving the other rounds completely undetermined, and search for high-probability characteristics by using the greedy algorithm presented in [10].

In summary, the first phase narrows the search space and gives us a good starting point for the second phase. Then, in the second phase, the actual characteristic is searched.

4   The Attack

The first thing to do when analyzing a cryptographic primitive is to find a promising point to attack. Thus, we explain the observations that have led to the attack on the finalization of ICEPOLE-128 using differential cryptanalysis in Sect. 4.1. After that, we discuss our first findings regarding forgeries in Sect. 4.2, and explain the trick leading to an improvement of the attack in Sect. 4.3. Finally, in Sect. 4.4, we show characteristics for 5 and 6 rounds of the ICEPOLE permutation which are not suitable for a forgery, but have a better probability than the best characteristics published by the designers [16].

4.1   Basic Attack Strategy

ICEPOLE uses a sponge-like mode of operation like several other CAESAR candidates including ASCON [9], KEYAK [6], or NORX [1,2]. When comparing those Sponge constructions with ICEPOLE, it is noticeable that the last permutation, which separates the last plaintext injection from the extraction of the tag, has much fewer rounds in the case of ICEPOLE compared to the others.

For more detail, we have a closer look at the number of permutation rounds during the three different stages for the proposals of ASCON, ICEPOLE, KEYAK, and NORX, which have the same security level of 128 bits. The number of rounds in each stage for these four primitives is given in Table 3. We can see that for ASCON and NORX, the number of rounds for data processing is reduced compared to finalization and initialization and for all three competitors of ICEPOLE, the finalization is equally strong as the initialization. In the case of ICEPOLE, the permutation used in the finalization has the same number of rounds as the permutation during the processing of data, and thus just half of the rounds of the permutation used during the initialization. So it is interesting to evaluate if the designers of the competitors of ICEPOLE have been overly conservative in the design of their respective finalization, or if their decision to invest more rounds can be justified.

Table 3. Permutation rounds for some sponge-like CAESAR candidates.

	Initialization	Data processing	Finalization
ASCON	12	6	12
ICEPOLE	12	6	6
KEYAK	12	12	12
NORX	8	4	8

4.2   Creating Forgeries

In this section, we first describe the principles of our attack on a high level. Afterwards, we discuss our preliminary results regarding suitable characteristics when just considering the 1024 bits of the ciphertext blocks to inject differences.

Attack Strategy. For creating forgeries with the help of differential characteristics, we have in principle two attack points in sponge-like constructions as ICEPOLE-128. We can either attack the data processing, or we can perform the attack on the finalization. In both cases, the key to a successful attack lies in the search for a suitable differential characteristic which holds with a high probability.

Figure 2 shows how a forgery during the processing of the data works. This approach requires a differential characteristic with differences only in those parts of the state that can be modified with message blocks, while the rest of the state has to remain free of differences. In other words, we search for a characteristic capable of producing collisions on the internal state. If we have found such a characteristic with input difference Δ_0 that holds with probability 2^{-x}, we can create a forgery which succeeds with probability 2^{-x} as follows: Assume we know a valid ciphertext-tag pair consisting of two ciphertext blocks $(c_0\|c_1, T)$. Then, the ciphertext-tag pair $(c_0 \oplus \Delta_0\|c_1, T)$ is valid with probability 2^{-x}. Thus, a valid forgery can be created with complexity 2^x.

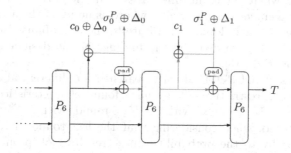

Fig. 2. Forgery during data processing.

The second option, attacking the finalization, is pictured in Fig. 3. In contrast to the previous attack, the requirements on a suitable characteristic can be relaxed. Here, we do not require a collision. It is sufficient that the difference Δ_1 for the tag T is known. The actual difference in the rest of the state does not matter in this attack. In other words, a forgery can be created from a known ciphertext-tag $(c_0\|c_1, T)$ by applying suitable differences to c_1 and T to get $(c_0\|c_1 \oplus \Delta_0, T \oplus \Delta_1)$.

In case of ICEPOLE, the permutation during the processing of the data and the finalization is equally strong. The requirements on suitable characteristics are less restrictive when attacking the finalization. Thus, attacks on the finalizations

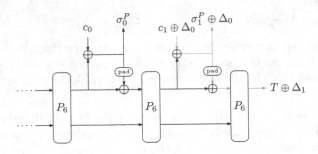

Fig. 3. Forgery during finalization.

are easier to achieve. In addition, the fact that the linear layer is located before the application of the S-boxes comes in handy. ICEPOLE has a state size of 1280 bits. For the generation of the tag, only 128 bits of the 1280 bits are extracted. The other bits do not influence the tag. Since the S-boxes are located at the end of the permutation, 128 of the 256 S-boxes of the last round have no influence on the tag and therefore, do not contribute to the probability of creating a forgery. Moreover, the other 128 S-boxes of the last round only contribute a single bit, which also has a positive effect on the total probability.

Suitable Characteristics. As discussed before, we need characteristics with a good probability, where the input differences lie in the part of the state that can be controlled by a ciphertext block, and where as many of the active S-boxes as possible lie in parts which do not contribute to the probability. However, before we present our results, we describe the findings of the designers [16] and the results by Huang et al. [13].

The designers of ICEPOLE already searched for differential characteristics without any special restrictions. They have found characteristics for 3 rounds with probability $2^{-18.4}$, 4 rounds with $2^{-52.8}$, 5 rounds with $2^{-186.2}$ and 6 rounds with $2^{-555.5}$. Indeed, when considering that the last round of ICEPOLE only contributes partially to the probability, these results look promising from the perspective of an attacker. However, as already observed by Huang et al. [13], these characteristics cannot be used for attacks on the cipher. They showed that if only 1024 bits of a message block are considered suitable for introducing differences, it is impossible to find a 3-round path with 9 active S-boxes in the form 4-1-4. Moreover, they show that the minimum number of active S-boxes in the first round is 2 in this case.

Our search for suitable characteristics supports their result. If we just consider the 1024 bits of the message block suitable for differences, we can create forgeries for 3 rounds with probability $2^{-25.3}$ and, for 4 rounds with a probability close to 2^{-128}. However, in the next section, we explain how we improved the probability for the 4-round attack to $2^{-60.3}$ by exploiting the padding rule of the last processed plaintext block.

4.3   Exploiting the Padding

ICEPOLE uses at most 1024-bit message blocks, which are padded to 1026 bits by appending a frame bit, which is 0 for the last plaintext block, followed by a single 1 and as many zeros until 1026 bits are reached. So using, for instance, a 1016-bit block and a 1024-bit block (where the last byte fulfills the padding rule applied to the 1016-bit block) virtually flips a bit in an otherwise unaccessible part of the state. By using this trick, we are able to use characteristics where only one S-box is active in the first round.

With these improved differential characteristics, we are able to create forgeries for ICEPOLE-128 with the finalization reduced to 3 (out of 6) rounds with probability $2^{-14.8}$, and for 4 rounds (out of 6) with probability $2^{-60.3}$. The characteristics for the 3-round attack can be found in Table 4, and for the 4-round attack in Table 5 of Appendix A.

The 3-round attack on ICEPOLE-128 has been verified using the reference implementation ICEPOLE128v1 submitted to CAESAR with a modified number of rounds for permutation P_6. We fixed a random key at the beginning and encrypted random 1024-bit messages (last byte of messages has to be equal to the padding 0x2) with random nonces to get 1024-bit ciphertexts. The forgeries are created by applying the difference shown in Table 4 to ciphertext and tag and discarding the last byte of the ciphertext. Removing the last byte of the ciphertext introduces a difference at bit 1026. Backed up by our experiments (2^{28} message-tag pairs), a forgery for round-reduced ICEPOLE-128, where the finalization is reduced to 3 out of 6 rounds, can be created with probability $2^{-11.7}$. For the 4-round attack, the probability is too low to be verified experimentally. However, parts of the used characteristic which have a high probability have been verified.

To introduce differences with the help of the padding, we can either extend or truncate known ciphertexts. As already discussed before, creating forgeries by truncating the last byte of the ciphertext only works if the last byte of the message before encryption equals the padding. Extending 1016-bit ciphertexts requires to guess 8 bits of the internal state correctly and hence decreases the probability by 2^{-8}. In the case of messages consisting of a fractional number of bytes, 1022-bit ciphertexts can be extended, leading to a decrease of 2^{-2}.

4.4   Characteristics for the Permutation

We also considered characteristics without any special restrictions. We have been able to improve the results published in the design documents. We have found a 5-round characteristic with an estimated probability of $2^{-104.5}$ and a 6-round characteristic with an estimated complexity of $2^{-258.4}$. The characteristics are given in Tables 6 and 7 of Appendix A. Both characteristics are a perceptible improvement over the characteristics given in the design document [16], which have a probability of $2^{-186.2}$ and $2^{-555.3}$, respectively.

5   Conclusion

In this work, we have analyzed the resistance of ICEPOLE-128 against forgery attacks. Our attacks work for versions of ICEPOLE-128 where the permutation used during the finalization is reduced to 4 (out of 6) rounds. This means that ICEPOLE-128 has a security margin of 2 rounds, which is lower than the 3 rounds expected by the designers [16].

Acknowledgments. The work has been supported in part by the Austrian Science Fund (project P26494-N15) and by the Austrian Research Promotion Agency (FFG) and the Styrian Business Promotion Agency (SFG) under grant number 836628 (SeCoS).

References

1. Aumasson, J., Jovanovic, P., Neves, S.: NORX. Submission to the CAESAR competition (2014). http://competitions.cr.yp.to/round1/norxv1.pdf
2. Aumasson, J.-P., Jovanovic, P., Neves, S.: NORX: parallel and scalable AEAD. In: Kutyłowski, M., Vaidya, J. (eds.) ICAIS 2014, Part II. LNCS, vol. 8713, pp. 19–36. Springer, Heidelberg (2014)
3. Bertoni, G., Daemen, J., Peeters, M., Assche, G.V.: Sponge functions. In: ECRYPT Hash Workshop 2007, May 2007
4. Bertoni, G., Daemen, J., Peeters, M., Van Assche, G.: On the indifferentiability of the sponge construction. In: Smart, N.P. (ed.) EUROCRYPT 2008. LNCS, vol. 4965, pp. 181–197. Springer, Heidelberg (2008)
5. Bertoni, G., Daemen, J., Peeters, M., Van Assche, G.: Duplexing the sponge: single-pass authenticated encryption and other applications. In: Miri, A., Vaudenay, S. (eds.) SAC 2011. LNCS, vol. 7118, pp. 320–337. Springer, Heidelberg (2012)
6. Bertoni, G., Daemen, J., Peeters, M., Van Assche, G., Van Keer, R.: Keyak. Submission to the CAESAR competition (2014). http://competitions.cr.yp.to/round1/keyakv1.pdf
7. Biham, E., Shamir, A.: Differential cryptanalysis of DES-like cryptosystems. J. Cryptology 4(1), 3–72 (1991)
8. De Cannière, C., Rechberger, C.: Finding SHA-1 characteristics: general results and applications. In: Lai, X., Chen, K. (eds.) ASIACRYPT 2006. LNCS, vol. 4284, pp. 1–20. Springer, Heidelberg (2006)
9. Dobraunig, C., Eichlseder, M., Mendel, F., Schläffer, M.: Ascon. Submission to the CAESAR competition (2014). http://competitions.cr.yp.to/round1/asconv1.pdf
10. Dobraunig, C., Mendel, F., Schläffer, M.: Differential cryptanalysis of siphash. In: Joux, A., Youssef, A. (eds.) SAC 2014. LNCS, vol. 8781, pp. 165–182. Springer, Heidelberg (2014)
11. Eichlseder, M., Mendel, F., Nad, T., Rijmen, V., Schläffer, M.: Linear propagation in efficient guess-and-determine attacks. In: Budaghyan, L., Tor Helleseth, M.G.P. (eds.) International Workshop on Coding and Cryptography, pp. 193–202 (2013)
12. Eichlseder, M., Mendel, F., Schläffer, M.: Branching heuristics in differential collision search with applications to SHA-512. In: Cid, C., Rechberger, C. (eds.) FSE 2014. LNCS, vol. 8540, pp. 473–488. Springer, Heidelberg (2015)

13. Huang, T., Tjuawinata, I., Wu, H.: Differential-linear cryptanalysis of ICEPOLE. In: Leander, G. (ed.) FSE 2015. LNCS, vol. 9054, pp. 243–263. Springer, Heidelberg (2015)
14. Mendel, F., Nad, T., Schläffer, M.: Finding SHA-2 characteristics: searching through a minefield of contradictions. In: Lee, D.H., Wang, X. (eds.) ASIACRYPT 2011. LNCS, vol. 7073, pp. 288–307. Springer, Heidelberg (2011)
15. Mendel, F., Nad, T., Schläffer, M.: Improving local collisions: new attacks on reduced SHA-256. In: Johansson, T., Nguyen, P.Q. (eds.) EUROCRYPT 2013. LNCS, vol. 7881, pp. 262–278. Springer, Heidelberg (2013)
16. Morawiecki, P., Gaj, K., Homsirikamol, E., Matusiewicz, K., Pieprzyk, J., Rogawski, M., Srebrny, M., Wójcik, M.: ICEPOLE. Submission to the CAESAR competition (2014). http://competitions.cr.yp.to/round1/icepolev1.pdf
17. Morawiecki, P., Gaj, K., Homsirikamol, E., Matusiewicz, K., Pieprzyk, J., Rogawski, M., Srebrny, M., Wójcik, M.: ICEPOLE: high-speed, hardware-oriented authenticated encryption. In: Batina, L., Robshaw, M. (eds.) CHES 2014. LNCS, vol. 8731, pp. 392–413. Springer, Heidelberg (2014)
18. The CAESAR committee: CAESAR: Competition for authenticated encryption: Security, applicability, and robustness (2014). http://competitions.cr.yp.to/caesar.html

A Differential Characteristics

Table 4. 3-round characteristic suitable for forgery with probability $2^{-14.8}$.

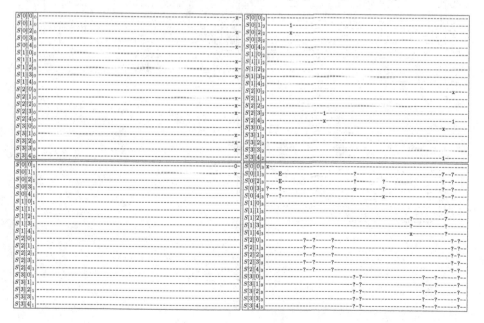

490   C. Dobraunig et al.

Table 5. 4-round characteristic suitable for forgery with probability $2^{-60.3}$.

Table 6. 5-round characteristic with probability $2^{-104.5}$.

Table 7. 6-round characteristic with probability $2^{-258.3}$.

```
S[0][0]0 x-------x--xxx--------x--x-x-----------------x
S[0][1]0 x--x-x--x--xx-------x--xx-----------------x
S[0][2]0 ---------x--------x-----x----x---------------x
S[0][3]0 x----x--x--x----x-----x----x--xx------------x
S[0][4]0 x-------xx-------x--xx--xx-----------------x
S[1][0]0 x-------x--x-x-----------------------------x
S[1][1]0 x------------------------------------------x
S[1][2]0 x---x--x----x--xx-----x---x--x----x---------x
S[1][3]0 ------------x--xxx-------xx--x--x------------x-x
S[1][4]0 ------xx--xxx--x--x-x--------x---x--xx-------x
S[2][0]0 ---------xx--x-x----------------------------x
S[2][1]0 x-------x--x--xx---------x--x-x-------------x
S[2][2]0 -------x-xx--x-x---------x-----------------x
S[2][3]0 x---x--x--x------x---xx--x-----x------------x
S[2][4]0 -------xx-----------------x----------------x
S[3][0]0 ------xx--xx------x----------x--------------x-x
S[3][1]0 x-------x--x-x--------x---x-------x--------x
S[3][2]0 x---x--xx--x-----x---x--x-----------------x
S[3][3]0 x-------x-x-----------x----x--x------------x
S[3][4]0 x---x--x--x--xx-------x--xx--x--x-----x----x

S[0][0]1 1------------------x----x------0----------
S[0][1]1 x--x-x----------------------x----x--------
S[0][2]1 x--x-x--------------x-----x--x-----------
S[0][3]1 x--x-x----------0--------x--x------------
S[0][4]1 1------------------x----x-------1--------
S[1][0]1 -------------------------x----------1----
S[1][1]1 --1----------------x----x------0--------
S[1][2]1 x--x-x--------0---------x--x------------
S[1][3]1 0------------------x----x----------------
S[1][4]1 x------------------x----x------0--------
S[2][0]1 --1----------------x----x----------------
S[2][1]1 1--0------------------x----x------------
S[2][2]1 x--x-x---------1--------x--x------------
S[2][3]1 ---x----------------x----0--------------
S[2][4]1 ---0------------------x----x------------
S[3][0]1 1------------------x----x----------------
S[3][1]1 0------------------x----x---------1-----
S[3][2]1 x--x-x--------------x----x--------------
S[3][3]1 x--x-0-----------------x----x----------
S[3][4]1 0--x-x------------------x----x----------

S[0][0]2 --0--------------------------------------
S[0][1]2 --x-------------------------------------
S[0][2]2 --x-------------------------------------
S[0][3]2 --x-------------------------------------
S[0][4]2 --0-------------------------------------
S[1][0]2 --x-------------------------------------
S[1][1]2 --1-------------------------------------
S[1][2]2 --1--x----------------------------------
S[1][3]2 --x-------------------------------------
S[1][4]2 --x-------------------------------------
S[2][0]2 --x-------------------------------------
S[2][1]2 --x-------------------------------------
S[2][2]2 --x-x-----------------------------------
S[2][3]2 --x-------------------------------------
S[2][4]2 --x-------------------------------------
S[3][0]2 --x-------------------------------------
S[3][1]2 --x-------------------------------------
S[3][2]2 --x-------------------------------------
S[3][3]2 --0-------------------------------------
S[3][4]2 --x-------------------------------------

S[0][0]3 ----------------------------------------
S[0][1]3 ----------------------------------------
S[0][2]3 ----------------------------------------
S[0][3]3 ----------------------------------------
S[0][4]3 ----------------------------------------
S[1][0]3 ----------------------------------------
S[1][1]3 -1--------------------------------------
S[1][2]3 --x-------------------------------------
S[1][3]3 ----------------------------------------
S[1][4]3 ----------------------------------------
S[2][0]3 ----------------------------------------
S[2][1]3 ----------------------------------------
S[2][2]3 ----------------------------------------
S[2][3]3 ----------------------------------------
S[2][4]3 ----------------------------------------
S[3][0]3 ----------------------------------------
S[3][1]3 ----------------------------------------
S[3][2]3 ----------------------------------------
S[3][3]3 ----------------------------------------
S[3][4]3 ----------------------------------------

S[0][0]4 ----------------------------------------------
S[0][1]4 ----------------1----------------------------
S[0][2]4 ---------------------------------------------
S[0][3]4 ---------------------------------------------
S[0][4]4 ---------------------------------------------
S[1][0]4 -------------------------------------x-------
S[1][1]4 ---------------------------------------------
S[1][2]4 ---------------------------------------------
S[1][3]4 ---------------------------------------------
S[1][4]4 -------------------------------------1-------
S[2][0]4 -----------------------------------------x--
S[2][1]4 ---------------------------------------------
S[2][2]4 ---------------------------------------------
S[2][3]4 ---------------------------------------------
S[2][4]4 ------------------------------------------1-
S[3][0]4 ----------------------------------1----------
S[3][1]4 --------------------------------------x------
S[3][2]4 ---------------------------------------------
S[3][3]4 ---------------------------------------------
S[3][4]4 ---------------------------------------------

S[0][0]5 x----------------------------1--------0-x-
S[0][1]5 x------1--------------------------------1-x-
S[0][2]5 x---x------------------------1----------x---
S[0][3]5 x----------------------------x--------------
S[0][4]5 ----------------------------------------
S[1][0]5 -----------------------------0---------1----
S[1][1]5 ---------------------------------1--x-x----
S[1][2]5 -------------------------1--------------
S[1][3]5 ----------------------------------0--------
S[1][4]5 -------------------------------x-----------
S[2][0]5 ------------------x---------------------
S[2][1]5 x------------------x--------------------
S[2][2]5 0-----------------x-------------------1----
S[2][3]5 x------------------x--------------------
S[2][4]5 ------------------0---------------------
S[3][0]5 1----------------------x1-----------x-------
S[3][1]5 x------------------------1--1---x-----1----
S[3][2]5 ------------------------x----------1-------
S[3][3]5 ------------------------x----------1-------
S[3][4]5 ------------------------1---------------

S[0][0]6 --------x---1--------x--x--1------------0----
S[0][1]6 -----1----------x----1--------x---------1--x1--
S[0][2]6 ------1x----11--1----x1-----------x-----x----
S[0][3]6 ----x-----x-xx--x--x-------1x-------1-1-----
S[0][4]6 --------0---------1------x-1--------x--------
S[1][0]6 ------------0-------------------xx--1-x---1----
S[1][1]6 ----------x1---------------x-------1--------
S[1][2]6 -------1------x----1--1------------x--------
S[1][3]6 ------1x-----------x--x--11------------x1---
S[1][4]6 ------x-----x-----------xx-----11-x-1----xx--
S[2][0]6 -------1-----x--------1-x--------------x--x-1-
S[2][1]6 -------x----x----1------x-----------x-----0--
S[2][2]6 ------------x-x--------------------x--1------
S[2][3]6 1-----1------1-1-1-0----------------x------1-
S[2][4]6 x------x-----x1x-x-x-------1--------0----1--1-1-
S[3][0]6 --------x--0-------x--0----1-x------x-111--1-x--1-
S[3][1]6 --------1-1--------1---x-----1--xxx---x----x--
S[3][2]6 -11----1--------------x------------x--------
S[3][3]6 -xx------x-------0----------0-1-1-----------
S[3][4]6 ---------1--x-----------1-x-x--x-1---1--------1----
```

Analysis of the CAESAR Candidate Silver

Jérémy Jean[1]([✉]), Yu Sasaki[1,2], and Lei Wang[1,3]

[1] Nanyang Technological University, Singapore, Singapore
JJean@ntu.edu.sg
[2] NTT Secure Platform Laboratories, Tokyo, Japan
sasaki.yu@lab.ntt.co.jp
[3] Shanghai Jiao Tong University, Shanghai, China
wanglei@cs.sjtu.edu.cn

Abstract. In this paper, we present the first third-party cryptanalysis against the authenticated encryption scheme Silver. In high-level, Silver builds a tweakable block cipher by tweaking AES-128 with a dedicated method and performs a similar computation as OCB3 to achieve 128-bit security for both of integrity and confidentiality in nonce-respecting model. Besides, by modifying the tag generation of OCB3, some robustness against nonce-repeating adversaries is claimed. We first present a forgery attack against 8 (out of 10) rounds with 2^{111} blocks of queries in the nonce-respecting model. The attack exploits a weakness of the dedicated AES tweaking method of Silver. Then, we present several attacks in the nonce-repeating model. Those include (1) a forgery against full Silver with $2^{49.46}$ blocks of queries which matches a conservative security claim by the designers, (2) a plaintext recovery against full Silver with a single query and (3) a key recovery against 8 rounds with 2^{111} blocks of queries. In particular, the plaintext recovery breaks the security claim by the designers. Considering that the current best key recovery for plain AES-128 is up to seven rounds, Silver lowers the security margin of AES due to its tweaking method. The attacks have been partially implemented and experimentally verified.

Keywords: Silver · CAESAR · Authenticated encryption · Forgery · Plaintext recovery · Key recovery

1 Introduction

An authenticated encryption is a symmetric-key cryptographic scheme that provides both of integrity and confidentiality at one time. Currently the CAESAR competition [1] is being conducted to determine a portfolio of authenticated encryptions, and research development for authenticated encryption deserves careful attention.

Security and efficiency are obviously important factors of authenticated encryption designs. Thanks to the AES-NI, which is a set of instructions in Intel's CPU, basing the scheme on block cipher AES [3] is a promising way of designing authenticated encryptions. On the other hand, the block size of AES (128 bits)

© Springer International Publishing Switzerland 2016
O. Dunkelman and L. Keliher (Eds.): SAC 2015, LNCS 9566, pp. 493–509, 2016.
DOI: 10.1007/978-3-319-31301-6_28

is often too small as a next-generation cryptographic scheme because security bound in many cases can only be proven up to a half of the block size. As a consequence, designing authenticated encryption schemes with large enough security while maintaining the efficiency advantages of AES-NI is a challenging topic.

One possible direction is designing some structure on top of AES to construct a tweakable block cipher [12]. The OCB mode designed by Rogaway [15] shows that a tweakable block cipher with n-bit block size enables to design an authenticated encryption with n-bit security. Several CAESAR candidates were designed in this line [5–7,13], including our target Silver [13].

As mentioned above, Silver [13] designed by Penazzi and Montes proposes a dedicated way of tweaking AES-128 to provide 128-bit security for both integrity and confidentiality in the nonce-respecting model. The way of tweaking AES-128 is a little bit complicated and thus a more careful security analysis is necessary. So far, no third-party analysis has been provided.

Silver provides some robustness against nonce repetition. The following argument is claimed for integrity and confidentiality against nonce-repeating adversaries: "*It is safe to assume that a forgery cannot be made with probability greater than, say, 2^{-50} for tags of length 128*", and "*any attack against Silver should be readily converted into an attack on AES-ECB. Thus, although there is loss of indistinguishability, the loss of confidentiality is not catastrophic, and it simply reduces to the loss one would be prepared to accept when using ECB.*" Thus, cryptanalysis in the nonce-repeating model is also important to examine the designers' claims.

Apparently a unique design feature of Silver is the way of tweaking AES-128. In each block, the tweak value only impacts to subkeys. Namely, when all the eleven subkeys are specified, computation between plaintext and ciphertext is exactly the same as plain AES. Subkeys in each block are generated with a key K, a nonce N and a tweak tw that depends on the block position. The algorithm is briefly explained as follows. First, $\alpha \leftarrow \text{AES}_K(N)$ is computed and then $(k_0, \ldots, k_{10}) \leftarrow \text{AES}^{KS}(K)$ and $(\alpha_0, \ldots, \alpha_{10}) \leftarrow \text{AES}^{KS}(\alpha)$ are computed, where AES^{KS} is the key schedule of AES-128. Additionally, an intermediate 128-bit value γ is computed from α. Intuitively, subkeys for the i-th message or associated data block are generated as follows (a complete specification appears in Sect. 2).

$$\begin{cases} sk_j \leftarrow k_j \oplus \alpha_j & \text{for } j = 0, 2, 3, 4, 6, 7, 8, 10, \\ sk_j \leftarrow k_j \oplus \alpha_j \oplus (\alpha + i \cdot \gamma) & \text{for } j = 1, 5, 9, \end{cases} \tag{1}$$

where '$+$' is the modular addition operated in 64 bits each. The multiplication by block index i introduces different impact among different block positions. The value $a + i \cdot \gamma$ is injected every four rounds to exploit the good diffusion property of 4-round AES.

Another unique design feature of Silver is the tag generation. Intuitively, the tag is the encryption of $\Sigma_A \oplus \Sigma_P \oplus \Sigma_C$. Here, Σ_A is the result of processing associate data which is similar to OCB3 [10] and Σ_P is the message checksum which is the same as OCB3. Then, Σ_C is computed as $\bigoplus_i (C_i + \alpha + i \cdot \gamma)$ with

C_i as the i-th ciphertext block. This additional checksum makes the essential difference from OCB3, and thus analyzing its security impact is important.

Our Contributions. We first present reduced-round analysis in the nonce-respecting model. We show a forgery against eight rounds with 2^{111} blocks of queries. Then, we explain several attacks in the nonce-repeating model. The first result is a forgery against full Silver with a complexity of $2^{49.46}$ blocks of queries. The second result is a plaintext recovery, i.e. breaking confidentiality of full Silver with a single query. The last result is a key recovery against eight rounds, which makes use of the above 8-round forgery as a tool and the bottleneck of the complexity is in that part, i.e. 2^{111} blocks of queries.

In general, differential cryptanalysis is hard to apply in the nonce-respecting model. To avoid this problem, our 8-round forgery in the nonce-respecting model exploits the internal (higher-order) difference [14], i.e. difference between the blocks of a single plaintext. As shown in Eq. (1), computations in different blocks only differ in the tweak value $i \cdot \gamma$, which is only injected in three subkeys. Thus, many subkeys have no difference ($\Delta sk_0 = 0$, $\Delta sk_2 = 0$ etc.). Such sparse tweak injections allow us to mount an internal differential cryptanalysis even in the nonce respecting model. By analyzing 256 consecutive blocks, the tweak value assumes the values $\gamma, 2\gamma, \ldots, 255\gamma$. Then, by forcing γ to have a single bit to one (and 127 zeros), an integral (8th-order differential) cryptanalysis [2,9] is applied.

Our forgery attack in the nonce-repeating model exploits the tag generation structure, which is the encryption of $\Sigma_P \oplus \Sigma_C$ (for empty associated data). Let n be the block size, i.e. $n = 128$ for Silver. We observe that the modular addition to compute Σ_C allows us to compute the sum of the x least significant bits (LSBs), where $x \in \{1, 2, \ldots, n\}$, independently of the $n - x$ most significant bits (MSBs). We first make $2^{n/3}$ queries under the same nonce to generate the first message block pair (P_1, P'_1) such that the $2n/3$ LSBs of P_1 and P'_1 are colliding and the $2n/3$ LSBs of C_1 and C'_1 are also colliding. The same is iterated until the $n/3$-th block. Now, we have $2^{n/3}$ possible combinations of plaintext blocks, in which any of them produces the same $2n/3$ LSBs of $\Sigma_P \oplus \Sigma_C$ and one pair will produce the same $n/3$ MSBs. With this pair of messages producing the same $\Sigma_P \oplus \Sigma_C$, a forgery can be mounted by the length extension attack, i.e. appending the same message block p to each of two messages. The number of queries is about $2^{n/3}$ and each query consists of $n/3$ blocks. Thus, the data complexity is about $n/3 \cdot 2^{n/3}$, which is $2^{48.08}$ for $n = 128$. By increasing the success probability and optimizing attack procedure, the complexity becomes $2^{49.46}$ message blocks with success probability 0.36, which is almost the same as the one for the ordinary birthday paradox. The designers claim a conservative 50-bit security against forgery in the nonce-repeating model. Here, the choice of 50 bits is very unclear without any reasoning. Our attack shows that their claim is no longer conservative, i.e. forgery in the nonce-repeating model can be mounted with a complexity at most 2^{50}.

Our plaintext-recovery attack exploits the weakness of the domain separation when the last message block is not full. For any ciphertext $C_1 \| \cdots \| C_{N-1} \| C_N$ for some N where $|C_N| < 128$, the corresponding last block of the plaintext P_N

is recovered with a single encryption query under the same nonce. Considering that the ECB mode never allows such a plaintext-recovery attack, it breaks the security claim by the designers, who state that *"any attack against Silver should be readily converted into an attack on AES-ECB"*.

Our key-recovery attack makes use of the fact that Σ_C in the tag generation is computed by modular addition. We first generate a tag collision with 1-block messages (and empty associated data) by making 2^{64} queries under the same nonce. For the colliding pair, we can directly observe the ciphertext difference ΔC. The tag collision indicates $\Delta \Sigma_P \oplus \Delta \Sigma_C = 0$, where $\Sigma_P = P$ for 1 block message, and thus we can recover $\Delta \Sigma_C$. For a 1-block message, Σ_C is computed as $\Sigma_C = C + (\alpha + \gamma)$. From ΔC and $\Delta \Sigma_C$, we recover the secret value $(\alpha + \gamma)$ similarly with the analysis by Lipmaa and Moriai [11] about XOR difference propagation through the modular addition. We notice that the current best key recovery attack on plain AES-128 [4] is up to seven rounds. In this sense, Silver lowers the security margin of AES-128 due to its tweaking method.

Our attacks exploit several design aspects of Silver such as the dedicated AES-128 tweaking method, tag generation structure, combination of XOR and modular addition, domain separation for special treatment, and so on. We believe that the presented analysis in this paper will bring useful feedback for the design of authenticated encryption schemes based on tweaked AES. Our results are summarized in Table 1.

Table 1. Summary of the cryptanalysis of Silver presented in this paper.

Model	Rounds	Type	#Queries	Reference
Nonce respecting	4	Forgery	2^{79}	Sect. 3.1
	8	Forgery	2^{111}	Sect. 3.2
Nonce repeating	Full	Forgery	$2^{49.46}$	Sect. 4.1
	Full	Plaintext recovery	1	Sect. 4.2
	8	Key recovery	2^{111}	Sect. 4.3

2   Description of Silver

Silver is an authenticated encryption scheme relying on a tweakable block cipher derived from the AES-128 block cipher E. We note that it can be regarded as an instance of the TWEAKEY framework introduced in [8] to construct tweakable block ciphers, where the tweaked values are injected by XORing together with the subkeys.

We describe here the design of Silver with the set of parameters recommended by the designers [13]. The original description being a bit hard to follow, we introduce new notations and figures to describe the cipher. Four inputs are processed by the encryption algorithm: a 128-bit public message number N, a 128-bit secret key k, possibly empty associated data A and a plaintext P. Those values are encrypted and authenticated in the form of a ciphertext C and a 128-bit tag T. From (N, A, C, T), the decryption algorithm produces the plaintext P if the tag T is verified, and $\perp$ otherwise.

Notations. We denote the byte length of P by b_P, b_A refers to the byte length of A, and $g = b_A \;||\; b_P$ encodes the two byte lengths of A and P in a 128-bit value. The XOR operation is denoted by $\oplus$. We represent the 64-bit most significant bits of a 128-bit value x by x^L, and its 64-bit less significant bits by x^R. We consider that $x = x^L \;||\; x^R$, where $||$ denotes the concatenation. More generally, we refer to the n least significant bits of x by $\lfloor x \rfloor_n$, and similarly to its n most significant by $\lceil x \rceil_n$. We denote the empty string by ϵ. Modular addition represented by $+$ is performed on 128-bit values in $(\mathbb{Z}/2^{64}\mathbb{Z}) \times (\mathbb{Z}/2^{64}\mathbb{Z})$, where the two 64-bit halves of the values are added independently. Finally, $\overline{x}$ denotes the value x with the least significant bit forced to 1, i.e. $\overline{x} = x \vee 1$.

Initialization. Before processing the inputs, Silver requires an initialization step with one block cipher call, and two calls to the AES-128 key scheduling algorithm. The block cipher call is parameterized by the secret key k and transforms N into an internal secret value that we denote α, i.e. $\alpha = E_k(N)$. Then, the AES-128 key schedule produces two sequences of eleven subkeys from both the secret k, and the value α: $(k_0, \ldots, k_{10})$ and $(\alpha_0, \ldots, \alpha_{10})$, respectively. The two sets of subkeys are used in the function denoted KS, which combines the subkeys with a 128-bit input value S: $KS(S) = (u_0, \ldots, u_{10})$, with:

$$u_0 = k_0 \oplus \alpha_1, \qquad u_4 = k_4 \oplus \alpha_4, \qquad u_8 = k_8 \oplus \alpha_8,$$
$$u_1 = k_1 \oplus (\alpha + S), \quad u_5 = k_5 \oplus \alpha_5 \oplus (\alpha + S), \quad u_9 = k_9 \oplus (\alpha + S),$$
$$u_2 = k_2 \oplus \alpha_2, \qquad u_6 = k_6 \oplus \alpha_6, \qquad u_{10} = k_{10} \oplus \alpha_{10}.$$
$$u_3 = k_3 \oplus \alpha_3, \qquad u_7 = k_7 \oplus \alpha_7,$$

Finally, two secret counters are also initialized: γ_A later used to process the associated data, and γ used during the plaintext encryption:

$$\gamma = \overline{\alpha_9^L} \;\Big|\Big|\; \overline{\alpha_9^R}, \qquad \gamma_A = \overline{\alpha_9^L} \;\Big|\Big|\; 0^{64}.$$

The tweakable block cipher used in Silver replaces the original AES-128 subkeys $(k_0, \ldots, k_{10})$ by the subkeys $KS(S)$ for a given tweak value S.

Associated Data. Similarly to several other authenticated encryption schemes [5–7,10], Silver first computes a MAC Σ_A on the associated data A, and then uses this value internally to authenticate A. The process is described on Fig. 1. First, A is divided into t blocks of 128 bits, possibly using the 10^* padding. Then, it applies the tweaked AES with subkeys $KS(i \cdot \gamma_A)$ independently on each block i, and XORs the t results to produce the checksum Σ_A.

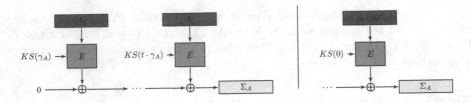

Fig. 1. Associated data. There are t blocks in A. If the last block is not full, pad it using the 10* padding. The checksum produced is Σ_A.

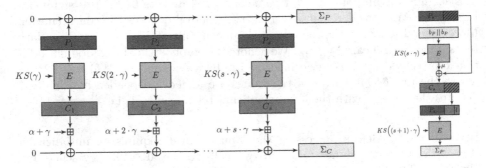

Fig. 2. Encryption. Σ_P is the checksum over the full input blocks, Σ_C is a checksum over the ciphertext blocks shifted by modular additions. The last partial block (if any) is processed with two calls to E.

Note that if the last block has been padded, the tweak input is 0 to produce subkeys $KS(0)$ for the last padded block.[1]

Encryption. To encrypt the plaintext P, we first divide it into blocks of 128 bits, say s blocks, the last one possibly non-full. The tweaked `AES-128` denoted E is used to transform each block P_i to ciphertext block C_i using the tweaked subkeys produced by $KS(i \cdot \gamma)$. Then, a checksum Σ_P is computed from the XOR of all the plaintext blocks, and a checksum Σ_C is computed from modularly masked value of the ciphertext blocks (see Fig. 2, left), namely:

$$\Sigma_P = \bigoplus_{i=1}^{s} P_i, \quad \text{and} \quad \Sigma_C = \bigoplus_{i=1}^{s}(C_i + \alpha + i \cdot \gamma).$$

In the event that the last block is not full and contains $0 < l < 16$ bytes (see Fig. 2, right), one first generates a mask μ from the encryption of $b_P \parallel b_P$ under subkeys $KS(s \cdot \gamma)$ and XORs it to the actual partial block P_s to produce the partial ciphertext block C_s. The unused bits from the mask μ are appended to P_s to produce a new 128-bit block, whose last byte is replaced by l, which is encrypted to produce another checksum $\Sigma_{P'}$.

---

[1] This special treatment of the last partial block, in particular the special tweak input value 0, will be later exploited in our forgery attacks in Sect. 3.

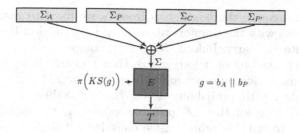

Fig. 3. Tag generation. The tag consists of the encryption of the XOR of the four checksums Σ_A, Σ_P, Σ_C and $\Sigma_{P'}$.

Tag Generation. To produce the tag T, one first XORs all the four (possibly null) checksums Σ_A, Σ_P, Σ_C and $\Sigma_{P'}$ to get Σ. Then, the tag T results from the encryption of Σ using E parameterized by the subkeys $\pi(KS(g))$, where the permutation π changes the order of the subkeys: $\pi = (2, 9, 3, 4, 6, 1, 7, 8, 10, 5, 0)$.

3   Nonce-Respecting Analysis

In this section, we propose an attack on reduced-round Silver against a nonce-respecting adversary. Our goal is to exploit the well-known integral property of the AES [2,3] by using the counter computation in the associated data. Ultimately, we show how to select associated data A comprised of 256 blocks $A = (A_1, \ldots, A_{255}, A_0)$ such that the checksum Σ_A equals zero. The following demonstrates that this behavior occurs with probability higher than 2^{-128} and can be used to replace A by A' producing the same $\Sigma_A = 0$, hence leading to a forgery attack.

We start by describing a simpler 4-round attack and later extend it to eight rounds.

3.1   Forgery on 4-Round Silver

In Sect. 2, we have introduced a counter value γ_A used to compute Σ_A from the associated data. We recall that γ_A is defined from a 128-bit value α_9 by $\gamma_A = \overline{\alpha_9^L} \,||\, 0^{64}$, in which $\overline{\alpha_9^L}$ overwrites the LSB of α_9^L with a single bit 1. Therefore, with probability 2^{-63}, $\gamma_A = 0^{63} 1 \,||\, 0^{64}$. If this holds, then the sequence of tweaks in the associated data computation with the non-full last block depicted on Fig. 4 would be:

$$\gamma_A = 0^{56}\, 00000001 \,||\, 0^{64},$$

$$2 \cdot \gamma_A = 0^{56}\, 00000010 \,||\, 0^{64},$$

$$3 \cdot \gamma_A = 0^{56}\, 00000011 \,||\, 0^{64},$$

$$\vdots$$

$$255 \cdot \gamma_A = 0^{56}\, 11111111 \,||\, 0^{64},$$

$$0 = 0^{56}\, 00000000 \,||\, 0^{64}.$$

Additionally, we require that eight bits from α are fixed to zero, so that the modular additions with the counter values $i \cdot \gamma_A$ performed in the KS function do not propagate any carry. Indeed, $KS(i \cdot \gamma_A)$ computes $i \cdot \gamma_A + \alpha$, and if the eight least significant bits of α^L are zeros, then no carry is propagated in the addition. This restriction on α occurs with probability 2^{-8}.

To summarize, with probability 2^{-71}, the 256 values $c_i = i \cdot \gamma_A + \alpha$ for $i \in [0, 255]$ take all the 256 possible values on the eight least significant bits of c_i^L. To use this property, we consider a 256-block associated data $A = A_1 || \cdots || A_{255} || A_0 = f(B)$, where:

$$A_0 = B,$$
$$\forall i \in \{1, \ldots, 255\}, \quad A_i = B \,||\, 10^7,$$

with B a random 120-bit value. This definition of A ensures that all full blocks are equal after the 10* padding (see Fig. 4).

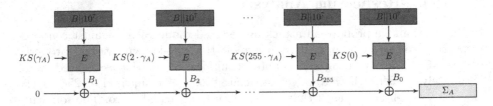

Fig. 4. We consider $t = 256$ blocks in A such that the 256 inputs to the block cipher calls are all equal.

In this context, the 256 equal blocks are processed with counter values carrying an integral-like property, which can be observed in the checksum value Σ_A. We recall that we consider a reduced variant of E where we apply only four rounds. We explain now why the resulting Σ_A equals zero by analyzing the propagation of the integral property in the different blocks. The following Fig. 5 shows the status of each byte in the state during the encryption across the 256 blocks. Namely, a byte marked C means that the same value appears for this byte in each of the 256 states, an A stands for a byte having the 256 different values in the 256 states, and a 0 means that the XOR of the bytes from the 256 states produces zero.

From the well-known integral property of the AES structure, the XOR of the 256 outputs produces zero in the 16 bytes after four rounds, with a difference introduced in the second subkey. As a result, the associated data of this particular A produces $\Sigma_A = 0$ with probability 2^{-71}.

Therefore, after querying the encryption oracle 2^{71} times with distinct nonce N, we expect that one query $(N, A, M = \epsilon)$ yields the ciphertext and tag pair (C, T) with the internal $\Sigma_A = 0$. Once this event happens, any subsequent

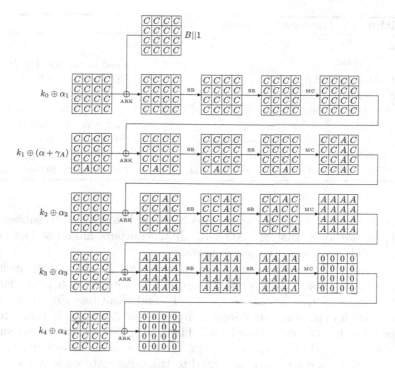

Fig. 5. Integral property in the associated data of Silver processed by the 4-round block cipher E.

decryption (N, A', C, T) query under the same nonce but with different associated data $A' = f(B')$ for a random 120-bit $B' \neq B$ would be a forgery. Indeed, the integral property still holds for the different input B', and the intermediate checksum Σ_A for the modified associated data would also be zero. Hence, the same tag has to be produced.

The procedure of the attack is described in Algorithm 1. We have experimentally verified that it works as expected assuming the event with probability 2^{-71} holds. The attack requires a total of queries of $2^{71+8} = 2^{79}$ blocks.

3.2   Forgery on 8-Round Silver

The idea behind the 8-round forgery attack on Silver is essentially the same as the one for four rounds, except that we make sure that the pre-added four rounds do not alter the integral property. We still want to find a 256-block associated data value such that $\Sigma_A = 0$. The main difference here is the choice of these blocks, as the integral property does not propagate over eight rounds.

However, on the nine subkeys added in the eight rounds, only two carry the tweak input, which is different for each block of associated data, namely, the second and the fifth subkeys. Hence, if we choose 256 blocks A_i such that the internal states after the second subkey addition are the same for all the

Algorithm 1 – Forgery for 4-round Silver.

1: $A \xleftarrow{\$} \{0,1\}^{120}$
2: $B \xleftarrow{\$} \{0,1\}^{120}$ ▷ Need to ensure $A \neq B$
3: $AD \leftarrow (A||10^7)^{255} \,||\, A$ ▷ $|AD| = b_A = 2^{12} - 1$
4: $AD' \leftarrow (B||10^7)^{255} \,||\, B$
5: **for** $i = 1 \rightarrow 2^{71}$ **do**
6:    $N_i \leftarrow \{0,1\}^{128}$ ▷ Pick a random nonce
7:    $(C = \epsilon, T_i) \leftarrow \mathrm{Silver}(N_i, AD, M = \epsilon)$ ▷ 4-round encryption query
8:    $P \leftarrow \mathrm{Silver}^{-1}(N_i, AD', C, T_i)$ ▷ 4-round decryption query
9:    **if** $P \neq \perp$ **then return** (N_i, AD', C, T_i) ▷ Success with probability 2^{-71}

256 blocks, then all the states would also be equal before the fifth subkey addition. Then, we know that the XOR of all the outputs after the last subkey addition yields zero, as in the previous case.

We are then left with the problem of choosing 256 blocks A_i such that the same state value X is reached after one-round encryption under subkeys $u_0 = k_0 \oplus \alpha_1$ and $u_1 = k_1 \oplus (\alpha + i \cdot \gamma_A)$. To find them (see Fig. 6), we pick a random value for this constant state X, apply all the 256 possible values to byte corresponding to the active byte in u_1. Then, we guess 32 bits of the subkeys u_0 (marked in gray) to partially decrypt the states. This results in 256 input states A_i, and each is partially encrypted to the same state value X by the first round. For each guess of 32 bits, the resulting sequence of 256 blocks A_i can be precomputed and stored along with the guessed value.

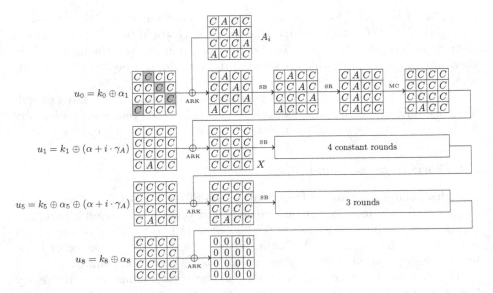

Fig. 6. Integral property in the associated data of Silver processed by the 8-round block cipher E. The four bytes marked in gray correspond to guessed values.

Similarly as the 4-round attack, we note that the last block A_{256} needs to be a partial block (e.g., it contains only 15 bytes) as we need the zero tweak value in the integral property. As before, this restricts the 255 other blocks on their last byte, but we emphasize that this can be handled easily as the last byte remains constant across all the blocks. The attack requires a total of queries of $2^{71+32+8} = 2^{111}$ blocks.

4   Nonce-Repeating Analysis

In this section, we propose to analyze the security of Silver with respect to the nonce-reuse scenario. In the original document, the designer make unclear statements about the expected security of Silver when the adversary reuses the nonce. Namely, they require the design to be used with non-repeating nonce, but still argue that the loss of security is not complete.

On the one hand, dealing with forgery, the designers claim that *"Silver has high forgery resistance even under nonce repetition."* They conjecture that a forgery cannot be made with probability greater than 2^{-50} for 128-bit tags. However, we show in Sect. 4.1 that 50-bit security is no longer conservative and it can actually match the upperbound of the security by showing how an adversary can forge a message in $2^{49.46}$ blocks of nonce-repeating queries.

On the other hand, about privacy, they write that *"any attack against Silver should be readily converted into an attack on AES-ECB. Thus, although there is loss of indistinguishability, the loss of confidentiality is not catastrophic, and it simply reduces to the loss one would be prepared to accept when using ECB"*. In the following Sect. 4.2, we show how a single nonce-repeating query can partially break the confidentiality of Silver by recovering part of the plaintext.

Finally, in Sect. 4.3, we show how to launch a key-recovery attack on 8-round Silver in 2^{111} blocks of queries, by extending the nonce-respecting forgery attack presented in the previous section.

4.1   Forgery Attack

Our forgery attack is mainly based on a *divide-and-conquer* approach to efficiently find two plaintext-ciphertexts (N, A, P, C, T) and (N, A, P', C', T') that have the same nonce and associated data, have the same plaintext length, and collide on the tag $T = T'$. Once such a pair is obtained, we choose a random one-block p, query $(N, A, P\|p)$ to receive $(C\|c, T'')$, and forge $(N, A, C'\|c, T'')$. From $T = T'$, it gives $\Sigma_A \oplus \Sigma_P \oplus \Sigma_C = \Sigma_A \oplus \Sigma_P' \oplus \Sigma_C'$, which is preserved after adding one more block p. Therefore the tag T'' is valid for $(N, A, C'\|c)$ and moreover the corresponding plaintext of forged decryption query is $P'\|p$.

Simple Attack. We now focus on how to find such a pair of plaintext-ciphertexts of Silver. The attack targets the mode used in Silver, and can be applied to any underlying block cipher. Therefore, we describe it in a general form with n being the block size in bits (e.g., $n = 128$ in the case of Silver).

To start with, we select a nonce N randomly, and use it as the nonce parameter for all the subsequent queries to Silver. Next, we construct a set of plaintexts which all collide on $\lfloor \Sigma_P \oplus \Sigma_C \rfloor_{2n/3}$. In detail, we find a pair of i-th plaintext-ciphertext blocks (P_i^*, C_i^*) and $(P_i^\$, C_i^\$)$, for $1 \leq i \leq n/3$, such that $\lfloor P_i^* \rfloor_{2n/3} = \lfloor P_i^\$ \rfloor_{2n/3}$ and $\lfloor C_i^* \rfloor_{2n/3} = \lfloor C_i^\$ \rfloor_{2n/3}$. This can be done by selecting $2^{n/3}$ plaintexts, whose i-th blocks P_i^j with $1 \leq j \leq 2^{n/3}$ have the same $2n/3$ LSBs and differ in $n/3$ MSBs; namely, $\lfloor P_i^j \rfloor_{2n/3} = \lfloor P_i^{j'} \rfloor_{2n/3}$ and $\lceil P_i^j \rceil_{n/3} \neq \lceil P_i^{j'} \rceil_{n/3}$ for any distinct $1 \leq j, j' \leq 2^{n/3}$, querying them under the nonce N to receive ciphertexts, and finding a pair of i-th ciphertext blocks C_i^j and $C_i^{j'}$ colliding on the $2n/3$ LSBs, i.e., $\lfloor C_i^j \rfloor_{2n/3} = \lfloor C_i^{j'} \rfloor_{2n/3}$. Their corresponding i-th plaintext blocks are selected as $(P_i^*, P_i^\$)$. After that, a set of plaintexts is constructed:

$$\left\{ P_1^{j_1} \| P_2^{j_2} \| \cdots \| P_{n/3}^{j_{n/3}} \ \Big| \ j_1, j_2, \ldots j_{n/3} \in \{*, \$\} \right\}.$$

It is trivial to get that all plaintexts in this set have the same $2n/3$ LSBs for both Σ_P and Σ_C. Finally, we query those $n/3$-block plaintexts with N as nonce and $A = \epsilon$ as associated data to Silver, and find a pair of plaintext-ciphertext that collide on the received tag. Note that if a pair of plaintexts also have the same $\lceil \Sigma_P \oplus \Sigma_C \rceil_{n/3}$, then the tags collide. Since there are in total $2^{n/3}$ plaintexts in the set, we can expect to obtain a pair colliding on tag with a good probability.

Here, one may think that trying $2^{n/6}$ plaintexts instead of trying $2^{n/3}$ plaintexts is sufficient to find a colliding pair on the $n/3$ MSBs. However this does not work. The important thing is that the $2^{n/3}$ plaintexts are not chosen randomly. They are generated by linearly combining $n/3$ paired message blocks, which prevents us from using classical birthday arguments.

Overall, the forgery attack on Silver in the nonce-repeating model is completed. The complexity for finding a collision for the first $n/3$ blocks requires $2^{n/3+1}$ queries and each query has $n/3$ blocks. The same complexity is required for finding the tag collision. In the end, the complexity amounts to $n/3 \cdot 2^{n/3+2}$ queried blocks, which equals $2^{50.08}$, and this slightly exceeds the claimed 50-bit security. In addition, considering the success probability of finding a collision in each block, more queries are required. In the following explanation, we show how to improve the data complexity.

Advanced Attack. The overall idea consists in generating two colliding message pairs in each of the first $n/6$ blocks, while the simple attack generated one pair in each of the first $n/3$ blocks. In the tag collision phase, we first have 2 choices in each block regarding which message pairs we pick. Namely, we have $2^{n/6}$ choices of $n/6$-block message-pairs chain. Then, for each of them, we can further consider $2^{n/6}$ combinations of messages. In total, $2^{n/6} \cdot 2^{n/6} = 2^{n/3}$ messages are examined for finding a tag collision, which is the same as the simple attack. With this effort, the complexity for generating collisions in each block is roughly $n/6 \cdot 2^{n/3+2}$, which stays unchanged from the simple attack, and the complexity for finding tag collision is reduced to $n/6 \cdot 2^{n/3+1}$, which is a half of the simple attack.

Hereafter, we evaluate the details with optimization particular for Silver, i.e. $n = 128$. First, to generate two colliding pairs in each block, we fix the 84 LSBs of the plaintext and try all the 2^{44} values in the 44 MSBs. Then, we examine the collision on the 84 LSBs of the ciphertext. According to the birthday paradox, we find a collision with probability 0.36 with 2^{42} messages. According to [16, Theorem 3.2], the probability of obtaining a collision for $(\log N)$-bit output function with trying $\theta \cdot N^{1/2}$ inputs is given by $1 - e^{-\frac{\theta^2}{2}}$. With 2^{44} messages, θ equals to 4, thus the success probability of finding a collision is $1 - e^{-8} \approx 0.999$. The probability of finding the second collision is almost the same, thus with probability $(1 - e^{-8})^2$, two colliding pairs can be obtained. We simultaneously apply this analysis for the first 22 blocks. Namely, each query consists of 22 blocks and the 84 LSBs are always fixed. This is iterated 2^{44} times by changing the value of 44 MSBs in all the blocks. In the end, the number of queries is $22 \cdot 2^{44} = 2^{48.46}$ message blocks and the probability of obtaining two colliding pairs in all the 22 blocks is $(1 - e^{-8})^{44} \approx 0.985$.

Generating a tag collision is rather simple, which is done as mentioned in the overview. We first choose which of colliding message pairs is used in each of the first 22 blocks, which yields 2^{22} choices of message-pair chains. For each of them, we further choose the message for each block, in which the 84 LSBs of Σ_P and Σ_C always take the same value. With 2^{22} messages, the probability for colliding 44 MSBs is 2^{-22}. Thus by examining all 2^{22} choices of 22-block message-pair chains, a tag collision can be generated. The number of queries for generating a tag collision is $22 \cdot 2^{44} \approx 2^{48.46}$ message blocks and the success probability is e^{-1}.

By combining the above two phases, the number of queries equals $2^{48.46} + 2^{48.46} = 2^{49.46}$ message blocks, and the success probability is $(1 - e^{-8})^{44} \cdot e^{-1} \approx 0.36$, which is almost the same as one for the ordinary birthday paradox. With this attack, we claim that 50-bit security against forgery in the nonce-repeating model is no longer conservative.

4.2 Breaking Confidentiality

First of all, we point out that Silver encrypts the last partial message block in the similar manner as the CTR mode, i.e. it generates the key stream and takes the XOR with the plaintext. Thus, the designers' claim only comparing the security of the ECB mode is strange. Comparison with the combination of the ECB and CTR modes seems more natural. Due to the property of CTR-mode like structure, recovering the plaintext of the last partial block with nonce-repeating queries is easy. Yet, we show another way of recovering plaintext which is particular to the computation structure in Silver.

Our observation to break confidentiality in Silver under nonce repetition essentially relies on a missing domain separation. In most ciphers, the cases of full block and partial block treatments are made distinct by using independent permutations by, for instance, injecting different tweak values in the corresponding block cipher calls. In Silver, this is not the case, and the adversary can use this at his advantage to learn internal values during an encryption.

Consider a short plaintext P of 15 bytes, a given nonce N and a secret key k. There is no associated data. To encrypt P, we observe that its length $b_P = 15$ does not fill one block, so we use the partial-block treatment (see Fig. 2, right). Namely, we first generate the mask $\mu = E_{KS(s \cdot \gamma)}(b_p \, \| \, b_p)$, where $s = 1$, and simply XOR its 15 most significant bytes to P to produce the corresponding ciphertext C. The subsequent operations related to the tag generation T are irrelevant here.

Once an adversary gets (N, C, T) with $|C| = 15$, he can recover the original plaintext P as long as he can recompute the internal mask value μ. We note that this is indeed possible using the nonce-repeating encryption query $(N, A = \epsilon, P = b_p \, \| \, b_p)$ since the same subkeys $KS(s \cdot \gamma)$ yield the same permutation to be used since $s = 1$.

Therefore, with a single encryption query under the same nonce, the adversary can break the confidentiality of C. We note that the same observation for longer P with b_p not a multiple of 16 can be conducted, but would only recover the last $b_p \pmod{16}$ bytes of P. The second query would simply consist of $s - 1$ random blocks, $s = \left\lfloor \frac{b_p}{16} \right\rfloor$ and the same last block $b_p \, \| \, b_p$ to use the same permutation to reveal μ.

4.3 Key-Recovery Attack on 8-Round Silver

The forgery attack on 8-round Silver in the nonce-respecting model discussed in Sect. 3.2, can be further extended into a key-recovery attack working on the same number of rounds in the nonce-repeating model. Recall that this attack recovers four bytes (marked in gray in Fig. 6) of $u_0 = k_0 \oplus \alpha_1$ for a nonce N. Then, if the value of α_1 for that nonce N can be recovered, which in turn determines four bytes of k_0, we can do a brute-force search to recover the other bytes of k_0, and therefore the key k. Therefore, we are left to find an algorithm to recover this α_1.

First, note that with respect to the target α_1, it is known that the 8 least significant bits of α and the 63 most significant bits of α_9^L are zeros. Hence, the number of candidate values for α_1 is essentially reduced to $2^{57=(128-8-63)}$. Below, we explain how to further reduce the number of its candidate values, and eventually to recover its value in the nonce-reuse model.

For a nonce N, let (N, A, P, C, T) and (N, A, P', C', T') be two plaintext-ciphertext pairs of Silver with N as the nonce and with the same associated data A. Moreover, both P and P' are one full-block long. We observe that if $T = T'$ holds, it implies $\Sigma_A \oplus \Sigma_P \oplus \Sigma_C = \Sigma_A \oplus \Sigma'_P \oplus \Sigma'_C$, which is equivalent to:

$$\Sigma_C \oplus \Sigma'_C = \left(C + (\alpha + \gamma) \right) \oplus \left(C' + (\alpha + \gamma) \right) = P \oplus P'.$$

From the known values of C, C' and $P \oplus P'$, we can recover partially the value of $\alpha + \gamma$. For the sake of simplicity, we denote $\alpha + \gamma$ as X and $P \oplus P'$ as Y, the i-th bit of C as $C[i]$, and the i-th carry bit for $C + X$ as $\mathtt{CR}[i]$. Similarly, we define $C'[i]$, $X[i]$, $Y[i]$ and $\text{CR}'[i]$. At the bit level, the computation of $\Sigma_C \oplus \Sigma_C'$ gives:

$$\Big(C[i] \oplus X[i] \oplus \text{CR}[i]\Big) \oplus \Big(C'[i] \oplus X[i] \oplus \text{CR}'[i]\Big) = Y[i],$$

which is equivalent to:

$$\text{CR}[i] \oplus \text{CR}'[i] = C[i] \oplus C'[i] \oplus Y[i].$$

Hence, we can compute the difference for each carry bit between $C + X$ and $C'+X$, that is $\text{CR}[i]\oplus\text{CR}'[i]$ for each i. Moreover, at the bit level, the computations of $C + X$ and $C' + X$ give:

$$\text{CR}[i + 1] = \Big(C[i] + X[i] + \text{CR}[i]\Big) \gg 1,$$

$$\text{CR}'[i + 1] = \Big(C'[i] + X[i] + \text{CR}'[i]\Big) \gg 1,$$

where $\gg$ denotes the right shift by one bit.

Observation 1. *If $C[i] = C'[i]$ and $P[i] \oplus P'[i] = 1$, then $X[i]$ can be computed as: $X[i] = C[i] \oplus \text{CR}[i + 1] \oplus \text{CR}'[i + 1]$.*

Proof. From $Y[i] = P[i] \oplus P'[i] = 1$ and $C[i] = C'[i]$, we have $\text{CR}[i] \oplus \text{CR}'[i] = 1$. Without loss of generality, we assume $\text{CR}[i] = 0$ and $\text{CR}'[i] = 1$ and distinguish two cases with $C[i] = 0$ and $C[i] = 1$ separately.

- **Case $C[i] = C'[i] = 0$.** We have that:

$$\text{CR}[i + 1] \oplus \text{CR}'[i + 1] = ((0 + X[i] + 0) \gg 1) \oplus ((0 + X[i] + 1) \gg 1),$$
$$= (X[i] + 1) \gg 1.$$

It is trivial to get that $\text{CR}[i + 1] \oplus \text{CR}'[i + 1] = 0$ implies $X[i] = 0$, and $\text{CR}[i+1]\oplus\text{CR}'[i+1] = 1$ implies $X[i] = 1$. Hence $X[i] = C[i]\oplus\text{CR}[i+1]\oplus\text{CR}'[i+1]$ holds.

- **Case $C[i] = C'[i] = 1$.** We have that:

$$\text{CR}[i + 1] \oplus \text{CR}'[i + 1] = ((1 + X[i] + 0) \gg 1) \oplus ((1 + X[i] + 1) \gg 1),$$
$$= ((X[i] + 1) \oplus (X[i] + 2)) \gg 1.$$

It is trivial to get that $\text{CR}[i + 1] \oplus \text{CR}'[i + 1] = 0$ implies $X[i] = 1$, and $\text{CR}[i+1]\oplus\text{CR}'[i+1] = 1$ implies $X[i] = 0$. Hence $X[i] = C[i]\oplus\text{CR}[i+1]\oplus\text{CR}'[i+1]$ holds. $\qquad\square$

Based on the above observation, for a nonce N, an algorithm of recovering partially its corresponding $\alpha + \gamma$ is as follows. The notations follow the above definitions.

1. Select two sets of 2^{64} distinct one full-block plaintext $\{P\}$ and $\{P'\}$ such that all plaintexts of $\{P\}$ (resp. $\{P'\}$) have the same value of P^R (resp. P'^R) and moreover $P^R \oplus P'^R = 1^{64}$ holds. Query all $(N, A = \epsilon, P)$s and $(N, A = \epsilon, P')$s to Silver and receive (C, T)s and (C', T')s, respectively. Find a pair (N, A, P, C, T) and (N, A, P', C', T') with $T = T'$.
2. Compute the difference of carry bits $\mathtt{CR} \oplus \mathtt{CR}'$ between $C + X$ and $C' + X$, where X refers to $\alpha + \gamma$.
3. Recover the bits $X[i]$s that satisfy $C[i] = C'[i]$ and $P[i] \oplus P'[i] = 1$.

We now evaluate the number of bits of X that can be recovered. Particularly, for each i with $1 \leq i \leq 64$, the condition $P[i] \oplus P'[i] = 1$ always holds, and $X[i]$ can be recovered as long as $C[i] = C'[i]$ holds. Therefore on average 32 bits of the right half X^R of X can be recovered. Then, by doing a similar attack again which forces $P^L \oplus P'^L = 1^{64}$ instead, we can recover 32 bits of the left half X^L of X. Combining together, we get to know (at least) 64 bits of X with a complexity of 2^{66}.

Putting everything together, the key-recovery algorithm for 8-round Silver in the nonce-repeating model is detailed as follows.

1. Launch the forgery attack in Sect. 3. Let N be the nonce used in the forged decryption query.
2. For this nonce N, launch the above attack algorithm to recover (at least) 64 bits of its corresponding $\alpha + \gamma$.
3. Guess the unknown 65 of α_9 (corresponding to the nonce N) exhaustively, then compute key schedule function of AES to get α, α_1 and $\alpha + \gamma$, and examine if all the conditions on them are satisfied or not. This recovers the correct value of α_9 and in turn α_1.
4. Compute 32 bits of k_0 (marked in grey in Fig. 6). Finally recover the other bits of k_0 by a brute-force search, which in turn recovers the key k.

The overall complexity is obviously dominated by Step 1, which is 2^{111} blocks for all queries.

5   Conclusion

In this paper, we have presented the first third-party security analysis of the CAESAR candidate Silver. For nonce-respecting adversaries, we show an 8-round forgery attack with 2^{111} blocks of queries. The attack exploits the sparse injection of the block tweak, which allows to exploit the internal difference between blocks. For nonce-repeating adversaries, we show practical forgery and plaintext-recovery attacks against full Silver and a key-recovery attack against eight rounds of Silver. Our plaintext-recovery attack shows that the security of Silver in the nonce-repeating model is much less than that of AES-ECB, which breaks the security claim made by the designers. Our key-recovery attack shows that the adversaries can attack more rounds against Silver than AES-128.

We believe that achieving 128-bit security based on AES-128 is very challenging but definitely worth trying. Proposing a new construction for tweaking

AES seems a good approach. Our results show that careful security analysis, or possibly security proofs, is strongly required when a new tweaking method is proposed. More security analysis on the other tweaked AES based designs are open.

Acknowledgement. Jérémy Jean is supported by the Singapore National Research Foundation Fellowship 2012 (NRF-NRFF2012-06). Lei Wang is supported by the Singapore National Research Foundation Fellowship 2012 (NRF-NRFF2012-06), Major State Basic Research Development Program (973 Plan) (2013CB338004), National Natural Science Foundation of China (61472250) and Innovation Plan of science and technology of Shanghai (14511100300).

References

1. Bernstein, D.: CAESAR Competition (2013). http://competitions.cr.yp.to/caesar.html
2. Daemen, J., Knudsen, L.R., Rijmen, V.: The block cipher SQUARE. In: Biham, E. (ed.) FSE 1997. LNCS, vol. 1267, pp. 149–165. Springer, Heidelberg (1997)
3. Daemen, J., Rijmen, V.: The Design of Rijndael: AES - The Advanced Encryption Standard. Springer, Heidelberg (2002)
4. Derbez, P., Fouque, P.-A., Jean, J.: Improved key recovery attacks on reduced-round AES in the single-key setting. In: Johansson, T., Nguyen, P.Q. (eds.) EUROCRYPT 2013. LNCS, vol. 7881, pp. 371–387. Springer, Heidelberg (2013)
5. Jean, J., Nikolić, I., Peyrin, T.: Deoxysv1.2 Submission to the CAESAR competition (2014)
6. Jean, J., Nikolić, I., Peyrin, T.: Joltikv1.2 Submission to the CAESAR competition (2014)
7. Jean, J., Nikolić, I., Peyrin, T.: Kiasuv1.2 Submission to the CAESAR competition (2014)
8. Jean, J., Nikolic, I., Peyrin, T.: Tweaks and keys for block ciphers: the TWEAKEY framework. In: Sarkar, P., Iwata, T. (eds.) Advances in Cryptology—ASIACRYPT 2014. LNCS, vol. 8874, pp. 274–288. Springer, Heidelberg (2014)
9. Knudsen, L.R., Wagner, D.: Integral cryptanalysis. In: Daemen, J., Rijmen, V. (eds.) FSE 2002. LNCS, vol. 2365, p. 112. Springer, Heidelberg (2002)
10. Krovetz, T., Rogaway, P.: The software performance of authenticated-encryption modes. In: Joux, A. (ed.) FSE 2011. LNCS, vol. 6733, pp. 306–327. Springer, Heidelberg (2011)
11. Lipmaa, H., Moriai, S.: Efficient algorithms for computing differential properties of addition. In: Matsui, M. (ed.) FSE 2001. LNCS, vol. 2355, p. 336. Springer, Heidelberg (2002)
12. Liskov, M., Rivest, R.L., Wagner, D.: Tweakable block ciphers. In: Yung, M. (ed.) CRYPTO 2002. LNCS, vol. 2442, p. 31. Springer, Heidelberg (2002)
13. Penazzi, D., Montes, M.: Silver v1. submitted to the CAESAR competition (2014)
14. Peyrin, T.: Improved differential attacks for ECHO and Grøstl. In: Rabin, T. (ed.) CRYPTO 2010. LNCS, vol. 6223, pp. 370–392. Springer, Heidelberg (2010)
15. Rogaway, P.: Efficient instantiations of tweakable blockciphers and refinements to modes OCB and PMAC. In: Lee, P.J. (ed.) ASIACRYPT 2004. LNCS, vol. 3329, pp. 16–31. Springer, Heidelberg (2004)
16. Vaudenay, S.: A Classical Introduction to Cryptography: Applications for Communications Security. Springer, Heidelberg (2006)

Cryptanalysis of the Authenticated Encryption Algorithm COFFE

Ivan Tjuawinata[✉], Tao Huang, and Hongjun Wu

Division of Mathematical Sciences, School of Physical and Mathematical Sciences,
Nanyang Technological University, Singapore, Singapore
S120015@e.ntu.edu.sg, {huangtao,wuhj}@ntu.edu.sg

Abstract. COFFE is a hash-based authenticated encryption scheme. In the original paper, it was claimed to have IND-CPA security and also ciphertext integrity even in nonce-misuse scenario. In this paper, we analyse the security of COFFE. Our attack shows that even under the assumption that the primitive hash function is ideal, a valid ciphertext can be forged with 2 enquiries with success probability close to 1. The motivation of the attack is to find a collision on the input of each of the hash calls in the COFFE instantiation. It can be done in two ways.

The first way is by modifying nonce and last message block size. Chosen appropriately, we can ensure two COFFE instantiations with different nonce and different last message block size can have exactly the same intermediate state value. This hence leads to a valid ciphertext to be generated. Another way is by considering two different COFFE instantiations with different message block size despite same key. In this case, we will use the existence of consecutive zero in the binary representation of π to achieve identical intermediate state value on two different COFFE instantiations. Having the state collisions, the forgery attack is then conducted by choosing two different plaintexts with appropriate nonce and tag size to query. Having this fact, without knowing the secret key, we can then validly encrypt another plaintext with probability equal to 1.

Keywords: COFFE · Authenticated cipher · Forgery attack

1 Introduction

Authenticated encryption is a symmetric encryption scheme aiming to provide authenticity at the same time as confidentiality to the message. Initially, Bellare and Namprempre proposed the authenticated encryption(AE) schemes by integrating an encryption scheme with an authentication scheme in 2000, [1]. In 2001, Krawczyk published a paper [8] that studies the possibility to solve this problem by applying the existing symmetric key cryptosystem and hash function one after another.

The difficulty of the general composition approach is although the security of the parts individually is well-studied, the application of one function may affect the security of the other. Furthermore, in implementation point of view, it is

© Springer International Publishing Switzerland 2016
O. Dunkelman and L. Keliher (Eds.): SAC 2015, LNCS 9566, pp. 510–526, 2016.
DOI: 10.1007/978-3-319-31301-6_29

not very efficient and error-prone considering it is required to have two different primitives, one for encryption, one for plaintext integrity.

To tackle the first difficulty, a lot of dedicated designs to simultaneously encrypt and authenticate the message have been proposed, among which the authenticated encryption mode is a commonly used design approach. Some examples of these mode of operations are IAPM [7], OCB [11], Jambu [13], GCM [5], CCM [4] and ELmD [3].

The consideration for the efficiency comes from the fact that encryption and authentication is done independently with each of their own primitive. So one way to solve this is to consider using the same primitive for both purposes. The initial direction that research goes was to construct a block-cipher based hash function for the authentication purpose such as the ones found in [9,10].

Another way to solve this problem is to purely use a hash function for both encryption and authentication purposes. Some of AE modes that is based on hash functions are OMD [2] and COFFE [6].

COFFE is a hash-function-based authenticated encryption scheme designed by Forler *et al.* It was published in ESC 2013 [6]. COFFE is designed to be secure for computationally constraint environment. As mentioned above, COFFE utilises a hash function for both encryption and authentication without introducing any block cipher primitive. According to [6], COFFE is one of the first authenticated encryption that is purely based on hash function. This alternative direction of constructing an authenticated encryption system is interesting for constructing a secure authenticated encryption.

The designers claim that COFFE is secure against chosen plaintext attack in nonce-respecting scenario. It is also claimed to have ciphertext-integrity even in nonce-misuse scenario. In particular, it is claimed that the ciphertext integrity of COFFE is at least strong as the indistinguishability of the hash function used. That is, forging a ciphertext with a valid tag should be as hard as finding collision in the underlying hash function. Furthermore, it also provides additional features. Firstly, it provides failure-friendly authenticity, that is, COFFE provides reasonable authenticity in the case of weaker underlying hash function. Secondly, it also provides side channel resistance under nonce-respecting scenario.

In this paper, we first analyse the design of COFFE. During the analysis we consider the scheme firstly under the nonce-repeating scenario. Instead of using any specific hash function for the underlying primitive, we analyse it on the generic construction case with an ideal underlying hash function. We show that under these settings, some instances of COFFE with particular parameters are vulnerable to distinguishing attack, ciphertext forgery attack, or related key recovery attack. Thus, the security claim of COFFE for these parameters does not hold.

The attacks come from the consideration that intermediate state values of two different COFFE instantiations can be made the same while having different inputs. The vulnerability comes from the fact that having most of the parameters to be variables, different set of parameters can be chosen and combined to create the collision. The attack starts by first trying to find a specific value for the parameters where this can happen. Having found these parameters, different approaches are made to exploit this discovery to launch either distinguishing attack, forgery attack, or key recovery attack. In this paper, we found that for the distinguishing and forgery attack, if we use the same secret key for all the instantiations, the success probability is close to 1.

The rest of this paper is structured as follows: The generic specification of COFFE is given in Sect. 2. Section 3 provides some analysis and observation of COFFE. Section 4 introduces the distinguishing attack. Section 5 provides two variants of ciphertext forgery attack. We proposed a related key recovery attack on Sect. 6. Lastly, Sect. 7 concludes the paper.

2   The COFFE Authenticated Cipher

The COFFE family of authenticated ciphers uses six parameters: key length, nonce length, block size, hash function input and output size, and tag length. We will briefly describe the specification of COFFE authenticated cipher. The full specification can be found in [6]. An overview of COFFE is provided in Fig. 1.

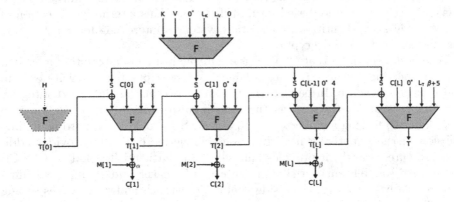

Fig. 1. General scheme of COFFE encryption and authentication (Fig. 2 of [6])

2.1   Notations

Throughout this paper, we will be using the following notations:

- $\mathcal{F}$: Underlying Hash function
  - γ: Input size for $\mathcal{F}$ assuming "one compression function invocation per hash function call"
  - δ: Output size for $\mathcal{F}$
- $\mathcal{L}_K$: Secret key length expressed in bits. The length of this string should be a multiple of a byte

- $\mathcal{L}_V$: Nonce length expressed in bits. The length of this string should be a multiple of a byte
- $\mathcal{L}_T$: Tag length expressed in bits. The length of this string should be a multiple of a byte, $\mathcal{L}_T \leq \delta$
- α: Message block size
- β: Last message block size, $\beta \leq \alpha \leq \delta$
- x: Bits for domain value. The number of byte for used for x follows the number of bytes needed to express $\beta + 5$ in bits.
- Let v be a binary string and b be a positive integer.
  - $|v|$: The length of v in bits.
  - $|v|_b$: A $b-$bit binary representation of v.
  - $[v]$: The length of v in byte.
  - $[v]_b$: A $b-$byte binary representation of v.
  - $\mathfrak{b}$: $[\beta + 5]$.
- $\mathcal{S}_1 || \mathcal{S}_2$: Concatenation of string $\mathcal{S}_1$ followed by $\mathcal{S}_2$.
- $\mathcal{S}_1 \bigoplus_\ell \mathcal{S}_2$: The ℓ- bit string obtained by XOR-ing the ℓ least significant bits of $\mathcal{S}_1$ and $\mathcal{S}_2$.
  - $\mathcal{S}_1 =_b \mathcal{S}_2$: The last b bits of both $\mathcal{S}_1$ and $\mathcal{S}_2$ is the same.
- $\mathcal{S}_1 || 0^\star || \mathcal{S}_2$: When clear the total length should be, say a bits, concatenate $\mathcal{S}_1$ with 0-bits then with $\mathcal{S}_2$ with the number of 0-bits being the difference between a and the total length of $\mathcal{S}_1$ and $\mathcal{S}_2$.
- $\mathcal{K}$: Secret key string
- $\mathcal{V}$: Nonce
- $\mathcal{L}$: The number of message blocks for the encryption
- $\mathcal{S}$: Session Key with length δ bits
- $\mathcal{H}$: Associated Data
- $\mathcal{M}[i], 1 \leq i \leq \mathcal{L}$: The i-th message block, an α bit string except for $\mathcal{M}[\mathcal{L}]$ having length β bits.
- $\mathcal{C}[0]$: The initial vector
- $\mathcal{C}[i], 1 \leq i \leq \mathcal{L}$: The i-th ciphertext block with the same length as $\mathcal{M}[i]$
- $\mathcal{T}[i], 0 \leq i \leq L$: Chaining values for the scheme each of which having length δ bits
- $\mathcal{T}$: Message Tag.

2.2 Associated Data Processing

The method of processing the associated data, $\mathcal{H}$, can be divided into three cases based on the length of the associated data.

- If the length of $\mathcal{H}$ is less than δ bits, it is appended by 1 followed by appropriate number of zeros to reach δ bits. This is defined as $\mathcal{T}[0]$ and a domain value x is defined to be 1.
- If the length of $\mathcal{H}$ is exactly δ bits, this is directly defined as $\mathcal{T}[0]$ while the domain value x is set to be 2.
- If the length of $\mathcal{H}$ is more than δ bits, feed $\mathcal{H}$ to $\mathcal{F}$ and the resulting hash output is used as the value of $\mathcal{T}[0]$ and x is defined as 3.

2.3 Initialization

There are two values that need to be computed in the initialization phase, S and $C[0]$. Firstly, the session key, S which is defined based on $\mathcal{K}, \mathcal{V}, \mathcal{L}_K, \mathcal{L}_V$, and $\mathfrak{b}$. The value of S is defined to be $\mathcal{F}(\mathcal{K}||\mathcal{V}||0^\star||\mathcal{L}_K||\mathcal{L}_V||[0]_\mathfrak{b})$. Note that here $0^\star$ is used to pad the string to make the length equals to γ.

Next, the constant $C[0]$ which depends only on the message block size α. $C[0]$ is defined to be the first $\frac{\alpha}{4}$ post-decimal values of π interpreted as a hexadecimal string. So for example, since the decimal values of π is $.14159\ldots$, if $\alpha = 16$, Then $C[0] = 0 \times 1415 = 0001010000010101$.

2.4 Processing Plaintext

Plaintext is encrypted to obtain the ciphertext after the generation of session key S, the initialization vector $C[0]$, initial chain value $T[0]$ and the domain value, x. The plaintext blocks are processed as follow:

$$T[1] = \mathcal{F}((S \bigoplus T[0]) \;||\; C[0] \;||\; 0^\star \;||\; [x]_\mathfrak{b})$$
$$C[1] = \mathcal{M}[1] \bigoplus_\alpha T[1]$$

for all blocks $\mathcal{M}[i], 2 \le i \le \mathcal{L} - 1\{$
$$T[i] = \mathcal{F}((S \bigoplus T[i-1]) \;||\; C[i-1] \;||\; 0^\star \;||\; [4]_\mathfrak{b})$$
$$C[i] = \mathcal{M}[i] \bigoplus_\alpha T[i]$$
$\}$

$$T[\mathcal{L}] = \mathcal{F}((S \bigoplus T[\mathcal{L}-1]) \;||\; C[\mathcal{L}-1] \;||\; 0^\star \;||\; [4]_\mathfrak{b})$$
$$C[\mathcal{L}] = \mathcal{M}[\mathcal{L}] \bigoplus_\beta T[\mathcal{L}].$$

2.5 Tag Generation

After the associated data and plaintext are processed, the $\mathcal{L}_T$-bit tag $\mathcal{T}$ is derived:

$$\mathcal{T} = \mathcal{F}((S \bigoplus T[\mathcal{L}]) \;||\; C[\mathcal{L}] \;||\; 0^\star \;||\; \mathcal{L}_T \;||\; \beta + 5)$$

The decryption is trivial and we omit it here. For the verification, only the $\mathcal{L}_T$ least significant bits of the tags are checked.

2.6 Security Goals of COFFE

COFFE is claimed to have the INT-CTXT (ciphertext integrity) and IND-CPA(indistinguishable under chosen plaintext attack) property under nonce-respecting scenario.

In particular, in Lemma 1 of [6], we have:

Lemma 1. *Let Π be a COFFE scheme as defined above with $\mathcal{F}$ as its underlying hash function. Then the advantage of adversary $\mathcal{A}$ under nonce-respecting*

*scenario with q queries and ℓ message blocks to the encryption oracle with time
bounded by t can be upper bounded by:*

$$\mathbf{Adv}_\Pi^{CPA}(q, \ell, t) \leq \frac{8\ell^2 + 3q^2}{2^n} + 2.\mathbf{Adv}_{\mathcal{F}}^{PRF\text{-}XRK}(q, \ell, t).$$

In other words, distinguishing COFFE from a random function with chosen
input under the bound of (q, ℓ, t) should be at least as hard as distinguishing $\mathcal{F}$
from a random function $\$: \{0, 1\}^\gamma \Rightarrow \{0, 1\}^\delta$.

Additionally, COFFE has some other security claim under different circumstances. Firstly, under the nonce-misuse scenario, it claimed that

– *"..., the integrity of the ciphertext does not depend on a nonce, but only on
the security of $\mathcal{F}$".*

In particular, in Lemma 2 of [6], we have:

Lemma 2. *Let Π be a **COFFE** scheme as defined above with $\mathcal{F}$ as its underlying hash function. Then in the nonce-ignoring adversary scenario with q queries
for ℓ message blocks and t times, we have*

$$\mathbf{Adv}_\Pi^{INT\text{-}CTXT}(q, \ell, t) \leq \frac{3\ell^2 + 2q^2}{2^\delta} + \frac{q}{2^{\mathcal{L}_T}} + \mathbf{Adv}_{\mathcal{F}}^{PRF}(q + \ell, O(t)).$$

This implies that the hardness of forging a ciphertext with a valid tag should
be at least as hard as distinguishing $\mathcal{F}$ from a random function from $\{0, 1\}^\gamma$ to
$\{0, 1\}^\delta$.

Secondly, COFFE also provides a failure-friendly authenticity. That is, under
a weaker assumption on the security of the underlying hash function $\mathcal{F}$, the
authenticity of the message is still kept.

Lastly, COFFE also provides a reasonable resistance against side channel
attack. This is so because *"for each encryption process, a new short term key is
derived from a nonce and the long term key"* [6].

3   Analysis on the COFFE Scheme

In our analysis, we will assume $\mathcal{F} : \{0, 1\}^* \Rightarrow \{0, 1\}^\delta$ to be an arbitrary ideal
hash function with γ being the largest possible length of the input to ensure
exactly one compression function invocation per hash function call. Here we are
assuming the possibility of the parameters to have length more than 255 bits.
In other words, it is possible that it requires more than 1 byte to represent
$\mathcal{L}_K, \mathcal{L}_V, \mathcal{L}_T$ in their binary format.

The first observation is about the input for the hash function call. Note that
since we are only considering concatenation, there is not always a way, given the
concatenated string, to uniquely determine the value for each strings before the
concatenation. For example, if $a||b = 11011$, it is possible for $a = 110, b = 11$ or
$a = 1, b = 1011$. This leads to the possibility that two different sets of strings to
be concatenated to the same string.

Observation 1. *For the input of any hash function call, due to the absence of separator between substrings and changeable elements lengths, it is possible to have two different sets of strings to be concatenated to the same string.*

On the following subsections, we analyse this observation further to find whether it is possible to utilise this to cause a collision in the intermediate state value of the COFFE. We first consider the case when we fix the message block size while allowing two different last message block sizes, β_1 and β_2, to be used. The analysis is focused on the case when $|\beta_1 - \beta_2|$ is a multiple of 256. The analysis on this can be found on Sect. 3.1. Next we also consider the possibility of having identical intermediate state values when we change the message block size, α, while keeping β fixed. The analysis is focused on how α should be chosen in such a way for the first message block encryption of both instantiations to have identical hash value output. This is discussed in Section 3.2.

3.1   Modification of β

We fix α and consider different values of β. In our next observation, with large enough α, it is possible to have $\beta_1 < \beta_2 \leq \alpha$ such that $\beta_2 - \beta_1$ is a multiple of 256. This implies that the last byte of the input of $\mathcal{F}$ in the tag generation for the two different plaintexts can be the same. As discussed above, however, we want the collision to happen in the whole input string for any $\mathcal{F}$ input. If both $\beta_1 + 5$ and $\beta_2 + 5$ require 2 bytes to represent in binary format, the second to last byte will never agree. So for collision to happen, we need $\beta_1 + 5 < 256, 256 \leq \beta_2 + 5 < 65536$ and $\beta_2 = \beta_1 + 256\rho$ for some integer $1 \leq \rho \leq 255$.

To further analyse this observation, we consider the note by the designers regarding the increase of number of byte required for the binary representations of the domain. In [6], it is stated that if $\beta + 5$ exceeds one byte, all domain representations in the current COFFE will be encoded as two-byte values instead of one. So this is important in our analysis on the possibility of exploring this observation to introduce a successful attack.

Note that in the message processing, assuming that γ is big enough, there are enough bits of the zero padding between $\mathcal{C}[i]$ and the domain values for the encryption to absorb the additional byte for the domain values in case $\beta + 5$ is increased from one byte to two bytes value. So the parts that need to be taken care of for this to happen are the session key generation and tag generation.

In the session key generation, we consider the last several bytes of the input of $\mathcal{F}$. Here we have the input to be ... $\| a \| \mathcal{L}_K \| \mathcal{L}_V \| 0$. Note that when we expand the domain value from 1- to 2-byte value, the domain value should still have the same value. So the second to last byte for the input must be 0. This gives us our next observation.

Observation 2. *To ensure that collision can occur when extending the domain from 1- to 2-byte value, the initial value of $\mathcal{L}_V$ must be a multiple of 256. This means that if the initial $\mathcal{L}_V$ is a 1-byte value, it must be 0, that is, no nonce in the first instance.*

Our primary goal in this section is to investigate the possibilities to have two different (key, nonce) pairs, $(\mathcal{K}_1, \mathcal{V}_1)$ and $(\mathcal{K}_2, \mathcal{V}_2)$ with lengths $\mathcal{L}_{K_1}, \mathcal{L}_{V_1}, \mathcal{L}_{K_2}, \mathcal{L}_{V_2}$ respectively such that

$$(\mathcal{K}_1 \parallel \mathcal{V}_1 \parallel 0^* \parallel \mathcal{L}_{K_1} \parallel \mathcal{L}_{V_1} \parallel [0]_1) = (\mathcal{K}_2 \parallel \mathcal{V}_2 \parallel 0^* \parallel \mathcal{L}_{K_2} \parallel \mathcal{L}_{V_2} \parallel [0]_2).$$

For simplicity, let

$$\mathcal{S}_1 = (\mathcal{K}_1 \parallel \mathcal{V}_1 \parallel 0^* \parallel \mathcal{L}_{K_1} \parallel \mathcal{L}_{V_1} \parallel [0]_1),$$
$$\mathcal{S}_2 = (\mathcal{K}_2 \parallel \mathcal{V}_2 \parallel 0^* \parallel \mathcal{L}_{K_2} \parallel \mathcal{L}_{V_2} \parallel [0]_2).$$

Here, we note that collision is indeed possible as illustrated by the following example: Let SHA-512 be our hash function. We can choose any 256-bit string, say $\mathcal{K}$, and set $\mathcal{K}_1 = \mathcal{K}_2 = \mathcal{K}$. Now we use any one bit value (0 or 1) as our $\mathcal{V}_2$ while $\mathcal{V}_1$ is set to be $\mathcal{V}_2$ appended by 255 zeros. Now if we use 472 zero paddings on the first string while using 727 bits for the second string we will have

$$\mathcal{S}_1 = \mathcal{S}_2 = (\mathcal{K} \parallel 1 \parallel 0^{255} \parallel 0^{472} \parallel [1]_1 \parallel [0]_1 \parallel [1]_1 \parallel [0]_1 \parallel [0]_1).$$

In the remaining of this section, we will try to analyse whether such collision is possible for other instances of COFFE. Here we analyse different cases of $\mathcal{S}_1$ on the possibility of having $\mathcal{S}_1 = \mathcal{S}_2$. The factors that we need to consider are the number of bytes required for $\mathcal{L}_{K_i}, \mathcal{L}_{V_i}$ and whether there is any zero paddings required. Note that if $\mathcal{L}_{K_i}$ or $\mathcal{L}_{V_i}$ is a 3-byte value, the value will be at least 65536 which is too big. To simplify our discussion, for this paper, we will only consider the key and nonce to have length whose binary format can be represented as at most a 2-byte value. Due to the big number of cases we need to consider and the similarity of the cases, we will just discuss one case as example and a full analysis of the other cases can be found in the appendix of the full version paper [12] while the Table 1 containing the conclusion is provided for reference.

I.3.c Case I.3.c.: $\mathcal{S}_1$ has no zero paddings, $[\mathcal{L}_{K_1}] = 1, [\mathcal{L}_{V_1}] = 2, [\mathcal{L}_{K_2}] = 1,$ $[\mathcal{L}_{V_2}] = 2.$

By Observation 2, $\mathcal{L}_{V_1} = 256\mathbf{b}$ and $\mathcal{L}_{K_1} = \mathbf{a}$ where $1 \leq \mathbf{a}, \mathbf{b} \leq 255$. Both $\mathbf{a}$ and $\mathbf{b}$ are nonzero because of the following reasons. First of all, since $\mathcal{L}_{K_1} = \mathbf{a}$, if $\mathbf{a} = 0$, then there is no secret key, in which case, no confidentiality for the message. So we can disregard the case when $\mathcal{L}_K = 0$. Next, since $[\mathcal{L}_{V_1}] = 2$, this should mean that $\mathcal{L}_{V_1} \geq 256$ since otherwise, $[\mathcal{L}_{V_1}] = 1$. So if $\mathbf{b} = 0$, this implies $\mathcal{L}_{V_1} = 0$ which violates the requirement $\mathcal{L}_{V_1} \geq 256$. Hence

$$\mathcal{S}_1 = (\mathcal{K}_1 \parallel \mathcal{V}_1 \parallel \mathbf{a} \parallel \mathbf{b} \parallel [0]_2).$$

Let $\mathcal{V}_1 = \mathcal{V}_1' \parallel \mathbf{d}$ where $\mathbf{d}$ is the last byte of $\mathcal{V}_1$. So

$$\mathcal{S}_1 = (\mathcal{K}_1 \parallel \mathcal{V}_1' \parallel \mathbf{d} \parallel \mathbf{a} \parallel \mathbf{b} \parallel [0]_2).$$

Consider the alternative string $\mathcal{S}_2$. Recall that here we want $\mathcal{S}_1 = \mathcal{S}_2$ where $\mathcal{S}_2$ has its domain value represented as a 2-bytes value. This implies that the $[0]_2$ in the last 2 bytes of S_1 must appear as the domain for S_2. So this implies that $\mathcal{L}_{K_2} = \mathbf{d}$ and $\mathcal{L}_{V_2} = 256\mathbf{a} + \mathbf{b}$. Let t' be the number of zero padding in S_2 where $t' \geq 0$. Equating S_1 with S_2, we have $\mathcal{K}_1 \, || \, V_1' = \mathcal{K}_2 \, || \, V_2 \, || \, 0^{t'}$. Now comparing the length of these substrings, we have $\mathbf{a} + 256\mathbf{b} - 8 = \mathbf{d} + 256\mathbf{a} + \mathbf{b} + t'$ or equivalently, $255(\mathbf{b} - \mathbf{a}) = \mathbf{d} + 8 + t'$. Consider the family:

$$\mathcal{F}_{(1,2),(1,2)} = \{(\mathbf{a}, \mathbf{b}, \mathbf{d}, t') : 1 \leq \mathbf{a}, \mathbf{b}, \mathbf{d} \leq 255, t' \geq 0, 255(\mathbf{b} - \mathbf{a}) = \mathbf{d} + 8 + t'\}.$$

Now we consider the feasibility of each element of $\mathcal{F}_{(1,2),(1,2)}$. Feasibility here means the possibilities of using these values as the parameters to have the collision. Let $(\mathbf{a}, \mathbf{b}, \mathbf{d}, t') \in \mathcal{F}_{(1,2),(1,2)}$. Note that the collision may not happen with probability 1 due to the case when $\mathcal{K}_1 \neq \mathcal{K}_2$. Note that since key is the first part of the collided string, this can only happen when $\mathcal{L}_{K_1} \neq \mathcal{L}_{K_2}$.

Before going on to the analysis, we have an assumption first. Suppose that $\mathcal{L}_{K_1} > \mathcal{L}_{K_2}$. Since $\mathcal{K}_1$ is the first $\mathcal{L}_{K_1}$ bits of S_1, $\mathcal{K}_2$ is the first $\mathcal{L}_{K_2}$ bits of S_2 and we need $S_1 = S_2$, the first $\mathcal{L}_{K_2}$ bits of $\mathcal{K}_1$ must be $\mathcal{K}_2$. Instead of assuming that this happens by chance, we will assume the following: The user has 2 different instantiations of COFFE scheme with different parameter and different key length. However, the keys chosen by the user are not independent. The longer key is an extension of the shorter key by a random secret string. We note that this assumption is only made for the related key setting attack and not for the general attack.

Based on this assumption, we then have the probability of $\mathcal{K}_1$ to have its first $\mathcal{L}_{K_2}$ bits to be the same as $\mathcal{K}_2$ is exactly 1.

Now back to our case, we have that $\mathcal{L}_{K_1} = \mathbf{a}$ and $\mathcal{L}_{K_2} = \mathbf{d}$. So the length difference of the two keys is $|\mathbf{a} - \mathbf{d}|$ bits. Now if $\mathbf{a} = \mathbf{d}$, then we have $\mathcal{K}_1 = \mathcal{K}_2$. Now the rest of the two strings are $V_1' \, || \, \mathbf{d} \, || \, \mathbf{a} \, || \, \mathbf{b} \, || \, [0]_2$ and $V_2 \, || \, 0^{t'} \, || \, \mathbf{d} \, || \, \mathbf{a} \, || \, \mathbf{b} \, || \, [0]_2$. So we have $V_1 = V_2 \, || \, 0^{t'} \, || \, \mathbf{d}$. Now since V_1 can be controlled by the attacker, we can easily set this to be true. So the probability of the two strings to collide is 1 if $\mathbf{a} = \mathbf{d}$.

Now consider when $\mathbf{a} \neq \mathbf{d}$, specifically, $\mathbf{a} > \mathbf{d}$. The other case can be analysed using exactly the same way. Now let $\mathcal{K}_1 = \mathcal{K}_2 \, || \, \mathcal{K}_1'$ where $\mathcal{K}_1'$ is the last $\mathbf{a} - \mathbf{d}$ bits of $\mathcal{K}_1$. We have $\mathcal{K}_1' \, || \, V_1' \, || \, \mathbf{d} \, || \, \mathbf{a} \, || \, \mathbf{b} \, || \, [0]_2$ and $V_2 \, || \, 0^{t'} \, || \, \mathbf{d} \, || \, \mathbf{a} \, || \, \mathbf{b} \, || \, [0]_2$ as the remaining part of the two strings truncating the first $\mathbf{d}$ bits. Thus, $\mathcal{K}_1' \, || \, V_1' = V_2 \, || \, 0^{t'}$. Note that since $1 \leq \mathbf{a}, \mathbf{d} \leq 255, \mathbf{a} - \mathbf{d} \leq 255, V_2$ has length $256\mathbf{a} + \mathbf{b} \geq 256$. So the entire $\mathcal{K}_1'$ is in V_2. In other words, for the two strings to collide, we need the last $\mathbf{a} - \mathbf{d}$ bits of V_2 must be equal to $\mathcal{K}_1'$. Since $\mathcal{K}_1'$ is supposed to be unknown, the probability of this collision is $2^{\mathbf{a} - \mathbf{d}}$. It is easy to see that the remaining substring can be set to collide with probability 1. So the probability of

Table 1. Session key generation input collision

$[\mathcal{L}_{K_1}]$	$[\mathcal{L}_{V_1}]$	t	$[\mathcal{L}_{K_2}]$	$[\mathcal{L}_{V_2}]$	t'	Collision?	Probability	Key restriction
1	1	Any	Any	Any	Any	No	0	Not Applicable
2	1	Any	1	1	Any	No	≈ 0	Not Applicable
2	1	0	2	1	Any	$\mathcal{F}_{(2,1),(2,1)}$	$2^{-(\mathcal{L}_{K_1}-\mathcal{L}_{K_2})}$	$\mathcal{K}_1 =_{(t'+8)} 0^{t'} \,\|\, \lfloor \frac{\mathcal{L}_{K_2}}{256} \rfloor$
2	1	$0<t<8$	2	1	Any	$\mathcal{F}_{p,(2,1),(2,1)}$	$2^{-(\mathcal{L}_{K_1}-\mathcal{L}_{K_2})}$	$\mathcal{K}_1 =_{(t'+8-t)} 0^{t'} \,\|\, \lfloor \frac{\mathcal{L}_{K_2}}{2^{8+t}} \rfloor$
2	1	Any	Any	2	Any	No	0	Not Applicable
1	2	Any	1	1	$255\mathcal{L}_{V_2}+t$	Yes	1	No
1	2	Any	2	1	Any	No	≈ 0	Not Applicable
1	2	0	1	2	Any	$\mathcal{F}_{(1,2),(1,2)}$	$2^{-|\mathcal{L}_{K_1}-\mathcal{L}_{K_2}|}$	No
1	2	$0<t<8$	1	2	Any	$\mathcal{F}_{p,(1,2),(1,2)}$	$2^{-|\mathcal{L}_{K_1}-\mathcal{L}_{K_2}|}$	No
1	2	0	2	2	Any	$\mathcal{F}_{(1,2),(2,2)}$	$2^{-(\mathcal{L}_{K_2}-\mathcal{L}_{K_1})}$	No
1	2	$0<t<16$	2	2	Any	$\mathcal{F}_{p,(1,2),(2,2)}$	$2^{-(\mathcal{L}_{K_2}-\mathcal{L}_{K_1})}$	No
2	2	Any	1	1	Any	No	0	Not Applicable
2	2	Any	2	1	$255\mathcal{L}_{V_2}+t$	Yes	1	No
2	2	Any	1	2	Any	No	≈ 0	Not Applicable
2	2	0	2	2	Any	$\mathcal{F}_{(2,2),(2,2)}$	$2^{-|\mathcal{L}_{K_1}-\mathcal{L}_{K_2}|}$	$\mathcal{K}_1 =_{\max(0,t'-(\mathcal{L}_{V_1}-8))} 0$
2	2	$0<t<8$	2	2	Any	$\mathcal{F}_{p,(2,2),(2,2)}$	$2^{-|\mathcal{L}_{K_1}-\mathcal{L}_{K_2}|}$	$\mathcal{K}_1 =_{\max(0,t'-(\mathcal{L}_{V_1}-(8-t)))} 0$

collision to happen is $2^{-(\mathbf{a}-\mathbf{d})}$. Using exactly the same analysis, we will see that when $\mathbf{d} > \mathbf{a}$, the probability of collision to happen is $2^{-(\mathbf{d}-\mathbf{a})}$. Hence, for any non-negative integer k, we can define a subfamily of $\mathcal{F}_{(1,2),(1,2)}$,

$$\mathcal{F}_{(1,2),(1,2),k} = \{(\mathbf{a}, \mathbf{b}, \mathbf{d}, t') \in \mathcal{F}_{(1,2),(1,2)} : |\mathbf{a} - \mathbf{d}| \leq k\}.$$

Then for any quadruplet $(\mathbf{a}, \mathbf{b}, \mathbf{d}, t') \in \mathcal{F}_{(1,2),(1,2),k}$ we take as parameter, the collision probability is at least 2^{-k}.

We remark that this probability is applicable for any choices of $\mathcal{K}_1$ and $\mathcal{K}_2$. This observation is essential in our attacks later to decide whether the attacks are only applicable to a family of key to any value of key with the given length.

We include in Table 1 the full list of conclusion of the session key generation analysis. Here we will use t to represent the zero padding for $\mathcal{S}_1$ and t' for $\mathcal{S}_2$. In the Collision column, No means a collision in this case is impossible, yes means a collision will always happen on any choices of the parameter values (with appropriate choice of key and nonce). Lastly, $\mathcal{F}_{(a,b),(c,d)}$ or $\mathcal{F}_{p,(a,b),(c,d)}$ is the family of parameter values that belongs to the respected case that collision is possible. The definition of each of the family can be found in the complete analysis of each case that is either can be found above or in the appendix of the full version paper [12].

Recall that in the tag generation, given $\mathcal{T}[\mathcal{L}], \mathcal{S}$, and $\mathcal{C}[\mathcal{L}]$, the input of $\mathcal{F}$ is

$$\left(\left(\mathcal{S} \bigoplus \mathcal{T}[\mathcal{L}]\right) \,\|\, \mathcal{C}[\mathcal{L}] \,\|\, 0^\star \,\|\, \mathcal{L}_T \,\|\, \beta + 5\right).$$

We note here that here we do not use the byte-aligned assumption in our analysis. The analysis can be restricted to a byte-aligned one by adding a restriction on the families to have some of the values to be divisible by 8. Here the values that are related to the remainder of any value divided by 256 must be divisible by 8. So for example, in the case when $[\mathcal{L}_K] = 2$, if $\mathcal{L}_K = (\mathbf{a} \parallel \mathbf{b})$, then we do not need $\mathbf{a}$ to be divisible by 8. We just need $\mathbf{b}$ to be divisible by 8. This will not change the existence of any of the families. However, it will certainly requires a bigger parameter value. For example, for the case when $[\mathcal{L}_{K_1}] = [\mathcal{L}_{V_1}] = [\mathcal{L}_{K_2}] = 2$ and $[\mathcal{L}_{V_2}] = 1$, if we want all the values to be byte aligned, the smallest parameters we can use is when $\mathcal{L}_{K_1} = \mathcal{L}_{K_2} = 256, \mathcal{L}_{V_2} = 8$, and $\mathcal{L}_{V_1} = 2048$. This leads to the input size for the hash function to be at least $2048 + 256 + 5 \times 8 = 2394$ bits. Here we set $V_2 = 128$ and $V_1 = V_2 \parallel 0^{2040}$ and the other settings to be the same as the previous example.

We move on to the tag generation when $\beta + 5$ changes from 1-byte value to 2-byte value. Note that the only possible source of this 1-byte value is from $\mathcal{L}_T$. So, in the second instantiation where $\beta + 5$ is changed to a 2-byte value, the tag length will be different from the initial one. In fact, the first tag length needs to be a 2-bytes value, say $a \parallel b$ and the second tag length needs to be a while the difference between the two βs needs to be $256 \times b$.

3.2   Modification of α

This section discusses a special case of the analysis in which the user has at least two instantiations of COFFE where they have different message block sizes but the same (or related) key. Since we are considering changing α, the one we really need to take care of is just the generation of $\mathcal{C}[0]$. This is because for any other place where α affects the system, it is generated by the previous chain in which we can truncate easily.

Recall that $\mathcal{C}[0]$ is the first $\frac{\alpha}{4}$ post decimal values of π interpreted as hexadecimal values. Suppose that we want the difference of the two block sizes, α_1 and α_2, to be k with α_2 being the larger value. Since we are assuming the ideality of $\mathcal{F}$, we want the input of $\mathcal{F}$ in this point for both instantiation to coincide. So in other words, if the initial vector of the first instantiation is the α_1 bit $\mathcal{C}_1$ and the second one to be the α_2 bit $\mathcal{C}_2$, the additional k bits of α_2 should be absorbed by the next substring of the input, which is the zero padding. Hence, the last k bits of $\mathcal{C}_2$ should all be zeros. In other words, the value of α_1 so that it can coincide with the positions in the post decimal values of π to have a consecutive 0^k bits. So for example, if we want $\alpha_2 = \alpha_1 + 8$, and α_1 and α_2 to be a multiple of 8, then we will need to wait until the 306-th decimal place to get the 8 bits of consecutive zeros. In this case, $\alpha_1 = 1224$ and $\alpha_2 = 1232$. The requirement that α_1 and α_2 are divisible by 8 comes if we are assuming that the design is byte-aligned. Note that different αs can be found along the places where we can find k consecutive zeros in the binary representation of π.

4   Distinguishing Attack

In this section, to form a distinguishing attack, we use the session key collision discussed in the previous section and the appendix of the full version paper [12].

Assuming that we have the same secret key, different nonce, and different number of byte of domain value but the same session key, as before, we assume that now the session key for each instantiation is the same, each uses the proper number of byte of domain value.

We set the parameters $\alpha, \beta_1, \beta_2, \mathcal{L}_{T_1}$ and $\mathcal{L}_{T_2}$ as follows:

1. $\beta_1 + 5 < 256 < \beta_2 + 5 \leq \alpha + 5 \leq \delta + 5$
2. $\beta_2 - \beta_1 = 256\rho$ for some positive integer ρ
3. $\mathcal{L}_{T_1} < 256 \leq \mathcal{L}_{T_2}$ where $\mathcal{L}_{T_2} = 256\mathcal{L}_{T_1} + \rho$.

Set the first plaintext to be a two-blocks message, $\mathcal{M}_1 = (\mathcal{MB}_1 \parallel \mathcal{MB}_2)$ such that $\mathcal{MB}_1$ has α bits and $\mathcal{MB}_2$ has β_1 bits with tag length set to be $\mathcal{L}_{T_2}$. Assume the ciphertext is $\mathcal{C}_1 = (\mathcal{CB}_1 \parallel \mathcal{CB}_2)$ with $\mathcal{T}_1$ as the tag.

The second message block, is then chosen to be $\mathcal{M}_2 = (\mathcal{MB}_1 \parallel \mathcal{MB}_2')$ such that $\mathcal{MB}_2'$ has β_2 bits. Here we will use the tag length to be $\mathcal{L}_{T_1}$. We also assume the ciphertext is $\mathcal{C}_2 = (\mathcal{CB}_1' \parallel \mathcal{CB}_2')$ with $\mathcal{T}_2$ as the tag.

As discussed above, since the first block of both message are the same, $\mathcal{MB}_1$, we should have $\mathcal{CB}_1 = \mathcal{CB}_1'$ and $\mathcal{T}[1]$ and $\mathcal{T}[2]$ should also be the same. Now remember that $\mathcal{MB}_2 \oplus \mathcal{CB}_2$ and $\mathcal{MB}_2' \oplus \mathcal{CB}_2'$ tells us the last β_1 and β_2 bits of $\mathcal{T}[2]$ respectively. So if $\mathcal{C}_1$ and $\mathcal{C}_2$ are both from COFFE instantiation, we must have $\mathcal{CB}_1 = \mathcal{CB}_1'$ and the last β_1 bits of $\mathcal{CB}_2$ and the last β_1 bits of $\mathcal{CB}_2'$ should agree. So we will guess that it is a COFFE instantiation instead of a random function if these requirements are met. Note that this can happen if it is a random function with probability $2^{-(\alpha+\beta_1)}$.

Recall that a distinguishing attack works as follows. An oracle randomly chooses whether it uses a random function or a COFFE instantiation with the given parameter. Then as an attacker, we can request for encryption for some plaintext. Then an adversary tries to decide whether the oracle uses a random function or a COFFE instantiation. The distinguishing attack described above has error probability 0 if we conclude that the oracle uses a random function. However, if we guess that the oracle uses a COFFE instantiation, there is a probability of $2^{-(\alpha+\beta_1)}$ of the function is actually a random function instead of COFFE. Note that since $\alpha \geq 256$ in our attack, the failure probability is at most 2^{-256} which is very small. Therefore, with 2 enquiries to the oracle with 4 message blocks, COFFE with ideal underlying hash function in nonce-respecting scenario can be distinguished with probability close to 1. So in these instantiations of COFFE, the security claim given in Lemma 1 is not satisfied.

5   Ciphertext Forgery Attack

In this section, we will propose two different ciphertext forgery attacks. The first attack is based on the observation on Subsect. 3.1. It exploits the possibility of having an identical intermediate state value for two different instantiations when we fix α while using different values of β. The detail of the attack can be found in Sect. 5.1. Similarly, Sect. 5.2 discusses the forgery attack based on the discussion on Subsect. 3.2. Here we try to forge a valid ciphertext in the case when there exists two different COFFE instantiations with same key for different message block size. Here both attacks require 2 enquiries and can forge a valid ciphertext with probability one. The success probability 1 is applicable whenever we assume for both instantiations, the secret key used is the same instead of one key being an extension of the other. Lastly, we will also discuss the possibility of combining the two forgery attacks. This can be found in Subsect. 5.3.

5.1   Forgery Attack with Constant Message Block Size

Take any $(\mathcal{K}_1, \mathcal{V}_1), (\mathcal{K}_2, \mathcal{V}_2)$ (key, nonce) pairs from the discussion session such that they generate the same session key, one with 1-byte domain value, the other with two. Let $\mathcal{S}_1$ be the input for session key generation with 1-byte domain value and $\mathcal{S}_2$ be the input for the session key generation with $2-$bytes domain value. Here we assume that the input for session key generation is chosen accordingly based on the number of bytes of domain value. Hence, after this point, we can ignore the secret key and nonce and we can just assume that for each instantiation, we are using the same session key and associated data.

Note that any full block plaintext-ciphertext pair leaks α least significant bits of the output of the hash function for a fixed input, while any β-bit block plaintext-ciphertext pair leaks only β least significant bits of it. So since $\beta \leq \alpha$, it is always more desirable to get a full-block plaintext-ciphertext pairs since they leak the output value more.

Here we set the parameters $\alpha, \beta_1, \beta_2, \mathcal{L}_{T_1}$ and $\mathcal{L}_{T_2}$ as described before in Sect. 4.

Next we define the first message $\mathcal{M}_1$, a $3-$block message $(\mathcal{MB}_1 \parallel \mathcal{MB}_2 \parallel \mathcal{MB}_3)$ such that $|\mathcal{MB}_1| = |\mathcal{MB}_2| = \alpha, |\mathcal{MB}_3| = \beta_2$. Let the ciphertext of this message be $\mathcal{C}_1 = (CB_1 \parallel CB_2 \parallel CB_3)$ with tag $\mathcal{T}_1$ with $\mathcal{L}_T$ set to any value. Here we can compute the values of CB_1 and CB_2 since $\mathcal{MB}_1 \bigoplus CB_1$ gives us the last α bits of $\mathcal{T}[1]$ and $\mathcal{M}_2 \bigoplus \mathcal{C}_2$ gives us the last α bits of $\mathcal{T}[2]$ which are essential in the attack.

We define our second message $\mathcal{M}_2$, a $2-$block message $(\mathcal{MB}_1 \parallel \mathcal{MB}_2')$ with the length of $\mathcal{MB}_2'$ to be β_1 bits and tag length to be $\mathcal{L}_{T_2}$. The first block is chosen to be exactly the same as before to ensure the value of $\mathcal{T}[1]$ and $\mathcal{T}[2]$ can be kept constant. Based on the previous message, the least α bits of both values are known. Suppose that the ciphertext of this plaintext is $\mathcal{C}_2 = (CB_1 \parallel CB_2')$ with tag $\mathcal{T}_2$.

Using the information we obtain, we generate the following valid ciphertext. Define another 2-block message $\mathcal{M}_3 = (\mathcal{MB}_1 \parallel \mathcal{MB}_2'')$. Here we set $\mathcal{MB}_1$ to be the same as the first block from the previous message blocks. This is again to ensure the value of $\mathcal{T}[1]$ and $\mathcal{T}[2]$ can be kept constant. We let the length of $\mathcal{MB}_2''$ to be β_2 and choose $\mathcal{MB}_2''$ such that $\mathcal{MB}_2'' \bigoplus_{\beta_2} \mathcal{T}[2] = \mathcal{C}_2 \parallel 0^{\beta_2 - \beta_1}$. Here, $\mathcal{MB}_2''$ can be calculated since we know the last α bits of $\mathcal{T}[2]$ and $\alpha > \beta_2$. Using this message, it is easy to see that the tag generation will have the same input as before, although $\mathcal{L}_T$ is now $\mathcal{L}_{T_1}$. So the tag for this ciphertext will be the last β_1 bits of $\mathcal{T}_2$.

This attack has success probability equal to the probability of the two strings used as the input session key generation to be the same. As we have discussed before, for some parameters such as the ones in case I.3 and case II.3, this can even be 1. In other words, in the case when the success probability is one, the attack above proves that the ciphertext integrity of this cipher does not satisfy the bound given in Lemma 2 even in an ideal hash function situation.

Note that here we use three COFFE instantiations for each attack (2 for enquiry and 1 for the guess), while in our discussion on session key generation collision, we only consider the collision for two (key, nonce) pairs. So the same attack cannot directly work for nonce-respecting scenario unless we can find three (key, nonce) pairs that collide to the same session key.

5.2 Forgery Attack with Dynamic Message Block Size

In this section, we are assuming the existence of two different instantiations of COFFE with different message block size but the same secret key and constant last message block size β. Now we pick $\alpha_1 < \alpha_2$ such that $\alpha_2 - \alpha_1 = k$. Next we find the valid size of α_1 and α_2 based on our discussion in the discussion section. Here since we assume constant last message block size, β, we assume $\beta \leq \alpha_1$. Since we are using constant last message block size, to get the same session key, we can consider the nonce-misuse scenario where we use the same key and nonce for both instantiations. Note that this means the tag length should still be kept the same.

First, we generate message, $\mathcal{M}_1 = (\mathcal{MB}_1 \| \mathcal{MB}_2)$ with $|\mathcal{MB}_1| = \alpha_2$ and $|\mathcal{MB}_2| = \beta$. Now assume that we get the ciphertext $\mathcal{C}_1 = (\mathcal{CB}_1 \| \mathcal{CB}_2)$ with tag $\mathcal{T}_1$. In this pair, our objective is to find the last α_2 bits of $\mathcal{T}[1]$ which can be obtained by calculating $\mathcal{MB}_1 \bigoplus \mathcal{CB}_1$.

We then consider the following message: $\mathcal{M}_2 = (\mathcal{MB}'_1 \| \mathcal{MB}'_2)$ with $|\mathcal{MB}'_1| = \alpha_2$ and $|\mathcal{MB}'_2| = \beta$. We further require the last k bits of $\mathcal{MB}'_2 \bigoplus_{\alpha_1} \mathcal{T}[1]$ are all zeros. Note that $\mathcal{MB}'_2$ can be generated easily with the knowledge of the last α_1 bits of $\mathcal{T}[1]$. Assume that the ciphertext is $\mathcal{C}_2 = (\mathcal{CB}'_1 \| \mathcal{CB}'_2)$ with tag $\mathcal{T}_2$. Here the last k bits of $\mathcal{CB}'_1$ are all zero and $\mathcal{CB}'_2 \bigoplus \mathcal{MB}'_2$ tells us the last β bits of $\mathcal{T}[2]$ when the first block of the message is $\mathcal{MB}'_1$.

Now having this information, we will proceed to our forgery attack. The message block that we will use is $\mathcal{M}_3 = (\mathcal{MB}''_1 \| \mathcal{MB}'_2)$. Here we have $|\mathcal{MB}''_1| = \alpha_1$ and $\mathcal{MB}''_1$ is chosen such that

$$\left(\mathcal{MB}''_1 \bigoplus_{\alpha_1} \mathcal{T}[1] \right) \| 0^k = \mathcal{CB}'_1.$$

We also note that the second block is exactly the same used in $\mathcal{M}_2$. Then it is easy to see that the ciphertext is $\mathcal{C}_3 = (\mathcal{CB}''_1 \| \mathcal{CB}'_2)$ where $\mathcal{CB}''_1 = \mathcal{MB}''_1 \bigoplus_{\alpha_1} \mathcal{T}[1]$. Furthermore, the tag is exactly $\mathcal{T}_2$.

This forgery attack requires 2 enquiries to the oracle with 4 message blocks. So this attack provides a family of instances of COFFE that cannot provide ciphertext integrity as claimed in Lemma 2 under nonce-misuse scenario.

5.3   Combination of the Existing Attacks

In our previous two subsections, we change one of the parameters (α, β) while letting the other constant. This is done to simplify the analysis. However, it is possible for us to combine both attacks to generate new attack, that is, we change α and β in the same time. Notice that by combining the two attacks, the "nonce-misuse" requirement is not a must anymore. As discussed in the constant message block size subsection, as long as we can find a triple of (key,nonce) pairs that generate the same session key, we can launch the attack in the nonce-respecting scenario.

6   Related Key Recovery Attack

Note that, in most of the attacks we have mentioned, we are assuming same secret key. In this section, we will discuss the case with two different instances of COFFE with different key length. As discussed in our observations, in this case we assume that the longer key is obtained by extending the shorter key with secret string. As we have discussed in the appendix of the full version paper [12], there are $(\mathcal{K}_1, \mathcal{V}_1), (\mathcal{K}_2, \mathcal{V}_2)$ pairs that leads to the same session key (with one of them using one-byte domain value while the other using two-byte value) with different key length. Here we will use the pairs with k-bits key length difference and all of the difference are all in the nonce of the corresponding shorter length key. Now assume that $|\mathcal{L}_{K_1}| > |\mathcal{L}_{K_2}|$.

We again choose the parameters $\alpha, \beta_1, \beta_2, \mathcal{L}_{T_1}$, and $\mathcal{L}_{T_2}$ as in Sect. 4. The attack here is an adaptation of the distinguishing attack we proposed earlier. We use the two messages $\mathcal{M}_1$ and $\mathcal{M}_2$ as described in Sect. 4. The difference here is that for $\mathcal{M}_2$ with two bytes domain value and shorter key length, we will enquire 2^k different blocks of it with different k most significant bits of $\mathcal{V}_2$. Note that if the k most significant bits of $\mathcal{V}_2$ coincide with the k-bit extension of the secret key, then $\mathcal{CB}_1 = \mathcal{CB}_1'$ and the last β_1 bits of $\mathcal{CB}_2$ and the last β_1 bits of $\mathcal{CB}_2'$ should agree. So by using this approach, we can guess the k-bits extension of the secret key with the same complexity as exhaustive search for a k-bits secret key.

As discussed in the distinguishing attack section, when we decide that the guessed k-bits is wrong, the probability that the k-bits is actually the correct extension key is 0. So there will not be a false negative. However, when the k-bits we guess is wrong, the probability of false positive is, as discussed in the distinguishing attack, $2^{\alpha+\beta_1}$ which is at least 2^{-255} which is negligible.

So for the related key recovery attack, to recover the k-bit extension of the secret key in nonce-respecting scenario, we will need $2^k + 1$ plaintext-ciphertext pairs with success probability approximately 1. Note that the exact same attack can be adapted to the case when $|\mathcal{K}_1| < |\mathcal{K}_2|$.

7 Conclusion

7.1 Attack Summary

From the discussion above, we see that the security claim for the nonce-misusing scenario is not met for many different parameters. The same attack can be adapted to give a distinguishing attack for the nonce-respecting scenario for some subfamilies of the parameters mentioned above.

Lastly, having two different instances of COFFE with different key length with the longer key being the extension of the shorter key may not be a good idea. This is because if the parameter used belongs to the family we have found earlier, the extension of the key can be recovered with exhaustive search in the same way as if the secret key is just k bits.

In conclusion, COFFE does not satisfy any of the two security claims for some of the parameters that we have discussed before. The problem arises from the fact that concatenation of strings cannot be inverted uniquely and hence giving the opportunity of having two different set of strings concatenated to the same resulting strings.

7.2 Lesson Learned

Here we see that the the forgery and distinguishing attacks are feasible due to the possibility to have different (key,nonce) pairs to generate the same session key. This can be fixed by fixing the space for every given parameters. If some parameters are variables (such as the message block size in COFFE), we should ensure that the values of the variables get authenticated so as to prevent the forgery attack.

References

1. Bellare, M., Namprempre, C.: Authenticated encryption: relations among notions and analysis of the generic composition paradigm. In: Okamoto, T. (ed.) ASIACRYPT 2000. LNCS, vol. 1976, pp. 531–545. Springer, Heidelberg (2000)
2. Cogliani, S., Maimut, D.S., Naccache, D., do Canto, R.P., Reyhanitabar, R., Vaudenay, S., Vizźar, D.: Offset Merkle-Damgård (OMD) version 1.0 A CAESAR proposal. In: CAESAR (2014). http://competitions.cr.yp.to/caesar-submissions.html
3. Datta, N., Nandi, M.: ELmD v1.0. In: CAESAR (2014). http://competitions.cr.yp.to/caesar-submissions.html
4. Dworkin, M.: Recommendation for BlockCipher Modes of Operation: The CCM Mode for Authentication and Confidentiality. NIST Special Publication 800–38C, May 2004
5. Dworkin, M.: Recommendation for Block Cipher Modes of Operation: Galois/Counter Mode(GCM) and GMAC. NIST Special Publication 800–38D, November 2007
6. Forler, C., McGrew, D., Lucks, S., Wenzel, J.: COFFE: Ciphertext Output Feedback Faithful Encryption authenticated encryption without a block cipher. In: Early Symmetric Crypto ESC (2013). https://eprint.iacr.org/2014/1003.pdf

7. Jutla, C.S.: Encryption modes with almost free message integrity. In: Pfitzmann, B. (ed.) EUROCRYPT 2001. LNCS, vol. 2045, pp. 529–544. Springer, Heidelberg (2001)
8. Krawczyk, H.: The order of encryption and authentication for protecting communications (or: how secure is SSL?). In: Kilian, J. (ed.) CRYPTO 2001. LNCS, vol. 2139, p. 310. Springer, Heidelberg (2001)
9. Meyer. C., Matyas. S.: Secure program load with manipulation detection code. In: Proceedings of Securicom, vol. 88, pp. 111–130 (1988)
10. Preneel, B., Govaerts, R., Vandewalle, J.: Hash functions based on block ciphers: a synthetic approach. In: Stinson, D.R. (ed.) CRYPTO 1993. LNCS, vol. 773, pp. 368–378. Springer, Heidelberg (1994)
11. Rogaway, P., Bellare, M., Black, J., Krovetz, T.: OCB: a block-cipher mode of operation for efficient authenticated encryption. In: Proceedings of the 8th ACM Conference on Computer and Communications Security, pp. 196–205 (2001)
12. Tjuawinata, I., Huang, T., Wu. H.: Cryptanalysis of the Authenticated Encryption Algorithm COFFE. Cryptology ePrint Archive, Report 2015/783 (2015). http://eprint.iacr.org/
13. Wu, H., Huang, T.: JAMBU Lightweight Authenticated Encryption Mode and AES-JAMBU. In: CAESAR (2014). http://competitions.cr.yp.to/caesar-submissions.html

Author Index